Marine Ecosystems and Biodiversity

Marine Ecosystems and Biodiversity

Editor: Suzy Bullock

Callisto Reference
www.callistoreference.com

Callisto Reference,
118-35 Queens Blvd., Suite 400,
Forest Hills, NY 11375, USA

Visit us on the World Wide Web at:
www.callistoreference.com

ISBN: 978-1-63239-943-4 (Hardback)

Printed in China.

Cataloging-in-Publication Data

Marine ecosystems and biodiversity / edited by Suzy Bullock.
p. cm.
Includes bibliographical references and index.
ISBN 978-1-63239-943-4
1. Marine ecology. 2. Marine biodiversity. 3. Marine sciences.
4. Oceanography. I. Bullock, Suzy.
QH541.5.S3 M37 2018
577.7--dc23

Table of Contents

Preface

Marine ecosystems are ecosystems found in the oceans and seas. This book on marine ecosystems studies new research trends with regard to this field. The marine ecosystem is the largest ecosystem of the planet and can be sub-classified into rocky shores, submarine canyons, cold seeps, etc. Research and study into the composition of ecosystems and their processes plays a key role in conservation and in upholding biodiversity on Earth. With state-of-the-art inputs by acclaimed experts of this field, this book targets students and professionals. For someone with an interest and eye for detail, this book covers the most significant topics in the field of marine ecosystems.

This book unites the global concepts and researches in an organized manner for a comprehensive understanding of the subject. It is a ripe text for all researchers, students, scientists or anyone else who is interested in acquiring a better knowledge of this dynamic field.

I extend my sincere thanks to the contributors for such eloquent research chapters. Finally, I thank my family for being a source of support and help.

Editor

1

Core Microbial Functional Activities in Ocean Environments Revealed by Global Metagenomic Profiling Analyses

Ari J. S. Ferreira[1], Rania Siam[1], João C. Setubal[2], Ahmed Moustafa[1], Ahmed Sayed[1], Felipe S. Chambergo[3], Adam S. Dawe[4], Mohamed A. Ghazy[1], Hazem Sharaf[1], Amged Ouf[1], Intikhab Alam[4], Alyaa M. Abdel-Haleem[1], Heikki Lehvaslaiho[4], Eman Ramadan[1], André Antunes[5], Ulrich Stingl[6], John A. C. Archer[4], Boris R. Jankovic[4], Mitchell Sogin[7], Vladimir B. Bajic[4], Hamza El-Dorry[1]*

1 Department of Biology and the Science and Technology Research Center, School of Sciences and Engineering, The American University in Cairo, Cairo, New Cairo, Egypt, **2** Departamento de Bioquímica, Instituto de Química, Universidade de São Paulo, São Paulo, Brazil, **3** Escola de Artes, Ciências e Humanidades, Universidade de São Paulo, São Paulo, Brazil, **4** Computational Bioscience Research Center, King Abdullah University of Science and Technology, Thuwal, Kingdom of Saudi Arabia, **5** Institute for Biotechnology and Bioengineering, Centre of Biological Engineering, University of Minho, Portugal, **6** Red Sea Research Center, King Abdullah University of Science and Technology, Thuwal, Kingdom of Saudi Arabia, **7** Josephine Bay Paul Center, Marine Biological Laboratory, Woods Hole, Massachusetts, United States of America

Abstract

Metagenomics-based functional profiling analysis is an effective means of gaining deeper insight into the composition of marine microbial populations and developing a better understanding of the interplay between the functional genome content of microbial communities and abiotic factors. Here we present a comprehensive analysis of 24 datasets covering surface and depth-related environments at 11 sites around the world's oceans. The complete datasets comprises approximately 12 million sequences, totaling 5,358 Mb. Based on profiling patterns of Clusters of Orthologous Groups (COGs) of proteins, a core set of reference photic and aphotic depth-related COGs, and a collection of COGs that are associated with extreme oxygen limitation were defined. Their inferred functions were utilized as indicators to characterize the distribution of light- and oxygen-related biological activities in marine environments. The results reveal that, while light level in the water column is a major determinant of phenotypic adaptation in marine microorganisms, oxygen concentration in the aphotic zone has a significant impact only in extremely hypoxic waters. Phylogenetic profiling of the reference photic/aphotic gene sets revealed a greater variety of source organisms in the aphotic zone, although the majority of individual photic and aphotic depth-related COGs are assigned to the same taxa across the different sites. This increase in phylogenetic and functional diversity of the core aphotic related COGs most probably reflects selection for the utilization of a broad range of alternate energy sources in the absence of light.

Editor: Fabiano Thompson, Universidade Federal do Rio de Janeiro, Brazil

Funding: This work was supported by King Abdullah University for Science and Technology Global Collaborative Partners (GCR) program. The funders had no role in study design, data collection and analysis, decision to publish, or preparation of the manuscript.

* E-mail: dorry@aucegypt.edu

Introduction

Microbial communities dominate the oceans and seas [1] and are involved in fundamental processes in the global ecosystem, synthesizing half of the photosynthetic biomass and producing significant amounts of oxygen [2]. They also play pivotal roles in carbon, nitrogen, sulfur and phosphorus cycles [3,4,5,6]. The diverse capabilities of marine microbes are the result of selection and adaptation to environments in which light level, oxygen concentration, nutrient availability, salinity, temperature and hydrostatic pressure all vary widely [7,8,9,10,11,12,13,14,15,16,17].

Acquisition and loss of genetic information, and mutational variation of gene regulatory networks, are known to shape the lifestyles of microorganisms including those that reside in marine environments [18,19,20,21,22,23,24,25]. However, some basic questions regarding these processes in marine environments remain unresolved. What metabolic and physiological capabilities enable microbes to thrive in marine habitats? How do biotic and abiotic factors, including human activities and climate change, affect their capacity to adapt to a specific environment? Thanks to recent developments in molecular biology, remote sensing and deep-sea sampling, these issues can now be addressed [26,27,28].

Metagenomic sampling of marine microbial communities at various sites has contributed much to our understanding of their biochemical and physiological diversity [7,8,9,10,11,12,13,14,29,30,31]. However, a comprehensive comparative analysis of publicly accessible sequences from a wide range of depth-related locations in oceans, which is necessary for a better understanding of the distribution and dynamics of microorganisms in marine ecosystems, has been lacking. In this study we report such an analysis, based on 24 metagenomic datasets from known depths at 11 sites representing a wide range of marine settings. These comprise all publicly available data from studies that generated sufficient sequence coverage to permit broad comparative analysis.

Four of the datasets come from samples collected at different depths in the Atlantis II Basin in the Central Red Sea (ATII), and are reported here for the first time. In addition, we included in our analysis three collections of datasets, each representing four different depths in the water column, from ALOHA Station in the North Pacific Subtropical Gyre [32], BATS Station in the Sargasso Sea [32], and Station 3 off the Northern Chilean coast near Iquique (Iquique) [31]; four datasets derived from samples obtained at depths of 1,000 m in the Central Basin of the Sea of Marmara [8], 4,000 m at ALOHA [9], 6,000 m within the Puerto Rico Trench (PRT) [11], and 50 m in the Mediterranean off Alicante in Southeastern Spain (Med.) [10]. We also selected for inclusion in our global study just four sites from the vast array of near-surface datasets acquired by the Global Ocean Sampling (GOS) expedition [7,13]. The entire collection comprises approximately 12.0 million sequences encompassing 5,358 Mb.

As a benchmark for understanding the genomic changes underlying microbial adaptation to different light and oxygen regimes, we assembled a core set of reference depth-related photic and aphotic COGs and defined a group of functional activities associated with extreme hypoxia, based on functional profiling of all 24 datasets mentioned above. These were then utilized to characterize biological activities in two specific oceanographic settings in which the normal relationships between light, oxygen and depth are disrupted: the Humboldt Current System (HCS) - a major upwelling zone - and the salinity-based two-layer flow system in the Sea of Marmara. Phylogenetic assignment of core COGs indicated that the same microbial taxa dominated the photic zone across all the sites considered, while aphotic zones show greater diversification and differentiation.

Materials and Methods

Sample collection, DNA isolation, and Whole-genome shotgun pyrosequencing

Water column samples from Atlantis II Basin in the Red Sea (21°13' N, 37°58' E) were obtained during Leg2 of the WHOI/KAUST/HCMR sampling cruise aboard the R/V Aegeo in April 2010. Permits to collect water samples from all the depths of Atlantis II basin in the Red Sea were issued by the Ministry of Defense, Kingdome of Saudi Arabia. All the field studies did not involve endangered or protected species. Detailed of samples collection and processing, environmental DNA isolation, whole-genome shotgun pyrosequencing, and establishing of ATII depth-related datasets for the four depths are presented in Methods S1.

Nucleotide sequence accession numbers

The ATII 454 metagenome has been deposited in the GenBank Sequence Read Archive with the following accession numbers: ATII 50 m: SRS598124; ATII 200 m: SRS598125; ATII 700 m: SRS598128; and ATII 1500 m: SRS598129.

ATII and publicly available datasets employed in comparative analysis

In addition to the four ATII depth-related datasets newly generated for this work, metagenomic datasets from ten other sites were obtained from public databases (fig. S1 and table S1 in file S1). Detailed information regarding the publically available datasets, 454 shotgun read simulation, sub-sampling, functional assignment using the eggNOG database, and analysis of differentially abundant COGs are presented in Methods S1.

Comparative analysis of datasets based on COG abundance

Because there were three depth-related metagenomic collections of datasets that could be used as reference (ALOHA, BATS and ATII), we elected to first screen for COGs that differ significantly in abundance among the various depths in each of the three water columns. This was achieved by means of Fisher's exact test ($p \leq 0.01$, FDR-corrected), using the exact Test function of the R bioconductor package edgeR [33], in order to identify common COGs that displayed significant differences in normalized abundance levels between at least two depths in one of the three water columns. 176 depth-related COGs were identified at this step and were employed for hierarchical clustering of all datasets (table S2 in file S1).

Determination of reference photic/aphotic global-core depth-related COGs

Datasets were initially classified as photic or aphotic based on publicly available PAR and oceanographic data. Subsequently the average normalized abundances of each COG were determined for the identified photic and aphotic groups of datasets, obtained by the hierarchical clustering of the 24 datasets based on the normalized abundances of the 176 depth-related COGs previously identified (table S3 in file S1).

Then, unequal variance conservative Student's t-test (Welch's test; $p \leq 0.05$, FDR-corrected) was employed to determine the COGs that presented statistically significant differences in the means of the two groups ($n = 12$ for each group). 82 COGs were found to vary significantly ($p \leq 0.0001$, FDR-corrected) in abundance between the photic and aphotic datasets and considered subsequently as the reference photic/aphotic global-core depth-related set of COGs. Out of those, 54 COGs were assigned as photic and 28 COGs were assigned as aphotic based on the means of the normalized abundance of each COGs between the photic and aphotic datasets (tables S4A and S4B in file S1).

Comparison of photic to aphotic depth-related COG abundance ratios in water columns

Using the reference photic/aphotic global-core depth-related set of COGs, the normalized abundance means of the reference 54 photic and 28 aphotic depth-related COGs were determined for each dataset. The ratio of photic to aphotic normalized abundance means of each dataset was then calculated and log2 transformed in order to be plotted. The average and mean standard deviations of the log2-transformed ratios were determined for photic and aphotic datasets from ATII, ALOHA and BATS to establish ratio ranges indicative of photic or aphotic genomic content.

Assessment of the effect of hypoxic environment on COG abundance

To evaluate the effect of low oxygen concentration on the genomic content of microbial communities, COGs that presented statistically significant changes in normalized abundance levels between datasets of the Iquique water column were initially identified following the same procedure performed for the selection of differentially abundant depth-related COGs in the three reference columns (ATII, ALOHA, BATS). Such analysis resulted in a list of 162 COGs from which the 82 global photic/

aphotic common-core depth-related COGs were subsequently excluded to avoid influence of sunlight intensity level on the identification of abundance changes specifically related to dissolved oxygen concentration (D.O.). Datasets were initially divided into two groups based on publicly available dissolved oxygen concentration (D.O.) data for the 24 datasets analyzed. Iquique datasets from 85 m, 110 m and 200 m depths (D.O.$<$ 20 μmol.kg-1) were considered as the low-oxygen standard ($n = 3$) to which all other remaining datasets ($n = 21$) were compared. Fisher's exact test was again employed to determine the COGs that were statistically different in abundance across the Iquique water column. COGs whose abundances had been shown to vary in association with light level were omitted from this analysis. Wilcoxon's test was then used to find the COGs whose normalized abundances varied in association with oxygen concentration. 18 COGs (table S5 in file S1) were found to vary significantly ($p \leq 0.05$, FDR-corrected) in abundance between low and high oxygen datasets.

Taxonomic distribution of photic/aphotic depth-related COGs among sites.

The taxonomic distribution analysis was performed on selected datasets representing euphotic and aphotic zones of the different oceanic sites sampled. Datasets obtained from the shallowest samples of the reference water columns (ATII, ALOHA and BATS) plus the four GOS and the Mediterranean ones were used as representative of the euphotic zone, whilst datasets from the deepest samples of the reference water columns and those from Marmara, PRT and ALOHA (4,000 m) corresponded to the aphotic zone in the analysis. Reads from each dataset mapped to the photic/aphotic core related COGs were subjected to BLASTX similarity search against the NCBI nr database using a maximum e-value threshold of 1e-05 [34]. A taxonoxomic identifier was assigned to each individual read based on the best BLAST hit using NCBI taxdump (version of March, 2013). Taxonomic counts were then summarized at the phylum level per COG for each analyzed dataset (tables S6A and S6B in file S1) by means of bash/perl scripts developed in-house. Eukarya or unidentified taxa were discarded from the analysis. Heatmaps were generated from the summarized raw counts for the core photic and aphotic depth-related COGs using the publicly available software environment R version 2.11.1 (www.r-project.org, RColorBrewer and gplots libraries).

Results and Discussion

Defining the reference depth-related environments

The metagenomic datasets employed in this work were derived from samples obtained at different sites in the world's oceans. Detailed information on sites and sampling methods is presented in figure S1, and table S1 in file S1. In order to avoid biases in the comparison of samples due to different sequencing methods and coverage, simulated 454 sequences were generated from datasets made up of assembled contigs (ALOHA 4,000 and GOS datasets) prior to extraction from all analyzed datasets of random subsamples of the same size as the smallest complete collection (BATS 500 m; for details, see Methods S1).

It is important to emphasize that the datasets analyzed here include all those based on sampling at multiple depths in water columns at single sites: ALOHA [32], BATS [32], Iquique [31], and ATII (this work). The Iquique site [31] lies within the Humboldt Current System (HCS), the most productive marine ecosystem in the global oceans [35], which dramatically affects environmental conditions at this location [35]. The HC flows along the West coast of South America and drives a major upwelling system that brings cold, nutrient-rich water toward the surface of the Eastern Tropical South Pacific off Northern Chile. This promotes the growth of a dense planktonic community and supports a high level of primary production [35]. Among the most notable effects is the efficient depletion of dissolved oxygen, from $>$200 μmol.kg^{-1} at the surface to 3 to 4 μmol.kg^{-1} at depths between 110 and 200 m, by aerobic heterotrophs [31,36]. In addition, due to dense growth of phytoplankton and the presence of organic particulates, the sunlit zone at the Iquique site is rather shallow compared to the other sites (fig. 1B). We therefore considered the water columns at ALOHA, BATS and Atlantis II, but not that at Iquique, as reference depth-related environments.

Clustering COGs according to depth

To identify known biochemical functions represented in the 24 datasets, we subjected each one to sequence similarity analysis using the eggNOG v2.0 database [37], and calculated the normalized abundances of the identified clusters of orthologous groups of proteins (COGs) [38]. COGs that exhibited statistically significant differences in abundance within the three reference water columns were retrieved individually using Fisher's exact test ($p \leq 0.01$, and FDR-corrected; for details see Methods S1). This resulted in 649, 413, and 463 COGs for ATII, ALOHA and BATS, respectively. Of these, 176 COGs were present at all three sites (table S2 in file S1).

We first performed hierarchical clustering on the normalized abundance data for these 176 COGs in the three reference columns. The major feature common to the ATII, ALOHA and BATS profiles is the division of the datasets into two depth-related groups, representing the surface/near-surface zone, and a deeper zone (fig. S2, and table S2 in file S1).

Next, we proceeded to a hierarchical clustering of the same 176-COG set in all 24 datasets. As expected for a generalized depth-related pattern, members of the set represented in the Sea of Marmara (1,000 m), ALOHA (4,000 m) and the Puerto Rico Trench (PRT, 6,000 m) samples clustered with the deeper water group, while those occurring in the four GOS surface-water and the 50 m Mediterranean Sea datasets fell into the surface/near-surface group (fig. 1A, and table S3 in file S1). This depth-related division reflects metabolic and physiological processes specific to the microbial communities residing in photic and aphotic zones. Note that the dividing line coincides with the bottom of the photic zones in the three columns, as determined by their PAR (photosynthetically active radiation) values (fig. 1B). Interestingly, inclusion of the Iquique datasets in the clustering analysis preserves the depth-related profile division found at the other ten sites. However, as shown in fig. 1A, all four datasets generated from the Iquique samples (50 m, 85 m, 110 m and 200 m) clustered with the aphotic group. Nevertheless, the apparently anomalous position of the datasets related to the Iquique column in the profile of the 11 sites (fig. 1A) is entirely compatible with the environmental conditions at the site [31] which, as mentioned above, differ quite significantly from those at ALOHA [12], BATS [39] and ATII sites. The PAR value at Iquique drops below 1% of its surface value at a depth of less than 50 m [31] (fig. 1B). If one defines this level as the lower limit of the photic zone, it is not surprising that all Iquique datasets cluster with the aphotic group.

The photic/aphotic global-core depth-related COGs

In order to identify specific photic/aphotic global-core depth-related COGs we compared the normalized abundance of the photic with the aphotic groups of datasets defined in the dendrogram shown in fig. 1A. Of the 176 COGs used in the

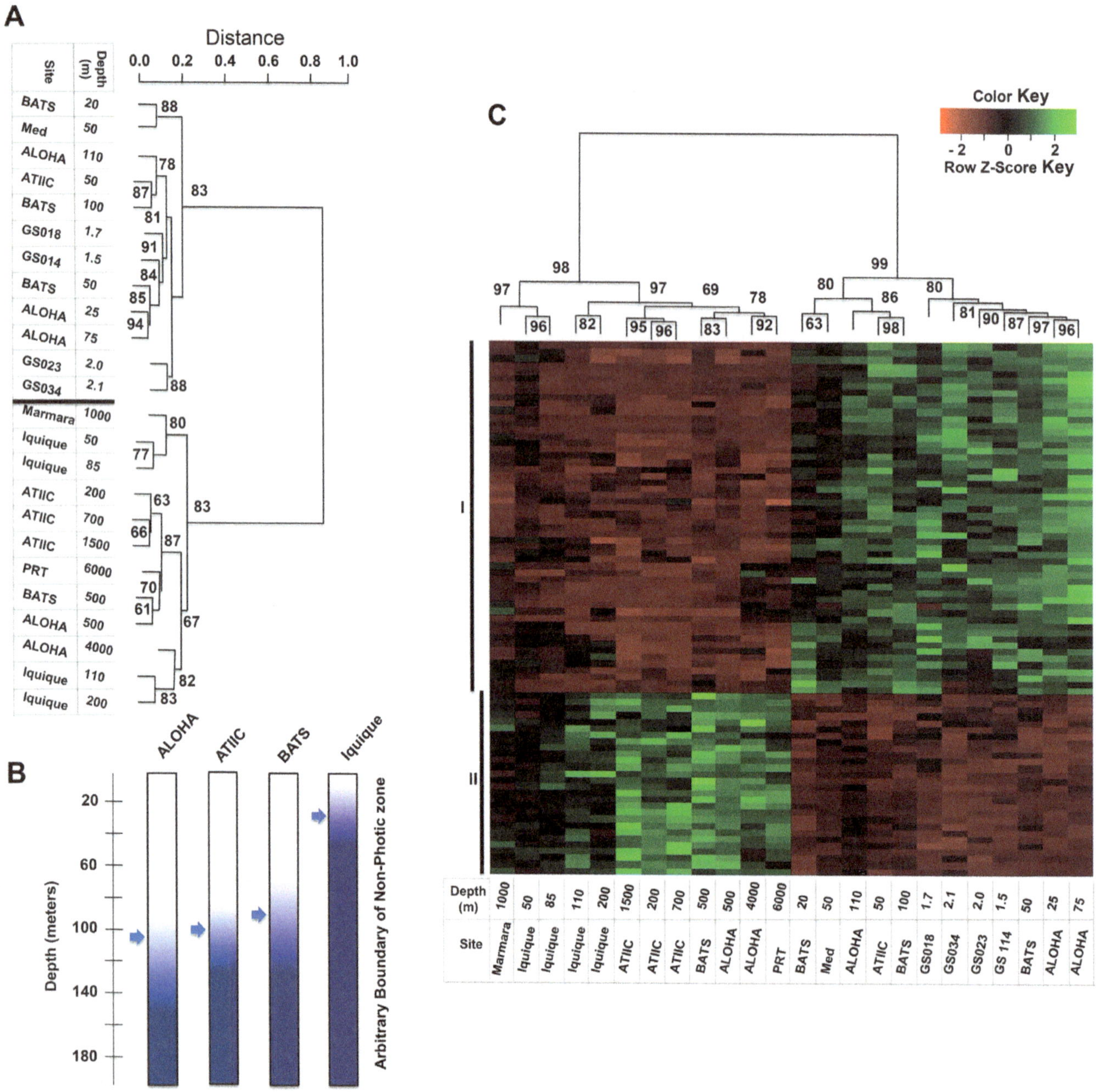

Figure 1. Clustering of the 176 depth-related COGs. (**A**) Hierarchical clustering of the 176 depth-related COGs in the 24 datasets. Clustering analysis is based on the normalized abundance profile of the 176 depth-related COGs that were shared by the three reference water columns (ATII, ALOHA and BATS) and significantly differed in abundance within at least one of them (details in Materials and Methods). The height indicates the relative distance between datasets. Bootstrap confidence values above 60 for the nodes are shown. The heatmap is shown in fig. S3A. (**B**) Location of the boundary between the photic and aphotic zones in each of the four water columns. The arrows indicate the depth at which PAR reaches 1% of the level at the surface. (**C**) Hierarchical clustering of the photic/aphotic global-core depth-related COGs. 82 COGs that showed statistically significant difference (Welch's test, FDR-corrected, p≤1E-04) in their normalized abundance between the photic and the aphotic groups of datasets were selected to establish a photic/aphotic global-core, depth-related reference set. 54 COGs had significantly higher abundance in the photic datasets (Group I, table S4A in file S1); contrary to the remaining 28 aphotic related COGs (Group II, table S4B in file S1). Bootstrap confidence values for the major nodes are shown. Heatmap coloring reflects the Z score of normalized abundances of each COG across all clustered datasets (details in Materials and Methods).

construction of this profile, a major subset of 82 COGs (19%) were found to show a statistically significant difference in abundance between photic and aphotic datasets ($p \leq 1e\text{-}04$ and FDR corrected; see details in Materials and Methods). Based on the mean normalized abundance for the whole set, 54 were considered photic related COGs and 28 were regarded as aphotic related COGs (fig. 1C, see also tables S4A and S4B in file S1). Bootstrapping evaluation indicated a high degree of confidence in the division of the photic and aphotic groups of the datasets presented in fig. 1A and 1C. This reflects the fact that functional adaptation to ambient light level is most probably one of the factors that strongly constrains the genetic variability of microorganisms in ocean environments. Thus, COGs related to photosynthesis, biosynthesis of light-harvesting pigments, assimilation of carbon dioxide by photosynthetic bacteria, as well as light-induced DNA repair and oxidative stress responses, were markedly more abundant in the photic than the aphotic datasets, as were COGs related to nitrogen fixation and phosphate metabolism. Conversely, in the aphotic datasets, COGs related to protein and amino acid catabolism, methane oxidation, sulfate assimilation and metabolism, selenocysteine metabolism and terpenoid biosynthesis were overrepresented, mirroring the exploitation of other energy sources in this zone.

It is important to note that previously detected differences in microbial gene abundances between the surface zone and the deep sea have been based on comparisons of limited metagenomic datasets [9,11,12]. In the present work all sufficiently extensive depth-related datasets available from different sites around the world's oceans (fig. S1, and table S1 in file S1) have been analyzed in a uniform fashion, allowing us to establish clearly defined sets of depth-related biological functions for the oceans as a whole. Such core functions shared between distinct sites around the global ocean can thus be used as a reliable diagnostic tool to analyze the distribution of light-related biological activities in sites that present atypical environmental conditions, as described below.

Abundance ratio of photic to aphotic depth-related COGs in the major upwelling zone of the Humboldt Current System and the salinity-based two-layer flow system in the Sea of Marmara

The environmental differences between the water columns at ALOHA, BATS and ATII relative to Iquique provide an excellent opportunity to explore the interplay between environmental factors - such as nutrient availability, light intensity, oxygen concentration - and the genomic changes that underlie microbial adaptation. When we examined the abundance ratio of the photic to the aphotic related COGs in all four water columns, we found, as expected, a well-defined correlation with the photic/aphotic boundary in the cases of ALOHA, BATS and ATII (fig. 2A) – high ratios of photic to aphotic related COG abundance in the photic samples and low ratios in aphotic ones. However, in the Iquique column the aphotic:photic ratio is merely 1.3 for the 50 m sample and 1.8 for the 85 m sample, whilst the mean ratio for the aphotic datasets of the reference columns is 3.8 (fig. 2A). The fact that microbial communities in this zone are genetically equipped for photosynthesis strongly suggests that the 50 m and 85 m levels are part of the mixed-water layer at this site.

To investigate this issue further, we inspected the ratios of photic to aphotic related COGs in all 24 datasets in relation to light levels. We selected one of the most striking positive functional associations with sunlight intensity as an example (correlation coefficient 0.89), the deoxyribodipyrimidine photolyase gene (COG0415), the product of which repairs DNA damage caused by exposure to ultraviolet light [40]. Figure 2B shows the abundance of this gene in each of the 24 datasets. It is overwhelmingly abundant in microbial communities that reside at all depth levels that clustered as photic in fig. 1A, whereas the gene is rarely found in those that reside in aphotic zones. Again, Iquique is an exception. Here, although the data from all four depths clustered with the aphotic group of datasets (fig. 1A), the abundance of the photolyase gene at 50 m and 85 m is nevertheless relatively high, while in the 110 m and 200 m datasets this drops to values comparable to those seen in aphotic zones (fig. 2B). Since the 50 m and 85 m levels at Iquique are most probably part of the mixed layer, microorganisms are presumably shuttled to the surface, where they are exposed to sunlight. Therefore, members of these communities are subjected to selective pressure to acquire and/or maintain biological functions that are crucial for survival in a sunlit environment. So, although its PAR value (fig. 1B) indicates that the 50 m layer at Iquique is already aphotic, the relative abundance of photic related COGs (fig. 2A) including photolyase (fig. 2B) argues that it belongs to a facultatively photic zone.

The data for the Sea of Marmara come from a depth of 1 km, which is unquestionably within the aphotic zone. The microbial community at this site was therefore expected to have low levels of the photolyase gene, comparable to those found in other aphotic zones. Paradoxically, however, the deep Marmara sample shows a relatively elevated abundance of this UV-protective gene (fig. 2B). Moreover, the ratio of aphotic to photic related COG abundances for the sample from the Sea of Marmara was found to be only 1.5, similar to that for the 50 m level in the Iquique column and much lower than the mean value for the aphotic datasets of the reference columns (fig. 2A). One possible explanation for this observation is provided by the hydrography and unusual circulation patterns in the Sea of Marmara [41], which receives an inflow of salty (around 3.9%) water from the Mediterranean Sea through the 65 m deep Strait of the Dardanelles. Owing to its density, this mainly euphotic water sinks to the bottom of the Marmara Basin [41]. Therefore, part of the microbial community in the 1-km deep aphotic Marmara environment is actually derived from the surface euphotic zone of the Mediterranean, and is continuously replenished. This would account for the unexpected presence of significant levels of photic genes at this site.

Effect of oxygen concentration on gene-content in marine microbial communities

Marine oxygen minimum zones (OMZs) occur naturally as a result of high oxygen consumption by heterotrophic microorganisms in nutrient-rich zones [35] like the upwelling system of the Humboldt Current (HC). They may also be induced by human activities such as fertilizer run-off and have a great environmental impact by creating zones dominated by microbial communities adapted to hypoxic conditions [42]. Although OMZs have important effects on the diversity, metabolism and physiology of microorganisms [43,44,45], the threshold oxygen concentration at which these begin to affect the content of microbial genomes in the ocean is not well defined. In order to clarify this issue, the three deepest Iquique datasets (85, 110 and 200 m) were chosen as reference hypoxic datasets because they represent the persistent OMZ present at this site [36] and comprise the datasets with the lowest oxygen concentrations among the 24 datasets studied in this work ($[O_2] = 11.0$, 4.0 and 3.2 $\mu mol.kg^{-1}$ at 85, 110 and 200 m, respectively). To avoid interference from any effects related to sunlight incidence, we removed members of the photic/aphotic global-core depth-related COGs from the list of 162 differentially abundant COGs that are statistically significant for the Iquique column. The normalized abundance medians of the remaining 149 COGs from these hypoxic Iquique samples were then compared with the corresponding values for the other 21 datasets (for details see Methods). As shown in fig. 3A (table S5 in file S1), 18 COGs in all were found to be significantly influenced by extreme oxygen limitation ($p <= 0.05$, FDR-corrected); 16 of these clustered in group I, and two in group II. Interestingly, approximately 43% of the COGs that are significantly more abundant in extremely hypoxic samples (Group I in fig. 3A) are related to nitrate metabolism.

Since the oxygen tension in the OMZ at the Iquique site is very low, the ability to use alternate final electron acceptors for the

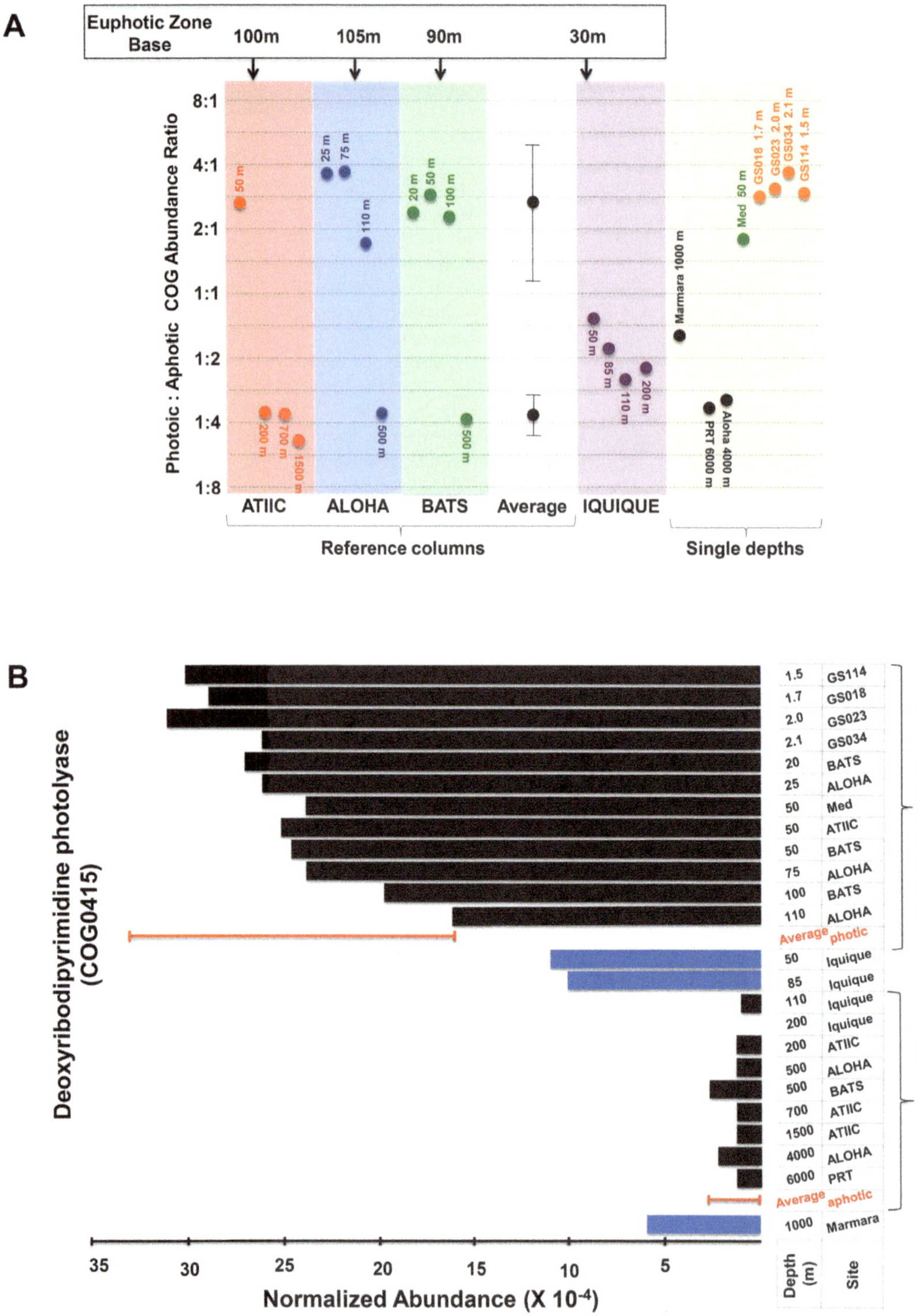

Figure 2. Abundance ratios of photic to aphotic related COGs. (**A**) Mean abundance ratios of photic to aphotic related COGs for datasets from the reference water columns, the Iquique water column, and from single-depth datasets. This analysis was initially determined using the 82-COG global-core, depth-related reference set, as described in Materials and Methods. Applying the ratios for the photic and aphotic datasets from the reference water columns, ratio ranges (mean±two SD) were established to indicate photic or aphotic functional genomic content enrichment. Mean abundance ratios for the Iquique water column and the single-depth datasets were calculated to verify their balance between photic and aphotic related biological functions. The arrows at the top of the diagram indicate the approximate depth of the boundary between photic and aphotic zones in each water column (fig. 1B). (**B**) Normalized abundance of the deoxyribodipyrimidine photolyase gene (COG0415; table S4A in file S1) in the 24 datasets obtained from the indicated depths. The average normalized abundance was calculated for the photic and for the aphotic groups of datasets. The bars represent two SD of the mean.

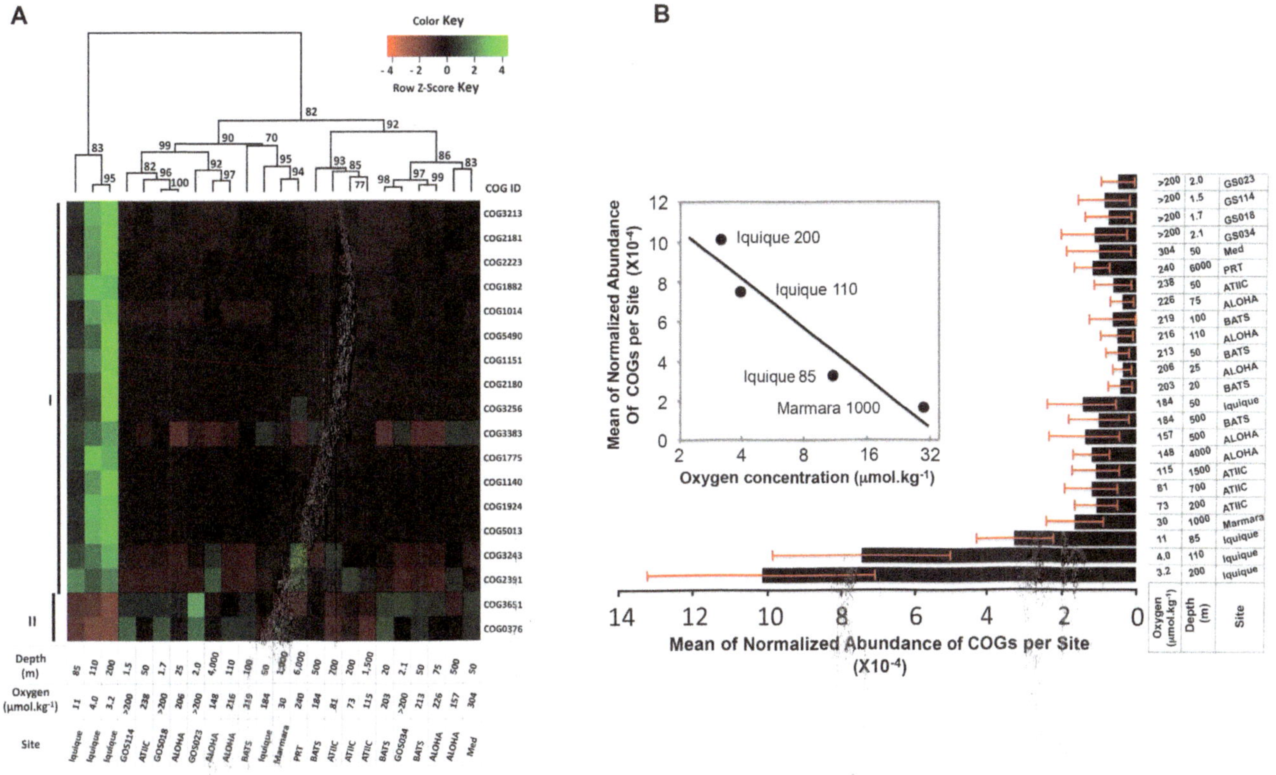

Figure 3. Effect of oxygen concentration on the functional genomic content of microbial communities represented in the 24 datasets. (A) Hierarchical clustering of COGs that significantly influenced by extreme oxygen limitation. Datasets analysis are based on the abundance profile of 18 COGs which significantly differed in abundance levels between an extremely hypoxic environment (Iquique 85-, 110 and 200 m deep, permanent OMZ) and environments with higher D.O. (Fisher's exact test, FDR-corrected, p≤0.05). Bootstrap confidence values for the nodes are shown. Heatmap coloring reflects the Z score of normalized abundances of each COG across all clustered datasets. Roman numbers on the left side of the figure present different groups of COGs as determined by abundance profile across the clustered datasets. (B) Mean abundances of the COGs in group I (figure 3A) as a function of the oxygen concentration. Inset shows the gradual increase in the abundance of these genes in datasets from Marmara 1000 m, Iquique 85 m, Iquique 110 m, and Iquique 200 m, as oxygen concentration drops from 30 μmol.kg^{-1} to 3.2 μmol.kg^{-1}.

replenishment of oxidative potential and energy production coupling becomes imperative for survival. This probably accounts for the fact that a number of the COGs overrepresented in this zone (COG2223, COG5013, COG1140, COG2181, COG2180, COG3256 and possibly COG3213) are related to nitrate reduction. The abundance of an enzyme involved in fermentation and specifically inactivated by oxygen (COG1882) was also significantly high. In addition, enzymes that are implicated in methanogenesis (COG1151) and pyruvate metabolism (COG1014) exhibited increased abundance in these datasets. Conversely, genes whose products require molecular oxygen for their activities - such as catalase (COG0376, fig. 3A, table S5 in file S1) - are poorly represented in this extremely hypoxic environment.

Analysis of the mean abundances of the COGs in group I as a function of the oxygen concentration reveals that these COGs are distinctively and overwhelmingly abundant in microbial communities from samples in which oxygen concentration was ≤ 11 μmol.kg^{-1} (fig. 3B). It is intriguing to note the gradual increase in the abundance of these genes in datasets as oxygen concentration drops from 30 μmol.kg^{-1} to 3.2 μmol.kg^{-1} (inset fig. 3B).

The fact that the relative abundances of genes in group I increases dramatically at oxygen concentrations of 11 μmol.kg^{-1} or less suggests that this level of oxygen represents the threshold at which selective effects on genome content in favor of anaerobic lifestyles set in.

Based on the content of the core sets of photic/aphotic depth-related COGs from all sites presented in fig. S3A, it appears that microbial communities living in photic zones share the same functions, regardless of site location. Thus, with respect to this zone, we do not see any specific branches clustering together that can be attributed to environmental conditions that are unique to any particular geographical location. The aphotic samples (based on PAR values), however, do display distinct branching profiles of genomic content that can be correlated with specific environmental conditions and oxygen concentrations. The three mixed-water layers, Iquique 50 and 85 m and Marmara 1 km (with aphotic:photic ratios of between 1.3 to 1.8; refer to fig. 2A), in which limitation of light and oxygen in these aphotic zones exert selective pressures, cluster together (fig. S3A, see also fig. S3B branch I), while the extremely hypoxic samples in the aphotic zone at Iquique (110 and 200 m) formed a separate cluster on the other side of the tree (fig. S3A, see also S3B branch II). The abundances of two COGs, photolyase (COG0415) and nitrate reductase (Alpha Subunit; COG5013), taken as representative of core photic and aphotic related COGs that contributed to the functional branching in these aphotic zone, are presented in fig. S3B. The results again argue that the concentration of oxygen at which selective pressure begins detectably to favor the acquisition of

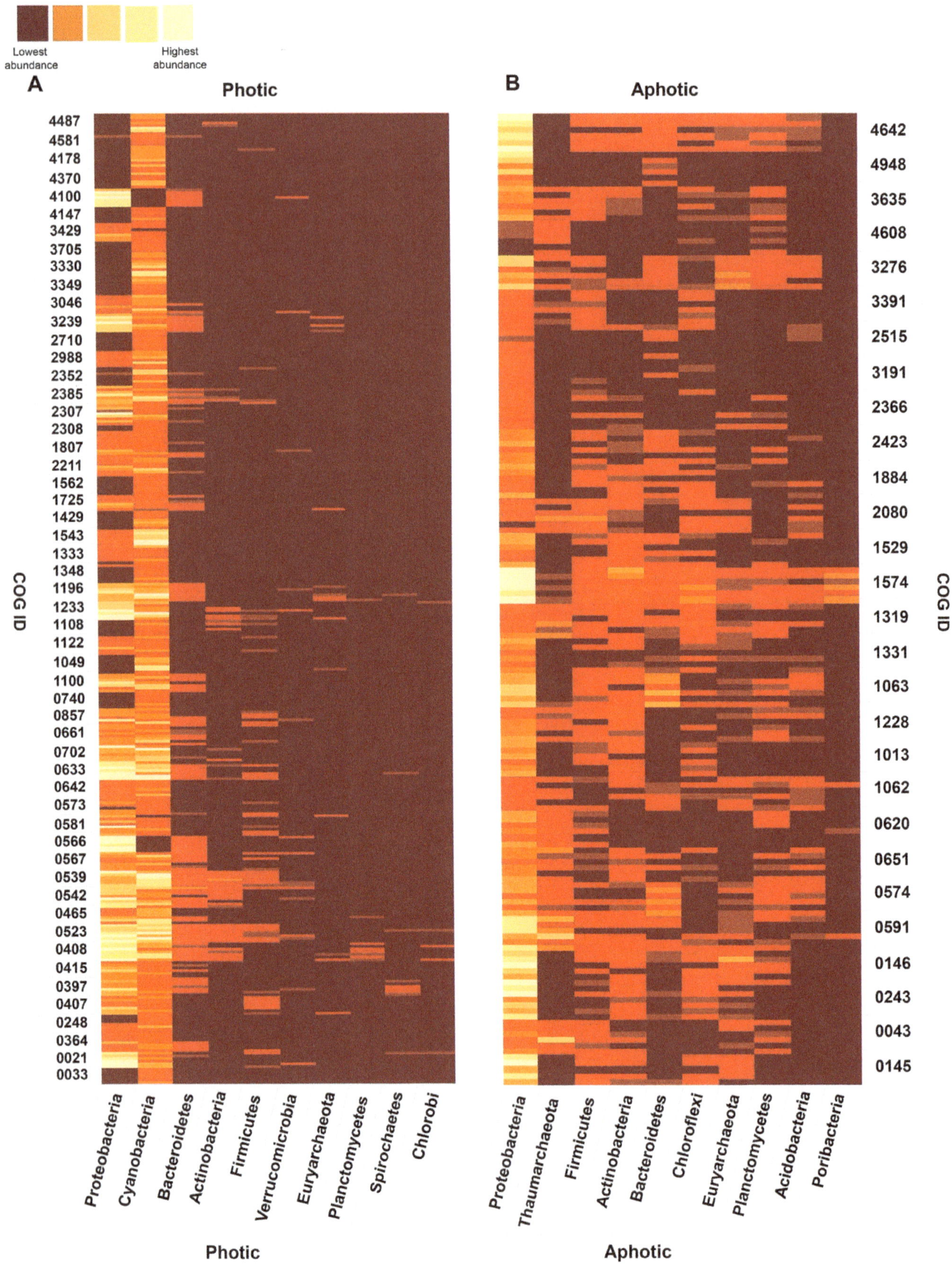
Lowest abundance
Highest abundance
A
Photic
B
Aphotic
COG ID
4487
4581
4178
4370
4100
4147
3429
3705
3330
3349
3046
3239
2710
2988
2352
2385
2307
2308
1807
2211
1562
1725
1429
1543
1333
1348
1196
1233
1108
1122
1049
1100
0740
0857
0661
0702
0633
0642
0573
0581
0566
0567
0539
0542
0465
0523
0408
0415
0397
0407
0248
0364
0021
0033
Proteobacteria
Cyanobacteria
Bacteroidetes
Actinobacteria
Firmicutes
Verrucomicrobia
Euryarchaeota
Planctomycetes
Spirochaetes
Chlorobi
Photic
4642
4948
3635
4608
3276
3391
2515
3191
2366
2423
1884
2080
1529
1574
1319
1331
1063
1228
1013
1062
0620
0651
0574
0591
0146
0243
0043
0145
COG ID
Proteobacteria
Thaumarchaeota
Firmicutes
Actinobacteria
Bacteroidetes
Chloroflexi
Euryarchaeota
Planctomycetes
Acidobacteria
Poribacteria
Aphotic

Figure 4. Taxonomical assignments of photic and aphotic depth-related COGs at the phylum level for the selected datasets. Reads assigned to COGs were compared to the NCBI nr database using BLASTX (E-value cutoff 1e-05) and the taxonomical identifiers of the best matches were retrieved. Subsequently, tables of relative frequency of taxa per COG per dataset were generated (tables S6A and S6B in file S1) and presented as heatmaps. The 10 phyla that have the highest average numbers of photic (**A**) and aphotic (**B**) COGs per site are presented. The taxa are sorted by the decreasing of their abundance. Datasets of the shallowest samples of the ALOHA, ATII and BATS water columns and single-depth datasets from GOS018, GOS023, GOS034, GOS114 and Mediterranean sites were used for the analysis of the photic related COGs, whereas the deepest datasets from the ALOHA, ATII and BATS water columns, and single-depth datasets from Marmara, PRT and ALOHA (4,000 m), were employed in the case of the aphotic related COGs. This analysis was based on a total of 36,938 sequences (24919 sequences are photic, and 12019 are aphotic).

adaptations for survival under hypoxic conditions lies below that of the Iquique 85 m sample (11 μmol.kg^{-1}), which is assigned to branch I.

Taxonomical distribution of photic and aphotic depth-related COGs

The establishment of collections of core photic and core aphotic depth-related COGs raises the question of how such functions are disseminated among phylogenetic clades present in the different environments covered by our analysis. To address this point, we generated taxonomic assignments, at the phylum level, for each core photic and aphotic depth-related COG based on the taxonomic identifiers of the respective reads (for details see Materials and Methods, and tables S6A and S6B in file S1). The results show that sequences related to core photic COGs derive from far fewer taxonomic clades than those of core aphotic ones. This reflects the predominance of certain taxa in most euphotic environments and the higher diversity of taxa found in deeper oceanic environments (fig. 4, and tables S6A and S6B in file S1). Moreover, the dominant taxon for any given COG is, with few exceptions, shared among all the sites analyzed.

The taxonomic distribution of the 54 core photic related COGs at the phylum level (fig. 4A, table S6A in file S1) reveals that the vast majority (91%) of them are most probably contributed by Proteobacteria (46%) and Cyanobacteria (45%). The taxonomic distribution of the 28 core aphotic related COGs at the phylum level is much more diverse than that observed for the photic functions. Although Proteobacteria are again predominant (50%), a total of ten phyla are represented at levels greater than 1%, compared to only five for the photic set (fig. 4B, table S6B in file S1).

The increased diversity of microbial communities in aphotic environments is perhaps the most striking result of the phylogenetic analysis of the core COGs. It can be attributed to adaptation to the wider variety (and lower abundance) of exploitable resources available in the aphotic zone, which contrasts with overwhelming importance of photosynthetic energy production in the photic zone. This interpretation is supported by the biochemical functions assigned to core photic/aphotic related COGs observed in our work.

In this study, using metagenomics-based functional profiling analysis, we have defined core sets of photic and aphotic depth-related biological activities that are shared by different sites distributed around the global oceans. These two sets of depth-related functional biological activities were used to provide insight into the interplay between light intensity - the most significant abiotic evolutionary force in the oceans - and the genetic and ecological differentiation of marine microbial communities. In addition, in conjunction with a set of functional biological activities related to extreme hypoxia, we provide a comprehensive picture of microbial biological activities in oceanographic settings in which light and oxygen imprint its effect in nominally aphotic zones. Moreover, our work highlights the importance of developing reference sets of functional biological activities that can be used as a tool for the diagnosis of the physiological and biochemical capabilities, and phylogenetic profiling, of marine microorganisms. This approach should be particularly helpful for investigations of the impact of anthropogenically induced environmental changes on marine ecosystems.

Supporting Information

Figure S1 Locations of the 11 sites and numbers of sequenced reads in the 24 datasets.

Figure S2 Hierarchical clustering of datasets from the three reference water columns (ATII, ALOHA, BATS). This analysis is based on the normalized abundance profile of 176 selected COGs that significantly differed in abundance within at least one of column (Fisher's exact test, FDR-corrected, $p \leq 0.01$) (table S2 in file S1). Heatmap coloring reflects the Z score of normalized abundances of each COG across clustered datasets. Roman numbers on the left side of the figure present different groups of COGs as determined by abundance profile across the clustered datasets.

Figure S3 Hierarchical clustering of the 176 depth-related COGs in the 24 datasets. (A) Heatmap of the datasets from 11 diverse marine sites (24 datasets). Refer to legend of figure 1A for details. Roman numbers on the left side of the figure present different groups of COGs as determined by abundance profile across the clustered datasets. The dendrogram of this figure is also presented in fig. 1A. (B) Distinct branching profiles of genomic content of the aphotic samples. The aphotic branch in figure S3A is presented together with the level of oxygen in each site and the normalized abundances of Deoxyribodipyrimidine photolyase (COG0415), and Nitrate reductase (Alpha Subunit; COG5013). The values of the normalized abundance of both COGs were obtained from table S4A and S5 in file S1.

File S1 Table S1, Sites locations, datasets, sampling and sequencing methods. Table S2, List of 176 depth-related COGs shared by the three reference columns, ATIIC, ALOHA and BATS. Fisher's exact test ($p \leq 0.01$, FDR corrected) was employed to identify COGs that presented a statistically significant difference in abundance between at least two depths of a reference column. COG cluster roman numbers refers to groups of COGs displaying similar normalized abundance profile. COGs are ordered as in figure S2. **Table S3, List of 176 depth-related COG set in all 24 datasets.** Fisher's exact test ($p \leq 0.01$, FDR corrected) was employed to identify the 176 depth-related COG set (table S2 in file S1) in all 24 datasets. Among these, the ones that were present in at least one dataset of each reference column were selected for subsequent comparative analysis of all datasets. COG cluster roman numbers refers to groups of COGs displaying similar abundance profile in the hierarchical clustering of all 24 datasets. COGs are ordered as in figures 1A and S3A. **Table S4, A. List of 54 photic global-**

core, depth-related COGs. Welch's test (p ≤ 1E-04, FDR corrected) was employed to identify COGs that presented a statistically significant difference in the mean abundance between the photic and the aphotic group of datasets (see details in Methods). Photic related COGs are presented in the same order as shown in group I Fig. 1C. **B. List of 28 aphotic global-core depth-related COGs.** Welch's test (p ≤ 1E-04, FDR corrected) was employed to identify COGs that presented a statistically significant difference in the mean abundance between the photic and the aphotic group of datasets (see details in Methods). Aphotic related COGs are presented in the same order as shown in group II Fig. 1C. **Table S5, List of 18 Oxygen Minimum Zone-related COGs.** Wilcoxon test (p ≤ 0.05, FDR corrected) was employed to identify COGs that significantly differed in abundance between the highly hypoxic group of datasets (Iquique 85-, 110- and 200-m deep samples) and the group of remaining datasets with higher oxygen concentration (details in Methods). COGs are presented in the same order as that in the heatmap (fig. 3A).**Table S6. A. Taxonomic distribution of the photic COGs at the phylum level.** Phylogenetic analysis was performed based on the NCBI taxonomic identifiers of the best matches to the nr database. For each COG, the rawcounts of assigned taxa is presented for eight different photic datasets: the shallowest samples of the reference columns, and five singledepth samples (details in Methods). The 10 phyla that have the highest average numbers of photic COGs per site are presented in fig. 4A. **B. Taxonomic distribution of the aphotic COGs at the phylum level.** Phylogenetic analysis was performed based on the NCBI taxonomic identifiers of the best matches to the nr database. For each COG, the rawcounts of assigned taxa is presented for six different aphotic datasets: the deepest samples of the reference columns and three single-depth samples. The 10 phyla that have the highest average numbers of aphotic COGs per site are presented in Figure 4B.

Acknowledgments

We thank Drs. Amy S. Bower, Stephen A. Swift, and Abdulaziz Al-Suwailem for offering technical assistants and important information regarding sample collection. We are grateful to Dr. Paul Hardy for critical reading of the manuscript.

Author Contributions

Conceived and designed the experiments: AJSF RS JCS AM HD. Performed the experiments: AJSF RS JCS FSC ER AMA HD. Analyzed the data: AJSF JCS HS AMA FSC US VBB HD. Contributed reagents/materials/analysis tools: AS ASD MAG AO IA HL AA JACA BRJ MS VBB. Wrote the paper: AJSF JCS HD.

References

1. Whitman WB, Coleman DC, Wiebe WJ (1998) Prokaryotes: the unseen majority. Proc Natl Acad Sci U S A 95: 6578–6583.
2. Field CB, Behrenfeld MJ, Randerson JT, Falkowski P (1998) Primary production of the biosphere: integrating terrestrial and oceanic components. Science 281: 237–240.
3. Arrigo KR (2005) Marine microorganisms and global nutrient cycles. Nature 437: 349–355.
4. Schafer H, Myronova N, Boden R (2010) Microbial degradation of dimethylsulphide and related C1-sulphur compounds: organisms and pathways controlling fluxes of sulphur in the biosphere. J Exp Bot 61: 315–334.
5. Jorgensen BB, Boetius A (2007) Feast and famine—microbial life in the deep-sea bed. Nat Rev Microbiol 5: 770–781.
6. Azam F, Malfatti F (2007) Microbial structuring of marine ecosystems. Nat Rev Microbiol 5: 782–791.
7. Yooseph S, Nealson KH, Rusch DB, McCrow JP, Dupont CL, et al. (2010) Genomic and functional adaptation in surface ocean planktonic prokaryotes. Nature 468: 60–66.
8. Quaiser A, Zivanovic Y, Moreira D, Lopez-Garcia P (2011) Comparative metagenomics of bathypelagic plankton and bottom sediment from the Sea of Marmara. ISME J 5: 285–304.
9. Konstantinidis KT, Braff J, Karl DM, DeLong EF (2009) Comparative metagenomic analysis of a microbial community residing at a depth of 4,000 meters at station ALOHA in the North Pacific subtropical gyre. Appl Environ Microbiol 75: 5345–5355.
10. Ghai R, Martin-Cuadrado AB, Molto AG, Heredia IG, Cabrera R, et al. (2010) Metagenome of the Mediterranean deep chlorophyll maximum studied by direct and fosmid library 454 pyrosequencing. ISME J 4: 1154–1166.
11. Eloe EA, Fadrosh DW, Novotny M, Zeigler Allen L, Kim M, et al. (2011) Going deeper: metagenome of a hadopelagic microbial community. PLoS One 6: e20388.
12. DeLong EF, Preston CM, Mincer T, Rich V, Hallam SJ, et al. (2006) Community genomics among stratified microbial assemblages in the ocean's interior. Science 311: 496–503.
13. Rusch DB, Halpern AL, Sutton G, Heidelberg KB, Williamson S, et al. (2007) The Sorcerer II Global Ocean Sampling expedition: northwest Atlantic through eastern tropical Pacific. PLoS Biol 5: e77.
14. Venter JC, Remington K, Heidelberg JF, Halpern AL, Rusch D, et al. (2004) Environmental genome shotgun sequencing of the Sargasso Sea. Science 304: 66–74.
15. Lipson DA, Haggerty JM, Srinivas A, Raab TK, Sathe S, et al. (2013) Metagenomic insights into anaerobic metabolism along an Arctic peat soil profile. PLoS One 8: e64659.
16. Varin T, Lovejoy C, Jungblut AD, Vincent WF, Corbeil J (2012) Metagenomic analysis of stress genes in microbial mat communities from Antarctica and the High Arctic. Appl Environ Microbiol 78: 549–559.
17. Gregoracci GB, Nascimento JR, Cabral AS, Paranhos R, Valentin JL, et al. (2012) Structuring of bacterioplankton diversity in a large tropical bay. PLoS One 7: e31408.
18. Ochman H, Lawrence JG, Groisman EA (2000) Lateral gene transfer and the nature of bacterial innovation. Nature 405: 299–304.
19. Kazazian HH Jr (2004) Mobile elements: drivers of genome evolution. Science 303: 1626–1632.
20. Gogarten JP, Townsend JP (2005) Horizontal gene transfer, genome innovation and evolution. Nat Rev Microbiol 3: 679–687.
21. Sobecky PA, Hazen TH (2009) Horizontal gene transfer and mobile genetic elements in marine systems. Methods Mol Biol 532: 435–453.
22. Balaji S, Babu MM, Aravind L (2007) Interplay between network structures, regulatory modes and sensing mechanisms of transcription factors in the transcriptional regulatory network of E. coli. J Mol Biol 372: 1108–1122.
23. Baumbach J (2010) On the power and limits of evolutionary conservation—unraveling bacterial gene regulatory networks. Nucleic Acids Res 38: 7877–7884.
24. Crombach A, Hogeweg P (2008) Evolution of evolvability in gene regulatory networks. PLoS Comput Biol 4: e1000112.
25. Maslov S, Krishna S, Pang TY, Sneppen K (2009) Toolbox model of evolution of prokaryotic metabolic networks and their regulation. Proc Natl Acad Sci U S A 106: 9743–9748.
26. Lutz RA, Falkowski PG (2012) Ocean science. A dive to Challenger Deep. Science 336: 301–302.
27. Giovannoni SJ, Vergin KL (2012) Seasonality in ocean microbial communities. Science 335: 671–676.
28. Kroger S, Law RJ (2005) Sensing the sea. Trends Biotechnol 23: 250–256.
29. Karl DM (2007) Microbial oceanography: paradigms, processes and promise. Nat Rev Microbiol 5: 759–769.
30. DeLong EF, Karl DM (2005) Genomic perspectives in microbial oceanography. Nature 437: 336–342.
31. Stewart FJ, Ulloa O, DeLong EF (2012) Microbial metatranscriptomics in a permanent marine oxygen minimum zone. Environ Microbiol 14: 23–40.
32. Martinez A, Tyson GW, Delong EF (2010) Widespread known and novel phosphonate utilization pathways in marine bacteria revealed by functional screening and metagenomic analyses. Environ Microbiol 12: 222–238.
33. Robinson MD, McCarthy DJ, Smyth GK (2010) edgeR: a Bioconductor package for differential expression analysis of digital gene expression data. Bioinformatics 26: 139–140.
34. Altschul SF, Madden TL, Schaffer AA, Zhang J, Zhang Z, et al. (1997) Gapped BLAST and PSI-BLAST: a new generation of protein database search programs. Nucleic Acids Res 25: 3389–3402.
35. Daneri G, Dellarossa V, Quinones R, Jacob B, Montero P, et al. (2000) Primary production and community respiration in Humboldt Current System off Chile and associated oceanic areas. Marine Ecology Progress Series 197: 41–49.

36. Revsbech NR, Larsen LH, Gundersen J, Dalsgaard T, Ulloa O, et al. (2009) Determination of ultra-low oxygen concentrations in oxygen minimum zones by the STOX sensor. Limnology and Oceanography: Mehods 7: 371–381.
37. Muller J, Szklarczyk D, Julien P, Letunic I, Roth A, et al. (2010) eggNOG v2.0: extending the evolutionary genealogy of genes with enhanced non-supervised orthologous groups, species and functional annotations. Nucleic Acids Res 38: D190–195.
38. Tatusov RL, Galperin MY, Natale DA, Koonin EV (2000) The COG database: a tool for genome-scale analysis of protein functions and evolution. Nucleic Acids Res 28: 33–36.
39. Steinberg DK, Carlson CA, Bates NR, Johnson RJ, Michaels AF, et al. (2001) Overview of the US JGOFS Bermuda Atlantic Time-series Study (BATS): a decade-scale look at ocean biology and biogeochemistry. Deep-Sea Research II 48: 1405–1447.
40. Osburne MS, Holmbeck BM, Frias-Lopez J, Steen R, Huang K, et al. (2010) UV hyper-resistance in Prochlorococcus MED4 results from a single base pair deletion just upstream of an operon encoding nudix hydrolase and photolyase. Environ Microbiol 12: 1978–1988.
41. Besiktepe ST, Sur IH, Ozsoy E, Abdul Latif M, Oguz T, et al. (1994) The circulation and hydrography of the Marmara Sea. Prog Oceanogr 34: 285–334.
42. Diaz RJ, Rosenberg R (2008) Spreading dead zones and consequences for marine ecosystems. Science 321: 926–929.
43. Bryant JA, Stewart FJ, Eppley JM, DeLong EF (2012) Microbial community phylogenetic and trait diversity declines with depth in a marine oxygen minimum zone. Ecology 93: 1659–1673.
44. Lam P, Kuypers MM (2011) Microbial nitrogen cycling processes in oxygen minimum zones. Ann Rev Mar Sci 3: 317–345.
45. Ulloa O, Canfield DE, DeLong EF, Letelier RM, Stewart FJ (2012) Microbial oceanography of anoxic oxygen minimum zones. Proc Natl Acad Sci U S A 109: 15996–16003.

2

Longer and Less Overlapping Food Webs in Anthropogenically Disturbed Marine Ecosystems: Confirmations from the Past

Fabiana Saporiti[1]*, Stuart Bearhop[2], Laura Silva[3], Damián G. Vales[3], Lisette Zenteno[1], Enrique A. Crespo[3], Alex Aguilar[1], Luis Cardona[1]

1 Department of Animal Biology and Institut de Recerca de la Biodiversitat (IRBio), Faculty of Biology, University of Barcelona, Barcelona, Spain, **2** Centre for Ecology and Conservation, School of Biosciences, University of Exeter, Cornwall Campus, Penryn, United Kingdom, **3** Laboratory of Marine Mammals, Centro Nacional Patagónico - Consejo Nacional de Investigaciones Científicas y Técnicas (CENPAT-CONICET), Puerto Madryn, Argentina

Abstract

The human exploitation of marine resources is characterised by the preferential removal of the largest species. Although this is expected to modify the structure of food webs, we have a relatively poor understanding of the potential consequences of such alteration. Here, we take advantage of a collection of ancient consumer tissues, using stable isotope analysis and SIBER to assess changes in the structure of coastal marine food webs in the South-western Atlantic through the second half of the Holocene as a result of the sequential exploitation of marine resources by hunter-gatherers, western sealers and modern fishermen. Samples were collected from shell middens and museums. Shells of both modern and archaeological intertidal herbivorous molluscs were used to reconstruct changes in the stable isotopic baseline, while modern and archaeological bones of the South American sea lion *Otaria flavescens*, South American fur seal *Arctocephalus australis* and Magellanic penguin *Spheniscus magellanicus* were used to analyse changes in the structure of the community of top predators. We found that ancient food webs were shorter, more redundant and more overlapping than current ones, both in northern-central Patagonia and southern Patagonia. These surprising results may be best explained by the huge impact of western sealing on pinnipeds during the fur trade period, rather than the impact of fishing on fish populations. As a consequence, the populations of pinnipeds at the end of the sealing period were likely well below the ecosystem's carrying capacity, which resulted in a release of intraspecific competition and a shift towards larger and higher trophic level prey. This in turn led to longer and less overlapping food webs.

Editor: Yan Ropert-Coudert, Institut Pluridisciplinaire Hubert Curien, France

Funding: Fundación BBVA funded this research as part of the project "Efectos de la explotación humana sobre depredadores apicales y la estructura de la red trófica del Mar Argentino durante los útimos 6000 años" (BIOCON08-194/09 2009-2011). Agencia Nacional de Promoción Científica y Tecnológica (Argentina) provided additional funding through the project "Análisis del uso de los recursos tróficos y su relación con cambios en la abundancia en tres predadores tope del Mar Argentino". FS has been supported by an FPU Fellowship granted by the Spanish Ministerio de Educación, Cultura y Deporte (AP 2009- 4573). Half of the cost of the publication of this article has been funded by the University of Barcelona. The funders had no role in study design, data collection and analysis, decision to publish, or preparation of the manuscript.

Competing Interests: The authors have declared that no competing interests exist.

* Email: fabiana.saporiti@ub.edu

Introduction

Human activities have altered most of the coastal marine ecosystems of the world over many centuries, causing reductions in population sizes, shifts in geographic ranges, and losses of diversity, biomass, and ecosystem functioning [1,2,3,4,5,6]. Both aboriginal and industrial exploitation of marine resources are characterised by the preferential removal of the largest species [3,7,8,9,10], a process thought to shorten size-structured marine food webs [4,11,12]. However, so far, megafaunal exploitation has generally resulted in population collapse and top predator rarity, rather than in true extinctions [13,14,15]. The difference between rarity and extinction is relevant, because the extinct species no longer belong to the local food web, whereas the scarce ones remain, although it has been argued that these can be considered functionally extinct [10,16].

To understand the actual relevance of both extinction and scarcity of formerly abundant megafauna species, we need to assess their role in the pristine ecosystems in which they have evolved. This can be achieved by studying marine protected areas (e.g. [17]) but the influence of the human disturbed matrix in which they are embedded is difficult to control. Furthermore, marine reserves are often too small to support viable populations of large top predators or encompass entire foraging ranges [18] and have not been protected for long enough to guarantee full recovery [19]. A second approach is through ecosystem modelling (e.g. [20,21]), but this can be hampered by the limited perspective of contemporary data [3], the difficulties of model parameterisation [22] and unknown changes in the diet and foraging behaviour of predators through time (e.g. [23,24]). A third approach consists of using ancient biological material, such as bones and shells, to reconstruct trophic relationships in a time span prior to the anthropogenic alteration of marine ecosystems [3].

The mechanisms and underlying causes of a number of cases of population collapses or extinctions have been clarified by analysing the fossil and zooarchaeological records (e.g. [25]). However, this method brings little information about past diets, trophic levels and food-web structure. The development of stable isotope analyses and quantitative methods for analysing the food-web topology [26,27,28] allows the reconstruction of past trophic relationships based on ancient biological material [29], since the stable isotope ratios in consumer tissues represent those of their diet in a predictable manner [30,31]. The marine ecosystems off Argentina have suffered major anthropogenic changes during the past two centuries. Firstly, European sealers locally extirpated the South American fur seal (*Arctocephalus australis*) during the 19th century and were at least in part responsible for a decline in South American sea lion populations (*Otaria flavescens*) during the first half of the 20th century [32,33,34,35,36]. When exploitation ceased most populations of this species had been reduced to <10% of the pre-exploitation numbers and their recovery did not begin until the early 1990s in Argentina, after several decades of stagnation [34,37,38]. As a result, the current populations of both species are still well below pre-exploitation numbers [34,39,40,41,42]. Furthermore, fisheries targeting large demersal fishes were established in Río de la Plata and northern Patagonia in the 1970s, causing a population decline in several large benthic predatory fish species [43,44,45,46]. Finally, the Argentine population of the Magellanic penguin (*Spheniscus magellanicus*) increased during the 20th century, both in number of individuals and geographic range [47], perhaps due to the decline of competitors and predators because of the exploitation of pinnipeds [48].

However, the European sealers and fishermen were not the first humans to exploit the marine resources off Argentina. Aboriginal hunter-gatherers began to exploit the local populations of fur seals, sea lions and other marine species during the middle Holocene [49,50,51,52,53,54]. Even though the general opinion is that the impact on the populations of pinnipeds was minimal [49,55]. More recently, Zangrando and colleagues [56] argued that human pressures on fur seals in the Beagle Channel during the late Holocene might have resulted in demographic and ecological changes, based on evidence from stable isotope analyses and the decreasing age and size of hunted individuals. However, these results remain inconclusive because potential variations of the stable isotope baseline through time were not accounted for [57]. The zooarchaeological record preserved in the hunter-gatherers middens offers an excellent opportunity to reconstruct the structure of ancient marine food webs (while accounting for potential changes in stable isotopic baselines) in the South-western Atlantic and compare it with that of modern food webs.

In this paper, we use the stable isotopes of carbon and nitrogen from the shells of mollusc shells to set the stable isotopic baseline and from bones of marine mammals and penguins to reconstruct the changes in the topology of coastal food webs from central-northern Patagonia and southern Patagonia through the second half of the Holocene. Through this, we aim to assess whether the structure of the food web has changed as a consequence of human exploitation during the studied period.

Material and Methods

Ethics Statement

Permits to collect modern samples (Table 1 and S1) were issued by the "Dirección de Fauna y Flora Silvestre", and the "Dirección de Areas Protegidas", both from the Province of Chubut. The zooarchaeological samples come from previous fieldwork carried out by Julieta Gómez Otero, Florencia Borrella, Martín Serrán and Lorena Peralta in Golfo San Matías and Península Valdés (central-northern Patagonia), Eduardo Moreno in Santa Cruz (southern Patagonia) and Ernesto Piana, Luis Orquera, Angie Tivoli and Francisco Zangrando in the Beagle Channel (Tierra del Fuego). All the samples used for this study come from stranded animals, died naturally (except for mussels and limpets), or from archaeological remains (shell middens). All specimen numbers and repository information are shown in Tables 2 and 3, and in Tables S2 and S3. Due to the low proportion of organic matter present in the archaeological shells, the entire valve was used to undertake isotopic analyses. However, samples of the same ages are available to allow the work to be reproducible.

Study site and sample collection

Bones and shells are commonly used in paleontological and archaeological isotopic studies because they contain organic remains and they are often abundant in archaeological deposits as well as in historic museum collections [24,58,59,60]. Furthermore, the proteins they contain integrate the diet over several years [61,62,63]. Accordingly, we determined the ratios of stable isotopes of carbon and nitrogen in the organic matrix from the shell of modern and archaeological intertidal mussels (*Aulacomya atra atra* and *Mytilus edulis*) and limpets (*Nacella magellanica*), and in modern and archaeological bone tissue from South American sea lions, South American fur seals and Magellanic penguins. Mussels and limpets were used to characterise the trophic level 1 (herbivores), which allows us to interpret whether shifts in the predators are more likely linked to prey switching or a change in the isotopic baseline. Both modern and archaeological samples were collected in two areas of the South-east coast of South America (Figure 1): central-northern Patagonia (Río Negro and Chubut provinces) and southern Patagonia (Santa Cruz and Tierra del Fuego provinces). These two areas differ in oceanographic, biogeographic and anthropogenic features, with central-northern Patagonia being less productive and more affected by anthropogenic impacts (industrial fishing) than southern Patagonia [64,65].

Nasal turbinates from modern marine mammal were sampled from specimens at the scientific collections of the *Centro Nacional Patagonico* (Puerto Madryn, Argentina) and *Museo Acatushún* (Ushuaia, Argentina) and analysed. Modern penguin bones from adult penguins found dead at breeding colonies distributed along central-northern Patagonia and southern Patagonia were also analysed. Modern mollusc samples were collected from December 2009 to February 2010 at three sites along the coastline of Argentina: two in the Río Negro province (central-northern Patagonia) and one in the Tierra del Fuego province (southern Patagonia) (Figure 1, Table 1).

Sex is often difficult to assess when only fragmented individuals are recovered from the zooarchaelogical record, despite being a major determiner in the foraging habits of fur seals, sea lions and penguins [47,66,67,68,69]. Accordingly, a sensitivity analysis was run using modern samples from the Río de la Plata region (between southern Brazil and Buenos Aires province) to assess the sensitivity of food web reconstruction using SIBER (see Data analysis). Blue mussels (*Mytilus edulis*) were collected in the Río de

Table 1. Samples used to reconstruct the modern food webs.

Species	Area	N
Herbivores		
Mytilus edulis	Río de la Plata	5
Aulacomya atra atra	Northern Patagonia	7
Aulacomya atra atra	Southern Patagonia	5
Siphonaria lessoni	Río de la Plata	5
Nacella magellanica	Northern Patagonia	5
Nacella magellanica	Southern Patagonia	5
Top Predators		
Arctocephalus australis	Río de la Plata	60
Arctocephalus australis	Northern Patagonia	29
Arctocephalus australis	Southern Patagonia	7
Otaria flavescens	Río de la Plata	19
Otaria flavescens	Northern Patagonia	36
Otaria flavescens	Southern Patagonia	41
Spheniscus magellanicus	Río de la Plata	20
Spheniscus magellanicus	Northern Patagonia	20
Spheniscus magellanicus	Southern Patagonia	40

la Plata region to represent ribbed mussel (*Aulacomya atra atra*), and the limpet *Siphonaria lessoni* to represent *Nacella magellanica* (Table 1).

Zooarchaeological bones (generally humeri, mandibles, coxal bones, ribs and femurs) and shell samples recovered from shell middens come from previous fieldwork carried out by Julieta Gómez Otero, Florencia Borrella, Martín Serrán and Lorena Peralta in Golfo de San Matías and Península Valdés (central-northern Patagonia), Eduardo Moreno in Santa Cruz (southern Patagonia) and Ernesto Piana, Luis Orquera, Angie Tivoli and Francisco Zangrando in the Beagle Channel (Tierra del Fuego) (Figure 1, Tables 2 and 3). The samples were dated in different laboratories and using different methods; in particular, in central-northern Patagonia, almost all dated samples were marine shells instead of charcoal. As CO_2 diffuses slowly from the atmosphere to the ocean, changes in the abundance of ^{14}C in sea water are delayed in comparison to changes in the atmosphere [70]. Such a lag is known as the reservoir effect and is the responsible for the difference between the radiocarbonic age of coal and shells from the same archaeological level. Reservoir effect data for the central-northern Patagonia region are emerging only recently [71], and they suggest a relatively small difference between marine and terrestrial ages on the basis of a few samples. Although the use of a correction for reservoir effect would be preferable [52], we refer to the conventional, uncalibrated radiocarbon dates for all the sites. This shortcoming should not be of great importance in these types of studies for which precise dates are not required, but large time intervals should suffice.

All samples were stored in a freezer at $-20°C$ until analysis.

Stable isotope analysis

Once in the laboratory, bone and shell samples were thawed and dried in an oven at 50°C, and ground to a fine powder with a mortar and pestle. Shell samples were pre-polished with sandpaper and with a diamond wheel drill to remove impurities. They were subsequently rinsed with distilled water and lipids were extracted in all samples with a chloroform/methanol (2:1) solution [72].

Since both bone and shell contain high concentrations of inorganic carbon, which may bias $\delta^{13}C$ values [73], they were divided into two aliquots. The first was decarbonised by soaking in 0.5 N (bone) or 1 N (shell) hydrochloric acid (HCl) until no more CO_2 was released [74]. Since the HCl treatment adversely affects $\delta^{15}N$ values [75], the second aliquot was not treated with HCl and used for $\delta^{15}N$ determination.

Dried powdered samples and secondary reference standards were combusted at 900°C, and analysed in a continuous flow isotope ratio mass spectrometer (Flash 1112 IRMS Delta C Series EA Thermo Finnigan). For the shell samples, a CO_2 absorbent for elemental analyses (CaO/NaOH) was employed to avoid the saturation of the spectrometer during the analysis of the non-acid washed samples, constituted by 90% $CaCO_3$. Stable isotope abundance is expressed in standard δ notation relative to carbonate Pee Dee Belemnite and atmospheric nitrogen. Analyses were performed at the Science and Technology Centre (CCiT) of the University of Barcelona.

Data analysis

Limpets and mussels were analysed together under the "herbivore" category, representing, in the same functional group, both benthic (limpets) and pelagic (mussels) compartments [76]. Each top predator species was analysed separately. Archaeological data were grouped into broad time intervals to enable quantitative reconstruction of food webs. Consequently in central-northern Patagonia all the archaeological samples were pooled together under a single time interval, from 5200 ± 70 yr ^{14}C BP until 380 ± 60 yr ^{14}C BP. This period is entitled the "Pre-contact period" since it predates the arrival of European settlers, opposed to the term post-contact used in archaeology [51,53,77]. Ancient samples from southern Patagonia were split in two different periods. The early aboriginal period (EAP) ranging from 6000 to 1100 years ago was characterised by very high oceanic primary

Table 2. Archaeological samples from northern-central Patagonia.

Species	N	Laboratory ID	Archaeological site	Age (^{14}C yr BP)	Reference	Repository information
Herbivores						
Aulacomya atra atra	3	A7-cC1, C4, C5	Los Abanicos 1	380±60	[105]	CENPAT, Puerto Madryn (Argentina)
Aulacomya atra atra	2	A8-cC2, C4	Las Ollas conchero 1	610±60, 640±60	[52]	CENPAT, Puerto Madryn (Argentina)
Aulacomya atra atra	5	A5-cC1, C2, C3, C4, C5.	Playas Las Lisas 2-perfil 1	2140±50	[52]	CENPAT, Puerto Madryn (Argentina)
Aulacomya atra atra	2	A11-cC3, C4	Cracker 8-Nivel 3	5200±70	[52]	CENPAT, Puerto Madryn (Argentina)
Nacella magellanica	5	A7-cL1, L2, L3 L4, L5.	Los Abanicos 1	380±60	[105]	CENPAT, Puerto Madryn (Argentina)
Nacella magellanica	5	A10-cL1, L2, L3, L4, L5.	Ecocentro Fogón 3	850±150	[52]	CENPAT, Puerto Madryn (Argentina)
Nacella magellanica	4	A4- cL1, L3, L4	Playa Las Lisas 2-conchero 2	2600±60	[52]	CENPAT, Puerto Madryn (Argentina)
Nacella magellanica	2	A11-cL1, L2	Cracker 8-Nivel 3	5200±70	[52]	CENPAT, Puerto Madryn (Argentina)
Predators						
Arctocephalus australis	3	F1 18, F1 19, FM1 11	Playa Unión-Barranca Norte	1040±70	Peralta, 2001 quoted in [52]	CENPAT, Puerto Madryn (Argentina)
Arctocephalus australis	2	90, 91	Bajada de los pescadores	2197±38	[51]	INCUAPA-UNCPBA, Buenos Aires F.D. (Argentina)
Otaria flavescens	2	36, 82	Los Abanicos 1	380±60	[105]	CENPAT, Puerto Madryn (Argentina)
Otaria flavescens	9	F1 17, FM1 13–17, FM1 19–21	Playa Unión-Barranca Norte	1040±70	Peralta, 2001 quoted in [52]	CENPAT, Puerto Madryn (Argentina)
Otaria flavescens	10	i1(61), i3(43), i4(44), i5(54), i15(C1), i18(89), i23, i24, M11, M13	Lobos*	1290±100	[106]	CENPAT, Puerto Madryn (Argentina)
Otaria flavescens	2	FSM - SRH Mont II OF costilla, OBS 13	Faro San Matías- Sondeo 6	1380±80	[51]	INCUAPA-UNCPBA, Buenos Aires F.D. (Argentina)
Otaria flavescens	1	OBS 137	Bajada de los pescadores	2197±38	[51]	INCUAPA-UNCPBA, Buenos Aires F.D. (Argentina)
Otaria flavescens	2	FSM - SRH Mont I OF cost px med, OBS 4	Faro San Matías, Sondeo 2	2910±90	[51]	INCUAPA-UNCPBA, Buenos Aires F.D. (Argentina)
Spheniscus magellanicus	4	7,72, 80, 82	Bajada de los pescadores	2197±38	[51]	INCUAPA-UNCPBA, Buenos Aires F.D. (Argentina)
Spheniscus magellanicus	3	OBS 26, FSM-S2N2 Obs 21, FSM-S2N3 Obs26-tibia derecha.	Faro San Matías, Sondeo 2	2910±90	[51]	INCUAPA-UNCPBA, Buenos Aires F.D. (Argentina)

*Lobos is a paleontological site, since here sea lions died naturally and massively due to a land slide.

productivity [78] and the prevalence of pinnipeds in the economy of hunter-gatherers [50]. Conversely, the late aboriginal period (LAP), ranging from 1000 to 100 years ago, was characterised by a much lower oceanic primary productivity [78] and a lower reliance of the hunter-gatherers economy on pinnipeds. Modern data were analysed separately, so five different food webs were analysed: two in central-northern Patagonia (one archaeological and one modern) and three in southern Patagonia (two archaeological and one modern).

SIBER (Stable Isotope Bayesian Ellipses in R) [79] was used to compute Bayesian Layman's metrics that summarised food web structures in each region and epoch. Only five of the six measures proposed by Layman et al. (2007) [27] were calculated. The δ^{15}N, δ^{13}C ranges (NR and CR, respectively) and the mean distance to centroid (CD) are measures of the total extent of spacing within the δ^{13}C–δ^{15}N bi-plot space and gives a measure of community niche width that is not particularly sensitive to sample size. Nitrogen range (NR) is the representation of the vertical structure of the web; carbon range (CR) gives us an idea of the trophic diversity at the base of the web, while the mean distance to centroid provides a measure of the average degree of trophic diversity within a food web [27]. Mean nearest neighbour distance (MNND) and the

Table 3. Archaeological samples from southern Patagonia.

Species	N	Laboratory ID	Archaeological site	Age (^{14}C yr BP)	Reference	Repository information
Herbivores						
Mytilus edulis	5	A17-cM1, M2, M3, M4, M5	Tunel VII	100±45	Piana et al., 1992 quoted in [107]	CENPAT, Puerto Madryn (Argentina)
Mytilus edulis	3	A15-cM1, M2, M3	Shamakush X, Capa E	500±100	Orquera and Piana, 1999 quoted in [50]	CENPAT, Puerto Madryn (Argentina)
Mytilus edulis	5	A19-cM1, M2, M3, M4, M5	Imiwaia I (M/K)	5940±50, 5750±170, 5840±45, 5710±50	Piana et al., 1992 quoted in [107]	CENPAT, Puerto Madryn (Argentina)
Nacella magellanica	5	A17-cL1, L2, L3, L4, L5	Tunel VII	100±45	Piana et al., 1992 quoted in [107]	CENPAT, Puerto Madryn (Argentina)
Nacella magellanica	2	A15-cL2, L4	Shamakush X, Capa E	500±100	Orquera and Piana, 1999 quoted in [50]	CENPAT, Puerto Madryn (Argentina)
Nacella magellanica	5	A19-cL1, L2, L3, L4, L5	Imiwaia I (M/K)	5940±50, 5750±170, 5840±45, 5710±50	Piana et al., 1992 quoted in [107]	CENPAT, Puerto Madryn (Argentina)
Top predators						
Arctocephalus australis	13	44331, 155288, 150329, 152253, 152439, 151607, 154656, 151575, 151912, 154284, 153887, 55456, 155447	Tunel VII, layer B	100±45	Piana et al., 1992 quoted in [107]	CADIC, Ushuaia (Argentina)
Arctocephalus australis	1	CV6 4/-45–50 cm	CV6 4/-45-50 cm	1190±60	[108]	INCUAPA-UNCPBA/ IMHICIHU, Buenos Aires F. D. (Argentina)
Arctocephalus australis	1	CdN2-0072	Cueva del Negro-cuadrícula 1 Nivel 2	1730±80	[109]	CENPAT, Puerto Madryn (Argentina)
Arctocephalus australis	3	37295, 37456, 37340	Tunel I, Capa X/α	2660±100,	[110]	CADIC, Ushuaia (Argentina)
				2690±80, 3030±90		
Arctocephalus australis	10	194047, 43247, 174498, 193261, 67319, 66397, 202083, 186854, 65989, 202401	Tunel I, Capa D	5000–4300	[110]	CADIC, Ushuaia (Argentina)
Arctocephalus australis	20	217933, 215241, 223614, 215940/215933, 53580, 68445, 64460, 213370, 215074, 58630, 190846, 63330, 213732, 52463, 226119, 69639, 189603, 216713, 212616/212653, 224151	Tunel I, Capa D/E	6400–5900	[110]	CADIC, Ushuaia (Argentina)
Otaria flavescens	2	43418, 154286	Tunel VII	100±45	Piana et al., 1992 quoted in [107]	CADIC, Ushuaia (Argentina)
Otaria flavescens	10	OF 2a, OF 2b, OF 3, OF 9, OF 12, OF 14, OF 15, OF 5, OF 7, OF 10	Kaiyawoteha III, Capa K	580±45	[111]	CADIC, Ushuaia (Argentina)
Otaria flavescens	1	CV6 4/-60–65 cm	CV6 4/-60-65 cm	1190±60	[108]	INCUAPA-UNCPBA/ IMHICIHU, Buenos Aires F. D. (Argentina)
Otaria flavescens	5	CV20 OF1, CV20 OF3, CV20 OF4, CV20 OF6, CV20 OF7	CV20	1256±50	[112]	CONICET, Río Gallegos/ CONICET-IMHICIHU, Buenos Aires (Argentina)
Otaria flavescens	9	30459, 33459, 33551, 33571, 33717, 34177, 34544, 34751, 33458	Tunel I, Capa D	5000–4300	[110]	CADIC, Ushuaia (Argentina)
Spheniscus magellanus	2	10030, 10100	Shamakush X, Capa E	500±100	Orquera and Piana, 1999 quoted in [50]	CADIC, Ushuaia (Argentina)
Spheniscus magellanus	3	pingüino 4a, pingüino 4b, pingüino 6,	Kaiyawoteha III, Capa K	580±45	[111]	CADIC, Ushuaia (Argentina)

Table 3. Cont.

Species	N	Laboratory ID	Archaeological site	Age (^{14}C yr BP)	Reference	Repository information
Spheniscus magellanus	4	12433, 12268, 10116, 10115	Mischiuen I, Capa C sup	890±90	Piana et al., 2004 quoted in [113]	CADIC, Ushuaia (Argentina)
Spheniscus magellanus	1	9255	Mischiuen I, Capa C inf	1060±85	Piana et al., 2004 quoted in [113]	CADIC, Ushuaia (Argentina)
Spheniscus magellanus	2	19098, 19264	Shamakush I, Capa D	1220±110, 940±110	[114]	CADIC, Ushuaia (Argentina)
Spheniscus magellanus	1	10122	Shamakush I, Capa C	1000 (ca.)	[114]	CADIC, Ushuaia (Argentina)
Spheniscus magellanus	4	3761, 4179, 3522, 3641	Mischiuen I, Capa F	4890±210, 4430±180	Piana et al., 2004 quoted in [113]	CADIC, Ushuaia (Argentina)
Spheniscus magellanus	3	1925, 26006, 27597	Imiwaia I (M/K)	5940±50, 5750±170, 5840±45, 5710±50	Piana et al., 1992 quoted in [107]	CADIC, Ushuaia (Argentina)

standard deviation of nearest neighbour distance (SDNND) reflect the relative position of species to each other within the niche space and can be used to estimate the extent of trophic redundancy.

Furthermore, the overlap (‰) among top predators' standard ellipses corrected for small sample size (SEA_C) was calculated in order to analyse resource partitioning among them over time. Comparison of metrics was based on 95% credibility intervals. All codes for SIBER analyses are contained in the package SIAR [80,81]. Finally, SIBER Bayesian ellipse areas (SEA_B) were calculated for the three top predators to measure their isotopic niche width. This approach is similar to a bootstrap, assigning measures of uncertainty, based on Markov-Chain Monte Carlo (MCMC) simulation to construct parameters of ellipses.

In addition, a simulation was performed to test the sensitivity of SIBER metrics to biases in the sex ratio of the top predators. A similar data set from Río de la Plata was used for this analysis (see Table S1). Three different scenarios were simulated: one where only females were included, one where only males were included and a third situation with a balanced sex ratio. These three simulated food webs should represent the three hypothetical and extreme situations emerging from the analysis of the zooarchaeological record, where the sex of the top predators was usually unknown.

Results

The structures of the five food webs analysed are shown in Figure 2. The $\delta^{15}N$ values of herbivores decreased from past to present (Table 4) both in central-northern Patagonia (Wilcoxon-Mann-Whitney test: $W = 3$, $p<0.001$) and in southern Patagonia (ANOVA: $F_{2,32} = 24.7$, $p<0.01$). Likewise the $\delta^{13}C$ values of ancient herbivores from southern Patagonia differed from the modern ones (ANOVA: $F_{2,32} = 5.42$, $p<0.01$) but changes were not statistically significant in central-northern Patagonia (t-test: $t = -0.09$, $df = 16.777$, $p = 0.93$). These results demonstrate that the isotopic baseline may change dramatically throughout time and allow us to properly interpret the structure of the ancient food webs.

The three food webs simulated for the sensitivity analysis did not differ in any of the Layman's metrics (Table 5). The areas of the Bayesian ellipses (SEA_B) estimated for the top predators did not either differ among scenarios either, except for fur seals, whose area was maximised when sex ratio was balanced. However, large differences were observed in the overlap between the standard ellipses corrected for small sample size (SEA_C) of fur seals and penguins: it ranged from zero when only male fur seals were considered, to 43.73% (fur seals) and 21.34% (penguins) when only females were included. This indicates that the lack of information on the actual sex ratio of ancient data sets is unlikely to bias the overall structure of the food web. Nonetheless caution is needed when interpreting patterns of niche overlap, as they may be sensitive to the sex ratio of the sample.

The horizontal structure of the food web (CR) did not vary throughout time in central-northern Patagonia but in southern Patagonia the carbon range is currently smaller than in the two past periods considered (Figure 3 and 4). The nitrogen range (NR), indicative of food chain length, increased from past to present in both regions, and did not differ between ancient food webs from southern Patagonia (Figure 3 and 4). Finally, the food webs from southern Patagonia were shorter than those from central-northern Patagonia during the late Holocene as well as in the present.

CD and MNND also changed throughout time in both regions, being smaller in the past. This indicates an increase in the trophic

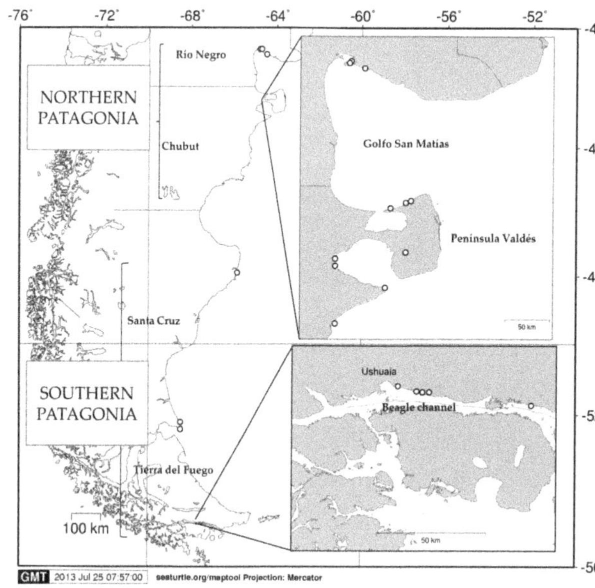

Figure 1. Study areas (central-northern Patagonia and southern Patagonia) and sampling approximate zones. Empty dots represent archaeological sites.

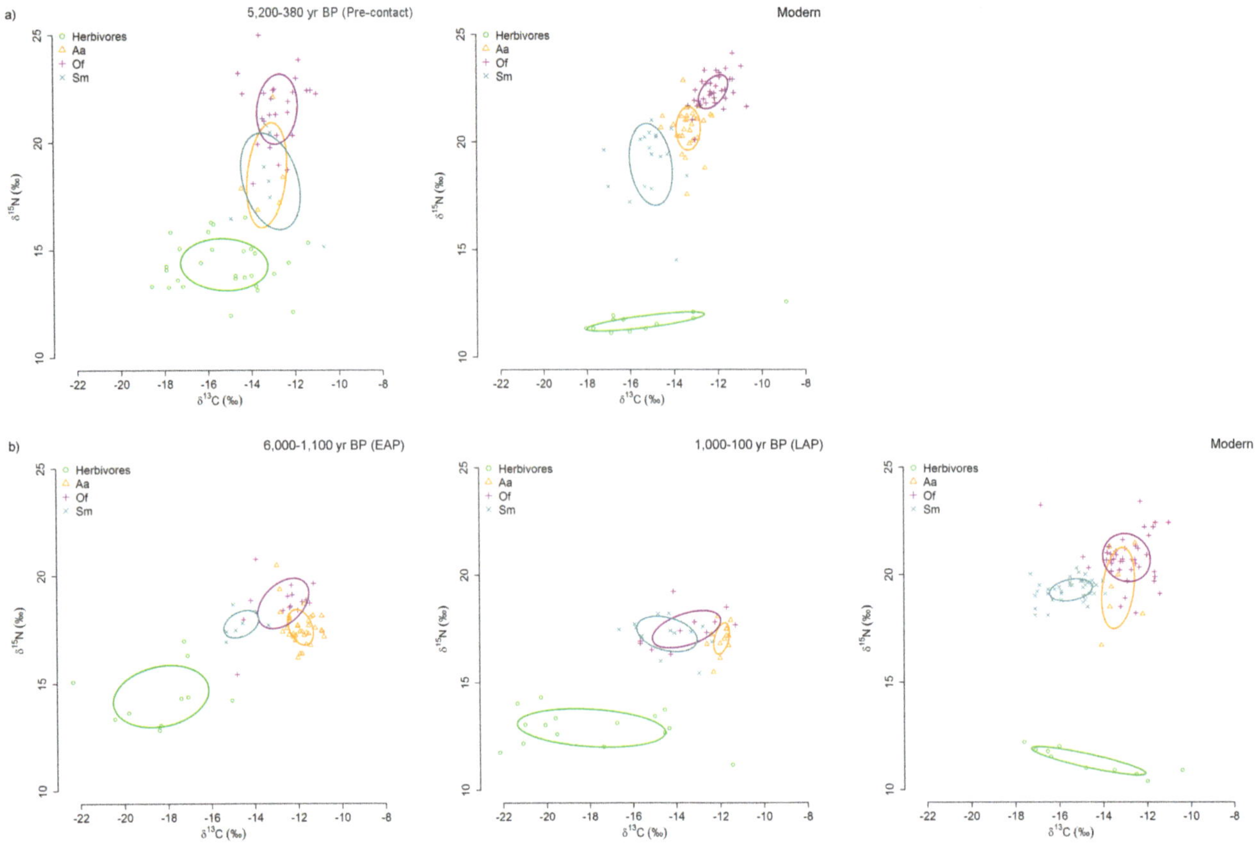

Figure 2. Isotopic niches/resource use areas of the species/functional groups described in the text and calculated with the standard ellipse areas corrected for small sample size (SEA_C) over time in the two geographical areas. a) Central-northern Patagonia and b) Southern Patagonia. Herbivores = mussels and limpets; Aa = South American fur seals; Of = South American sea lions; Sm = Magellanic penguins.

diversity and a decrease in the trophic redundancy in modern food webs. Regarding the three species of air predators the areas of the standard ellipses did not overlap in the modern $\delta^{13}C$–$\delta^{15}N$ bi-plot space of central-northern Patagonia (Table 6 and Figure 2). Conversely, the topology of the three species within the Pre-contact $\delta^{13}C$–$\delta^{15}N$ bi-plot space of the same region departed dramatically from expectations, as the standard ellipses of all the species overlapped one another, especially those of Magellanic penguins and fur seals, in sharp contrast with the low overlap currently observed (Table 6). The topology of the three species in the $\delta^{13}C$–$\delta^{15}N$ bi-plot space of southern Patagonia also changed over time, as the overlap among the ellipses of the three species was almost zero in the EAP period (except between sea lions and fur seals where it was 15.38%), increased between penguins and sea lions in the LAP period and is currently about 34% between fur seals and sea lions. Interestingly, the SIBER ellipse areas (SEA_B) calculated for each species (Table 7) exhibited a decreasing trend throughout time in central-northern Patagonia, although it was statistically significant only for sea lions. Conversely in southern Patagonia, the sea lions' ellipse area has been constant over time, whereas that of fur seals showed a trend to increase through time and that of penguins slightly increased from the EAP period to the LAP period to then slightly decreased in the present food web (Table 7).

All the results of the stable isotopes analyses described in the above are publicly available in Table S1–S3.

Discussion

The study of the zooarchaeological record using stable isotope analysis certainly offers a window to explore the structure of ancient food webs, but it is not free from problems. Historical changes in the isotopic baseline are a major shortcoming in retrospective studies using stable isotope analysis to assess changes in trophic level and food web structure [57]. Although widely recognised as a confusing factor, previous studies often assumed the temporal stability of the isotopic baseline [82,83,84,85,86,87,88,89], but the results reported here demostrate that changes can be dramatic. Post [76] suggested the use of filter feeding mussels and surface-grazing snails as proxies for the baselines of pelagic and littoral aquatic food webs respectively and Bailey and colleagues [88] and Casey and Post [57] recognised the potential of mollusc shells to reconstruct the isotopic baseline in retrospective studies, since their organic matrix is encased within mineral crystals and hence, preserved [90]. The present study demonstrates the potential of the method and offers new perspectives using material that is widely available in paleontological and archaeological collections.

Another drawback from retrospective studies is the uneven distribution of specimens across space and time, which forced us to pool samples of disparate radiocarbon age to reconstruct ancient communities. Uncertainty about the gender of most specimens is another setback of the zooarchaeological record, as sexual secondary characters can seldom be observed. The sensitivity analysis conducted here confirmed the robustness of Layman's

Table 4. Mean and standard deviation of nitrogen and carbon stable-isotope in central-northern and southern Patagonia.

Central-northern Patagonia				$\delta^{15}N$ (‰)		$\delta^{13}C$ (‰)	
Group	**Tissue**	**N**		**Mean (±SD)**		**Mean (±SD)**	
		Pre-contact	**Modern**	**Pre-contact**	**Modern**	**Pre-contact**	**Modern**
Herbivores	Shell	28	12	14.30 (±1.19)	11.61 (±0.42)	−15.25 (±1.97)	−15.32 (±2.57)
Top Predators							
Arctocephalus australis	Bone	5	29	18.45 (±2.12)	20.59(±0.99)	−13.26 (±0.80)	−13.32 (±0.54)
Otaria flavescens	Bone	26	36	21.52 (±1.60)	22.30 (±0.78)	−12.77 (±0.91)	−12.17 (±0.66)
Spheniscus magellanicus	Bone	7	20	18.19(±2.05)	18.93 (±1.85)	−13.10 (±1.25)	−15.03 (±0.94)

Southern Patagonia					$\delta^{15}N$ (‰)			$\delta^{13}C$ (‰)		
Group	**Tissue**	**N**			**Mean (±SD)**			**Mean (±SD)**		
		EAP	**LAP**	**Modern**	**EAP**	**LAP**	**Modern**	**EAP**	**LAP**	**Modern**
Herbivores	Shell	10	15	10	14.39 (±1.36)	12.87 (±0.86)	11.32 (±0.62)	−18.32 (±2.07)	−17.93 (±3.31)	−14.69 (±2.46)
Top Predators										
Arctocephalus australis	Bone	35	13	7	17.58 (±0.81)	17.02 (±0.69)	19.34 (±1.72)	−11.86 (±0.58)	−11.92 (±0.34)	−13.31 (±0.69)
Otaria flavescens	Bone	15	12	41	18.70 (±1.13)	17.47 (±0.82)	20.76 (±1.10)	−12.66 (±1.13)	−13.52 (±1.48)	−12.92 (±1.07)
Spheniscus magellanicus	Bone	7	13	40	17.75 (±0.58)	17.25 (±0.79)	19.27 (±0.51)	−14.61 (±0.74)	−14.43 (±1.33)	−15.46 (±0.96)

Table 5. Probability values of Layman's metrics in three simulated model food webs for Río de la Plata region, including only females, males or both sexes in the samples of predators.

	NR	CR	CD	MNND	SDNND
♀♀	8.01 (7.00–9.04)	6.03 (4.46–7.64)	3.14 (2.77–3.51)	2.95 (2.50–3.41)	2.43 (1.55–3.35)
♂♂	8.32 (7.82–8.82)	6.16 (4.75–7.55)	3.44 (3.13–3.76)	3.23 (2.90–3.55)	2.50 (1.80–3.22)
♀♂	8.22 (7.55–8.87)	6.03 (4.57–7.42)	3.25 (2.93–3.57)	3.14 (2.75–3.55)	2.39 (1.69–3.10)

NR = nitrogen range; CR = carbon range; CD = mean distance to centroid; MNND = mean nearest neighbour distance; SDNND = standard deviation of nearest neighbour distance. For more details see *Data analysis* in Material and Methods.

metrics to changes in sex ratios of penguins, sea lions and fur seals. Conversely, the overlap between the Bayesian ellipses was highly sensitive to the sex ratio of those species and hence caution is needed when interpreting those results.

The overall evidence reported here indicates that modern marine food webs of central-northern and southern Patagonia are longer (NR) than the ancient ones (Figs 2–4). Such a conclusion is in sharp contrast with the idea that human exploitation has shortened food webs because of the preferential removal of top predators [4]. Certainly, increased scarcity of large, marine species may have forced fisheries to target smaller species, but there is no reason why surviving top predators had to experience a similar shift. Marine predators are limited by the size of the prey they can consume and draw their energy from a very limited range of trophic levels, in contrast with fisheries [91]. Furthermore, it should be noted that the impact of sealing on the populations of fur seals and sea lions was much larger [39,40] than that of fishing on the populations of hake, squid and anchovies [20].

Surviving sea lions and fur seals off Argentina have certainly been under carrying capacity during the second half of the 20th century, when the population was less than 10% of the original numbers [39]. This reduced intraspecific competition and led to a major dietary shift in favour of benthic, larger prey with a higher trophic level [24]. Only recently, as the population of sea lions is rebuilding and approaching carrying capacity, sea lions experience resource limitation again and intraspecific competition forces them to increase the consumption of smaller and less profitable prey [92]. Information about the historical dietary changes of South American fur seals is restricted to the last three decades in Río de la Plata and reveals no major dietary changes [93]. However, the population was dramatically reduced during the first half of the 20th century and evidence from other fur seal species in the South-western Atlantic Ocean [94] suggest that this group of smaller and more pelagic pinnipeds also forage at a higher trophic level after severe population declines.

Trophic diversity, measured as the mean distance to the centroid in the isotopic space (CD) also increases in the present food webs, while redundancy (MNND) decreases thus revealing a higher trophic overlap in the past (Figs 2–4). This was confirmed by the isotopic niche overlap of the three air-breathing top predators, although they differed dramatically in body size and mouth diameter (Table 6). A similar scenario has been reported for waters off Peru, where sea lions, fur seals and penguins primarily rely on the large population of anchoveta *Engraulis ringens*, supported by intense oceanic productivity [95,96,97]. Similarly, a high trophic overlap has often been reported for other wasp-waisted ecosystems in upwelling regions [98]. Currently, a wasp-waist ecosystem supported by amphipods exists off southern Patagonia [99], but amphipods are not consumed directly by fur seals and sea lions. Perhaps a similar wasp-waisted ecosystem supported by small schooling fish might have existed off Patagonia during the Holocene, where primary productivity was usually higher than now [78] and hence it could have supported a larger population of small schooling fish and squids.

However, there are at least two reasons to believe that this is not the explanation for the high levels of overlap and redundancy of the ancient food webs reported here. Firstly, the large population of fur seals inhabiting the highly productive Río de la Plata rely heavily on anchovies and squids [93], but the much smaller sea

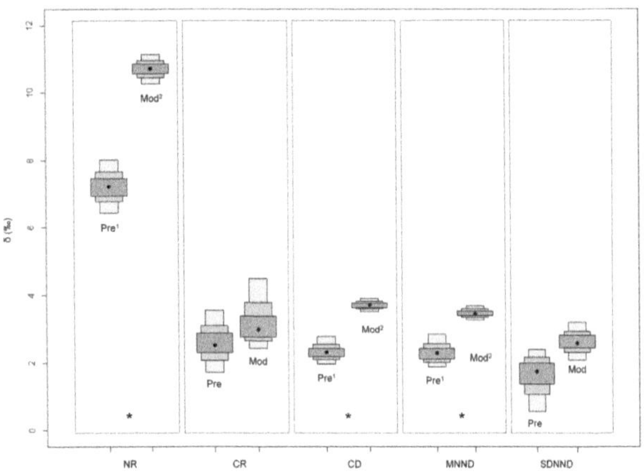

Figure 3. Probability values of Layman's metrics over the time in the central-northern Patagonia. Metrics that present differences over the time are indicated by an asterisk and superscripts. Pre = pre-contact period; Mod = modern period.

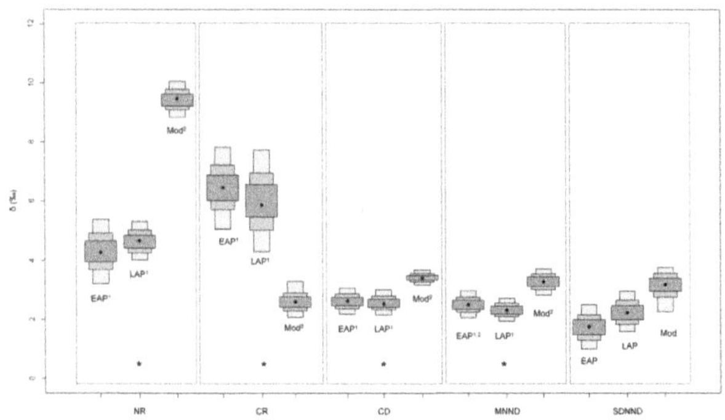

Figure 4. Probability values of Layman's metrics over the time in the southern Patagonia. Metrics that present differences over the time are indicated by an asterisk ad superscripts. Groups sharing the same superscript number (1, 2) are not significantly different. EAP = early aboriginal period; LAP = late aboriginal period; Mod = modern period.

Table 6. Overlap as a percentage of standard ellipses corrected for small samples (SEA_c) of the three top predators described in the text.

Central-northern Patagonia											
PRE-CONTACT	**Aa**	**Of**	**Sm**	**MODERN**	**Aa**	**Of**	**Sm**				
Aa	1	11.37	86.30	**Aa**	1	0.00	0.00				
Of	16.42	1	5.05	**Of**	0.00	1	0.00				
Sm	64.56	2.62	1	**Sm**	0.00	0.00	1				
Southern Patagonia											
EAP	**Aa**	**Of**	**Sm**	**LAP**	**Aa**	**Of**	**Sm**	**MODERN**	**Aa**	**Of**	**Sm**
Aa	1	15.38	0.00	**Aa**	1	0.00	0.00	**Aa**	1	34.19	0.00
Of	5.60	1	0.05	**Of**	0.00	1	53.24	**Of**	38.42	1	0.00
Sm	0.00	0.13	1	**Sm**	0.00	58.28	1	**Sm**	0.00	0.00	1

The tables should be read horizontally, as each number in the cell refers to the percentage of overlap of the area of the group indicated in each row (e.g. 11.37% is the percentage of the ellipses of fur seals that are overlapped with the ellipses of the sea lions, while 16.42% is the percentage of the ellipses of the sea lions that are overlapped with the fur seals). Aa = *Arctocephalus australis*; Of = *Otaria flavescens*; Sm = *Spheniscus magellanicus*.

lion population consumes primarily demersal fishes [100]. Secondly, marine productivity during the late aboriginal period off southern Patagonia was much lower than during the early aboriginal period and closer to actual productivity levels [78], but the structure of the trophic web was indistinguishable from that observed in the same region during the early aboriginal period. Thus, a direct link between productivity, predator diet and the structure of trophic overlap is probably unclear.

Empirical evidence has indicated that marine environments with lower annual temperature variability have smaller predator-prey mass ratios and, consequently, longer food chains [101]. It can be argued that changes in sea water temperature throughout the second half of the Holocene may explain the difference observed between the ancient and modern marine food webs in the South-western Atlantic. However, available evidence rules out major changes in sea surface temperature during the considered time span [51,55,102]. Furthermore, there is an inverse relationship between the length of the current food webs and the seasonal variability in the sea surface temperature in the South-western Atlantic [103], a pattern opposite to that reported by Jennings and Warr (2003)[101]. Finally, the difference in the length of the food chain observed between the two areas in both periods is smaller than the difference between periods in both areas.

Thus, human disturbance stands as the most likely reason for the differences in the structure of the food web reported here. Even so, human disturbance is not the only factor explaining the length of marine food webs, as those from southern Patagonia were always shorter and more overlapping than the central-northern Patagonia ones. This might be a consequence of the latitudinal decrease in species diversity reported in the South-western Atlantic [64,104] and was also probably the reason in the late Holocene, since it is unlikely that the impact of the hunter-gatherers inhabiting central-northern Patagonia on marine resources was larger than that of the hunter-gatherers inhabiting southern Patagonia [51,52,53].

In conclusion, this study strongly support the hypothesis that selective exploitation of marine ecosystems, targeting primarily top predators, leads to longer and less overlapping food webs, if top predators are not extinct but survive well below carrying capacity. The situation might be different if human exploitation targeted primarily intermediate trophic levels as a result of the legal protection of top predators.

Furthermore, this study demonstrates the necessity to reconstruct the isotopic baseline in retrospective studies and how this can be achieved by analysing the organic matter encased into the shell of molluscs.

Table 7. Areas of the estimated Bayesian ellipses of the species/functional groups over the time in central-northern and southern Patagonia.

Central-northern Patagonia	**Herbivores**	**Aa**	**Of**	**Sm**
Pre-contact	7.4 (4.9–10.3)	6.1 (1.9–11.7)	4.7 (3.0–6.6)[1]	7.7 (3.1–13.6)
Modern	3.9 (2.0–6.3)	1.9 (1.3–2.6)	1.6 (1.1–2.1)[2]	5.5 (3.3–8.1)
Southern Patagonia				
EAP	8.6 (4.1–14.2)	1.6 (1.1–2.1)	4.0 (2.1–6.1)	2.2 (0.9–3.8)
LAP	9.0 (5.0–13.9)	1.3 (0.7–2.0)	3.9 (2.0–6.2)	3.5 (1.8–5.5)
Modern	4.1 (1.9–6.7)	4.4 (1.8–7.8)	3.8 (2.7–5.0)	1.7 (1.2–2.2)

Values are indicated as the mean and the 95% of credibility interval in parentheses. Metrics that present differences between ages are indicated by superscripts. Aa = *Arctocephalus australis*; Of = *Otaria flavescens*; Sm = *Spheniscus magellanicus*.

Supporting Information

Table S1 Modern nitrogen and carbon stable-isotope ratios in Río de la Plata and adjoining areas. These data have been used to perform the sensitivity test for the method.

Table S2 Modern and archaeological nitrogen and carbon stable-isotope ratios in central-northern Patagonia. Table shows the list of samples, grouped according to their historical period (modern and pre-contact), and $\delta^{13}C$ and $\delta^{15}N$ values.

Table S3 Modern and archaeological nitrogen and carbon stable-isotope ratios in southern Patagonia. Table shows the list of samples, grouped according to their historical period (modern, LAP and EAP), and $\delta^{13}C$ and $\delta^{15}N$ values.

Acknowledgments

Authors are grateful to Pilar Rubio and Pilar Texidor of the Centres Científics i Tecnològics (CCiT-UB) of the University of Barcelona, where the stable isotope analyses were carried on.

Authors also thank the logistical support provided by the Centro Nacional Patagónico (CENPAT-CONICET) and all the archaeologists who provided the ancient samples: Julieta Gómez Otero, Florencia Borrella, Martín Serrán and Lorena Peralta in Golfo de San Matías and Península Valdés (central-northern Patagonia), Eduardo Moreno, Luis Borrero, Juan Bautista Belardi and Florencia Borella in Santa Cruz (southern Patagonia) and Ernesto Piana, Luis Orquera, Angie Tivoli and Francisco Zangrando in the Beagle Channel (Tierra del Fuego). Finally, the authors wish to acknowledge Maptool program for maps presented in this paper. Maptool is a product of SEATURTLE.ORG (information is available at www.seaturtle.org).

Author Contributions

Conceived and designed the experiments: FS LC EAC AA. Performed the experiments: FS DV LS LZ. Analyzed the data: FS SB LC. Contributed reagents/materials/analysis tools: AA LC SB. Contributed to the writing of the manuscript: FS LC SB.

References

1. Caddy JF (1998) How pervasive is "fishing down marine food webs"? Science 282: 1383.
2. Devine JA, Baker KD, Haedrich RL (2006) Fisheries: deep-sea fishes qualify as endangered. Nature 439: 29.
3. Jackson JB, Kirby MX, Berger WH, Bjorndal KA, Botsford LW, et al. (2001) Historical overfishing and the recent collapse of coastal ecosystems. Science 293: 629–637.
4. Pauly D, Christensen V, Dalsgaard J, Froese R, Torres FJ (1998) Fishing down marine food webs. Science 279: 860–863.
5. Halpern BS, Walbridge S, Selkoe KA, Kappel CV, Micheli F, et al. (2008) A global map of human impact on marine ecosystems. Science 319: 948–952.
6. Rivadeneira MM, Santoro CM, Marquet PA (2010) Reconstructing the history of human impacts on coastal biodiversity in Chile: constraints and opportunities. Aquatic Conservation: Marine and Freshwater Ecosystems 20: 74–82.
7. Christensen V, Gue S, Heymans JJ, Walters CJ, Watson R, et al. (2003) Hundred-year decline of North Atlantic predatory fishes. Fish and fisheries 4: 1–24.
8. Lewison RL, Freeman SA, Crowder LB (2004) Quantifying the effects of fisheries on threatened species: the impact of pelagic longlines on loggerhead and leatherback sea turtles. Ecology Letters 7: 221–231.
9. Myers RA, Worm B (2003) Rapid worldwide depletion of predatory fish communities. Nature 423: 280–283.
10. Lotze HK, Coll M, Dunne JA (2011) Historical Changes in Marine Resources, Food-web Structure and Ecosystem Functioning in the Adriatic Sea, Mediterranean. Ecosystems 14: 198–222.
11. Post DM (2002) The long and short of food-chain length. Trends in Ecology & Evolution 17: 269–277.
12. Vander Zanden MJ, Fetzer WW (2007) Global patterns of aquatic food chain length. Oikos 116: 1378–1388.
13. McClenachan L, Cooper AB, Carpenter KE, Dulvy NK (2012) Extinction risk and bottlenecks in the conservation of charismatic marine species. Conservation Letters 5: 73–80.
14. Worm B, Barbier EB, Beaumont N, Duffy JE, Folke C, et al. (2006) Impacts of biodiversity loss on ocean ecosystem services. Science 314: 787–790.
15. Dulvy NK, Pinnegar JK, Reynolds JD (2009) Holocene extinctions in the sea. In: Turvey ST, editor. Holocene Extinctions. Oxford: Oxford University Press. pp. 129–150.
16. Sala E (2004) The Past and Present Topology and Structure of Mediterranean Subtidal Rocky-shore Food Webs. Ecosystems 7: 333–340.
17. Libralato S, Coll M, Tempesta M, Santojanni A, Spoto M, et al. (2010) Food-web traits of protected and exploited areas of the Adriatic Sea. Biological Conservation 143: 2182–2194.
18. Kramer DL, Chapman MR (1999) Implications of fish home range size and relocation for marine reserve function. Environmental Biology of Fishes 55: 65–79.
19. García-Rubies A, Hereu B, Zabala M (2013) Long-Term Recovery Patterns and Limited Spillover of Large Predatory Fish in a Mediterranean MPA. PLoS ONE 8: e73922.
20. Koen-Alonso M, Yodzis P (2005) Multispecies modelling of some components of the marine community of northern and central Patagonia, Argentina. Canadian Journal of Fisheries and Aquatic Sciences 62: 1490–1512.
21. Coll M, Navarro J, Olson RJ, Christensen V (2012) Assessing the trophic position and ecological role of squids in marine ecosystems by means of food-web models. Deep Sea Research Part II: Topical Studies in Oceanography: 1–16.
22. Schweder T (2006) The Scenario Barents Sea study: a case of minimal realistic modelling to compare management strategies for marine ecosystems. In: Boyd I, Wanless S, Camphuysen CJ, editors. Top predators in marine ecosystems Their role in monitoring and management: Cambridge University Press. pp. 310–323.
23. Chamberlain CP, Waldbauer JR, Fox-Dobbs K, Newsome SD, Koch PL, et al. (2005) Pleistocene to recent dietary shifts in California condors. Proceedings of the National Academy of Sciences of the United States of America 102: 16707–16711.
24. Drago M, Crespo E, Aguilar A, Cardona L, García N, et al. (2009) Historic diet change of the South American sea lion in Patagonia as revealed by isotopic analysis. Marine Ecology Progress Series 384: 273–286.
25. Anderson A (2008) Short and sometimes sharp: human impacts on marine resources in the archaeology and history of South Polynesia. In: Erlandson J, Rick TC, editors. Human impacts on ancient marine ecosystems: a global perspective. Berkley: University of California Press. pp. 21–42.
26. Bearhop S, Adams CE, Waldron S, Fuller RA, Susan EA, et al. (2004) Determining trophic niche width: a novel approach using stable isotope analysis. Journal of Animal Ecology 73: 1007–1012.
27. Layman CA, Arrington A, Montaña CG, Post DM (2007) Can stable isotope ratios provide for community-wide measures of trophic structure? Comment. Ecology 88: 42–48.
28. Schmidt SN, Olden JD, Solomon CT, Vander Zanden MJ (2007) Quantitative approaches to the analysis of stable isotope food web data. Ecology 88: 2793–2802.
29. Schoeninger MJ, DeNiro MJ (1984) Nitrogen and carbon isotopic composition of bone collagen from marine and terrestrial animals. Geochimica et Cosmochimica Acta 48: 625–639.
30. DeNiro MJ, Epstein S (1978) Influence of diet on the distribution of carbon isotope in animals. Geochimica et Cosmochimica Acta 42: 495–506.
31. DeNiro MJ, Epstein S (1981) Influence of diet on the distribution of nitrogen isotopes in animals. Geochimica et Cosmochimica Acta 45: 41–351.
32. Arnould JPY (2008) Southern fur seals *Arctocephalus spp*. In: Perrin WF, Würsig B, Thewissen JGM, editors. Encyclopedia of Marine Mammals. 2nd ed. San Diego: Academic Press. pp. 1079–1084.
33. Godoy JC (1963) Fauna Silvestre. Buenos Aires: Consejo Federal de Inversiones.
34. Schiavini A, Crespo EA, Szapkievich V (2004) Status of the population of South American sea lion (*Otaria flavescens* Shaw, 1800) in southern Argentina. Mammalian Biology 69: 108–118.
35. Vaz-Ferreira R (1982) *Arctocephalus australis*, South American fur seal. Mammals in the sea. Rome: FAO Fisheries Series. pp. 497–508.
36. Vaz-Ferreira R (1982) *Otaria flavescens* (Shaw) South American sea lion. Mammals in the sea: FAO Fisheries Series. pp. 477–495.
37. Crespo EA, Pedraza SN (1991) Estado actual y tendencia de la población de lobos marinos de un pelo (*Otaria flavescens*) en el litoral norpatagónico. Ecología Austral 1: 87–95.
38. Reyes L, Crespo EA, Szapkievich V (1999) Distribution and population size of the Southern sea lion (*Otaria flavescens*) in central and southern Chubut, Patagonia, Argentina. Marine Mammal Science 15: 478–493.

39. Dans SL, Crespo EA, Pedraza SN, Koen Alonso M (2004) Recovery of the South American sea lion (*Otaria flavescens*) population in northern Patagonia. canadian Journal of Fisheries and Aquatic Sciences 61: 1681–1690.
40. Túnez JI, Cappozzo HL, Cassini MH (2008) Natural and anthropogenic factors associated with the distribution of South American sea lion along the Atlantic coast. Hydrobiologia 598: 191–202.
41. Túnez JI, Cappozzo HL, Cassini MH (2008) Regional factors associated with the distribution of South American fur seals along the Atlantic coast of South America. ICES Journal of Marine Science 65: 1733–1738.
42. Páez E (2006) Situación de la administración del recurso lobos y leones marinos en Uruguay. In: Menafra R, Rodríguez-Gallego L, Scarabino F, Conde F, editors. Bases para la conservación y el manejo de la costa uruguaya. Montevideo: Vida Silvestre, Sociedad Uruguaya para la Conservación de la Naturaleza. pp. 577–583.
43. Jaureguizar AJ, Milessi AC (2008) Assessing the sources of the fishing down marine food web process in the Argentinean-Uruguayan Common Fishing Zone. Production 72: 25–36.
44. Dans SL, Koen Alonso M, Crespo EA, Pedraza SN, García NA (2003) Interactions between marine mammals and high seas fisheries in Patagonia: an integrated approach. In: Gales N, Hindell M, Kirkwood R, editors. Marine mammals: fisheries, tourism and management issues. Collingwood: CSIRO Publishing. pp. 100–115.
45. Vasconcellos M, Csirke J (2011) B6. Southwest Atlantic. FAO-Review of the state of world marine fishery resources. Rome: FAO Fisheries and Aquaculture Technical Paper No. 569. pp. 93–105.
46. Milessi AC, Jaureguizar AJ (2013) Evolución temporal del nivel trófi co medio de los desembarques en la Zona Común de Pesca Argentino-Uruguaya años 1989-2010. Frente Marítimo 23: 83–93.
47. Schiavini A, Yorio P, Gandini P, Raya Rey A, Boersma PD (2005) Los pingüinos de las costas argentinas: estado poblacional y conservación. El hornero 20: 5–23.
48. Boersma DP, Stokes D, Yorio P (1990) Reproductive variability and historical change of Magellan penguin (*Spheniscus magellanicus*) at Punta Tombo, Argentina. In: Davis LS, Darby JT, editors. Penguin biology. San Diego: Academic Press. pp. 15–43.
49. Orquera LA, Legoupil D, Piana EL (2011) Littoral adaptation at the southern end of South America. Quaternary International 239: 61–69.
50. Tivoli AM, Zangrando AF (2011) Subsistence variations and landscape use among maritime hunter-gatherers. A zooarchaeological analysis from the Beagle Channel (Tierra del Fuego, Argentina). Journal of Archaeological Science 38: 1148–1156.
51. Favier Dubois CM, Borella F, Tykot RH (2009) Explorando tendencias en el uso humano del espacio y los recursos en el litoral rionegrino durante el Holoceno tardío y medio. In: Salemme MC, Santiago M, Alvarez M, Piana EL, Vázquez M, et al., editors. Holocene. Ushuaia: Editorial Utopía. pp. 985–998.
52. Gómez Otero J (2006) Dieta, uso del espacio y evolución en sociedades cazadoras-recolectoras de la costa centro- septentrional de Patagonia durante el Holoceno medio y tardío. [PhD thesis]. Buenos Aires: Universidad de Buenos Aires.
53. Gómez Otero J (2007) Isótopos estables, dieta y uso del espacio en la costa atlántica centro-septentrional y el valle inferior del rio Chubut (Patagonia, Argentina). In: Morello F, Martinic M, Prieto A, Bahamond G, editors. Arqueología de Fuego-Patagonia Levantando piedras, desenterrando huesosy develando arcanos. Punta Arenas, Chile: CEQUA. pp. 151–161.
54. Moreno E (2008) Arqueología y etnohistoria de la costa patagónica central en el Holoceno tardío.: Fondo Editorial Provincial, Secretaría de Cultura del Chubut.
55. Saporiti F, Bala LO, Crespo EA, Gómez Otero J, Zangrando AFJ, et al. (2013) Changing patterns of marine resource exploitation by hunter-gatherers throughout the late Holocene of Argentina are uncorrelated to sea surface temperature. Quaternary International 299: 108–115.
56. Zangrando FA, Panarello H, Piana EL (2013) Zooarchaeological and Stable Isotopic Assessments on Pinniped-human Relations in the Beagle Channel (Tierra del Fuego, Southern South America). International Journal of Osteoarchaeology 24: 231–244.
57. Casey MM, Post DM (2011) The problem of isotopic baseline: Reconstructing the diet and trophic position of fossil animals. Earth-Science Reviews 106: 131–148.
58. Newsome SD, Clementz MT, Koch PL (2010) Using stable isotope biogeochemistry to study marine mammal ecology. Marine Mammal Science 26: 509–572.
59. O'Donnell TH, Macko SA, Chou J, Davis-Hartten KL, Wehmiller JF (2003) Analysis of δ13C, δ15N, and δ34S in organic matter from the biominerals of modern and fossil Mercenaria spp. Organic Geochemistry 34: 165–183.
60. Versteegh EAA, Gillikin DP, Dehairs F (2011) Analysis of $\delta^{15}N$ values in mollusk shell organic matrix by elemental analysis/isotope ratio mass spectrometry without acidification: an evaluation and effects of long-term preservation. Rapid communications in mass spectrometry: RCM 25: 675–680.
61. Hilderbrand GV, Farley SD, Robbins CT, Hanley TA, Titus K, et al. (1996) Use of stable isotopes to determine diets of living and extinct bears. Canadian Journal of Fisheries and Aquatic Sciences 74: 2080–2088.
62. Hobson KA, Clark RG (1992) Assessing avian diets using stable isotopes I: turnover of ^{13}C in tissues. The Condor 94: 181–188.
63. Jones D, Quitmyer I (1996) Marking time with bivalve shells: oxygen isotopes and season of annual increment formation. Palaios 11: 340–346.
64. Cousseau MB, Perrotta RG (2000) Peces marinos de Argentina. biología, distribución, pesca. Mar del Plata: INIDEP.
65. Piola A, Falabella V (2009) El Mar Patagónico; Falabella V, Claudio C, Croxal J, editors. Buenos Aires: Wildlife Conservation Society y BirdLife International.
66. Koen Alonso M, Crespo EA, Pedraza SN, García N, Coscarella MA (2000) Food habits of the South American sea lion, Otaria flavescens, off Patagonia, Argentina. Fishery bulletin 98: 250–263.
67. Page B, McKenzie J, Goldsworthy SD (2005) Dietary resource partitioning among sympatric New Zealand and Australian fur seals. Marine Ecology Progress Series 293: 283–302.
68. Drago M, Cardona L, Crespo EA, Aguilar A (2009) Ontogenic dietary changes in South American sea lions. Journal of Zoology 279: 251–261.
69. Silva L, Saporit F, Vales D, Tavares M, Gandini P, et al. (2014) Differences in diet composition and foraging patterns between sexes of the Magellanic penguin (*Spheniscus magellanicus*) during the non-breeding period as revealed by $\delta^{13}C$ and $\delta^{15}N$ values in feathers and bone. Marine Biology 161: 1195–1206.
70. Mangerud J (1972) Radiocarbon dating of marine shells, including a discussion of apparent ages of Recent shells from Norway. Boreas 1: 143–172.
71. Favier Dubois CM (2009) Valores de efecto reservorio marino para los últimos 5.000 años obtenidos en concheros de la costa atlántica norpatagónica (Golfo san Matías, Argentina). Magallania 37: 139–147.
72. Bligh EG, Dyer WJ (1959) A rapid method of total lipid extraction and purification. Canadian Journal of Biochemistry and Physiology 37: 911–917.
73. Lorrain A, Savoye N, Chauvaud L, Paulet Y-M, Naulet N (2003) Decarbonation and preservation method for the analysis of organic C and N contents and stable isotope ratios of low-carbonated suspended particulate material. Analytica Chimica Acta 491: 125–133.
74. Newsome SD, Koch PL, Etnier MA, Aurioles-Gamboa D (2006) Using carbon and nitrogen isotope values to investigate maternal strategies in northeast Pacific otariids. Marine Mammal Science 22: 556–572.
75. Bunn SE, Loneragan NR, Kempster MA (1995) Effects of acid washing on stable isotope ratios of C and N in penaeid shrimp and seagrass: implications for food-web studies using multiple stable isotopes. Limnology and Oceanography 40: 622–625.
76. Post DM (2002) Using stable isotopes to estimate trophic position: models, methods, and assumptions. Ecology 83: 703–718.
77. Moreno E, Zangrando AF, Tessone A, Castro A (2011) Isótopos estables, fauna y tecnología en el estudio de los cazadores-recolectores de la costa norte de santa cruz. MAGALLANIA, (Chile) 39: 265–276.
78. Saporiti F, Bala LO, Gómez Otero J, Crespo EA, Piana E, et al. (unpublished) Paleoindian pinniped exploitation in South America was driven by oceanic productivity Quaternary International.
79. Jackson AL, Inger R, Parnell AC, Bearhop S (2011) Comparing isotopic niche widths among and within communities: SIBER - Stable Isotope Bayesian Ellipses in R. The Journal of animal ecology 80: 595–602.
80. Parnell A, Jackson A (2013) SIAR: Stable Isotope Analysis in R.
81. Parnell AC, Inger R, Bearhop S, Jackson AL (2010) Source partitioning using stable isotopes: coping with too much variation. PloS one 5: e9672.
82. Burton R, Snodgrass J, Gifford-Gonzalez D, Guilderson T, Brown T, et al. (2001) Holocene changes in the ecology of northern fur seals: insights from stable isotopes and archaeofauna. Oecologia 128: 107–115.
83. Wainright SC, Fogarty MJ, Greenfield RC, Fry B (1993) Long-term changes in the Georges Bank food web: trends in stable isotopic compositions of fish scales. Marine Biology 115: 481–493.
84. Emslie SD, Patterson WP (2007) Abrupt recent shift in $d^{13}C$ and $d^{15}N$ values in Adélie penguin eggshell in Antarctica. Proceedings of the National Academy of Sciences of the United States of America 104: 11666–11669.
85. Becker BH, Beissinger SR (2006) Centennial Decline in the Trophic Level of an Endangered Seabird after Fisheries Decline. Conservation Biology 20: 470–479.
86. Christensen J, Richardson K (2008) Stable isotope evidence of long-term changes in the North Sea food web structure. Marine Ecology Progress Series 368: 1–8.
87. Corbett A, Causey D, Clementz M, Koch PL, Doroff A, et al. (2008) Aleut hunters, sea otters, and sea cows. In: Rick TC, Erlandson JM, editors. Human impacts on ancient marine ecosystems. Berkley: University of California Press. pp. 43–75.
88. Bailey G, Barrett J, Craig O, Milner N (2008) Historical ecology of the North Sea basin: an archaeological perspective and some problems of metodology. In: Rick TC, Erlandson JM, editors. Human impacts on ancient marine ecosystems: University of California Press. pp. 215–242.
89. Barrett J, Johnstone C, Harland J, Van Neer W, Ervynck A, et al. (2008) Detecting the medieval cod trade: a new method and first results. Journal of Archaeological Science 35: 850–861.
90. Crenshaw MA (1980) Mechanisms of shell formation and dissolution. In: Rhoads DC, Lutz RA, editors. Skeletal growth of aquatic organisms. New York: Plenum Press. pp. 115–132.
91. Trites A, Christensen V, Pauly D (2006) Effects of fisheries on ecosystems: just another top predator? In: Boyd I, Wanless S, Camphuysen C, editors. Top Predators in Marine Ecosystems: Cambridge University Press. pp. 11–27.

92. Drago M, Cardona L, Crespo EA, Grandi M, Aguilar A (2010) Reduction of skull size in South American sea lions reveals density-dependent growth during population recovery. Marine Ecology Progress Series 420: 253–261.
93. Vales DG, Saporiti F, Cardona L, De Oliveira LR, Dos Santos RA, et al. (2013) Intensive fishing has not forced dietary change in the South American fur seal Arctophoca(= Arctocephalus)australis off Río de la Plata and adjoining areas. Aquatic Conservation: Marine and Freshwater Ecosystems: n/a-n/a.
94. Hanson N, Wurster C, Bird M, Reid K, Boyd I (2009) Intrinsic and extrinsic forcing in life histories: patterns of growth and stable isotopes in male Antarctic fur seal teeth. Marine Ecology Progress Series 388: 263–272.
95. Muck P, Fuentes H (1987) Sea lion and fur seal predation on the Peruvian anchoveta, 1953 to 1982. In: Pauly D, Tsukuyama I, editors. The peruvian anchoveta and its upwelling ecosystem: three decades of change: ICLARM studies and Reviews. pp. 351.
96. Soto KH, Trites AW, Arias-Schreiber M (2004) The effects of prey availability on pup mortality and the timing of birth of South American sea lions (Otaria flavescens) in Peru. Journal of Zoology 264: 419–428.
97. Herling C, Culik BM, Hennicke JC (2005) Diet of the Humboldt penguin (Spheniscus humboldti) in northern and southern Chile. Marine Biology 147: 13–25.
98. Cury P, Bakun A, Crawford RJM, Jarre A, Quiñones RA, et al. (2000) Small pelagics in upwelling systems: patterns of interaction and structural changes in "wasp-waist" ecosystems. ICES Journal of Marine Science 57: 603–618.
99. Padovani LN, Viñas MD, Sánchez F, Mianzan H (2012) Amphipod-supported food web: Themisto gaudichaudii, a key food resource for fishes in the southern Patagonian Shelf. Journal of Sea Research 67: 85–90.
100. Riet-Sapriza FG, Costa DP, Franco-Trecu V, Marín Y, Chocca J, et al. (2013) Foraging behavior of lactating South American sea lions (Otaria flavescens) and spatial–temporal resource overlap with the Uruguayan fisheries. Deep Sea Research Part II: Topical Studies in Oceanography 88–89: 106–119.
101. Jennings S, Warr KJ (2003) Smaller predator–prey body size ratios in longer food chains. Proceedings of the Royal Society Biological Sciences 1413–1417.
102. Obelic B, Álvarez A, Argullós J, Piana EL (1998) Determination of water palaeotemperature in the Beagle Channel (Argentina) during the last 6000 yr through stable isotope composition of Mytilus edulis shells. Quaternary of South America and Antarctic Peninsula 11: 47–71.
103. Rivas AL (2010) Spatial and temporal variability of satellite-derived sea surface temperature in the southwestern Atlantic Ocean. Continental Shelf Research 30: 752–760.
104. Balech E, Ehrlich MD (2008) Esquema biogeográfico del Mar Argentino. Revista de investigación y desarrollo pesquero 19: 45–75.
105. Gómez Otero J, Suárez F (1999) Lobos marinos y guanacos: análisis arqueofaunístico de un fogón hallado en la costa del Golfo de San Matías, Península Valdés (Chubut). Resúmenes del XIII Congreso Nacional de Arqueología Argentina: Universidad de Córdoba. pp. 396.
106. Serrán M, Centeno N, Weiler N, Gómez Otero J (2008) Massive death of pinnipeds 1200 years ago: taphonomic history of the "Lobos site" (Golfo Nuevo, Patagonia, Argentina). Quaternary International 183: 135–142.
107. Schiavini A (1993) Los lobos marinos como recurso para cazadores-recolectores marinos: el caso de Tierra del Fuego. Latin American Antiquity 4: 346–366.
108. L'Heureux G, Franco NV (2002) Ocupaciones humanas en el área de cabo Vírgenes (Pcia de Santa Cruz, Argentina): el sitio Cabo Vírgenes 6. Anales del Instituto de la Patagonia (Serie Ciencias Humanas) 30.
109. Beretta M, Corinaldessi L, Castro A (2011) Recursos marinos vs. recursos terrestres: análisis arqueofaunístico en el sitio Cueva del Negro, costa norte de Santa Cruz. Arqueología 17: 137–159.
110. Orquera L, Piana E (1987) Composición tipológica y datos tecnomorfológicos y tecnofuncionales de los conjuntos arqueológicos del sitio Túnel I (Tierra del Fuego, República Argentina). Relaciones de la Sociedad Argentina de Antropología 17: 201–239.
111. Piana E, Vázquez M, Ceraso A (2007) Kaiyawoteha. Informe de excavación. Ushuaia.
112. Belardi JB, Carballo Marina F, L'Heureux GL (2011) Nuevos resultados arqueológicos en Cabo Vírgenes (Santa Cruz, Argentina): el sitio Cabo Vírgenes 20. Magallania, (Chile) 39: 279–292.
113. Tivoli AM (2010) Las aves en la organización socioeconómica de cazadores-recolectores-pescadores del extremo sur sudamericano. Buenos Aires: Universidad de Buenos Aires.
114. Orquera L, Piana E (1997) El sitio Shamakush I. Relaciones. Relaciones de la Sociedad Argentina de Antropologia 21: 215–265.

Assessing the Impact of Human Activities on British Columbia's Estuaries

Carolyn K. Robb*

School of Environment and Management, Royal Roads University, Victoria, British Columbia, Canada

Abstract

The world's marine and coastal ecosystems are under threat and single-sector management efforts have failed to address those threats. Scientific consensus suggests that management should evolve to focus on ecosystems and their human, ecological, and physical components. Estuaries are recognized globally as one of the world's most productive and most threatened ecosystems and many estuarine areas in British Columbia (BC) have been lost or degraded. To help prioritize activities and areas for regional management efforts, spatial information on human activities that adversely affect BC's estuaries was compiled. Using statistical analyses, estuaries were assigned to groups facing related threats that could benefit from similar management. The results show that estuaries in the most populated marine ecosections have the highest biological importance but also the highest impacts and the lowest levels of protection. This research is timely, as it will inform ongoing marine planning, land acquisition, and stewardship efforts in BC.

Editor: Simon Thrush, University of Auckland, New Zealand

Funding: Living Oceans Society (www.livingoceans.org), an environmental non-profit organization and the author's employer, provided the computer and GIS software used in the completion of this research. No other funding assistance was received. The funders had no role in study design, data collection and analysis, decision to publish, or preparation of the manuscript.

Competing Interests: This research was completed as part of the requirements for a Master of Science in Environment and Management from Royal Roads University in Victoria, British Columbia, Canada. While a student at Royal Roads University, the author also worked as the GIS Analyst for Living Oceans Society (LOS), a marine conservation organization based in British Columbia, Canada.

* E-mail: ckrobb18@hotmail.com

Introduction

Estuaries are one of the world's most biologically productive ecosystems [1,2,3], providing ecosystem services valued as high as $22,000 US per hectare per year [4]. In British Columbia, Canada (BC), calculations have shown that estuarine habitats sequester a minimum of 180,000 tonnes of carbon annually [5] and Kelsey [6] estimated that 80% of all coastal wildlife use estuaries. However, anthropogenic threats have made estuaries one of the most degraded ecosystems on earth [3,7] and these threats are expected to increase in prevalence and severity as coastal populations continue to expand [7].

Many estuarine threats originate within the watersheds that drain through the estuaries [7,8]. The conversion of land for agriculture, forestry, or residential and industrial development causes habitat loss or fragmentation [7,9]. These changes in land use can trigger an increase in sediment, debris, nutrient, and pollutant levels in estuaries that smothers or entangles organisms or causes eutrophication, hypoxia, and anoxia [1,3,7,10,11]. Development that expands impervious surfaces within the watershed can increase water flow into estuaries, particularly during storm events [7], while dams impede water inputs [1,7,9]. As the climate warms, increased precipitation and more frequent storm surges are also anticipated in BC [9]. All of these hydrological changes can disrupt water filtration [7] and alter estuarine productivity and trophic structure [12,13].

Other threats originate in the marine environment. Development and industrial activities within estuaries, such as log-handling, aquaculture, dredging, and vessel traffic, can result in habitat loss or degradation, environmental contamination, shoreline erosion, the re-suspension of sequestered carbon, and the invasion of alien species [1,7,9,10,12]. Commercial and recreational fishing can alter the composition of estuarine species through the preferential removal of target species [7,12]. Shoreline armouring to stabilize and protect coastal developments limits the ability of estuaries to retreat inland as the climate warms and sea levels rise and may therefore cause habitat and shorebird losses [1,9,13,14]. Higher sea temperatures may increase the severity of eutrophication in estuarine salt marshes [15]. Lastly, the acidification of estuarine waters may inhibit the growth of shelled organisms [16] and increase the bioavailability of trace metals from pollution [17].

The continuing degradation of estuarine habitats has led to a variety of research efforts. In BC, the Pacific Estuary Conservation Program (PECP), a partnership of government and non-government organizations, has mapped the boundaries of 442 estuaries and designed regional scale projects to prioritize estuaries for conservation based on their biological importance [2]. Further research has found that 38% of BC estuaries had economic tenures within the intertidal area [18]. Merrifield *et al.* [1] recently completed a regional assessment of threats to estuaries along the west coast of the United States which face similar threats to BC estuaries [19]. The researchers found that only 16% of surveyed estuaries faced few or no anthropogenic threats while 25% of estuaries had moderate levels of all assessed threats [1]. However, a systematic investigation of human activities affecting BC estuaries has not yet been completed.

In BC, the management of estuarine areas and activities requires the coordination of several government agencies. Subtidal and intertidal areas are typically public land while upland areas can be publicly or privately owned. The BC government regulates the use of public lands by issuing tenures but the ongoing activities of some industries with tenures, such as aquaculture, are managed federally. Federal jurisdiction also pertains to all organisms and most activities within the water column, areas designated for port facilities, and activities on or beneath the seabed in offshore areas [20]. In public areas, the BC government and several federal agencies have established protected areas that, to varying degrees, shield estuarine habitat from human activities [21]. In private areas, the PECP works to secure and rehabilitate important estuarine habitats through direct land purchases [2,22].

Recent reports attribute estuary degradation to a failure of management efforts that have focused predominantly on individual industrial sectors [23,24,25] without examining the cumulative impacts of activities [24,26,27,28] or the ecosystem services on which the activities rely [29]. There is a growing consensus that a broader management focus, termed ecosystem-based management (EBM), is necessary to tackle the problems facing marine and coastal environments [23,24,26,27,30,31,32]. EBM is place-based and considers both the spatial variations of, and the interactions between, the ecological, physical, and human elements of ecosystems [24,31]. In Canada, Fisheries and Oceans Canada (DFO) has developed a framework for estuarine, coastal, and marine management following the principles of EBM [23].

To implement EBM effectively, ecosystem elements must be mapped and the impacts of human activities assessed [23,24,28,33]. This research uses geographic information systems (GIS) to investigate the spatial distribution of human activities that affect BC estuaries, incorporating threats relevant to BC that have not been considered previously. Cluster analysis is used to categorize estuaries based on the presence and magnitude of threat variables. Unlike in previous research [1], the spatial range of influence of each threat is incorporated to generate a more realistic spatial representation of each threat [33], the cluster analysis uses a non-hierarchical approach to allow the use of non-normal variables, and results are compared to assessments of each estuary's biological importance and conservation status. By identifying estuaries facing similar threats, this work gives context to previous prioritization efforts that focused solely on ecological factors [2], informs regional planning efforts, and helps prioritize limited conservation and research resources to ensure vital ecosystem services are sustained.

Methodology

Study Area

BC's estuaries, comprising less than three percent of the coastline, form the study area for this research (Figure 1). I used estuaries and associated watersheds as the basic spatial units for the analysis, referred to as "estuary-watershed systems", because they are considered the complete geographic boundaries for estuarine management [8,34].

Estuary and Watershed Data Preparation

In a desktop GIS, I overlaid spatial data for watersheds and estuaries (Table S1) to identify the watershed associated with each estuary (Figure 1). I confirmed that overlapping estuaries and watersheds shared the same primary waterway using attribute information found in both datasets. Estuaries that could not be linked to a unique watershed were removed. I added attribute information on importance for waterbirds [2] and carbon sequestration [5] to each estuary based on estuary name. I also overlaid the estuary-watershed systems with spatial data on legislated protected areas and privately conserved lands (Table S1) to determine the count and total area of conservation designations within each system. Lastly, I overlaid the systems with BC's marine ecosections, which are planning areas with similar physical and biological characteristics, to determine the average estuarine importance and conservation values by ecosection (Table 1).

Threat Data Preparation

To identify the primary anthropogenic threats facing BC estuaries, I reviewed the relevant literature, focusing on research on estuaries and estuarine habitats, such as salt marshes, seagrass beds, and sand flats, from BC and the Pacific coast of North America (Table 2). Spatial data for the primary threats (Table S2) were chosen based on their relevance to estuaries, spatial coverage, quality, resolution, and currency and used to create a comprehensive spatial layer for each threat along the BC coast. While gaps did exist (Table S3), data were available for the majority of the identified threats.

Several spatial analysis methodologies were used to compile the data (Table S2) for further statistical analysis. I classified land use data into agricultural or urban areas, combining the urban areas with railroads and road networks. I identified armoured shorelines by combining dikes and manmade shoreline segments. I selected the locations of freshwater diversions from a point dataset of obstructions. To identify areas that had been clearcut over the past decade, I extracted polygons with a stand age of 10 or less from a provincial dataset of harvest depletion. I located sources of point and nonpoint watershed pollution from data on pulp mills, mines, and areas closed due to contamination. Lastly, I calculated the percent change in modeled future precipitation using data on current (1961–1990 normal) and future (2080's CGSM A2x scenario) precipitation from ClimateBC [36] and the methodology employed by Merrifield *et al.* [1].

The compilation of marine threat data (Table S2) was guided by the best available data identified by the BC Marine Conservation Analysis (BCMCA) [37]. I obtained provincial tenure data to identify public areas leased to marine-related industrial activities and confirmed site usage with site-specific finfish aquaculture production data from DFO or by visually confirming the presence of infrastructure, such as aquaculture pens or log storage sites, using Google Earth 6.1 [38] satellite imagery. I merged the tenures to create features for shoreline development and aquaculture. I transformed the tenures for marinas, commercial wharves, and ferry terminals into points and combined them with BCMCA datasets representing marinas, anchorages, and tow boat reserves. I obtained the point locations of dredging operations and extracted data for large vessels from a gridded dataset of vessel traffic (Table S2). Smaller vessels were not well represented in this dataset so I used linear data on recreational boating routes to represent the distribution of small vessels. I combined fisheries datasets to create features for commercial and recreational fisheries. To retain the catch values available in the commercial fishing datasets, the values for all fisheries were converted to kg and summed. No high resolution datasets were available for projected ocean acidification, sea surface temperature, or sea level rise within BC's coastal zone (Table S3).

Threats within Estuaries

The maximum spatial range of influence of each activity was determined from zone values specified by Ban *et al.* [33] and Quigley and Hall [39] (Table S2). Incorporating zones more

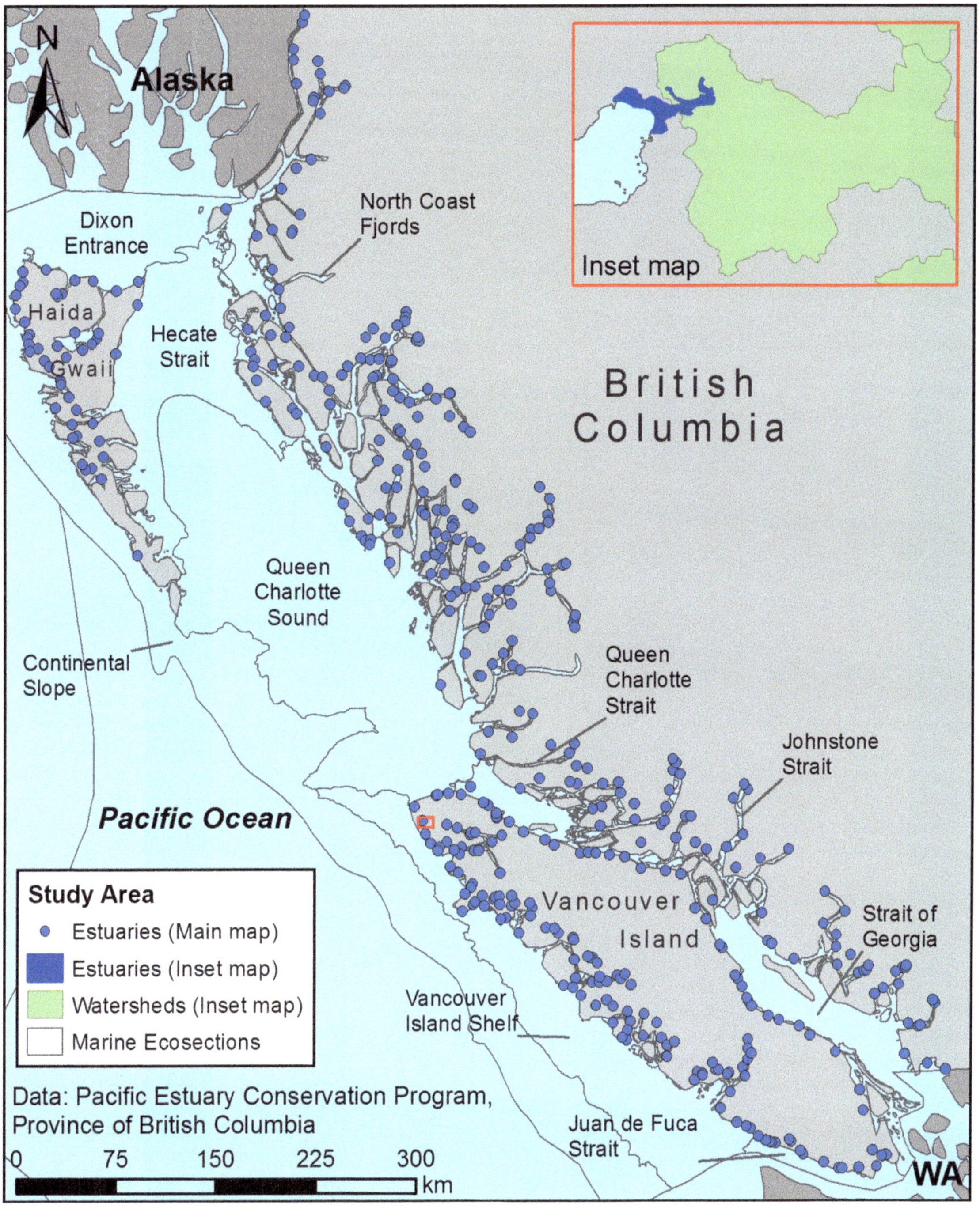

Figure 1. Study area. Description: This research investigates the presence of threats within BC estuaries, which are found along the BC coastline from the southern border with Washington to the northern border with Alaska. The inset map shows a typical estuary and watershed.

realistically represents the full spatial extent of each threat [33] and ensures that all relevant threats are considered, though the spatial footprint of associated infrastructure may not overlap directly with an estuary or watershed. I buffered each threat by the zone value to create polygons representing the larger zone of influence. Spatial ranges were not calculated for features with low data resolution [33], features where the zone of impact was not known, or features that already incorporated the zone of impact (Table S2). The spatial range was calculated for large vessel traffic, despite the low data resolution, due to the wide influence of this feature and because the original data had been fit to a coastline that did not incorporate the mapped estuaries in all areas.

To determine the impact of the threats within each estuary-watershed system, I overlaid the buffered threats with the estuaries and watersheds. To standardize for variation in the size of the systems, I calculated the proportion of each system affected by the activity. For point features, I determined the density of points within each estuary-watershed system. For a few gridded features, information on level of use was available and used to better represent their impact spatially. Because the impact of vessel

Table 1. Average conservation importance, protection level, and threat status of BC estuaries, grouped by marine ecosection.

Marine Ecosection	Importance for Waterbirds	Count of Threats	Count of Protected Areas	Proportion of Area Protected (%)	Average Watershed Area (ha)	Average Estuarine Area (ha)
Continental Slope	3.4	6.2	1.6	57.9%	4,328.3	17.0
Dixon Entrance	3.2	6.1	1.2	42.4%	6,646.5	140.0
Hecate Strait	3.5	7.0	1.3	22.4%	3,859.1	55.7
Johnstone Strait	3.6	8.9	0.8	12.8%	5,724.1	64.0
Juan de Fuca Strait	3.7	9.1	2.3	11.5%	5,019.6	25.1
North Coast Fjords	3.6	6.6	1.3	27.8%	6,514.0	192.7
Queen Charlotte Sound	3.5	5.2	1.3	20.3%	4,101.3	29.4
Queen Charlotte Strait	3.3	9.3	0.9	3.1%	5,542.8	125.4
Strait of Georgia	3.2	11.0	2.4	4.2%	7,132.2	783.3
Vancouver Island Shelf	3.4	8.9	1.0	13.3%	5,255.5	45.9

traffic, such as ship wake, can extend into an estuary from outside, I spatially joined the vessel traffic data with the estuaries and summed the number of vessel hours for all overlapping grid cells to calculate the total impact. For commercial fishing, I intersected the grid of fisheries catch data (Table S2) with the estuary polygons and summed the total catch of the overlapping grid cells proportional to the area of overlap to estimate the catch taken within each estuary. I calculated the average precipitation change for each estuary-watershed system using zonal statistics, which examined and combined the values of cells in the gridded precipitation dataset that fell within the estuary-watershed systems.

I counted the number of threats affecting each estuary and determined the proportion of the estuary and watershed area covered by those activities. The average values were compared among estuaries in different marine ecosections (Table 1) and estuaries with different conservation status and biological importance.

Statistical Analyses

I ran cluster analyses using Systat 13 [40] to group the estuaries based on the presence and magnitude of the threats (Table 3). Pairwise correlations between threats were examined to ensure the included threats were independent and considered equal in the analysis. Most of the threat variables had negatively skewed distributions and were log-transformed to minimize the impact of a few scattered higher values [28]. To determine the optimal number of clusters, I performed preliminary analyses with hierarchical clustering and the Ward method. I used non-hierarchical K-medians for the final cluster analysis because the method is resistant to non-normal variables and outliers [41]. Euclidean distance was used to determine the distance of each site from the centers of each cluster.

I mapped the results and generated spatial statistics for the clusters, including the average number of threats and the proportion of each estuary affected by each threat (Tables 3 and 4). I also analyzed the distribution of the clusters among ecosections, sites with a conservation designation, and estuaries considered biologically important (Table 4).

Results

Importance and Conservation Status of Estuaries

The average importance of estuaries for waterbirds was similar among all ecosections but priority carbon sequestration sites were only found within the Strait of Georgia and North Coast Fjords ecosections. Federally and provincially legislated protected areas overlapped 58.5% of estuary-watershed systems, predominantly in the North Coast Fjords ecosection. On average, 21% of estuarine and watershed area was covered by a legislated protected area, with the highest proportions in north coast ecosections. A further 27 estuary-watershed systems, primarily on the south coast, overlapped a privately conserved area, with an average of 1.4% of the estuarine and watershed area protected. In total, 61% of estuary-watershed systems had at least one private or legislated protected area.

Threats in Estuaries

Of 442 possible estuaries, 376 were joined to adjacent watersheds and included in the analysis. The literature review uncovered 20 threat categories and, based on the available data, 16 threat variables were created (Table S2). A spatial dataset showing the threat distribution within each estuary is available from the author. Change in future precipitation was the only threat to overlap with all estuary-watershed systems but all systems overlapped at least two threats (Figure 2). No system overlapped all of the threats, though two systems in the Strait of Georgia ecosection were affected by 15 threats. Overall, there was an average of 7.9 threats per estuary-watershed system but systems within the Strait of Georgia ecosection averaged 11.0 threats and four other ecosections, also on the south coast, had a higher than average number of threats per estuary-watershed system (Table 1).

Estuary-watershed systems with the highest importance to waterbirds had the highest counts of threats. Systems with a protected area, either legislated or privately conserved, faced an average of 7.6 threats. Estuary-watershed systems with less area protected faced a greater average number of threats and those facing the fewest threats were the most protected. Systems with a legislated protected area faced fewer average threats than those with a privately conserved area.

Table 2. The incidence of estuarine threats within the literature relevant to BC estuaries and estuarine habitats, such as salt marshes, seagrass beds, and sand flats.

	Key sources used to identify threats to estuaries and estuarine habitats						
Estuarine threats	**[9]**	**[8]**	**[19]**	**[7]**	**[1]**	**[2]**	**[35]**
Terrestrial							
Development in watershed	X	X	X	X	X	X	
Agriculture in watershed	X	X	X	X	X		
Forestry in watershed		X	X	X	X		
Watershed pollution	X	X	X	X	X	X	X
Freshwater diversions	X	X	X	X	X	X	X
Shoreline armouring in estuary	X	X	X	X	X		X
Impaired waterways		X	X	X	X	X	
Debris/litter				X		X	X
Marine							
Development of estuary shoreline	X	X	X	X	X	X	X
Port facilities in estuary	X	X		X	X		X
Dredging	X	X	X	X			X
Marine pollution				X	X	X	X
Aquaculture	X		X	X	X		X
Resource extraction				X		X	X
Introduced species	X	X	X	X		X	X
Recreation		X		X			
Increased atmospheric carbon dioxide							
Sea level rise	X	X		X	X	X	X
Sea surface temperature		X					X
Ocean acidification							X
Change in future precipitation	X	X			X		
Ecosystem focus (location)	Est* (BC)	SM (WC US)	Est (WC NA)	Est (global)	Est (WC US)	Est (BC)	SM (CA)

*Abbreviations: Est = estuary; SM = salt marsh; BC = British Columbia; WC = west coast; US = United States of America; NA = North America; CA = California.

Table 3. Average values of the threats affecting BC estuaries, grouped into the three clusters generated by the K-medians cluster analysis.

Cluster	Urban (%)	Tenures (%)	Shoreline Armouring (m/ha)	Aquaculture (%)	Ports (%)	Dredging (%)	Watershed Pollution (%)	Agriculture (%)
1	67.8%	5.4%	5.7	1.7%	61.8%	0.6%	12.6%	0.2%
2	92.1%	7.8%	9.0	11.4%	21.4%	0.5%	32.0%	0.4%
3	100.0%	17.8%	38.2	18.6%	72.4%	7.7%	87.8%	14.5%

Cluster	Freshwater Diversion (count/ha)	Forestry (%)	Contamination Closures (%)	Commercial Fishing (catch in kg)	Recreational Fishing (%)	Large Vessel Traffic (hours)	Small Vessel Traffic (%)	Future Precipitation (% change)
1	0.000010	15.4%	3.7%	890.8	19.2%	605.1	75.4%	19.2
2	0.000028	70.3%	3.1%	1,697.9	58.4%	3,650.7	67.5%	16.9
3	0.000075	52.8%	66.3%	23,345.3	37.8%	19,439.6	81.9%	15.8

Table 4. Average conservation importance, protection level, and threat status of BC estuaries, grouped into the three clusters generated by the K-medians cluster analysis.

Cluster	Number of Estuaries	Importance for Waterbirds	Priority Sites for Carbon Sequestration	Count of Threats	Count of Protected Areas	Proportion of Area Protected (%)	Average Estuarine Area (ha)	Average Watershed Area (ha)
1	180	3.6	0	6.6	1.3	34.4%	88.7	5,918.8
2	127	3.5	2	8.0	0.9	9.5%	163.6	6,034.4
3	69	3.1	6	11.1	2.1	6.7%	521.8	6,129.8

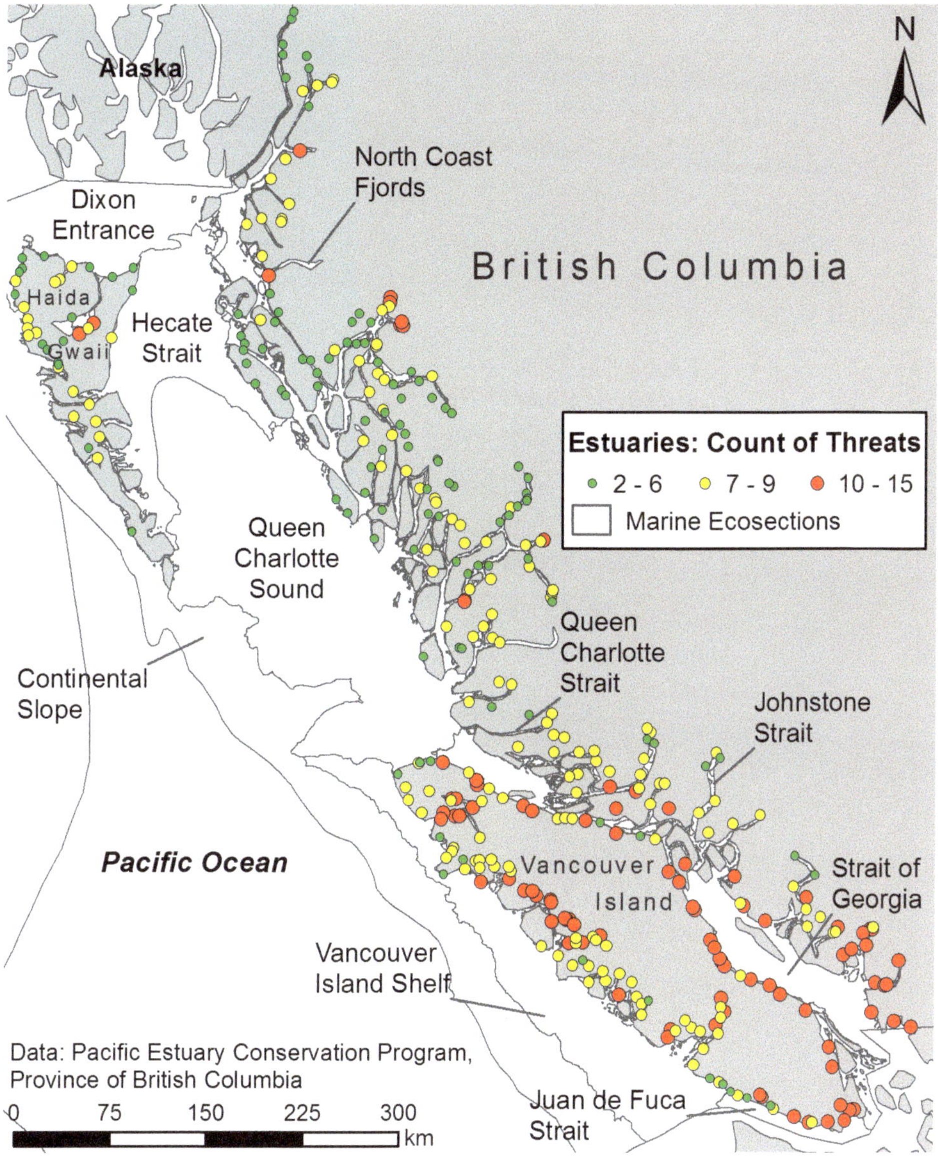

Figure 2. Count of threats by estuary-watershed system. Description: Estuary-watershed systems are shown classified based on the number of the 16 possible threats found within their bounds. The classification uses Jenks natural breaks which maximizes the variance between the classes. The number of threats faced by an estuary is greatest in the more heavily populated south coast ecosections.

Cluster Analysis

The cluster analysis revealed three optimal clusters (Figure 3). Cluster 3, representing estuaries with the greatest importance for waterbirds and carbon sequestration (Table 4), was the cluster with the greatest impact from most of the threat variables (Table 3). These 69 estuaries were found primarily in the south coast ecosections, dominated by the Strait of Georgia ecosection. Cluster 3 had the greatest average number of protected areas within each estuary, but the smallest average proportion of area protected. Cluster 2 represented 127 estuaries moderately affected by most threat variables and the highest impact from forestry and recreational fishing. The majority of estuaries in this cluster were found on the south coast in the less populated Vancouver Island Shelf ecosection. Cluster 1 faced minimal impacts from most threats but had a higher than average future change in precipitation. Cluster 1 also had the most estuaries, located predominantly within the North Coast Fjords ecosection. Based on average area conserved, these estuaries were also the most protected.

Discussion

A variety of local approaches have been important for estuary conservation, including the creation of legislated protected areas, the purchase of sites for private conservation, and the establishment of estuary management plans. A conservation designation was found within 61% of the estuary-watershed systems but systems in the north coast ecosections had greater protection than their southern counterparts (Table 1). The biased placement of legislated conservation designations in less threatened areas further from human settlements, and therefore less costly to protect, matches global trends for terrestrial protected areas [42]. Estuary-watershed systems with privately conserved areas faced a greater number of threats than those with legislated protected areas. As

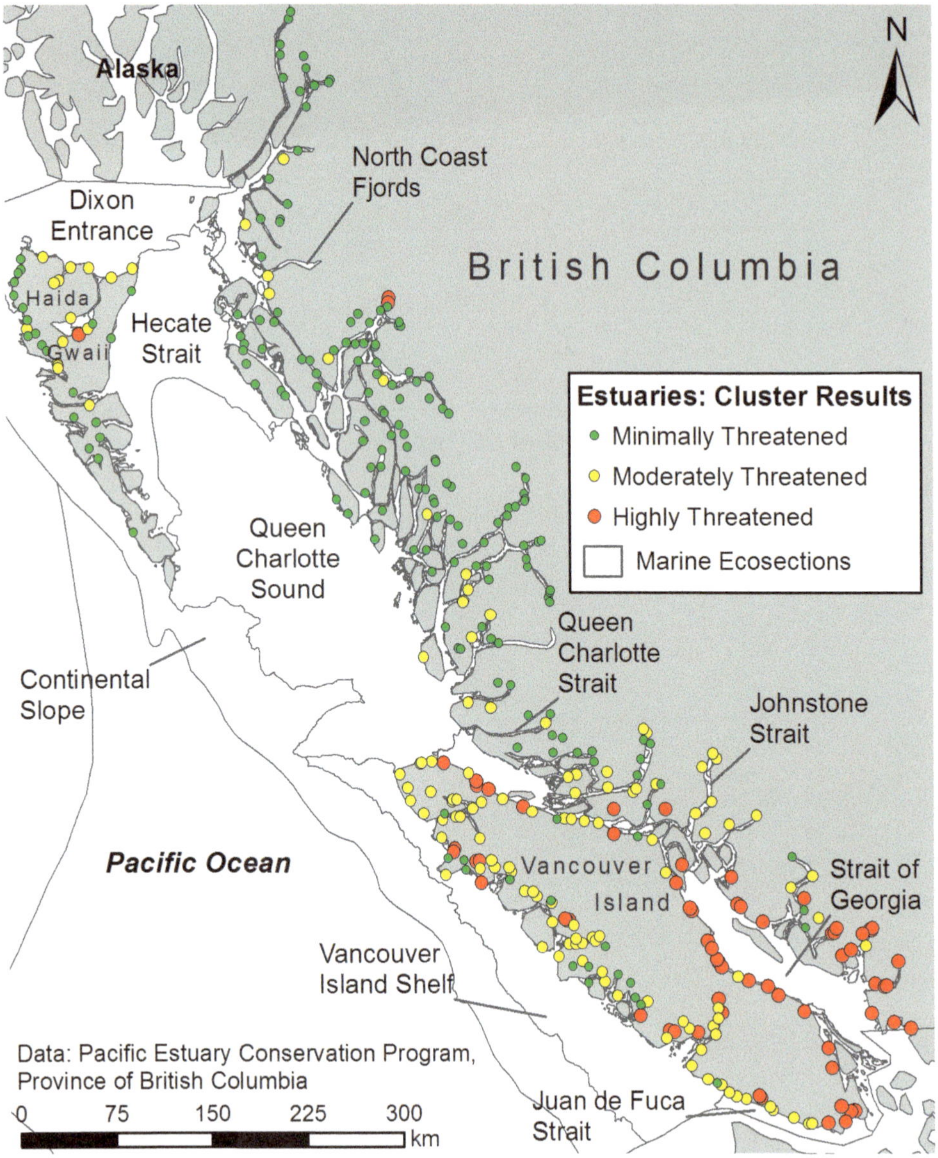

Figure 3. Distribution of K-medians cluster results. Description: Estuary-watershed systems are shown distributed into clusters that resulted from the K-medians cluster analysis based on the presence of 16 threat variables. Similar to the count of threats, the clusters highlight a north-south gradient, with highly threatened estuaries primarily observed in the south coast ecosections and minimally threatened estuaries predominantly observed in the north coast ecosections.

Brooks *et al.* [42] note, people most value "what is close to them". Privately protected areas, often negotiated by the PECP with landowners, are found primarily along the south coast, where the greatest concentration of humans, and industrial activities, is found [10,17].

In 2002, a review of estuary management plans in BC identified only nine completed plans, predominantly for estuaries on the south coast [43]. Since that time, planning has been completed for two additional estuaries but some of these plans, particularly those with limited stakeholder support, have taken up to 20 years to finalize [43] and some have not been fully implemented (K. Moore, Environment Canada, personal communication). With uneven estuarine protection along the coast and management plans for less than 2% of mapped estuaries, it is apparent that local efforts alone will not be sufficient to conserve estuarine habitats effectively into the future.

To ensure the maintenance of biodiversity, a regional approach based on the principles of EBM offers a more comprehensive, efficient, and cost-effective alternative to site-specific protection and planning [30,44]. The PECP's efforts to map and rank estuaries began the work of looking at BC's estuaries through a regional lens. However, the PECP lacked the comprehensive anthropogenic data necessary to complete the ranking [2]. Identifying clusters of estuaries that face similar threats can highlight priority activities and areas where coordinated management could achieve a higher return on investment.

Highly Threatened Estuaries

Areas with high diversity or importance and high cumulative impacts are considered conservation priorities to protect species and habitats that would otherwise be lost to habitat modification [28,42]. Identified as highly important for waterbirds and carbon sequestration, the highly threatened estuaries in Cluster 3 qualify as conservation priorities (Table 4). These estuaries were located predominantly within the heavily populated south coast ecosections (Figure 3), which correlates with results from previous research [18,33] as well as the simpler count of threats by estuary (Table 1 and Figure 2).

Protected areas that prohibit human activities are a prominent means of conserving estuaries [23] but the systems in Cluster 3 had the lowest average proportion of area protected (Table 4). Many of those estuaries are located near major urban centers, where the financial costs and logistical challenges associated with establishing and ensuring compliance of conservation designations are higher [45]. Protecting estuaries in highly populated areas will require stakeholder engagement, a component of EBM, to ensure that all relevant ecosystems and activities are considered, that priorities are established collectively, and that efforts are not duplicated [30,46]. However, only the conservation of representative sites across the coast will ensure the continued viability of estuarine ecosystems and help Canada meet its international commitment to preserve 10% of each ecological region in the marine and coastal realm [46].

Moderately Threatened Estuaries

Cluster 2 faced a moderate level of most threats (Table 3) and contained estuaries located primarily in the less populated ecosections of the south coast (Figure 3). These systems had the greatest impact from recreational fishing and forestry, suggesting that coordinating estuarine management with these two sectors might alleviate significant stresses. Regional planning is a means of engaging industries whose activities adversely affect an ecosystem from afar [47], such as the terrestrial activities like agriculture, forestry, and watershed development that affect estuaries in Clusters 2 and 3. Smith *et al.* [47] suggest that a system of local plans for highly used areas nested within regional plans might offer the most efficient solution. This suggestion meshes with recommendations from the Royal Society of Canada [48], who proposed that planning for DFO's large ocean management areas should be tied to planning for the smaller coastal management areas because the areas are physically and biologically integrated.

Minimally Threatened Estuaries

While some prioritize action in highly threatened areas, there are also calls to minimize conflict by proactively protecting areas required by few industries [49,50]. Over 40% of the estuary-watershed systems assessed, represented by Cluster 1, were minimally affected (Table 3). This result is encouraging, suggesting that many of BC's estuaries are relatively pristine. In comparison, research done for estuaries along the more heavily populated west coast of the United States found that only 16% were minimally affected [1]. The systems of Cluster 1 were located primarily in the lightly populated north coast ecosections (Figure 3). These sites faced a slightly greater average increase in future precipitation but because of the relatively low levels of other anthropogenic threats, the estuaries may be best poised to respond to the corresponding increase in water flow [51].

While the estuaries of Cluster 1 were the least affected by the combination of all threats, they did show moderate impacts from a few threats (Table 3). These estuaries are primarily located on the north coast, away from human settlements. However, small human populations do not guarantee minimal anthropogenic impacts [28] and some threats, such as forestry, shipping, and climate change, can influence even remote estuaries. Threats that impact all of the clusters are examples of regional issues that are likely managed most efficiently at a broad scale [11].

Limitations in Data and Analysis

Cluster analysis is an inherently subjective statistical methodology relying on researcher judgment to gauge the integrity of the results. A strong theoretical foundation for the analysis helps exclude irrelevant variables that might distort results [41] and the extensive literature review of anthropogenic threats (Table 2) ensured only well-researched variables were incorporated. However, the lack of fine scale or historical spatial data on marine activities and climate change is a noted impediment to conservation and research efforts [17,44].

Two specific data gaps should be noted. Observations of invasive species, one of the most important stresses in estuarine ecosystems [9,19], have not been catalogued across the BC coast (D. Buffett, Ducks Unlimited Canada, personal communication). Instead, two vectors of invasive species, aquaculture and large vessel traffic [10], were used and the results can help identify highly disturbed estuaries most susceptible to invasion [8]. Perhaps the most important data gap is projected sea level rise. The shoreline armouring feature represents estuaries with barriers preventing adaptation to higher sea levels but does not consider spatial variations in elevation. The vulnerability of BC shorelines to sea level rise has been approximated by the BC Government using coarse elevation data and the physical parameters of shoreline segments [17]. However, better projections of the impact of SLR on estuaries require high resolution elevation data that are not currently available for much of the coast [17].

Limitations in Implementation

Following the approach taken by Merrifield *et al.* [1], this analysis incorporated threats found throughout estuary-watershed systems. Estuaries link terrestrial watersheds with marine systems and the influences of the two realms both enhance [52] and threaten [3,7] the ecosystem services provided by estuaries. However, while terrestrial, coastal, and marine environments are integrated, they are often not governed harmoniously. Considering the entire land to sea continuum is a more comprehensive approach to conservation and management advocated by both government agencies [23,27] and researchers [7,8,34,47,53], despite the challenges associated with the large area and number of stakeholders involved. A failure to utilize EBM and explicitly consider threats originating across the land-sea continuum hinders effective estuary management and can undercut the success of protected areas [30]. Conservation efforts require the coordinated involvement of all relevant government bodies to facilitate successful integrated and adaptive planning [7,21,23,34]. This will be particularly relevant for the estuaries in Clusters 2 and 3, which face a suite of significant marine and terrestrial threats. In BC, improved coordination may be underway as the BC government and First Nations are developing an EBM plan that considers the cumulative impacts of human activities (www.mappocean.org) and federal agencies are collaborating on a network of marine protected areas [46].

Priorities for Management and Conservation

Land purchases such as those coordinated by the PECP play an integral role in the conservation of estuaries. To a certain extent, these purchases will likely always be done in an ad hoc manner, taking advantage of the generosity of landowners adjacent to estuaries. However, planning at a regional scale using biological and socioeconomic data can help ensure that management and conservation efforts are proactive, scientifically defensible, maximize available resources, and minimize impacts on local resource users [2,25,54]. A comprehensive network approach to protected areas can also help guarantee that protected areas fully realize their benefits [50]. To conserve estuarine ecosystem services, a network of protected areas must ensure that contributing sites include replications across the coast, maintain connectivity, and are adequately sized and protected [46].

Estuarine conservation and management would also benefit from further research into the cumulative impacts of estuarine threats. This analysis determined the specific threats facing BC's estuaries but more information is required to understand how threats interact, in particular for the highly threatened estuaries in Clusters 2 and 3. Ban *et al.* [33] began this work for BC by extrapolating expert assessments from California on the vulnerability of marine habitats to different activities [35]. However, information on the level of impact of each threat is also required and is currently only available for a few of the regional scale datasets, a shortcoming identified in other jurisdictions [1]. The BCMCA has worked with industry representatives in the marine environment to validate the accuracy of spatial data [37] and continuing that work to generate a better representation of site usage and impact is an important next step.

Austin *et al.* [9] declared climate change to be the "foremost threat to biodiversity" in BC but a lack of spatial data (Table S3) make it challenging to incorporate the relevant threats. The BC estuaries identified as priorities for carbon sequestration [5] were shown to be highly threatened (Table 4). Protecting pristine carbon sequestration sites and restoring degraded areas through an EBM process has been referred to as "one of the strongest win-win mitigation efforts known today" [32] because safeguarding carbon sequestration requires maintaining estuarine structure and function, facilitating the continued provision of the other ecosystem services that estuaries provide. Emerging information on the capacity of estuaries to sequester carbon, along with better accounting methodologies, may bring renewed impetus to improve and fund coastal management from both the public and private sectors [22,32,42,55].

Conclusions

Estuaries are an important component of BC's biodiversity and the ecosystem services they provide are well established. However, estuary degradation continues, in part due to a historical lack of integrated management and the investment required to develop and implement individual estuary management plans. Regional planning based on EBM that involves stakeholders across the entire land-sea continuum is required to prioritize sites for protection or restoration and ensure conservation benefits are realized.

By working at a broad scale and clustering estuaries based on their predominant threats, this analysis assesses human activities within BC's estuaries scientifically and reveals that estuaries deemed most important for waterfowl habitat and carbon sequestration disproportionately face the highest threats. The results also highlight a north-south gradient of threats, with minimally threatened estuaries located predominantly on the north coast and highly threatened estuaries primarily on the south coast. This information can be combined with the PECP's ecological ranking of estuaries to guide threat mitigation in marine, terrestrial, and coastal areas. Continuing to broaden our understanding of estuaries, including the cumulative impacts they face and the amount of carbon they sequester, will be essential to ensuring the sustained delivery of their vital ecosystems services.

Acknowledgments

I am grateful to the organizations that made their data available for this study, as well as the individuals who provided support and feedback, including Jonathan Moran, Kathleen Moore, Dan Buffett, Karin Bodtker, and two anonymous reviewers.

Author Contributions

Conceived and designed the experiments: CKR. Performed the experiments: CKR. Analyzed the data: CKR. Wrote the paper: CKR.

References

1. Merrifield MS, Hines E, Liu X, Beck MW (2011) Building regional threat-based networks for estuaries in the Western United States. PLoS One 6(2): 1–10.
2. Ryder JL, Kenyon JK, Buffett D, Moore K, Ceh M, et al. (2007) An integrated biophysical assessment of estuarine habitats in British Columbia to assist regional conservation planning. Delta BC: Canadian Wildlife Service, Pacific and Yukon Region.
3. Edgar GJ, Barrett NS, Braddon DJ, Last PR (2000) The conservation significance of estuaries: A classification of Tasmanian estuaries using ecological, physical and demographic attributes as a case study. Biological Conservation 92: 383–397.
4. Costanza R, d'Arge R, de Groot R, Farber S, Grasso M, et al. (2007) The value of the world's ecosystem services and natural capital. Nature 387: 253–260.
5. Campbell CR (2010) Blue carbon – British Columbia: The case for the conservation and enhancement of estuarine processes and sediments in BC. Victoria BC: Sierra Club of BC.
6. Kelsey E (1995) The Pacific Estuary Conservation Program. Victoria BC: Pacific Estuary Conservation Program.
7. Kennish MJ (2002) Environmental threats and environmental future of estuaries. Environmental Conservation 29(1): 78–107.
8. Callaway GC, Zedler JB (2009) Conserving the diverse marshes of the Pacific coast. In Batzer DP, Baldwin AH, editors. Wetland habitat of North America: Ecology and conservation concerns. London: University of California Press. 285–307.
9. Austin MA, Buffett DA, Nicolson DJ, Scudder GGE, Stevens V (2008) Taking nature's pulse: The status of biodiversity in British Columbia. Victoria BC: Biodiversity BC.
10. Environment Canada (1999) Draft Canada's National Programme of Action (NPA) for the protection of the marine environment from land-based activities. Ottawa, ON: Environment Canada.
11. Leschine TM, Ferriss BE, Bell KP, Bartz KK, MacWilliams S, et al. (2003) Challenges and strategies for better use of scientific information in the management of coastal estuaries. Estuaries 26(4B): 1189–1204.
12. Crossland CJ, Baird D, Ducrotoy JP, Lindebloom H (2005) The coastal zone – a domain of global interactions. In: Crossland, CJ, Kremer HH, Lindebloom HJ, Marshall Crossland JI, Le Tissier MDA, editors. Coastal fluxes in the anthropocene. New York: Springer-Verlag Berlin Heidelberg. 1–37.
13. Scavia D, Field JC, Boesch DF, Buddemeier RW, Burkett V, et al. (2002). Climate change impacts on U.S. coastal and marine ecosystems. Estuaries 25(2): 149–164.
14. Galbraith H, Jones B, Park R, Clough J, Herrod-Julius S, et al. (2002) Global climate change and sea level rise: Potential losses of intertidal habitat for shorebirds. Waterbirds 25(2): 173–183.
15. Nicolas D, Chaalali A, Drouineau H, Lobry J, Uriarte A, et al. (2011) Impact of global warming on European tidal estuaries: Some evidence of northward migration of estuarine fish species. Regional Environmental Change 11: 639–649.

16. Feely RA, Alin SR, Newton J, Sabine CL, Warner M, et al. (2010) The combined effects of ocean acidification, mixing, and respiration on pH and carbonate saturation in an urbanized estuary. Estuarine, Coastal and Shelf Science 88: 442–449.
17. Okey TA, Alidina HM, Lo V, Montenegro A, Jessen S (2012) Climate change impacts and vulnerabilities in Canada's Pacific marine ecosystems. Vancouver BC: CPAWS BC and WWF-Canada.
18. Ryder JL, Ceh M, Buffett D, Moore K (2006) Economic and conservation tenures in the intertidal areas of BC estuaries. In: Gilkeson L et al., editors. Alive and inseparable: British Columbia's coastal environment 2006. Victoria: Province of British Columbia. 35–41.
19. Emmett R, Llansó R, Newton J, Thom R, Hornberger CM, et al. (2000) Geographic signatures of North American West Coast estuaries. Estuaries 23(6): 765–792.
20. DFO (2008) The role of the provincial and territorial governments in the oceans sector. Ottawa ON: Fisheries and Oceans Canada.
21. Robb CK, Bodtker KM, Wright K, Lash J (2011) Commercial fisheries closures in marine protected areas on Canada's Pacific coast: The exception, not the rule. Marine Policy 35: 309–316.
22. Herr D, Pidgeon E, Laffoley D (2012) Blue carbon policy framework: Based on the discussion of the International Blue Carbon Policy Working Group. Gland Switzerland: International Union for Conservation of Nature.
23. DFO (2002) Policy and operational framework for integrated management of estuarine, coastal and marine environments in Canada. Ottawa ON: Fisheries and Oceans Canada.
24. Crowder L, Norse E (2008) Essential ecological insights for marine ecosystem-based management and marine spatial planning. Marine Policy 32: 772–778.
25. Lotze HK, Lenihan HS, Bourque BJ, Bradbury RH, Cooke RG, et al. (2006) Depletion, degradation and recovery potential of estuaries and coastal seas. Science 312: 1806–1809.
26. McLeod KL, Lubchenco J, Palumbi S, Rosenberg AA (2005) Scientific consensus statement on marine ecosystem-based management.
27. DFO (2005) Canada's oceans action plan: For present and future generations. Ottawa ON: Fisheries and Oceans Canada.
28. Halpern BS, Walbridge S, Selkoe KA, Kappel CV, Micheli F, et al. (2008) A global map of human impact on marine ecosystems. Science 319: 948–952.
29. Millennium Ecosystem Assessment (2005) Ecosystems and human well-being: A framework for assessment. Nairobi Kenya: United Nations Environment Programme.
30. Agardy T, di Sciara GN, Christie P (2011) Mind the gap: Addressing the shortcomings of marine protected areas through large scale marine spatial planning. Marine Policy 35: 226–232.
31. Douvere F (2008) The importance of marine spatial planning in advancing ecosystem-based sea use management. Marine Policy 32: 762–771.
32. Nelleman C, Corcoran E, Duarte CM, Valdés L, DeYoung C, et al. (2009) Blue carbon: A rapid response assessment. Arendal Norway: United Nations Environment Programme.
33. Ban NC, Alidina HM, Ardron JA (2010) Cumulative impact mapping: Advances, relevance and limitations to marine management and conservation, using Canada's Pacific waters as a case study. Marine Policy 34: 876–886.
34. Imperial MT, Hennessey TM (1996) The ecosystem-based approach to managing estuaries: An assessment of the National Estuary Program. Coastal Management 24: 115–139.
35. Teck SJ, Halpern BS, Kappel CV, Micheli F, Selkoe KA, et al. (2010) Using expert judgment to estimate marine ecosystem vulnerability in the California Current. Ecological Applications 20(5): 1402–1416.
36. Wang T, Hamann A, Spittlehouse DL, Murdock TQ (2012) ClimateWNA – high-resolution spatial climate data for western North America. Journal of Applied Meteorological and Climatology 51: 16–29.
37. Ban NC, Bodtker KM, Nicolson D, Robb CK, Royle K, et al. (2013) Setting the stage for marine spatial planning: Ecological and social data collation and analyses in Canada's Pacific waters. Marine Policy 39: 11–20.
38. Google (2012) Google Earth, v. 6.1. Mountainview CA: Google, Inc.
39. Quigley MP, Hall JA (1999) Recovery of macrobenthic communities after maintenance dredging in the Blyth Estuary, north-east England. Aquatic Conservation: Marine Freshwater Ecosystems 9: 63–73.
40. Systat Software (2010) Systat, v 13. Chicago IL: Systat Software, Inc.
41. Ketchen DJ Jr, Shook CL (1996) The application of cluster analysis in strategic management research: An analysis and critique. Strategic Management Journal 17: 441–458.
42. Joppa LN, Pfaff A (2009) High and far: Biases in the location of protected areas. PLoS One 4(12): 1–6.
43. Williams GL, Langer OE (2002) Review of estuary management plans in British Columbia. Vancouver BC: Fisheries and Oceans Canada.
44. Gleason MG, Merrifield MS, Cook C, Davenport A, Shaw R (2006) Assessing gaps in marine conservation in California. Frontiers in Ecology and the Environment 4(5): 249–258.
45. Balmford A, Gaston KJ, Blyth S, James A, Kapos V (2003) Global variation in terrestrial conservation costs, conservation benefits, and unmet conservation needs. Proc Natl Acad Sci U S A 100(3): 1046–1050.
46. Government of Canada (2011) National framework for Canada's network of marine protected areas. Ottawa ON: Fisheries and Oceans Canada.
47. Smith HD, Maes F, Stojanovic TA, Ballinger RC (2011) The integration of land and marine spatial planning. Journal of Coastal Conservation 15: 291–303.
48. Hutchings JA, Côté IM, Dodson JJ, Fleming IA, Jennings S, et al. (2012) Sustaining Canadian marine biodiversity: Responding to the challenges posed by climate change, fisheries, and aquaculture. Ottawa ON: Royal Society of Canada.
49. Brooks TM, Mittermeier RA, da Fonseca GAB, Gerlach J, Hoffmann M, et al. (2006) Global biodiversity conservation priorities. Science 313(5783): 58–61.
50. Jessen S, Chan K, Côté I, Dearden P, De Santo E, et al. (2011) Science-based guidelines for MPAs and MPA networks in Canada. Vancouver BC: CPAWS.
51. Baustian JJ, Mendelssohn IA, Hester MW (2012) Vegetation's importance in regulating surface elevation in a coastal salt marsh facing elevated rates of sea level rise. Global Change Biology 18: 3377–3382.
52. Barbier EB (2012) Progress and challenges in valuing coastal and marine ecosystem services. Review of Environmental Economics and Policy 6(1): 1–19.
53. Sloan NA, Vance-Borland K, Ray GC (2007) Fallen between the cracks: Conservation linking land and sea. Conservation Biology 21(4): 897–898.
54. Klein CJ, Chan A, Kircher L, Cundiff AJ, Gardner N, et al. (2008) Striking a balance between biodiversity conservation and socioeconomic viability in the design of marine protected areas. Conservation Biology 22(3): 691–700.
55. Luck GW, Chan KMA, Fay JP (2009) Protecting ecosystem services and biodiversity in the world's watersheds. Conservation Letters 2: 179–188.

4

Overfishing and the Replacement of Demersal Finfish by Shellfish: An Example from the English Channel

Carlotta Molfese[1]*, Doug Beare[2], Jason M. Hall-Spencer[1]

1 Marine Biology and Ecology Research Centre, Plymouth University, Plymouth, United Kingdom, 2 Worldfish, Batu Maung, Penang, Malaysia

Abstract

The worldwide depletion of major fish stocks through intensive industrial fishing is thought to have profoundly altered the trophic structure of marine ecosystems. Here we assess changes in the trophic structure of the English Channel marine ecosystem using a 90-year time-series (1920–2010) of commercial fishery landings. Our analysis was based on estimates of the mean trophic level (mTL) of annual landings and the Fishing-in-Balance index (FiB). Food webs of the Channel ecosystem have been altered, as shown by a significant decline in the mTL of fishery landings whilst increases in the FiB index suggest increased fishing effort and fishery expansion. Large, high trophic level species (e.g. spurdog, cod, ling) have been increasingly replaced by smaller, low trophic level fish (e.g. small spotted catsharks) and invertebrates (e.g. scallops, crabs and lobster). Declining trophic levels in fisheries catches have occurred worldwide, with fish catches progressively being replaced by invertebrates. We argue that a network of fisheries closures would help rebalance the trophic status of the Channel and allow regeneration of marine ecosystems.

Editor: David Mark Bailey, University of Glasgow, United Kingdom

Funding: The authors have no support or funding to report.

Competing Interests: The authors have declared that no competing interests exist.

* Email: carlotta0911@gmail.com

Introduction

Effects of overfishing on marine trophic structure

The field of historical marine ecology has introduced a different perspective to our understanding of marine ecosystems; it has revealed that overfishing has had profound effects on coastal ecosystems worldwide for centuries [1,2]. The historical response to overfishing is an increase in fishing effort, an expansion to new and deeper grounds and a shift to new target species [3]. In the last decade, fisheries have shifted towards smaller, lower-trophic level species as large predatory species with a higher economic value had been depleted [4]. This phenomenon, known as "fishing down marine food webs" was first described by [5] in 1998: they demonstrated a decline in the trophic level of global fisheries landings from 3.3 units in the early 1950s to 3.1 in 1994. Studies performed independently from commercial catch data on smaller, regional scales over the last decades have shown even more rapid declines in trophic level (Table 1).

Fisheries typically remove top predators first and as a result their direct competitors and prey are able to prosper, affecting the overall productivity and ecological stability of the ecosystem [1]. Severe declines in the populations of major predator species have now been reported around the world [6,7]. Overexploitation of a species can have cascading effects and have the potential to trigger regime shifts altering the ecological function of marine systems [8,9]. In many instances, the decline of finfish species has been followed by an increase in their invertebrate prey [10,11] and although new and economically viable fisheries have developed for these new target species, concerns have been raised about their long-term sustainability as well as shifts towards homogenized, simplified ecosystems [12,13].

In the present study, we used a 90-year dataset of international catch statistics from the English Channel marine ecosystem, a region that has numerous important fishing ports and where finfish landings now make up a far smaller proportion of the catch than they did historically (Figure 1). This dataset spans a period of intensive fishing which we use to assess whether there has been a trend for 'fishing down' food webs in a region where it has not been reported before. Finally, we discuss the way forwards to improve fisheries sustainability using area closures to aid recovery of marine ecosystems.

The English Channel

The English Channel ('La Manche' in French) is a narrow strait between England and France (Figure 1). It covers 75,000 km^2 and narrows to *ca* 30 km wide at its easternmost point; the Channel is relatively shallow, with an average depth of around 100 m in the west gradually decreasing to 40 m depth in the east [25]. The western Channel is influenced greatly by Atlantic water while the eastern part receives large freshwater inputs from coastal rivers, this gradation from oceanic to neritic waters forms a biogeographical transition zone with a variety of both boreal/cold temperate and warm temperate organisms [26].

Today the UK and France account for 98% of the total landings from the Channel [27]: around 25,884 t and 68,732 t of finfish and 26,605 t and 48,871 t of invertebrate were landed in 2010 in the UK and France respectively. Fishing has exerted pressure in these waters since the Middle Ages [25,26] but at the turn of the 20th century fishing effort increased substantially due to the advent of motorised fishing vessels. Monitoring of fish populations has revealed reductions in mean length and length at maturity of demersal communities [28]. Moreover, large and slow-growing

Table 1. Instances of "Fishing Down Marine Food Web" across the globe, showing rates of decline in mean trophic level (mTL).

Country/Area	Period	mTL decline	Source
Cuba EEZ	1960–1995	0.10 decade^{-1}	[14]
Canada (West and East coast)	1950–1997 and 1873–1996	0.03–0.1 decade^{-1}	[15]
Celtic Sea	1982–2000	0.04 year^{-1} (ICES catch data) and 0.03 year^{-1} (scientific survey)	[16]
Thailand	1965–1997	0.05–0.09 decade^{-1}	[17]
Iceland	1918–1999	0.06 decade^{-1}	[18]
Chile	1979–1999	0.175 decade^{-1}	[19]
Greece	1950–2001	0.02 decade^{-1}	[20]
Indian States and Union Territories	1950–2000	0.058 decade^{-1}	[21]
Argentinean-Uruguayan Common Fishing Zone (AUCFZ)	1989–2003	0.03 year^{-1}	[22]
Portugal	1970–2006	0.005 year^{-1}	[23]
Brazil	1978–2000	0.16 decade^{-1}	[24]

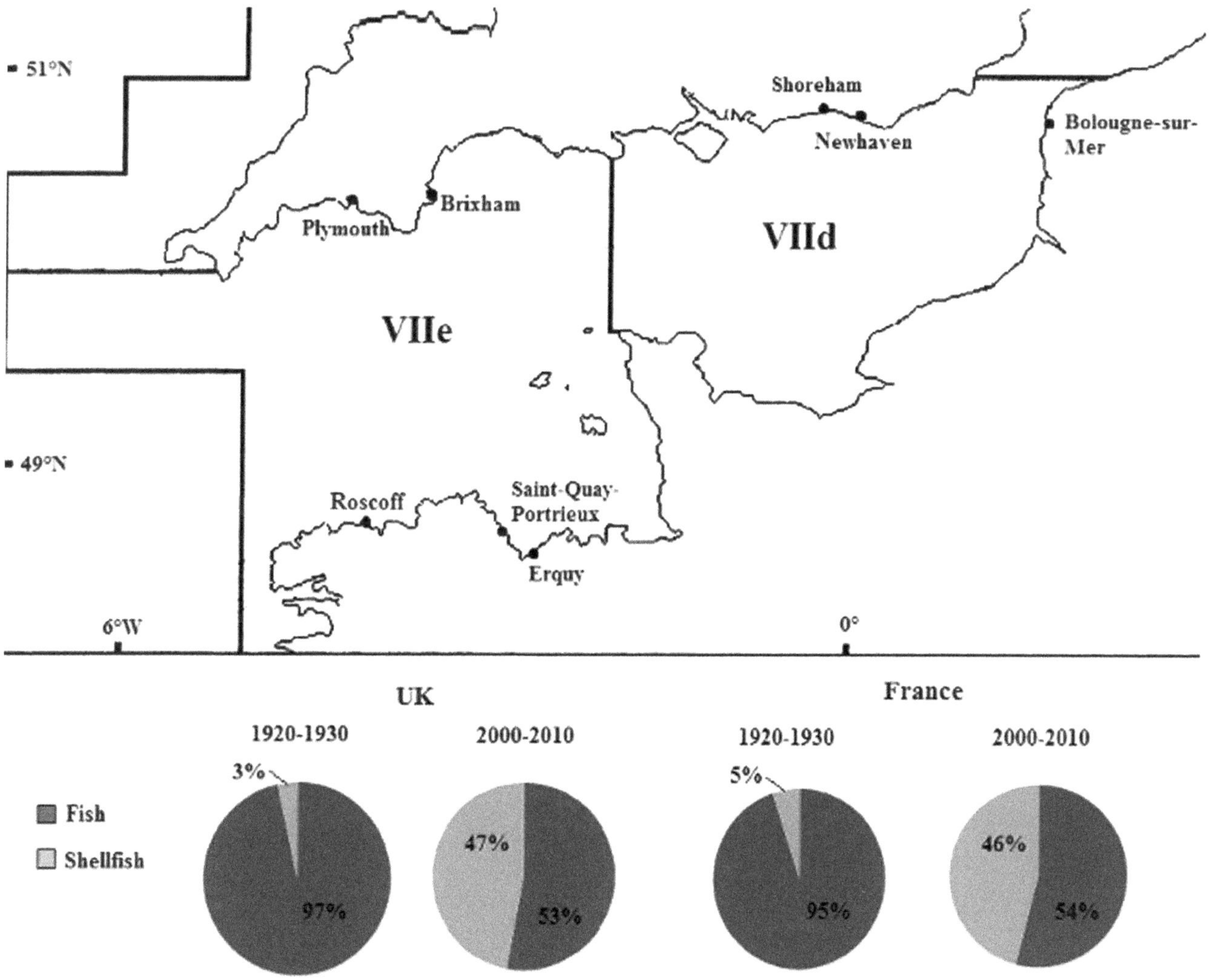

Figure 1. Major English Channel fishing ports by landings value in 2010, ICES areas VIIe and VIId. Data sourced from MMO and France AgriMer. Pie charts show the proportions of fish and shellfish landed by the UK and French fishing fleets for the period 1920–1930 and 2000–2010.

Table 2. Finfish species included in our analysis with respective trophic levels (TL).

Common Name	Scientific Name	TL	Common Name	Scientific Name	TL
Witch	*Glyptocephalus cynoglossus*	3.1	Houndsharks, smoothhounds nei	Triakidae	3.9[a]
European flounder	*Platichthys flesus*	3.2	Red gurnard	*Chelidonichthys cuculus*	3.9
Lemon sole	*Microstomus kitt*	3.2	Small-eyed ray	*Raja microocellata*	3.9
Red mullet	*Mullus barbatus*	3.2	Blonde ray	*Raja brachyura*	4
Sand sole	*Pegusa lascaris*	3.2	Nursehound	*Scyliorhinus stellaris*	4
Common dab	*Limanda limanda*	3.3	Dogfish etc.	*Squalus* spp.	4.1[a]
Common sole	*Solea solea*	3.3	Dogfish sharks nei	Squalidae	4.1[a]
European plaice	*Pleuronectes platessa*	3.3	Haddock	*Melanogrammus aeglefinus*	4.1
Mullets nei	Mullidae	3.3[a]	Various sharks nei	Selachimorpha (Pleurotremata)	4.1[a]
Striped red mullet (= Surmullet)	*Mullus surmuletus*	3.4	Blue shark	*Prionace glauca*	4.2
Grey gurnard	*Eutrigla gurnardus*	3.6	Megrim	*Lepidorhombus whiffiagonis*	4.2
Small-spotted catshark	*Scyliorhinus canicula*	3.6	Pollack	*Pollachius pollachius*	4.2
Pouting	*Trisopterus luscus*	3.7	Tope shark	*Galeorhinus galeus*	4.2
Saithe	*Pollachius virens*	3.7	European conger	*Conger conger*	4.3
Spotted ray	*Raja montagui*	3.7	Ling	*Molva molva*	4.3
Turbot	*Scophthalmus maximus*	3.7	Spurdog	*Squalus acanthias*	4.3
Brill	*Scophthalmus rhombus*	3.8	Atlantic cod	*Gadus morhua*	4.4
Groundfishes nei	Osteichthyes	3.8[a]	European Hake	*Merluccius merluccius*	4.4
Gurnards, searobins	*Triglidae*	3.8[a]	Whiting	*Merlangius merlangus*	4.4
Raja rays nei	*Raja* spp.	3.8[a]	Monkfish	*Lophius piscatorius*	4.5
Smooth-hound	*Mustelus mustelus*	3.8	Atlantic halibut	*Hippoglossus hippoglossus*	4.5
Thornback ray	*Raja clavata*	3.8	John dory	*Zeus faber*	4.5
Dogfishes and hounds nei	Squalidae, Scyliorhinidae	3.9[a]	Monkfish nei	*Lophius* spp.	4.5[a]

nei: not elsewhere included.
[a]Figure represent the mean TL value of known species belonging to that taxonomic group within UK waters. Species included were obtained from the UK Fisheries Statistics list [37], see Table S2.

species have decreased dramatically in the region over the last century including angel shark (*Squatina squatina*, Squatinidae) and common skate (*Raja batis*, Rajidae) which is now commercially extinct in the Channel; whereas small, commercially undesirable species such as the small-spotted catshark (*Scyliorhinus canicula*, Scyliorhinidae) have increased in abundance [29,30].

Table 3. Invertebrate species included in our analysis with respective trophic levels (TL).

Common Name	Scientific Name	TL	Common Name	Scientific Name	TL
Blue mussel	*Mytilus edulis*	2	Velvet swimcrab	*Necora puber*	2.6
European flat oyster	*Ostrea edulis*	2	Common prawn	*Palaemon serratus*	2.7
Great Atlantic scallop	*Pecten maximus*	2	Marine crabs nei	Brachyura	2.8[a]
Pacific cupped oyster	*Crassostrea gigas*	2	Norway lobster	*Nephrops norvegicus*	2.9
Periwinkles nei	*Littorina* spp.	2[a]	Whelk	*Buccinum undatum*	3.1
Common edible cockle	*Cardium edule*	2.1	Common shrimp	*Crangon crangon*	3.2
Queen scallop	*Aequipecten opercularis*	2.1	Cuttlefish, bobtail squids nei	*Sepiidae, Sepiolidae*	3.5[a]
Spinous spider crab	*Maja squinado*	2.3	Common cuttlefish	*Sepia officinalis*	3.6
Variuos shellfish	Mollusca, Crustacea, Echinodermata	2.4[a]	Octopuses, etc. nei	Octopodidae	3.6[a]
Edible crab	*Cancer pagurus*	2.6	Various squids nei	Loliginidae, Ommastrephidae	4[a]
European lobster	*Homarus gammarus*	2.6	Common squids nei	*Loligo* spp.	4.2[a]

nei: not elsewhere included.
[a]Figure represent the mean TL value of known species belonging to that taxonomic group within UK waters. Species included were obtained from the UK Fisheries Statistics list [37], see Table S2.

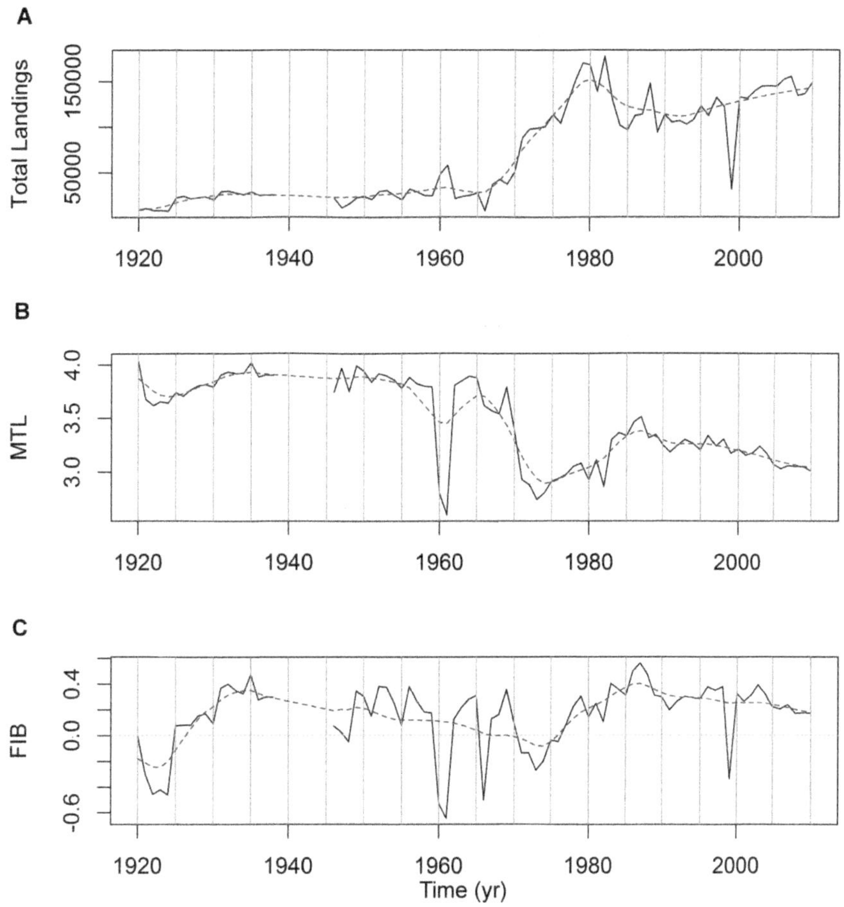

Figure 2. ICES data for the English Channel on landings, mTL and FiB index for 1920–2010. Analysis excludes pelagic species. (A) Annual landings from the English Channel. (B) Changes in the mTL over time. (C) Changes in the FiB index over time. The blue dashed line is a smoothing function, "supsmu" [50] available as standard with the R software package [51].

Materials and Methods

Fishing down marine food webs

The trophic level (TL) of an organism denotes its position within a foodweb and it can be estimated from diet observations, nitrogen isotope measurements or models of trophic fluxes from the equation:

$$TL_i = 1 + \sum_j \left(TL_j \times DC_{ij} \right) \quad (1)$$

Where TL_j is the 'non-integer' trophic level of prey j and DC_{ij} is the fraction of j in the diet of i [31]. The Marine Trophic Index (MTI), which corresponds to the mTL of fishery landings, was developed to describe the structure of an ecosystem resulting from a fishery-induced depletion of its components and can be computed for any year y:

$$MTI = mTL_y = \frac{\sum_i \left(TL_i \times Y_{iy} \right)}{\sum_i Y_{iy}} \quad (2)$$

Table 4. Granger causality testes at lag 1.

Model	Res. D.f.	D.f.	F-value	P(>F)
'FiB causes mTL'				
FiB ~ lags(fib) + lags(mTL)	80			
FiB ~ lags(fib)	81	−1	1.024	0.315
'mTL causes FiB'				
mTL ~ lag(mTL) + lag(FiB)	80			
mTL ~ lag(mTL)	81	−1	3.933	0.050

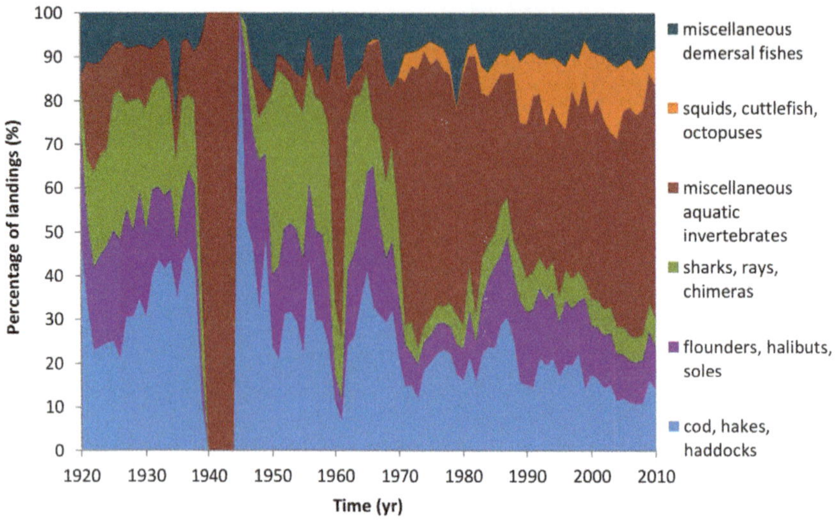

Figure 3. ICES data on changes in catch composition for the English Channel 1920–2010. Species grouped into ISSCAAP categories.

Where Y_{iy} is the catch of species i in year y, and TL_i is defined as in Equation 1 [32].

Additionally, [31] developed an index to assess if changes in the mTL were compensated by changes in catches. This is because biological production is higher at lower TL, hence an inevitable consequence of 'fishing down the food web' will, ironically, be greater biological production potentially being available to the fisheries, and any decline in TL should be accompanied by an 'ecologically appropriate' increase in the overall biomass of catches. Hence, the Fishing-in-Balance (FiB) index:

- Will maintain a value of zero when a decrease or increase in TL is accompanied by an ecologically balanced increase or decrease in catches;
- Will increase (>0) if bottom-up effects have occurred or if the fishery has expanded beyond its traditional ground;
- Will decrease (<0) if the fishery has contracted geographically or if it has taken so much biological productivity from the ecosystem that has impaired its natural functioning.

It is defined for any year y by:

$$FiB_y = \log\left[Y_y \times \left(\frac{1}{TE}\right)^{TLy}\right] - \log\left[Y_0 \times \left(\frac{1}{TE}\right)^{TL0}\right] \quad (3)$$

Where Y_y is the catch at year y; TL_y is the trophic level of the catch in year y; Y_0 and TL_0 are the catch and trophic level of the catch at the beginning of the series analysed and TE is the energy transfer efficiency between TL estimated to be 0.1 (= 10%) in several marine ecosystem studies [32].

International fishery landings

Catch statistics for the English Channel were obtained from ICES (International Council for the Exploration of the Sea) [33] and comprised two datasets; Historical catch statistics (1903–1949) and Official catch statistics (1950–2010). The English Channel consists of West and East Channel areas defined as divisions VIIe and VIId respectively (Figure 1). The statistics represent the live weight equivalent of landings and so do not include discards. In our analyses we excluded all pelagic species as previous studies in the Western English Channel have shown a very strong climatic influence that affects their abundance and distribution in the Channel [29,34] and would therefore complicate the interpretation of the analysis. Landings reported as <1 t were omitted, as well as data collected prior to 1920 and between 1939–1945, since the older dataset excluded shellfish catches and there were data gaps during the World Wars.

Table 2 and 3 lists the 68 taxa analysed in our study and the respective TL obtained from FishBase [35] and the 'Sea Around Us' database [36]. For taxa reported at levels coarser than species, or under a general category (e.g. 'sharks etc.', 'various shellfish'), a list of all marine species caught in UK waters was obtained from the Marine Management Organization (MMO) [37] and was used to derive an average trophic level for such categories. Tables S1 and S2 list the aggregated taxa and average trophic levels estimated for such groups. Species were also grouped into ISSCAAP categories (International Standard Statistical Classification of Aquatic Animals and Plants) to evaluate changes in landings composition over time. Table S3 shows taxa belonging to each ISSCAAP category.

Results

There was a clear increase in landings from the English Channel between 1920 and 2010 (Figure 2A). These increased gradually from 9,146 t in 1920 to 50,924 t in 1970 and escalated rapidly to a maximum weight of 177,793 t in 1982 t. These however fell abruptly to 96,783 t in 1985 and have stabilized at around 130,000–150,000 t in the last decade (Figure 2A).

The mTL of fish landed from the English Channel has declined from 4.0 in 1920 to 3.0 in 2010 (Figure 2B) and there is a marked negative correlation of −0.8 between mTL and total landings. Confirmatory statistical tests are difficult to support because of autocorrelation and non-linear trends. 'Differencing' the data is one way to reduce or remove the effects of autocorrelation, and when this was done the series were still negatively and significantly correlated with a coefficient of −0.30 (Pearson's Product-Moment correlation, $t = -2.7$, $df = 80$, $p = 0.0065$). Between 1920 and 1970 both mTL and total landings underwent relatively little change but after the 1970s as catches increased mTL declined, the period of highest catches (1971–1982) corresponding to mTL values of 2.7–3.1. In the following years, catches declined considerably and

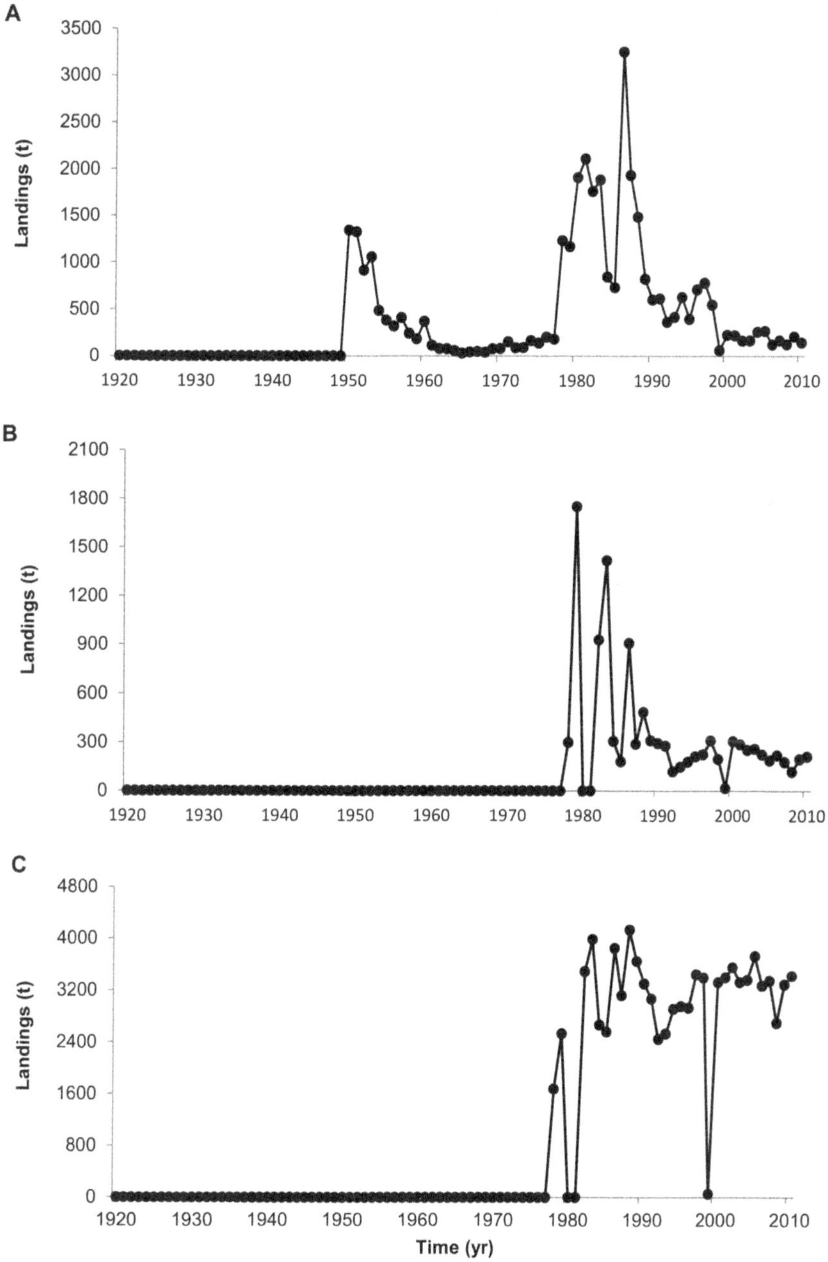

Figure 4. Annual landings of selected elasmobranch species. (A) Spurdog, (B) Tope shark, and (C) Small-spotted catshark.

mTL has continued to fall and was at 3.0 in 2010. An overall increase in the FiB index was also detected (Figure 2C), suggesting that the decline in mTL has indeed been compensated by increased catches as a consequence of either a geographic expansion of the fishery or increased productivity of the Channel.

We decided to use the concept of 'Granger causality' [38] to decide if there was any statistical evidence that mTL had 'caused' the change in FiB. We tested both whether mTL caused FiB and the reverse (whether FiB causes mTL) and the results are presented in Table 4. Clearly there is statistical evidence that changes in the mean trophic level 'cause' changes in the fishing balance index ($p = 0.050$) while, gratifyingly, the reverse is untrue ($p = 0.315$). For further statistical tests see Supporting Information S1.

The composition of landings has also undergone changes over time with regards to both higher and lower trophic level species (Figure 3). The contribution of higher trophic level species to UK and France fisheries landings has decreased considerably in recent years. The group 'sharks, rays, chimeras' declined markedly from 34% in 1926 to 6.0% in 2010 with spurdog and tope shark (*Galeorhinus galeus*, Triakidae) landings declining considerably after the 1980s while small-spotted catshark landings increasing significantly (Figure 4). Similarly, the contribution of the 'cods, haddock, hakes' group has declined from 48% in 1920 to 14% in 2010. The most remarkable declines in landings have occurred for Atlantic cod (*Gadus morhua*, Gadidae), ling (*Molva molva*, Gadidae) and European hake (*Merluccius merluccius*, Merlucciidae) (Figure 5). Landings of 'flounders, halibuts, soles' and 'miscellaneous demersal fishes' has changed relatively little over the whole time-series (Figure 3).

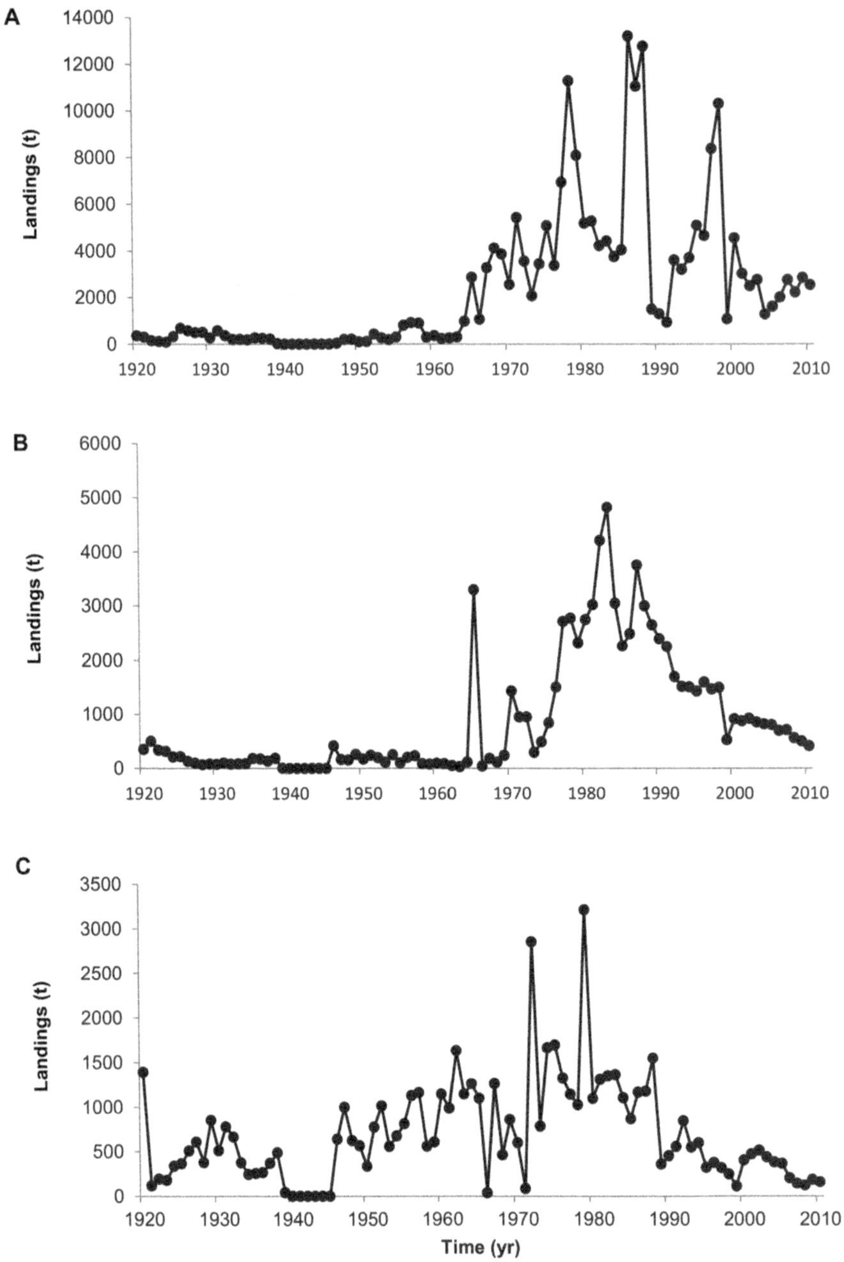

Figure 5. Annual landings of selected gadoid species. (A) Atlantic cod, (B) Ling, (C) European hake.

As for invertebrate species, a marked increase in landings from the English Channel is evident with the 'miscellaneous aquatic invertebrates' and 'squids, cuttlefish, octopuses' groups accounting to more than half of the total landings since the 1970s (Figure 3). In particular, landings of edible crab (*Cancer pagurus*, Cancridae), European lobster (*Homarus gammarus*, Nephropidae) and Great Atlantic scallop (*Pecten maximus*, Pectinidae) have increased considerably (Figure 6).

Discussion

It is clear from our analyses that fishing pressure has caused significant changes to food webs of the English Channel over the past 90 years. The mean Trophic Level of English Channel landings has fallen by 0.1 unit per decade, one of the fastest rates reported among other heavily fished regions of the world providing yet more evidence that 'fishing down food webs' (*sensu* [5]) is a worldwide phenomenon (Table 1). The FiB index suggests that, either a geographic expansion of the Channel fishery or an increase in the productivity of the region has compensated for declining mTL with increased catches during two quite distinct periods in the time-series: 1925–1970 and 1980–2010. The former corresponds to the period of rapid industrialization and expansion of fishing documented by [28] and [39] for the English Channel and the UK respectively. Increases in landings were only

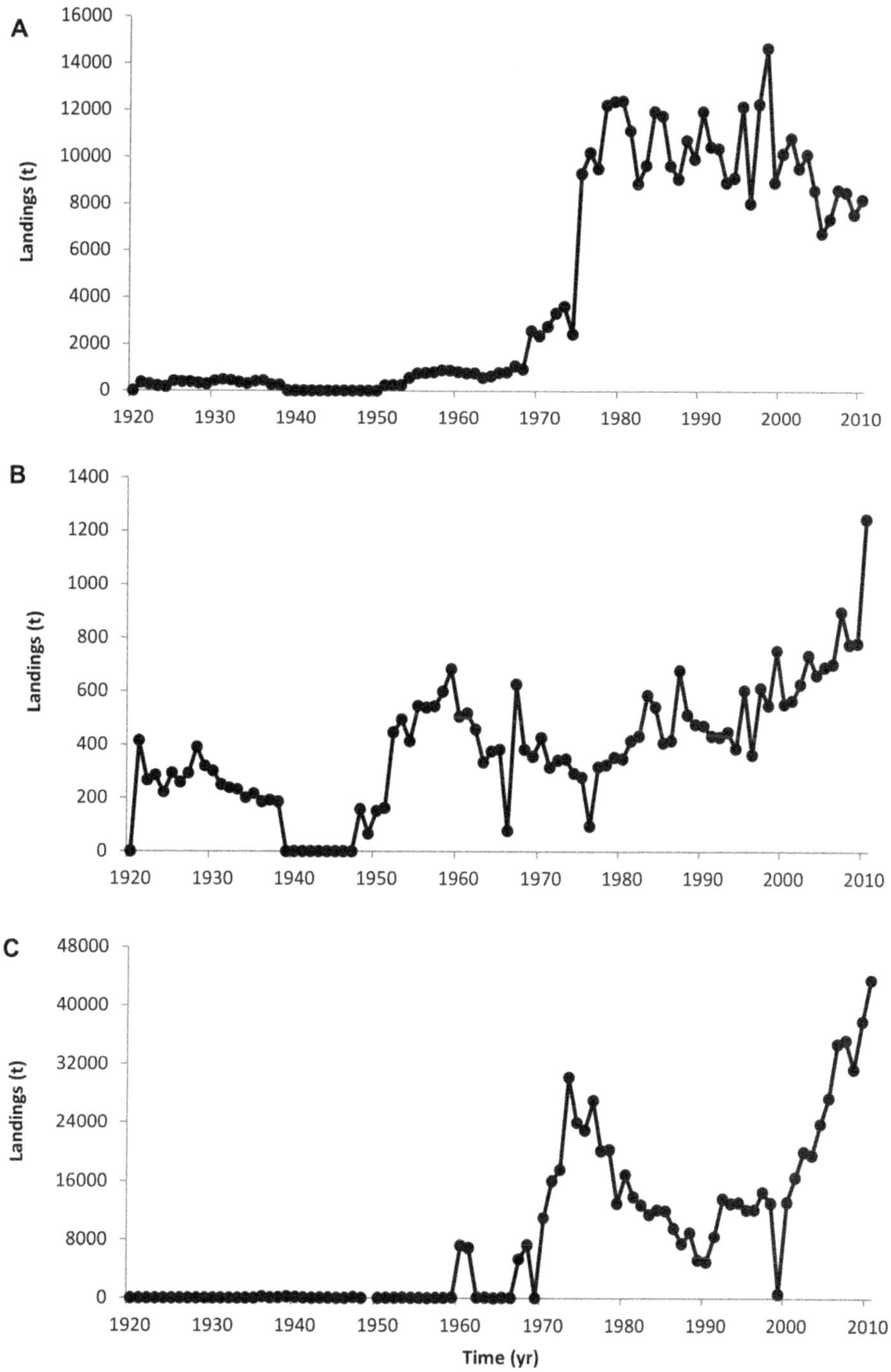

Figure 6. Annual landings of selected shellfish species. (A) Edible crab, (B) European lobster and (C) Great Atlantic scallop.

maintained thanks to increased fishing efficiency and expansion into distant and deeper grounds, but stocks were declining long before the 1980s. Historical evidence reveals that signs of overfishing in UK waters were already apparent at the end of the 19th century [40] and concerns regarding declining stocks date back at least to 1863 [3]. The second period matches an increase in water temperatures and overall productivity of the North East Atlantic during which stocks of cold-water species such as cod and haddock have seen a dramatic decline [41]. However, the declining trend of the FiB index in the last decade of the time series suggests that the factor that compensated for declining mTL in the second period is now fading away.

The landings data show a decrease in high trophic level species, such as gadoid fish and elasmobranchs, and an increase in low trophic level species such as invertebrates (Figure 3). The pattern is strikingly similar to that found elsewhere around the UK [11] and the rest of the world [10,18,22].

It is now well known that fishing removed populations of species that were common around Britain and Ireland a century ago. Large and long-lived species of elasmobranch, such as the common skate and the angel shark, have proved to be particularly vulnerable as they have low fecundity rates and are slow to mature [42,43]. Conversely, small-spotted catshark have had a major increase in English Channel landings, this species has a very high rate of discard survival and matures earlier than larger species [28,30]. Overall, English Channel elasmobranch landings have been declining steadily since the 1950s, with dramatic declines in spurdog, tope shark and thornback ray (*Raja clavata*, Rajidae).

When fishing pressure eased off during World Wars I and II stocks of demersal fish such as 'cod, hake and haddock' built up around the UK but recent decades of overfishing have brought these stocks to historic lows [39,44]. The contribution of the 'miscellaneous demersal' group to total landings has changed little over time but landings of high trophic level species within this group, such as European conger (*Conger conger*, Congridae) and monk fish (*Lophius piscatorious*, Lophiidae), have declined since the 1960s as also documented in the Celtic Sea by [16].

The removal of high trophic level species can have a 'cascading effect' on lower trophic levels [1,8]. In the English Channel, the proportion of lower trophic level species have increased over the years with invertebrate species accounting to more than 50% of the total landings since 1970s (Figure 3). The replacement of finfish species by invertebrates is a phenomenon which has been documented in many marine regions around the world and it has generally been followed by the development of new economically viable invertebrate fisheries [10,11]. Despite their initial profitability, an increasing percentage of these new fisheries are already overexploited, collapsed or closed [12] and concerns have been raised where trawls and dredges are used to catch invertebrates as they degrade habitat complexity and species diversity [13].

Socioeconomic factors are also known to influence the composition of landings. These include consumers' income and preferences, catching restrictions, fuel prices and technological innovations. These factors are in turn reflected in the price placed on a certain product. Generally as a resource become scarcer its price increases. In the Celtic Sea and in Portugal high trophic level species have become more expensive than low trophic level ones [16,23] suggesting that the supply of those species has diminished but their demand is still high. Indeed the UK is a net importer of fish. 588,000 t of fish (excluding shellfish) were imported in 2010, with cod accounting for the highest import at 101,400 t [37]. Two-thirds of the total supply of haddock (60,300 t) is also derived from Iceland and Norway and in a lesser quantity also ling, hake and pollack are imported [37]. These are all fish species with catch restrictions in the UK but their landings have never fully met their allocated quotas. Conversely, in 2010 the UK exported 98,000 t of shellfish including 15,300 t of crab, 14,500 t of scallops and 16,500 t of shrimps and prawns [37]. However, [45] showed an increase in the price of low trophic level species globally in the past 50 years as a consequence of markets adjusting to the scarcity of target species by increasing the value of less desirable species, essentially creating 'perverse incentives' to unsustainable fishing. What is particularly compelling about the current study is that we've shown that by fishing or forcing down the food web, fishermen potentially create more biomass for themselves to harvest but there are many implications which remain to be investigated in detail of what seems a rather 'perverse incentive'.

The recent recognition of a global fishery crisis has induced many governments and fishery scientists to question current strategies used in fisheries management. Since the 1950s, the concept of Maximum Sustainable Yield (MSY) and annual catch allowances has been at the foundation of all fisheries management [4]. However, disregard of scientific advices over the years has caused the deterioration of many fish stocks and has rendered the single-species approach highly ineffective [46]. As this study has demonstrated the impact of fisheries is much more pervasive than previously thought. Fisheries can affect the physical structure of an ecosystem as well as its ecological stability and productivity by depriving it of its components [1,11]. The complexity of marine ecosystems compels a more holistic management approach that takes into account as many interactions as possible and whose priority is the integrity of the ecosystem from which the resource is extracted from [4].

There have been moves to improve the ecological status of the English Channel. In 2012, the UK Government approved 31 of 127 conservation areas recommended by an extensive stakeholder consultation, representing 3.7% of the marine area under English jurisdiction [47]. Removal of towed demersal fishing pressure from within such sites would allow habitat recovery and removal of all fishing effort within these areas could enhance adjacent fisheries through "spillover" or larval dispersal [48]. The Lyme Bay MPA in the English Channel is a great example of how an exploited ecosystem can regenerate and provide benefits to dependent fishing communities in only few years if closed to damaging fishing gear [49].

Conclusion

The community-level changes observed in the English Channel reflect those that have occurred in other heavily-fished systems around Europe and the rest of the world [5,10,42]. The use of the Marine Trophic Index (MTI) and the Fisheries-in-Balance Index (FiB) on this long-term data series have helped expose a major shift from demersal fish to shellfish landings in the English Channel as a consequence of an unsustainable fishing practice fuelled by 'perverse economic incentives'. These trends may be reversed by removing fishing pressure from within a network of closed areas and by implementing more rigid management measures including decommissioning schemes and reduction in fishing effort.

Acknowledgments

We thank Neal Gray of the Marine Management Organisation for help with sourcing data. The Consultative Group on International Agricultural Research (CGIAR) program on Climate Change, Agriculture and Food Security (CCAFs) is also owed a debt of thanks.

Author Contributions

Conceived and designed the experiments: CM. Analyzed the data: DB. Wrote the paper: CM JHS.

References

1. Jackson JBC, Kirby MX, Berger WH, Bjorndal KA, Botsford LW, et al. (2001) Historical overfishing and the recent collapse of coastal ecosystems. Science 293: 629–638.
2. Robinson LA, Frid CLJ (2008) Historical Marine Ecology: Examining the Role of Fisheries in Changes in North Sea Benthos. Ambio 37(5): 363–371.
3. Roberts CM (2007) The Unnatural History of the Sea. The past and future of humanity and fishing. Washington DC: Island Press. 435p.
4. Pauly D, Maclean J (2003) In a perfect ocean: the state of fisheries and ecosystems in the North Atlantic Ocean. Washington DC: Island Press. 208 p.
5. Pauly D, Christensen V, Dalsgaard J, Froese R, Torres F (1998) Fishing Down Marine Food Webs. Science 279: 860–86.
6. Christensen V, Guénette S, Heymans JJ, Walters CJ, Watson R, et al. (2003) Hundred-year decline of North Atlantic predatory fishes. Fish Fish 4: 1–24.
7. Myers RA, Worm B (2005) Extinction, survival or recovery of large predatory fishes. Phil. Trans. R. Soc. B. 360: 13–20.
8. Pinnegar JK, Polunin NVC, Francour P, Badalamenti F, Chemello R, et al. (2000) Trophic cascades in benthic marine ecosystems: lessons for fisheries and protected-area management. Environ. Conserv. 27: 179–200.
9. Frank KT, Petrie B, Fisher JAD, Leggett WC (2011) Transient dynamics of an altered large marine ecosystem. Nature 477: 86–89.
10. Worm B, Myers RA (2003) Meta-analysis of cod-shrimp interactions reveals top-down control in oceanic food webs. Ecology 84: 162–173.
11. Thurstan RH, Roberts CM (2010) Ecological Meltdown in the Firth of Clyde, Scotland: Two Centuries of Change in a Coastal Marine Ecosystem. PLoS ONE 5(7): e11767.
12. Anderson SC, Flemming JM, Watson R, Lotze HK (2011) Rapid Global Expansion of Invertebrate Fisheries: Trends, Drivers, and Ecosystem Effects. PLoS ONE 6(3): e14735.
13. Howarth LM, Roberts CM, Thurstan RH, Stewart BD (2013) The unintended consequences of simplifying the sea: making the case for complexity. Fish Fish doi: 10.1111/faf.12041.
14. Baisre JA (2000) Chronicles of Cuban marine fisheries (1935–1995). Trend analysis and fisheries potential. FAO Fisheries Technical Paper 394, 26 p.
15. Pauly D, Palomares ML, Froese R, Sa-a P, Vakily M, et al. (2001) Fishing down Canadian aquatic food webs. Can. J. Fish. Aquat. Sci. 58: 51–62.
16. Pinnegar JK, Jennings S, O'brien CM, Polunin NVC (2002) Long-term changes in the trophic level of the Celtic Sea fish community and fish market price distribution. J Appl Ecol 39: 377–390.
17. Pauly D, Chuenpagdee R (2003) Development of fisheries in the Gulf of Thailand Large Marine Ecosystem: analysis of the unplanned experiment. In: Hempel G, Sherman K, editors. Large Marine Ecosystems of the World: Trend in exploitation, Protection and Research. Elsevier, the Netherlands. 337–354.
18. Valtýsson HP, Pauly D (2003) Fishing down the food web: an Icelandic case study. In: Guðmundsson E and Valtysson HP, editors. Proc. Conference on Competitiveness within the Global Fisheries. University of Akureyri, Akureyri. 12–24.
19. Arancibа H, Neira S (2005) Long-term changes in the mean trophic level of Central Chile fishery landings. Sci Mar 69 (2): 295–300.
20. Stergiou K.I (2005) Fisheries impact on trophic levels: long-term trends in Hellenic waters. In: Papathanassiou E and Zanetos A, editors. State of the Hellenic Marine Environment, Hellenic Centre for Marine Research, Athens, Greece. 326–329.
21. Bhathal B, Pauly D (2008) 'Fishing down marine food webs' and spatial expansion of coastal fisheries in India, 1950–2000. Fish Res 91: 26–34.
22. Jaureguizar AJ, Milessi AC (2008) Assessing the sources of the fishing down marine food web process in the Argentinean-Uruguayan Common Fishing Zone. Sci Mar 72(1): 25–36.
23. Baeta F, Costa MJ, Cabral H (2009) Changes in the trophic level of Portuguese landings and fish market price variation in the last decades. Fish Res 97: 216–222.
24. Freire KMF, Pauly D (2010) Fishing down Brazilian marine food webs, with emphasis on the east Brazil large marine ecosystem. Fish Res 105: 57–62.
25. Stanford R, Pitcher TJ (2004) Ecosystem simulations of the English Channel: climate and trade-offs. Fish Centre Res Reports 12 (3): 1–103.
26. Dauvin JC (2012) Are the eastern and western basins of the English Channel two separate ecosystems? Mar Pollut Bull 64: 463–471.
27. Engelhard GH (n.d.). CHARM III project. Available: http://www.charm-project.org/en/themesen/human-activities/fishing-activities/138-an-integrated-view-on-the-french-and-british-fisheries-in-the-channel. Accessed 5 February 2013.
28. Genner MJ, Kendall M, Sims DW (2001) Archiving and analysis of the MBA bottom trawl and benthic survey data: Unravelling fishing efforts from climate change. Report to the Ministry of Agriculture, Fisheries and Food, London: 1–36.
29. Hawkins SJ, Southward AJ, Genner MJ (2003) Detection of environmental change in a marine ecosystem – evidence from the Western English Channel. Sci Total Enviro 310: 245–56.
30. McHugh M, Sims DW, Partridge JC, Genner MJ (2011) A century later: Long-term change of an inshore temperate marine fish assemblage. J Sea Res 65: 187–194.
31. Pauly D, Palomares ML (2000) Approaches for dealing with three sources of bias when studying the fishing down marine food web phenomenon. In: Durand F, editors. Fishing down the Mediterranean food webs? CIESM Workshop Series No. 12, Kerkyra, Greece. 61–66.
32. Kleisner K, Pauly D (2011) The Marine Trophic Index (MTI), the Fishing in Balance Index (FiB) and the spatial expansion of fisheries. In: Christensen V, Lai S, Palomares MLD, Zeller D, Pauly D, editors. The State of Biodiversity and Fisheries in Regional Seas. Fisheries Centre Research Report 19(3): 41–44.
33. ICES (2010) ICES Catch Statistics 1903–1949. Available: http://www.ices.dk/fish/CATChSTATISTICS.asp. Accessed 12 September 2012.
34. Southward AJ, Boalch GT and Maddock L (1988) Fluctuations in the herring and pilchard fisheries of Devon and Cornwall linked to change in climate since the 16th century. J Mar Biol Assoc UK 68: 423–445.
35. Froese R, Pauly D (2008) FishBase. WorldWideWeb Electronic Publication www.fishbase.org, version 09/2008. Accessed 21 October 2012.
36. Pew Charitable Trust (1999) Sea Around Us Project. Available: http://www.seaaroundus.org/topic/species. Accessed on 12 November 2012.
37. Marine Management Organization (2010) Fisheries Statistics. Available: http://www.marinemanagement.org.uk/fisheries/statistics/annual_additional.htm. Accessed 06 December 2012.
38. Granger CWJ (1969) Investigating Causal Relations by Econometric Models and Cross-spectral Methods. Econometrica 37(3): 424–438. doi:10.2307/1912791. JSTOR 1912791.
39. Thurstan RH, Brockington S, Roberts CM (2010) The effects of 118 years of industrial fishing on UK bottom trawl fisheries. Nat Commun 1 doi: 10.1038/ncomms1013.
40. Garstang W (1900) The impoverishment of the sea. J Mar Biol Assoc UK 6: 1–69.
41. Heath MR, Neat FC, Pinnegar JK, Reid DG, Sims DW, Wright PJ (2012) Review of climate change impacts on marine fish and shellfish around the UK and Ireland. Aquatic Conserv: Mar. Freshw. Ecosyst. 22: 337–367.
42. Walker PA, Hislop JRG (1998) Sensitive skates or resilient rays? Spatial and temporal shifts in ray species composition in the central and north-western North Sea between 1930 and the present day. ICES J Mar Sci 55: 392–402.
43. Rogers SI, Ellis JR (2000) Changes in the demersal fish assemblages of British coastal waters during the 20th century. ICES J Mar Sci 57: 866–881.
44. Beare D, Hölker F, Engelhard GH, McKenzie E, Reid DG (2010) An unintended experiment in fisheries science: a marine area protected by war results in Mexican waves in fish numbers-at-age. Naturwissenschaften 97(9): 797–808.
45. Sumalia UR (1998) Markets and the fishing down marine food webs phenomenon. EC Fish Coop Bulletin, 11: 25–28.
46. O'Leary BC, Smart JCR, Neale FC, Hawkins JP, Newman S, et al. (2011) Fisheries mismanagement. Mar Pollut Bull 62(12): 2642–2648.
47. Jones PJS (2012) Marine Protected areas in the UK: challenges in combining top-down and bottom-up approaches to governance. Environ Conserv 39 (3): 248–258.
48. Gell FR, Roberts CM (2003) Benefits beyond boundaries: the fishery effects of marine reserves. Trends Ecol Evol 18(9): 448–455.
49. Sheehan EV, Stevens TF, Gall SC, Cousens SL, Attrill MJ (2013) Recovery of a Temperate Reef Assemblage in a Marine Protected Area following the Exclusion of Towed Demersal Fishing. PLoS ONE 8(12): e83883. doi:10.1371/journal.pone.0083883.
50. Friedman JH (1984) A variable span scatterplot smoother. Laboratory for Computational Statistics, Stanford University Technical Report No. 5.
51. R Core Team (2013) R: A language and environment for statistical computing. Available: http://www.R-project.org/R Foundation for Statistical Computing, Vienna, Austria.

5

How Can We Identify and Communicate the Ecological Value of Deep-Sea Ecosystem Services?

Niels Jobstvogt[1,2]*, Michael Townsend[3], Ursula Witte[1], Nick Hanley[4]

1 Oceanlab, University of Aberdeen, Aberdeen, United Kingdom, 2 Aberdeen Centre for Environmental Sustainability (ACES), University of Aberdeen, Aberdeen, United Kingdom, 3 National Institute of Water and Atmospheric Research (NIWA), Hamilton, New Zealand, 4 Department of Geography and Sustainable Development, University of St Andrews, St Andrews, United Kingdom

Abstract

Submarine canyons are considered biodiversity hotspots which have been identified for their important roles in connecting the deep sea with shallower waters. To date, a huge gap exists between the high importance that scientists associate with deep-sea ecosystem services and the communication of this knowledge to decision makers and to the wider public, who remain largely ignorant of the importance of these services. The connectivity and complexity of marine ecosystems makes knowledge transfer very challenging, and new communication tools are necessary to increase understanding of ecological values beyond the science community. We show how the Ecosystem Principles Approach, a method that explains the importance of ocean processes via easily understandable ecological principles, might overcome this challenge for deep-sea ecosystem services. Scientists were asked to help develop a list of clear and concise ecosystem principles for the functioning of submarine canyons through a Delphi process to facilitate future transfers of ecological knowledge. These ecosystem principles describe ecosystem processes, link such processes to ecosystem services, and provide spatial and temporal information on the connectivity between deep and shallow waters. They also elucidate unique characteristics of submarine canyons. Our Ecosystem Principles Approach was successful in integrating ecological information into the ecosystem services assessment process. It therefore has a high potential to be the next step towards a wider implementation of ecological values in marine planning. We believe that successful communication of ecological knowledge is the key to a wider public support for ocean conservation, and that this endeavour has to be driven by scientists in their own interest as major deep-sea stakeholders.

Editor: James P. Meador, Northwest Fisheries Science Center, NOAA Fisheries, United States of America

Funding: This research project was funded by MASTS (Marine Alliance for Science and Technology for Scotland; URL: www.masts.ac.uk). MASTS is funded by the Scottish Funding Council (grant reference HR09011) and contributing institutions. Townsend's involvement was funded by NIWA (National Institute of Water and AtmosphericResearch; URL: www.niwa.co.nz) under the Coasts and OceansResearch Programme 3 (2013/14 SCI). The funders had no role in study design, data collection and analysis, decision to publish, or preparation of the manuscript.

Competing Interests: The authors have declared that no competing interests exist.

* Email: niels.jobstvogt@abdn.ac.uk

Introduction

The concept of ecosystem services (ES) has inspired a movement away from conservation for the sake of nature's inherent value to one that explicitly identifies, links and communicates the benefits of conservation to human wellbeing [1–3]. The endeavour of describing, quantifying and valuing the economic benefits that nature provides to society through ES has been identified as a powerful tool to make ecosystems count in cost-benefit analysis for environmental decision making [1,4]. Throughout this paper, however, the term 'value' is used in a broader sense, as a holistic concept which can include social, ecological and economic values. This broadening of the concept of value is needed because for the remotest places on earth like the deep sea, it is particularly challenging to make direct links between changes in system functioning and effects on the delivery of final ES (and thus on human well-being) [5,6].

The deep sea accounts for nearly 91% of the world's oceans with depths ranging from 200 m to almost 11,000 m. Despite its remoteness and size, its ecosystems are far from being unaffected by anthropogenic impacts such as fishing, climate change, and pollution [7–10]. To date many knowledge gaps remain around the functioning of deep-sea ecosystems. This is partially explained through the high costs, difficulties, and risks that are associated with deep-sea research. This lack of ecological knowledge means that we also know very little about the social and economic value of protecting the deep sea. By identifying and quantifying the ES benefits provided by the deep sea it is likely that appreciation for these benefits will change. This should lead to a larger emphasis on mitigating anthropogenic impacts in the oceans.

The major challenges of accounting for deep-sea ES stem from most people's lack of awareness about the deep-sea environment, and from the prevalence of intermediate services relative to easier-to-appreciate final services. Intermediate services in this paper refer to the indirect services that the ecosystem provides, such as habitat provision and nutrient cycling (the Millennium Ecosystem Assessment [1] refers to this category as supporting services). Intermediate services are the functional basis of the final services supplied by the system (Figure 1). The final services are considered as the ecosystem's contribution to human well-being [11] and include the ES categories of provisioning (e.g. commercial fish species), regulating (e.g. waste absorption and detoxification) and

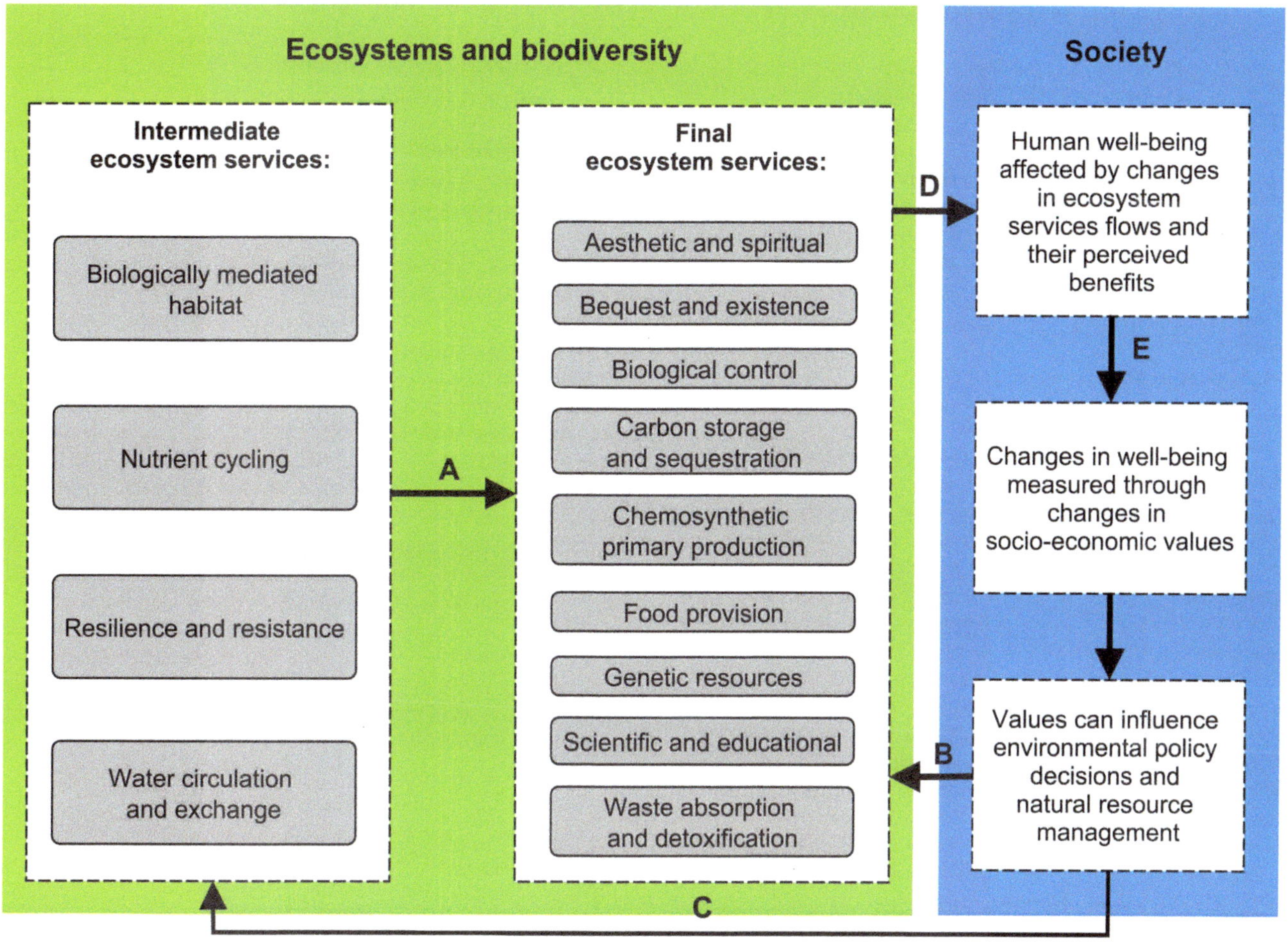

Figure 1. The ecosystem services framework for the example of deep-sea ecosystem services (ES). Environmental policies can either influence the management of final ES directly (arrow B) or indirectly via the intermediate ES (arrow C). The latter requires a sufficient understanding of the dependencies between intermediate and final ES (arrow A). Our understanding for the benefits provided by deep-sea ES (arrow D) and the values associated with them (arrow E) is currently very limited. The framework was simplified from [53] and adapted to the ES used for this expert consultation.

cultural (e.g. aesthetic values) [1]. There is a need to improve the integration of intermediate services and the processes that sustain them into the way in which we assess ES [12], since this ecological understanding is essential for demonstrating how human well-being ultimately depends on ecological processes and biodiversity [13,14].

In this paper we analyse the links between ES and the underlying processes that sustain them. This analysis was undertaken with the help of an expert panel of deep-sea scientists, focussing on submarine canyons. Within the deep sea, submarine canyons are an important ecosystem, which host unique reservoirs of biodiversity [15–18]. Along with much of the deep sea, they remain mostly underexplored and unprotected [9,19,20]. We selected this deep-sea ecosystem to test the Ecosystem Principles Approach (EPA) [21,22]. The EPA has recently been developed as a way of incorporating and translating ecological knowledge into easily understandable 'units' of information ('ecosystem principles') suitable for a wide range of audiences and thus for use in an ecosystem management context. The focus is on known and broadly-accepted information, with scientists from a wide field of expertise condensing this knowledge into principles that explain the linkages between ES, and their dependencies on underlying processes. The ecosystem principles also provide marine managers with qualitative information on temporal, spatial, and causal dependencies of ES flows. In the New Zealand case study by Townsend and colleagues [22], the EPA highlighted the importance of accounting for intermediate services in marine management, which were often provided by different geographical areas relative to the location at which final services were taken into account in the ES assessment. Economic theory suggests that intermediate ES can and should be valued through the final services that they support and the resultant direct benefits to people [40]. Indeed, ES are often perceived as a purely economic concept [23], but they also have social and ecological values, which when integrated with economic values produce a more holistic ecosystem assessment that can better inform natural resource management decisions [24]. This dominance of economic approaches and monetary valuation of ES stems from the often-felt pressure among the nature conservation sector to "speak the same language" as business and policy sectors in order to make conservation count [25–29].

In contrast, the EPA's advantage lies where economic valuation reaches its limits, by offering a more holistic picture which might help non-experts to better understand the high ecological value that scientists associate with the deep sea. Links between ecosystem processes and ES are less well known for the deep sea. For this reason it may be more helpful to highlight the links between deep-sea ES and processes themselves, rather than presenting decision makers with a set of economic values that are likely to underestimate the ecosystems' holistic value due to the omission of important ecological aspects of the deep sea [23]. In this paper we test the applicability of the EPA to little known and remote deep-sea ecosystems, such as submarine canyons, and demonstrate how the approach can provide decision makers with an accessible knowledge base for conservation decisions, despite some deficiencies in scientific data and associated uncertainty. We further discuss the approach's utility for expert consultation and cross-disciplinary knowledge transfer.

Methodology

2.1 Case study area: The Nazaré Canyon

The Nazaré Canyon on the Portuguese continental margin (Figure 2), also described as "Europe's Grand Canyon" [16], was chosen as the case study area of this paper to test the applicability of the EPA for deep-sea ecosystems. The Nazaré Canyon has attracted scientific interest due to its habitat heterogeneity and is considered to be a biodiversity hotspot [16]. Like other submarine canyon ecosystems it plays an important role in transportation processes at the continental margins [30,31]. The canyon is shallowest at 1 km off the Portuguese coast (50 m depth) and with a total length of 210 km it extends into the Iberian abyssal plain where it reaches depths of over 4,900 m [32].

2.2.1 Sampling and survey structure. Twenty-three researchers with knowledge of the Nazaré Canyon, covering a broad range of disciplines such as ecology, biology, microbiology, biogeochemistry, geography, geomorphology, geology, sedimentology and oceanography, were invited to participate in a HERMIONE (Hotspot Ecosystem Research and Man's Impact On European Seas project; URL: www.eu-hermione.net; last access July 2013) workshop in September 2012 and in an online pre-workshop survey (Figure 3). These opportunities were used to gather ideas and feedback for the two main surveys that followed. We used an email-based Delphi process to gather structured information by consensus from the invited expert panel in two consecutive rounds of surveying (post-workshop survey I and II; Figure 3). The Delphi process was originally developed as an interactive forecasting technique, where an expert panel goes through iterative survey rounds. The group results of each round are fed back to participants who are able to adapt their responses in the next survey round. The main idea of the Delphi process is to lead the group towards a consensus through the indirect exchange of information via a process coordinator. This process allowed us to subsequently include experts' requests for changes and additional information into the post-workshop surveys (Figure 3). The Delphi process makes it less likely that some researchers dominate the discussion and the outcomes, by maintaining anonymity throughout the communication process, thus avoiding the potential peer pressure of an expert workshop setting (further detail on the Delphi process in [33]). The survey questionnaires are available on request from the corresponding author.

2.2.2 Ethics statement. The nature of this research did not require ethical approval according to the University of Aberdeen Research Ethics Framework (*University of Aberdeen Research Ethical Review Checklist, Appendix A,* pp. 26–28; URL: www.abdn.ac.uk/documents/research-governance-framework-appendix4.pdf; last access March 2013). All study participants were recruited using an opt-in strategy and therefore consent was not explicitly recorded. Workshop and surveys did not include any sensitive personal questions. We asked participants to state their age, years of research experience, and field of studies; answering these questions was optional. Participants were provided with information on study objectives, sponsors, the participants' role, survey and workshop durations, potential benefits to the participants, summarised methodology, destination of gathered data and research results, the potential science impact and a contact address for further questions. Throughout the post-workshop surveys participants were identifiable via their email addresses. However, data on stated opinions and personal information was stored anonymously and kept confidential at all times. We chose email as the preferred communication method to facilitate the exchange of information during the Delphi phase of the expert consultation.

2.3 Submarine canyon ecosystem services

The experts helped to identify ES that were either perceived as less important or not relevant for the submarine canyon based on Table 1. Subsequently, the ES 'genetic resources', 'biological control', 'aesthetic and spiritual', 'scientific and educational', and 'chemosynthetic primary production' were excluded from Table 1 as less important relative to the other deep-sea ES. Accordingly eight ES (Table 1) were taken forward as a focus for the development of ecosystem principles.

2.4 Ecosystem Principles Approach

One of the main goals of the expert consultation was to develop a list of submarine canyon ecosystem principles, which could then be linked to ES. As an initial step, a review on the submarine canyon literature identified relatively well-explored ecosystem processes and relationships. The review findings were then discussed in the expert workshop and principles added or refined according to experts' suggestions. The following paragraph shows how the concept of ecosystem principles was described to canyon experts: "[An ecosystem principle] explicitly defines a key element

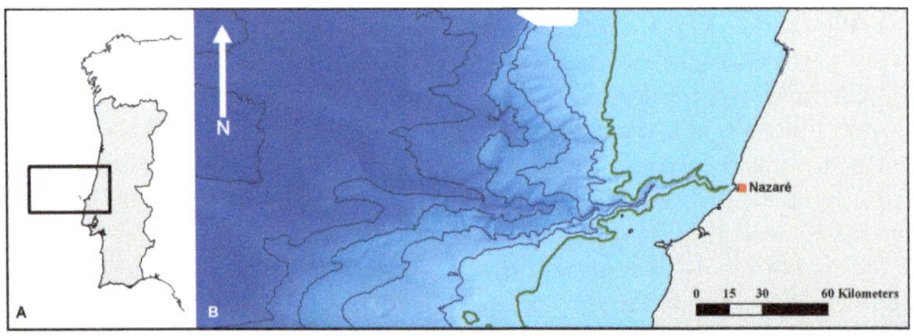

Figure 2. Nazaré Canyon. (**A**) Overview map of Portugal and the Nazaré Canyon area. (**B**) Nazaré Canyon bathymetry map with the Portuguese coastline to the east. Contour lines (blue) at 1000 m intervals; the 200 m depth contour, indicating the shelf edge, is marked in green. Data courtesy of Instituto Hidrografico, Lisbon and National Oceanography Centre, Southampton.

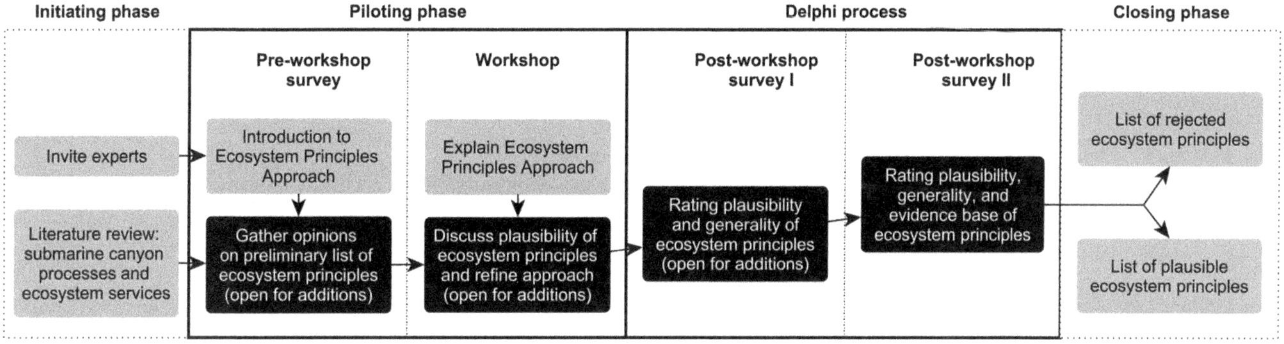

Figure 3. The survey phases of the submarine canyon expert consultation. Survey steps where experts were directly involved are highlighted as black boxes.

of how we expect the ecological system to operate" [22]. The workshop invitation summarised the EPA and included the methodology paper by Townsend and colleagues [22] as preparation for the workshop. During the workshop we provided further information in the form of a presentation that helped to increase participants' familiarity with the EPA. Principles were excluded from the initial list after each of the two consecutive post-workshop surveys (Delphi process; Figure 3) when fewer than 50% of the experts agreed with the plausibility of the principle. Experts were also able to propose new principles or suggest changes to the list of principles that was identified from the literature and subsequently refined throughout the piloting and Delphi phases (Figure 3).

We also asked the expert panel to categorise ecosystem principles according to their level of generality. The following categories were available: (i) general deep-sea principle, (ii) general canyon principle, (iii) shelf-incising canyon specific principles, to (iv) Nazaré Canyon-specific principle. The option with the highest frequency was then presented as the group vote in the subsequent survey.

Only in the second stage of the Delphi process were experts asked to distinguish their rating based on evidence on the one side and their expert view (as individuals) on the other side. The evidence base was rated on a five-point scale from 'very poor' to 'very good'. For the presentation of the group result, this was then divided into three categories of good, intermediate or poor evidence, according to the average group scores. These evidence scores had no influence on the decision to include or exclude any principle, but were introduced to separate personal opinion from the levels of evidence that existed in support of the principle.

During the workshop experts stressed the importance of the connectivity function of submarine canyons at the continental

Table 1. Submarine canyon ecosystem services.

Ecosystem services	Descriptions
Provisioning services:	
Carbon sequestration and storage	The uptake, storage, and burial of organic material within the canyon.
Food provision	The provision of marine organisms for human consumption.
Genetic resources and chemical compounds*	The use of canyon organisms in biotechnological, pharmaceutical, or industrial applications.
Regulating services:	
Biological control*	The control of diseases and invasive species.
Waste absorption and detoxification	The burial, decomposition and transformation of waste within the canyon ecosystem.
Cultural services:	
Aesthetic and spiritual*	The canyon ecosystem aesthetic and spiritual or inspirational source for religion, arts, movies, documentaries, books and folklore.
Bequest and existence	Safeguarding the canyon ecosystem for future generations and for the existence of marine species.
Scientific and educational*	The cognitive use of the canyon ecosystem for science and education.
Intermediate services:	
Biologically mediated habitat	Canyon habitats formed by marine organisms that provide nursery and refuge sites for other marine life.
Nutrient cycling	The storage and recycling of nutrients by canyon organisms.
Chemosynthetic primary production*	Primary productivity that is not dependent on energy from the sun.
Resilience and resistance	The amount of disturbance that the canyon ecosystem can cope with and its ability to regenerate after disturbance.
Water circulation and exchange	The currents, such as up-and down-welling, dense shelf water cascading, and mixing of water masses.

Services are grouped into four categories: provisioning, regulating, cultural and intermediate.
Listed items taken from [5,7,36,52] with alterations.
*Deep-sea ES that were not taken forward for the development of submarine canyon principles.

margin and we therefore chose 'water circulation and exchange' as an example to demonstrate how ecosystem principles link the supply of ES with their underlying processes. In the final results section, those ecosystem principles which in their description indicated a relationship between the ES 'water circulation and exchange' and other services were linked. The ecosystem principles relevant to this particular ES were assigned by the authors of this paper.

Results

3.1 Sample characteristics

The workshop was attended by 14 deep-sea researchers, 11 of whom had completed the pre-workshop pilot survey (Figure 3). All 14 workshop attendees were invited to participate in the two stages of the Delphi process (Figure 3), which 11 did for the first round and 10 for the second round. The average survey participant had 21 years of research experience and the survey covered academics from senior professors to PhD researchers, with male and female researchers equally represented.

3.2 Ecosystem principles

Over the course of the Delphi process, 21 ecosystem principles were identified from the literature and were then assessed and refined by the expert group (Tables 2 and 3). To highlight the nature of ecosystem principles, we present principle P1 as an example, which was rated to be plausible by all experts: 'canyons host a large number of different habitats and as a result increase species diversity at a regional scale' (for further principle descriptions we refer the reader to Tables 2 and 3). Four principles were discarded, whereas 17 principles were rated as plausible. Ten ecosystem principles fell into the category 'general submarine canyon principles', five into category 'general deep-sea principles', and two into 'shelf-incising canyon specific principles' (Tables 2 and 3). This indicated that the majority of ecosystem principles were at an appropriate level to describe processes and linkages between ES for submarine canyons in general and that they can be readily transferred to other canyons. The Delphi process had the expected effect of driving opinions closer towards consensus. Seven of the ten experts who participated in Delphi-rounds I and II (Figure 3) were closer to the group rating after the second Delphi-round.

Comparing the ratings of evidence and plausibility, we recognised that the existence of supporting evidence was not necessarily a requisite for an ecosystem principle to be plausible. Seven principles obtained intermediate evidence scores and for P16 and P17 evidence was rated as poor. However, this lack of evidence did not translate into a lack of plausibility. It was therefore an advantage to separate the two ratings from each other to distinguish between the experts' opinions and their evidence-based judgments. However, plausibility was clearly lower overall when evidence ratings were poor (Tables 2 and 3).

Developing principles to link food availability with biodiversity was challenging, and none were rated as plausible (P18 and P20). Experts had strong concerns of oversimplification when it came to the type, quality and amount of organic matter as a source of food and how changes of those parameters affected biodiversity. For biodiversity there was again a concern of over-simplification by omitting information on the spatial scale of biodiversity. In the same way, geographical scale mattered to experts, and lack of information on depth ranges and exact geographical position was criticised. The rating on generality provided a preliminary solution for implementing information on the geographical transferability of principles. Giving experts the chance to express uncertainty about the generality of principles as well as disentangling opinions about the rating of evidence allowed them to express their expectations for submarine canyons based on their research experience, and to transfer widely accepted knowledge from other ecosystems.

3.3 Linking principles and services

Many of the principles in Tables 2 and 3 have the capacity to provide information on *where* and *when* principles are likely to operate: for the principles included in Figure 4, particularly P8 and P7 reflect these spatio-temporal components. Other principles such as P3 and P12 explain *how* certain ES are provided and go into more detail on the processes involved. Principles like P17 and P16 that address effects of high biodiversity on ecosystem processes are capable of linking a broader set of ES such as 'carbon storage', 'food provision', 'bequest and existence' and 'waste absorption'. The ecosystem principles associated with 'biologically mediated habitat' were mainly thought to have an effect on biodiversity (e.g. P1 and P13) and to indirectly affect final ES such as 'food provision' and 'bequest and existence values' (P5).

For sustainable ecosystem management it can be equally important to understand the processes and principles that are involved in the provision of ES, as it is to understand the social and economic benefits of those services. We used 'water circulation and exchange' as an example to showcase how ecosystem principles explain links between ecological processes and ES (Figure 4; see also Table S1 for further details). 'Water circulation and exchange' has an important connectivity function in the submarine canyon (P11) and upwelling effects can lead to enhanced 'nutrient cycling' and as a result enhance productivity (P7 and P6; Figure 4). Further, 'nutrient cycling' might be important as an intermediate service for two different final ES, 'food provision' and 'bequest and existence values', because it can enhance fish abundance (P7). The 'bequest and existence' value can arise through the value that people tend to hold for iconic species (including fish), whereas 'food provision' relies on the abundance of commercially important fish as a consumptive resource. Trophic relationships, enhanced biomass, maintenance of deep-sea organisms (including non-iconic and non-commercial species) are important processes that sustain 'bequest and existence' as well as 'food provision' (P3, P4, P9; Figure 4) and should therefore be considered for management purposes. For 'carbon storage' and 'waste absorption' ecosystem management might be more concerned with other processes such as the transportation of organic and inorganic material, means of transportation, sedimentation rates, storage time, and burial processes that are important in parts of the submarine canyon (P3, P8, P12; Figure 4). How 'water circulation and exchange' is linked to 'resilience and resistance' as well as 'biologically mediated habitats' could not be resolved through the ecosystem principles developed in our workshop. This might be an indication that either too little evidence exists to support any ecosystem principles or that the links with processes that sustain these two ES are too complex to be described in the simplified form of ecosystem principles.

Discussion

The deep-sea case study for the Nazaré Canyon resulted in new insights on how to address the difficulties of assessing marine ES for ecosystem management purposes, especially when uncertainty is high due to lack of scientific data. In times where the demands on deep-sea resources are increasing, and scientific data on the potential impacts on marine biodiversity is scarce [5,34–36],

Table 2. Submarine canyon ecosystem principles with expert ratings on their plausibility and evidence base.

ID	Ecosystem principles	Plausibility	Evidence (mean score ± SE)
	General submarine canyon principles:		
P1	Canyons host a large number of different habitats and as a result increase species diversity at a regional scale.	100%	GOOD (3.8±0.2)
P2	The canyon topography tends to have a focusing or channelling effect for sediment and organic material.	100%	GOOD (3.8±0.2)
P3	The strength of large scale transportation events varies and occurrence ranges from a yearly to decadal pattern. They can be triggered by storms, high sediment load in the water column, cooling and increasing salinity of surface waters, or slope failures.	100%	GOOD (3.9±0.3)
P4	The transport of organic material from shallower waters to the deep seabed, which is mainly driven by large scale transportation events, is an important source of food for deep-sea organisms.	100%	MEDIUM (3.3±0.3)
P5	Canyons can serve as fish feeding ground, refuge and nursery area and therefore often show higher abundance of fish than their surroundings.	90%	MEDIUM (2.8±0.3)
P6	Canyons can enhance the mixing of water masses and as a result influence the exchange of nutrients, heat and salt between the shelf and the deep sea.	90%	MEDIUM (3.4±0.4)
P7	The canyon topography affects up- and down-welling of water masses at the continental margin. Upwelling events around the canyon head enhances productivity locally; as a result fish abundance can be higher.	90%	MEDIUM (3.3±0.3)
P8	By transporting large amounts of organic material from the shelf into deeper waters, canyons act as temporary stores of sediment and carbon. It can take decades or even centuries until the transported material reaches the abyssal plain, where it is then deposited on geological time scales.	80%	GOOD (3.8±0.2)
P9	Food quantity and quality tends to be higher within some canyon areas compared to the surrounding slope. This can enhance the biomass of the benthic and pelagic fauna.	80%	MEDIUM (3.3±0.4)
P10	Many species that are found in canyons are not found on the slope. They are therefore contributing to regional diversity.	80%	MEDIUM (3.1±0.4)

ID = principle identification number. The plausibility rating: ten experts participated in the full rating process (i.e. 100% = 10 experts). The evidence rating (1–5 from 'very good' to 'very poor'): poor (mean score <2.5), medium (2.5≤ mean score <3.5) and good (mean score ≥3.5); SE = standard error.

approaches such as the Ecosystem Principles Approach (EPA) are crucial to draw the link between the ecological and socio-economic dimensions of the ecosystem. Currently this linkage is poorly understood, contributing to the under-valuation of deep-sea ecosystems which is likely to undermine conservation efforts. We briefly outline the utility of expert consultation under these circumstances and reflect on the ability of the EPA to integrate more ecology into the assessment of ES and into the decision-making process. We first discuss how the EPA can help to communicate the overall importance of deep-sea ecosystems for the provision of ecosystem services. Further, we explain how the EPA can improve marine ecosystem-based management by promoting the inclusion of information on ecosystem processes into an ecosystem services assessment.

4.1 Communicating ecological values

The EPA has the ability to broaden access to ecological knowledge so that decision makers are not dependent on science advisors alone, but can take informed decisions on the basis of simplified ecological knowledge made available to them [37,38].

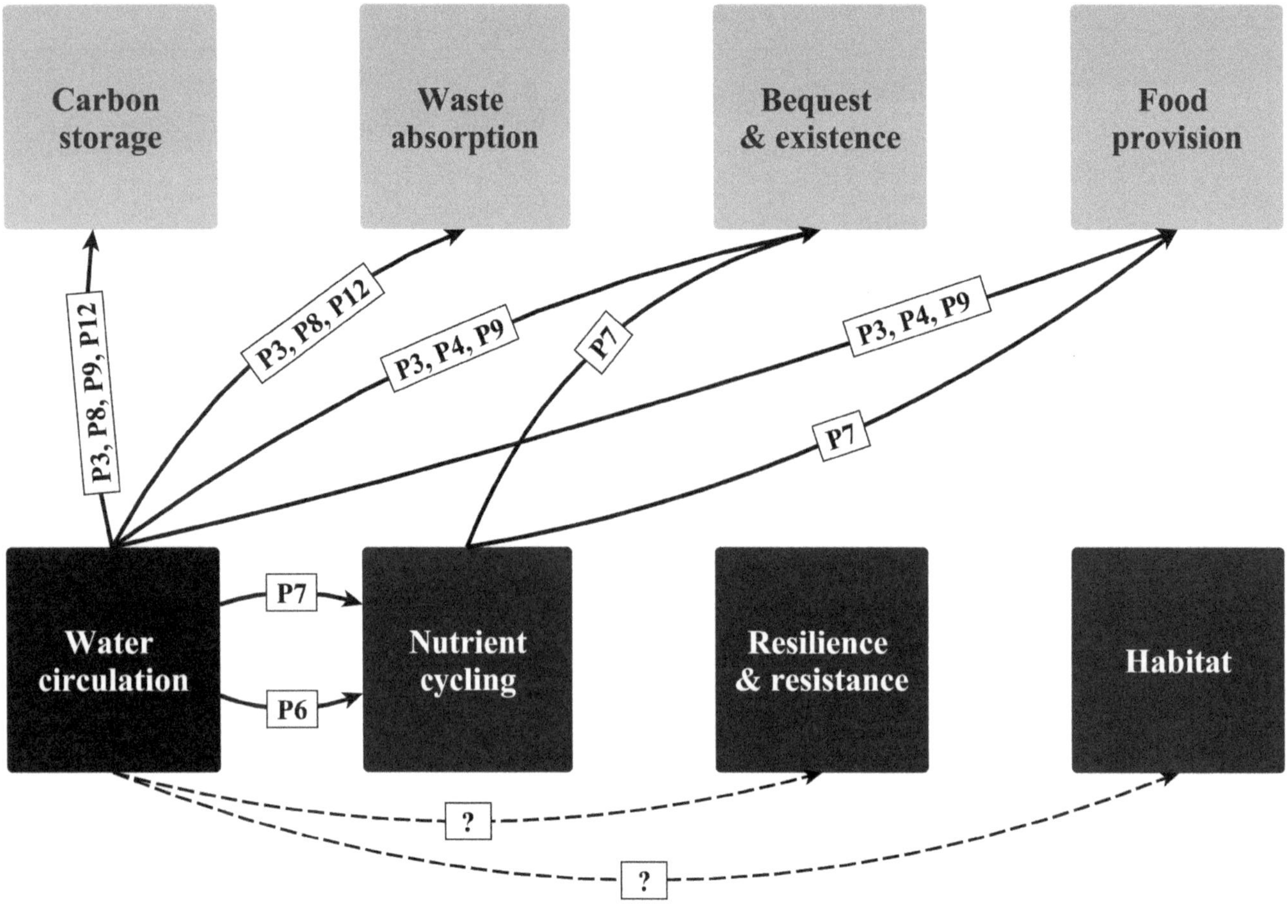

Figure 4. Links between 'water circulation' (black box) and other canyon ecosystem services explained through ecosystem principles. The intermediate services are in the lower half (dark grey and black boxes) and final services in the upper half (light grey boxes) of the diagram. Principles are indicated as arrows with their respective ID (cf. explanation in Table S1). Research gaps highlighted as question marks with dotted lines. Principles unrelated to 'water circulation' were omitted from this figure.

This broadening of access could also assist in increasing information about deep-sea ES amongst the general public. The process of developing ecosystem principles for the submarine canyon environment made clear that experts were able and willing to make predictions on connections between ES, generalisations on important canyon processes, and to link canyon characteristics to effects on ES. Also, while there remain research gaps and uncertainties, the list of ecosystem principles presented in Tables 2 and 3 provides the best available science knowledge to date, presented in easily understandable units of information. The list includes spatial and temporal information, as well as information on how the principles influence the supply of ES. We found that it was very important for the participating scientists to distinguish between their opinions (i.e. plausibility of the principles) and the (less subjective) rating of the existing scientific evidence base to inform principles. It was not imperative for principles to have a good evidence base, but rather to be generally accepted as plausible (cf. P5, P17 and P16). Herein lies a predictive strength of the EPA, in backing up the uncertainty associated with deep-sea science with a consensus-based approach, thereby decreasing uncertainty about ES linkages. The generality of principles was equally important, accounting for concerns that some ecosystem principles were valid on a larger scale than others i.e. 'general deep-sea principles' or 'shelf-incising canyon specific principles'. This additional type of information is crucial to highlight the ability of the EPA to transfer principles to other submarine canyons or even other deep-sea ecosystems. The majority of experts rated the principles as either very broadly applicable to the deep sea or to submarine canyons in general, irrespective of the type of canyon. The broad applicability of principles was thought to be an effect of reducing the complexity of ecological information.

We share the view of one workshop participant who stated that it will be difficult to determine when the list of ecosystem principles is complete. New evidence, the inclusion of researchers with different academic backgrounds, and assigning more time to the task might increase the number of principles on the list. Thus including a broad range of disciplines into the principle development process is crucial. Nonetheless, there exists an asymptotic relationship between effort expended and the number of principles, where spending more time on identifying and reviewing submarine canyon principles might increase the detail of principles, but not their utility for management decisions. The EPA's utility lies not in providing large amounts of detail, but in providing meaningful, concise information to better understand the overall functioning of the ecosystem in conjunction with the ES it provides. The EPA is based on what we know today and the ecosystem principles in this paper cover a broad range of topics, with further workshops or surveys being likely to provide diminishing returns of new principles to our established list.

There are many aspects of the deep sea that the science community remains uncertain about, but the EPA helped to focus and assimilate known information with the underlying ecosystem processes that are better understood and agreed on. However, the fact that some ecosystem principles were discarded (Table 3) indicated that there remain gaps in understanding on how canyon biodiversity is influenced by current regimes, and by different types and quality of organic material, and also the importance of recruitment processes between deeper and shallower waters. The high specificity and complexity of these and other processes might not allow us to develop these processes as ecosystem principles at the current time.

4.2 Utility of the Ecosystem Principles Approach

The deep sea is hard to sample and poorly understood, yet we were able to draw on experts' knowledge and condense what they know about submarine canyons, one of the deep-sea's biodiversity hotspots. We showed that the EPA, in combination with a Delphi process, can be a useful tool for working at the fringes of our current knowledge, using collective expert opinions to evaluate and arbitrate on the content of ecological understanding. Through this process we can also highlight where knowledge and hence research gaps lie, and where funding is needed. The EPA might be seen as a balancing act between the precautionary principle on the one side and economic reasoning on the other. The precautionary principle as framed in Rio in 1992 states that lack of scientific certainty shall not be used as an excuse to postpone actions that might prevent environmental degradation [39]. Lubchenco [37] lists guidance for decision-making under uncertainty as one of the roles that science should play in society. This might include reliance on more subjective approaches such as the EPA to support more holistic decision-making in marine resource management until we have a greater body of scientific evidence to prove or disprove what researchers have outlined as ecosystem principles.

The principles for 'water circulation and exchange' (Figure 4) demonstrate the EPA's ability to provide information on the ecological value of the ecosystem, and how these are linked to the kinds of final ES which economists are likely to value in monetary terms. The approach does not focus on final ES alone, but provides information at multiple levels without losing sight of the indirect impacts on ecosystems through intermediate services and underlying processes, and of the multiple connections between ecosystems. By linking processes with services through ecosystem principles, we draw the attention towards the network character of ecosystems. The highly interlinked nature of this network means that it is actually far from straightforward to categorise and separate services for ES valuation, especially given the multitude of connections within the marine environment. This is so irrespective of whether social, ecological, or economic definitions of 'value' are assessed. Presenting the information on final and intermediate services together with their underlying processes in a network style can better inform future management scenarios with more realistic ecological information than assessments that are limited to final ES alone.

Current ES valuation frameworks suggest that intermediate services should be valued only in terms of the final services that they support and not be included directly in a valuation of ES flows to avoid double-counting of their social or economic value [25,40–42]. The resulting requirement for effective management is that underlying processes and linkages are sufficiently well understood [23] and the ES they support are provided within the managed area [5]. However, the spatial and temporal distances between marine intermediate and final services can span millennia and act on a global scale, as is the case with the ocean nutrient cycle [35,43]. Marine ecosystems are highly connected systems with many processes being important for the provision of intermediate and final ES and crossing ecosystem boundaries [5]. Hence, if intermediate and final ES are spatially separated, the chances are that recommendations which focus only on final ES will be poor for marine resource management [22,44]. The field of ES valuation has its roots in terrestrial ecosystems where ecosystems with their services and processes are less open than in marine ecosystems [45,46]; valuation approaches might have to be adapted for the marine environment to account for its higher connectivity. Also, to capture the holistic value of intermediate services, we would have to successfully value all final ES (including cultural ES values and other non-marketed ES), which is still more of a research aspiration rather than a currently-achievable outcome. A failure to recognise the contribution of intermediate services for the ES valuation in ecosystems where they dominate, such as the deep sea, will likely lead to misguided policy decisions [47].

The EPA should be seen as an addition to baseline ecological research and economic ES valuation, rather than as a substitute for either. The EPA's advantage lies where monetary valuation reaches its limits, in highlighting links between ES and their underlying processes, and in linking intermediate services with final ES. While economic ES values can help set marine management priorities that are socially and economically desirable (Daily et al., 2009), the EPA focusses on the ecological 'value' of the ecosystem and can provide important information on how such management priorities can be achieved. Where economic values require empirical links to well-being and monetary quantification, ecological 'value' is more focussed on the importance of the ecosystem processes and characteristics that lead to such economically valued benefits being produced.

4.3 Future research opportunities and lessons learnt

The EPA might not only enhance the availability of ecological information and its uptake by decision makers, but can also improve how research results are shared across disciplines. Inter- as well as trans-disciplinary collaborations are complicated by the existence of language barriers. The use of different key terms or jargon restricts access to the pool of knowledge to only a small number of experts. The lack of frameworks that translate research findings into understandable and meaningful formats has been described as one of the major reasons why information might not reach beyond disciplinary boundaries [48,49]. Different methodologies, attitudes and perceptions between disciplines might further decrease the flow of scientific evidence [38,48]. Thus to allow economists, geologists, biologists, oceanographers and other disciplines involved in marine science to share information it would be beneficial to work on a global matrix of ecosystem principles similar to the ES valuation databases provided by the Marine Ecosystem Services Partnership (MESP; URL: www.marineecosystemservices.org; last access August 2013) which gathers studies on monetary ES values. In contrast to the MESP database, the EPA would be able to add to the evidence base not just on economic and social, but also on ecological values. Also, a more extensive dataset on ecosystem principles for marine ecosystems would increase the chances that more complex management scenarios could be developed, such as in Bayesian belief networks (BBNs), which depend heavily on the availability of baseline information on ecosystem processes, even though BBNs are able to deal with knowledge gaps when expert knowledge is available [50]. Other fields that are using approaches like habitat mapping as well as biological value mapping might benefit from

Table 3. Submarine canyon ecosystem principles continued from Table 2.

ID	Ecosystem principles	Plausibility	Evidence (mean score ± SE)
	Shelf-incising canyon specific principles:		
P11	Canyons function as major transport pathways between the shelf and the deep sea.	100%	GOOD (4.0±0.3)
P12	Sediment, organic material, and pollutants that are transported alongshore get trapped by the canyon and transported down the canyon slope.	90%	MEDIUM (3.4±0.3)
	General deep-sea principles:		
P13	Areas with reef forming or habitat creating organisms can support higher diversity than their surroundings. These habitats are most common on hard substrates, such as areas with steep slopes, rocks, boulders, vertical walls, or overhangs.	100%	GOOD (4.1±0.3)
P14	The biomass of invertebrates living in and on the seafloor can constitute an important food source for commercially important deep-sea species.	100%	GOOD (3.7±0.3)
P15	The organisms inhabiting soft substrates play a major role for the recycling of nutrients. The process is largely dominated by bacteria, but is to a smaller extent also attributed to the animals living in and on the sediment.	70%	GOOD (3.5±0.3)
P16	Higher biodiversity can support higher rates of ecosystem processes.	70%	POOR (2.4±0.5)
P17	Higher biodiversity increases the insurance value of an ecosystem by increasing the likelihood that the ecosystem is able to provide the same ecosystem functions after an ecosystem impact occurred.	60%	POOR (2.0±0.4)
	Rejected ecosystem principles*:		
P18	Diversity tends to be lower in areas with high food availability.	10%	Not assessed
P19	Space and resource occupancy by native species can decrease invasion risk.	30%	POOR (1.4±0.2)
P20	Where strong bottom currents are common, food availability and substrate characteristics become less important and current speed becomes the main driver for species abundance and diversity.	40%	Not assessed
P21	The disturbances caused by strong bottom currents keep species diversity and abundance at low levels.	40%	POOR (2.3±0.4)

ID = principle identification number. The plausibility rating: ten experts participated in the full rating process (i.e. 100% = 10 experts). The evidence rating (1–5 from 'very good' to 'very poor'): poor (mean score <2.5), medium (2.5≤mean score <3.5) and good (mean score ≥3.5); SE = standard error.
*Principles P18–P21 were rejected by the majority of experts, i.e. their plausibility was below 50%.

the EPA as well, given that spatial ecosystem principles were developed [51].

Using the EPA it should be possible to provide more precise temporal and spatial information on ecosystem principles, and to develop management strategies based on the list of principles and evidence on social and economic values and resulting management priorities. Showing the EPA's potential to improve people's understanding of ecosystem functioning was beyond the scope of this study, but future research involving the wider public and decision makers would be beneficial to test the effect of such simplified ecological knowledge on their decisions.

Three insights in particular emerge from our interdisciplinary workshop, which might improve future marine conservation initiatives and their acceptance:

(1) To further the field of marine ES valuation it would be beneficial to acknowledge that the traditional approaches to ES valuation, which have their roots in terrestrial research, might not be easily transferable to a highly linked marine

environment. Marine ecosystem boundaries are much less clearly defined than in terrestrial environments, and ES flows are less easily traceable. We might therefore need different approaches to valuing ES flows in the marine environment. The EPA is but one potential approach to improve integration of ecological values with social and economic values.

(2) The precautionary principle demands that we are cautious with our exploitation of the environment, but in the same time that management recommendations are made on a timely basis to the best of our knowledge, without postponing decisions for indefinite time until more certainty has been gained. The marine science community should more willingly embrace its important societal role in providing recommendations for nature conservation management with the support of social science approaches.

(3) We propose greater transparency in decisions on the conservation importance of marine areas. It should be possible to enhance understanding of the social, ecological as well as the economic values of certain areas, and to justify their protection, by providing easy understandable information on marine ES and how they relate to underlying ecosystem processes.

Acknowledgments

We are deeply thankful to the workshop and survey participants for their time and input, and to Dr Rene van der Wal, Dr Nicolas Krucien, and Prof Graham Pierce (University of Aberdeen) for constructive feedback on early drafts of this paper. We also like to thank Dr Veerle Huvenne (National Oceanography Centre) and the Instituto Hidrografico for the Nazaré Canyon bathymetry data, and the conference organizers for the workshop opportunity during the HERMIONE Annual Meeting 2012. We also thank two anonymous reviewers for their comments.

Author Contributions

Conceived and designed the experiments: NJ NH UW MT. Performed the experiments: NJ. Analyzed the data: NJ MT. Contributed reagents/materials/analysis tools: MT. Wrote the paper: NJ MT NH UW.

References

1. MEA (2005) Millenium Ecosystem Assessment: ecosystems and human well-being synthesis report. Washington DC: Island Press, 141.
2. Salles J-M (2011) Valuing biodiversity and ecosystem services: Why put economic values on Nature? Comptes rendus biologies 334: 469–482.
3. Lele S, Springate-baginski O, Lakerveld R, Deb D, Dash P (n.d.) Ecosystem services: origins, contributions, pitfalls, and alternatives. Conservation & Society: 1–46.
4. Hanley N, Barbier E (2009) Pricing Nature: Cost-Benefit Analysis and Environmental Policy. Cheltenham: Edward Elgar, 353.
5. Armstrong CW, Foley NS, Tinch R, Van den Hove S (2012) Services from the deep: Steps towards valuation of deep sea goods and services. Ecosystem Services 2: 2–13.
6. Jobstvogt N, Hanley N, Hynes S, Kenter J, Witte U (2014) Twenty thousand sterling under the sea: Estimating the value of protecting deep-sea biodiversity. Ecological Economics 97: 10–19.
7. Van den Hove S, Moreau V (2007) Deep-sea biodiversity and ecosystems: A scoping report on their socio-economy, management and governance. Cambridge: UNEP-WCMC, 84.
8. Benn AR, Weaver PP, Billet DSM, Hove S Van Den, Murdock AP, et al. (2010) Human activities on the deep seafloor in the North East Atlantic: an assessment of spatial extent. PloS one 5(9): 1–15.
9. Ramirez-Llodra E, Tyler PA, Baker MC, Bergstad OA, Clark MR, et al. (2011) Man and the Last Great Wilderness: Human Impact on the Deep Sea. PLoS ONE 6(7): 1–25.
10. Jones DOB, Yool A, Wei C-L, Henson SA, Ruhl HA, et al. (2013) Global reductions in seafloor biomass in response to climate change. Global change biology: 1–12.
11. Haines-Young R, Potschin M (2013) Common International Classification of Ecosystem Services (CICES): Consultation on Version 4, August–December 2012. Nottingham, 34.
12. Turner RK, Paavola J, Cooper P, Farber S, Jessamy V, et al. (2003) Valuing nature: lessons learned and future research directions. Ecological Economics 46: 493–510.
13. Cardinale BJ, Duffy JE, Gonzalez A, Hooper DU, Perrings C, et al. (2012) Biodiversity loss and its impact on humanity. Nature 486: 59–67.
14. Mace GM, Norris K, Fitter AH (2012) Biodiversity and ecosystem services: a multilayered relationship. Trends in ecology & evolution 27: 19–26.
15. Stiles ML, Ylitalo-Ward H, Faure P, Hirshfield MF (2007) There's no place like home: Deep seafloor ecosystems of New England and the mid-Atlantic. Oceana, 40.
16. Tyler P, Amaro T, Arzola R, Cunha M, De Stigter H, et al. (2009) Europe's Grand Canyon: Nazaré Submarine Canyon. Oceanography 22: 46–57.
17. Danovaro R, Company JB, Corinaldesi C, D'Onghia G, Galil B, et al. (2010) Deep-sea biodiversity in the Mediterranean Sea: the known, the unknown, and the unknowable. PloS one 5(8): 1–25.
18. Auster PJ, Moore J, Heinonen KB, Watling L (2005) A habitat classification scheme for seamount landscapes: assessing the functional role of deep-water corals as fish habitat. In: Freiwald A, Roberts JM, editors. Cold-Water corals and ecosystems. Berlin Heidelberg: Springer, 761–769.
19. Glover AG, Smith CR (2003) The deep-sea floor ecosystem: current status and prospects of anthropogenic change by the year 2025. Environmental Conservation 30: 219–241.
20. Würtz M (2012) Mediterranean submarine canyons: Ecology and governance. Gland, Switzerland and Malaga, Spain: IUCN, 216.
21. Townsend M, Thrush S (2010) Ecosystem functioning, goods and services in the coastal environment. Prepared by the National Institute of Water and Atmospheric Research for Auckland Regional Council. Auckland Regional Council Technical Report 2010/033.
22. Townsend M, Thrush S, Carbines M (2011) Simplifying the complex: an "Ecosystem Principles Approach" to goods and services management in marine coastal ecosystems. Marine Ecology Progress Series 434: 291–301.
23. Cornell S (2010) Valuing ecosystem benefits in a dynamic world. Climate Research 45: 261–272.
24. De Groot RS, Wilson M a, Boumans RM. (2002) A typology for the classification, description and valuation of ecosystem functions, goods and services. Ecological Economics 41: 393–408.
25. Fisher B, Turner K, Zylstra M, Brouwer R, De Groot R, et al. (2008) Ecosystem services and economic theory: integration for policy-relevant research. Ecological Applications 18: 2050–2067.
26. Brink P, Mazza L, Badura T, Kettunen M, Withana S (2012) Nature and its role in the transition to a green economy. The Economics of Ecosystems & Biodiversity (TEEB), 72.
27. Kallis G, Gómez-Baggethun E, Zografos C (2013) To value or not to value? That is not the question. Ecological Economics 94: 97–105.
28. Balmford A, Fisher B, Green RE, Naidoo R, Strassburg B, et al. (2010) Bringing Ecosystem Services into the Real World: An Operational Framework for Assessing the Economic Consequences of Losing Wild Nature. Environmental and Resource Economics 48: 161–175.
29. Peterson MJ, Hall DM, Feldpausch-Parker AM, Peterson TR (2010) Obscuring ecosystem function with application of the ecosystem services concept. Conservation biology: the journal of the Society for Conservation Biology 24: 113–119.
30. De Leo FC, Smith CR, Rowden AA, Bowden DA, Clark MR (2010) Submarine canyons: hotspots of benthic biomass and productivity in the deep sea. Proceedings Biological sciences/The Royal Society 277: 2783–2792.
31. Harris PT, Whiteway T (2011) Global distribution of large submarine canyons: Geomorphic differences between active and passive continental margins. Marine Geology 285: 69–86.
32. Lastras G, Arzola RG, Masson DG, Wynn RB, Huvenne V a I, et al. (2009) Geomorphology and sedimentary features in the Central Portuguese submarine canyons, Western Iberian margin. Geomorphology 103: 310–329.
33. Linstone HA, Turoff M (2002) The Delphi Method: Techniques and Applications. New Jersey: New Jersey Institute of Technology, 616.
34. Mengerink KJ, Van Dover CL, Ardron J, Baker M, Escobar-Briones E, et al. (2014) A Call for Deep-Ocean Stewardship. Science 344: 696–698.
35. Thurber AR, Sweetman AK, Narayanaswamy BE, Jones DOB, Ingels J, et al. (2013) Ecosystem function and services provided by the deep sea. Biogeosciences Discussions 10: 18193–18240.
36. Armstrong CW, Foley N, Tinch R, Van den Hove S (2010) Ecosystem goods and services of the deep sea. Deliverable D6.2 HERMIONE Project, 68.

37. Lubchenco J (1998) Entering the Century of the Environment: A New Social Contract for Science. Science 279: 491–497.
38. Rudd MA, Lawton RN (2013) Scientists' prioritization of global coastal research questions. Marine Policy 39: 101–111.
39. Myers N (2002) The Precautionary Principle Puts Values First. Bulletin of Science, Technology & Society 22: 210–219.
40. UK NEA (2011) UK NEA: Synthesis of the key findings. UK National Ecosystem Assessment, Cambridge, UNEP-WCMC, 87.
41. Boyd J, Banzhaf S (2007) What are ecosystem services? The need for standardized environmental accounting units. Ecological Economics 63: 616–626.
42. Wallace KJ (2007) Classification of ecosystem services: Problems and solutions. Biological Conservation 139: 235–246.
43. Roberts CM, Andelman S, Branch G, Bustamante RH, Carlos Castilla J, et al. (2003) Ecological criteria for evaluating candidate sites for marine reserves. Ecological Applications 13: 199–214.
44. McLeod K, Leslie H (2009) Ecosystem-based management for the oceans. Washington, DC: Island Press, 368.
45. Hawkins SJ (2004) Scaling up: the role of species and habitat patches in functioning of coastal ecosystems. Aquatic Conservation: Marine and Freshwater Ecosystems 14: 217–219.
46. Raffaelli D (2006) Biodiversity and ecosystem functioning: issues of scale and trophic complexity. Marine Ecology Progress Series 311: 285–294.
47. Johnston RJ, Russell M (2011) An operational structure for clarity in ecosystem service values. Ecological Economics 70: 2243–2249.
48. Davies AL, Colombo S, Hanley N (2013) Improving the application of long-term ecology in conservation and land management. Journal of Applied Ecology 51: 63–70.
49. Peters DPC (2010) Accessible ecology: synthesis of the long, deep, and broad. Trends in Ecology & Evolution 25: 592–601.
50. Campbell D, Chilton S, Clark S, Fitzsimmons C, Gazzola P, et al. (2012) Bayesian Belief Networks as an Interdisciplinary Marine Governance and Policy Tool. Valuing Nature Network Report, 27.
51. Townsend M, Thrush SF, Lohrer AM, Hewitt JE, Lundquist CJ, et al. (2014) Overcoming the challenges of data scarcity in mapping marine ecosystem service potential. Ecosystem Services 8: 44–55.
52. Beaumont N, Townsend M, Mangi S, Austen MC (2006) Marine biodiversity: an economic valuation. Building the evidence base for the Marine Bill. Department for Environment Food and Rural Affairs, 64.
53. Mace GM, Bateman I, Albon S, Balmford A, Brown C, et al. (2011) UK NEA Technical Report: Conceptual Framework and Methodology. UK National Ecosystem Assessment: Technical Report. Cambridge: UK National Ecosystem Assessment, UNEP-WCMC, 11–26.

6

Anthropogenic Litter in Urban Freshwater Ecosystems: Distribution and Microbial Interactions

Timothy Hoellein*, Miguel Rojas, Adam Pink, Joseph Gasior, John Kelly

Department of Biology, Loyola University Chicago, Chicago, Illinois, United States of America

Abstract

Accumulation of anthropogenic litter (i.e. garbage; AL) and its ecosystem effects in marine environments are well documented. Rivers receive AL from terrestrial habitats and represent a major source of AL to marine environments, but AL is rarely studied within freshwater ecosystems. Our objectives were to 1) quantify AL density in urban freshwaters, 2) compare AL abundance among freshwater, terrestrial, and marine ecosystems, and 3) characterize the activity and composition of AL biofilms in freshwater habitats. We quantified AL from the Chicago River and Chicago's Lake Michigan shoreline, and found that AL abundance in Chicago freshwater ecosystems was comparable to previously reported data for marine and terrestrial ecosystems, although AL density and composition differed among habitats. To assess microbial interactions with AL, we incubated AL and natural substrates in 3 freshwater ecosystems, quantified biofilm metabolism as gross primary production (GPP) and community respiration (CR), and characterized biofilm bacterial community composition via high-throughput sequencing of 16S rRNA genes. The main driver of biofilm community composition was incubation location (e.g., river vs pond), but there were some significant differences in biofilm composition and metabolism among substrates. For example, biofilms on organic substrates (cardboard and leaves) had lower GPP than hard substrates (glass, plastic, aluminum and tiles). In addition, bacterial communities on organic substrates were distinct in composition from those on hard substrates, with higher relative abundances of bacteria associated with cellulose decomposition. Finally, we used our results to develop a conceptual diagram designed to unite the study of AL in terrestrial and freshwater environments with the well-established field of marine debris research. We suggest this broad perspective will be useful for future studies which synthesize AL sources, ecosystem effects, and fate across multiple ecosystem types, and will benefit management and reduction of global AL accumulations.

Editor: Stefan Bertilsson, Uppsala University, Sweden

Funding: Loyola University Chicago supported undergraduate researchers Adam Pink and Joseph Gasior through summer research fellowships from the Center for Urban Environment Research and Policy (CUERP) and the Department of Biology, respectively. Miguel Rojas's work on this project was supported by an NSF Graduate Research Fellowship (NSF-DGE-1257295). The funders had no role in study design, data collection and analysis, decision to publish, or preparation of the manuscript.

Competing Interests: The authors have declared that no competing interests exist.

* Email: thoellein@luc.edu

Introduction

Accumulation of anthropogenic litter (i.e. garbage; AL) in marine environments has received increased attention in the popular press and scientific literature in recent years [1–3]. Researchers have documented large AL accumulations in pelagic gyres of the Pacific and Atlantic Oceans, which have earned such titles as "Pacific trash vortex" and "plastic soup" [3,4]. Images of AL accumulations on beaches, and ingestion of plastic and styrofoam by fish, marine birds and turtles have also increased public awareness [5–7]. Finally, the release of toxic compounds and microscopic particles via plastic decomposition in the environment has garnered attention [8,9].

Current research on AL in the ocean (i.e. marine debris) identifies two primary sources: direct inputs from boats and anglers, and inputs from terrestrial and riverine ecosystems [10]. However, the patterns of AL abundance, retention, transformation and transport within freshwater ecosystems are unknown, and are likely a significant component of the global AL 'life cycle'. For example, because rivers can transport AL between terrestrial and marine environments, AL retention or transformation in rivers can attenuate its delivery to downstream ecosystems. From a management perspective, freshwaters may serve as easier collection sites for AL relative to marine coastal and pelagic zones. Therefore, empirical analyses of AL pools, fluxes, and interactions with biota in freshwaters are needed to understand its ecosystem effects and to develop management strategies for the global concern of AL accumulation [11].

When AL enters aquatic habitats it will become rapidly colonized by microbial biofilms composed of bacteria, fungi, and algae in an extracellular mucilaginous matrix. Biofilms are complex, taxonomically diverse communities which support a wide range of metabolic activities [12]. Biofilms can play important roles in nutrient cycling within aquatic habitats [13], and they can also serve as an important food source for higher trophic levels, such as invertebrates and fish [14]. Therefore, biofilm colonization often serves as the initial biotic interaction of solid substrates in aquatic environments, and this interaction can be ecologically significant.

Substrate composition can have a strong effect on biofilm community composition and activity [15,16]. In addition, differences in substrate quality (e.g., nitrogen content) and mobility drive rates of nutrient demand, metabolism, and community structure within biofilms [16,17]. Recent evidence indicates microplastics in the open ocean support microbial biofilms that are distinct in composition from the microbial community of surrounding waters [18]. The pattern suggests that the plastic surfaces selected for a unique microbial community, which may carry out different metabolic pathways than those in the open water, including plastic decomposition. However, patterns of biological activity and community composition of biofilms colonizing AL have not previously been measured in freshwaters.

Our objectives for this study were to 1) measure AL abundance in urban freshwaters, 2) compare AL abundance across terrestrial, freshwater, and marine ecosystems, and 3) measure biofilm activity and bacterial community composition on natural and AL substrates in freshwater ecosystems.

Methods

Anthropogenic litter collection

We sampled three 70–100-m reaches in the North Branch of the Chicago River and adjacent riparian zone, on 3 different days in summer 2011. Study reaches were in the Bunker Hill Forest Preserve in Chicago, IL (latitude 41.999273, longitude −87.779868), and were spaced approximately 250-m apart. The North Branch of the Chicago River watershed area is 110 km^2, and the land-use is largely urban and suburban with very little agriculture [19]. In each reach, we removed all AL from the river benthos (width = 18–23 m), and all AL from one bank on the riparian zone within 10-m of the water's edge. We also collected AL from three 400-m reaches of Lake Michigan beaches adjacent to the lakeshore campus of Loyola University Chicago (LUC), located in the Rogers Park neighborhood of Chicago, IL (latitude 42.002541, longitude −87.656464). We sampled the beaches on 3 separate days in summer 2011 by establishing a 400-m transect parallel to the shore and collecting all AL within 50-m from the water line. AL <1 cm was not collected, so this analysis does not include microplastic particles. In the laboratory, we removed dirt from the AL, sorted it by category, and let it air dry for several days. We weighed and measured surface area of each item from its length and mean width, and calculated AL density as the number, mass, and surface area of AL per area searched (No. m^{-2}, g m^{-2}, and AL cm^2 m^{-2}). Field studies did not involve endangered or protected species at any of the locations of our study. The Forest Preserve District of Cook County issued permission for research at the North Branch of the Chicago River. No permission was required for litter collection at the Lake Michigan beach. The Land Management Committee issued permission for research at the pond on private land at the Loyola University Retreat and Ecology Campus (LUREC; see below). Future permissions for work at that site can contact the campus office at (815) 338–1032.

Anthropogenic litter density in freshwater vs marine ecosystems

We compared our results to AL abundance from published studies conducted in marine beaches. Comparisons among AL studies are complicated by variability in the units used to quantify AL abundance and the categories used to report AL types. AL density in the literature is reported as relative proportion (%), the number of items collected along a transect (No. m^{-1}), or the number of items per area searched (No. m^{-2}). AL density by mass (i.e, g m^{-2} or g m^{-1}) is infrequently reported [1,20]. Studies often classify AL types according to different schemes, which are either based upon material composition (e.g., metal, plastic, and glass), identity (e.g., cigarettes, plastic bags, tires), or function (e.g., fishing, food-related, and construction) [1,21,22]. To facilitate comparison among our data and literature values, we first compared AL abundance and mass (No. m^{-1} and g m^{-1}) of individual AL categories between our sites and studies from marine beaches which used similar units, categories, and did not include microplastic abundance. These included sites in Brazil [23], New Jersey USA [24], South Africa [25], and Oman [26]. The majority of AL studies report abundance as the relative proportion of different categories present. We compared relative amounts of AL categories across habitats including city blocks [27], marine beach [28], lake beach (this study), riparian zone (this study), shallow marine benthos [1], river benthos (this study), and marine offshore benthos [29]. The selected studies reflect the global nature of AL research, however, we note the selected sites are examples of each habitat, and do not represent mean AL composition of the different ecosystems.

Litter incubation and biofilm activity

To measure biofilm activity and community composition on AL and natural surfaces, we selected 4 of the most common types of AL (i.e. glass, aluminum, plastic and cardboard) and used leaves and ceramic tiles (a common surrogate for rocks) to represent natural substrates. We categorized leaf and cardboard as organic substrates and glass, aluminum, tile and plastic (a synthetic organic material) as hard substrates. Polypropylene mesh bags (38×20 cm) with 5 mm openings (Cady Bag Company, Pearson GA) were used to incubate AL and natural substrates in the North Branch of the Chicago River, a 1.4 hectare pond at LUREC in Woodstock, IL (latitude 42.288834, longitude −88.366064), and in the artificial stream facility at LUC. Artificial streams were re-circulating chambers with a paddle wheel, where channel width is 14.0 cm and total flowpath length is 2.0 m [30]. Streams were refilled with 60 L of tap water that had been allowed to dechlorinate for a minimum of 2 d prior to adding to streams, and water level was marked and maintained throughout the study. On the outside of each bag we attached 3 unglazed porcelain tiles and 3 glass tiles (tile size = 25 cm^2). We drilled a 0.3 cm^2 hole in each tile using a diamond drill bit and attached the tiles to the bags using cable ties. We cut 6×5.5 cm^2 pieces from aluminum cans from cola beverages and 20 oz. plastic drinking bottles (polyethylene terephthalate) and attached these to the outside of litterbags with cable ties (N = 3 of each type). Cans and bottles were collected from recycling containers on campus and thoroughly rinsed. Aluminum was situated such that the surface that was previously inside the can was exposed for biofilm growth (i.e. the non-painted surface). Inside the bags, we placed 3 pieces of corrugated, non-waxed or colored cardboard (170 cm^2) and 3 naturally senesced red maple (*Acer rubrum*) leaves. The cardboard and leaves were separated from each other inside the bag using cable ties. Leaves and cardboard were situated inside the bags, which may have reduced primary production and grazing relative to substrates inside the bag.

Five replicate bags were attached to the benthos of the North Branch Chicago River and LUREC pond in early summer 2012. Bags were incubated from May 7–August 13, 2012 (98 days) in the river and June 1–July 20, 2012 (50 days) in the LUREC pond. In the artificial stream, we placed 25 replicate bags starting on June 8, 2012. We collected 3 replicate bags (1 bag from each artificial stream) on days 7, 27, 34, 48, and 52. When bags were removed from the river and pond, individual substrates were placed in 160 mL specimen containers filled with site water. We collected an additional 6–8 L of unfiltered site water for measurements of GPP and CR. Substrates and site water were kept on ice until returned to the lab, and metabolism measurements were started on the same day, within 2–4 hours of collection. For substrates from artificial streams, we placed substrates directly in specimen containers and began metabolism measurements straightaway.

We measured CR and net primary production (NPP) using the light/dark method [16,31]. Each specimen container was filled with unfiltered stream water of known dissolved oxygen (DO) concentration and a single substrate (3 replicates of each substrate; N = 18 for the pond and artificial streams, N = 12 for the river). Containers were capped so no air bubbles were present. All containers were placed in an environmental chamber at 22–23°C with constant light. Final DO and incubation time were recorded after 2–4 h in the light. The containers were refilled with fresh site water and incubated in an environmental chamber in the dark for 2–4 h, after which final DO and time were recorded. Three 'blank' containers with site water only were incubated for both NPP and CR measurements to account for metabolism of water column organisms or abiotic changes in DO. Running GPP and CR in sequence could affect results due to cumulative bottle effects, but these were minimized by the short duration of each measurement period. All substrates were immediately frozen at −20°C after metabolism measurements were completed. We calculated metabolism as change in DO in the light (NPP) and dark (CR), and gross primary production (GPP) as NPP - CR (units: $\mu g\ O_2\ cm^{-2}\ h^{-1}$). For leaves and cardboard, we measured surface area by tracing each piece onto paper. The pieces were cut-out and weighed. We weighed paper of known surface area, and used simple linear regression to develop a mass to surface area conversion [15].

Biofilm community composition

Biofilm was scraped from each substrate incubated in the pond, river, and artificial stream using a sterile razor blade and collected in 2 mL microcentrifuge tubes. For the artificial stream biofilms, we collected biofilm only from substrates sampled on the final incubation date. Genomic DNA was isolated from each biofilm sample (N = 18 in the artificial stream and pond, N = 12 in the river) with the PowerBiofilm DNA Isolation Kit (MoBio Laboratories, Inc., Carlsbad, CA USA) according to manufacturer's instructions. Isolated DNA was placed in 100 µL of a 20 mM Tris solution and stored at −20°C. DNA from each sample was sent to Argonne National Lab for massively parallel, paired-end sequencing of partial 16S rRNA genes using the Illumina MiSeq platform. PCR amplification was performed using primers 515F and 806R to amplify the V4 region of the 16S rRNA gene [32]. PCR amplification failed for 2 samples (1 aluminum and 1 glass substrate from the pond) due to low DNA yield from low biomass biofilms, so no sequencing data were obtained for these samples. Sequencing was conducted as described previously [32]. Sequences were processed using MOTHUR v.1.30.1 [33] as described previously [34]. Briefly, any sequences with ambiguities or homopolymers longer than 8 bases were removed from the data set. Sequences were aligned using the SILVA-compatible alignment database available within MOTHUR. Sequences were trimmed to a uniform length of 253 base pairs and chimeric sequences were removed using Uchime [35]. We classified sequences using the MOTHUR-formatted version of the RDP training set (v.9) and any unknown, chloroplast, mitochondrial, archaeal and eukaryotic sequences were removed. Sequences were clustered into operational taxonomic units (OTUs) based on 97% sequence identity. To avoid biases associated with uneven numbers of sequences across samples, the entire dataset was randomly subsampled to 17,000 sequences per sample.

Data analysis

We used 1-way ANOVA to compare AL abundance across 3 ecosystem types: Chicago River benthic zone, riparian zone, and Lake Michigan beach. For substrates incubated in the Chicago River and LUREC pond, we used a 1-way ANOVA to compare GPP and CR across AL types. 2-way repeated measures ANOVA (RM-ANOVA) was used to compare GPP and CR across substrates and dates during incubation in the artificial streams.

The compositions of the bacterial communities from individual samples were compared by calculating dissimilarities based on the theta index [36] and visualizing the resulting distance matrix using non-metric multidimensional scaling (nMDS) run within MOTHUR. The significance of differences in theta index scores between sites and substrates was assessed by ANOSIM run within MOTHUR. ANOSIM reports R statistics and p values for comparisons of pairs of groups. Briefly, R = 0 when there are no differences between groups, R approaches 1 as groups become more distinct, R = 1 when all samples in different groups are more dissimilar to each other than any samples in the same group, and the p value indicates the statistical significance of the R statistic [37]. Bacterial families making the most significant contributions to differences in composition between sites and substrates were identified by SIMPER analysis run in Primer V.5 (Primer-E Ltd., Plymouth, United Kingdom). We used two-way ANOVA to assess the effects of sites and substrates on the total number of OTUs identified, the relative abundance of the bacterial families identified by SIMPER, and on the most abundant bacterial phyla. MOTHUR was also used to calculate Inverse Simpson and Shannon (H′) diversity indices [38]. We used one-way ANOVA to compare diversity among substrates in each habitat type. All ANOVAs were completed in SYSTAT 13.0 (Systat, Inc. Chicago, IL), with significant ANOVA ($p<0.05$) followed by Tukey's multiple comparison tests.

Data Sharing

All of the sequence data analyzed in this paper can be downloaded from the National Center for Biotechnology Information (NCBI) Sequence Read Archive (SRA) with accession number SRP042298.

Results

Anthropogenic litter abundance across freshwater ecosystem types

Lake Michigan beaches had significantly less AL relative to the Chicago River riparian and benthic zones when AL abundance was expressed as number of items, mass, and surface area (ANOVA $p \leq 0.012$; Figure 1A, B, and C). There was no difference in AL abundance between the riparian and benthic zones of the Chicago River. Across all three ecosystem types, the dominant types of AL were plastic, paper, and glass (Figure 1D, E, F). Plastic was more abundant in the riparian zone than the beach (ANOVA $p = 0.048$), but there was no difference in the abundance of paper among locations (ANOVA $p = 0.684$; Figure 1D, E). Glass was more abundant in the river benthos than the riparian or beach areas (ANOVA $p = 0.001$; Figure 1F).

Anthropogenic litter abundance across terrestrial and aquatic ecosystems

The abundance and density of AL at our study sites fall within the range reported in the literature, although values for Lake Michigan beaches were lower than values from two marine sites which used similar methodology and AL categories (Figure 2).

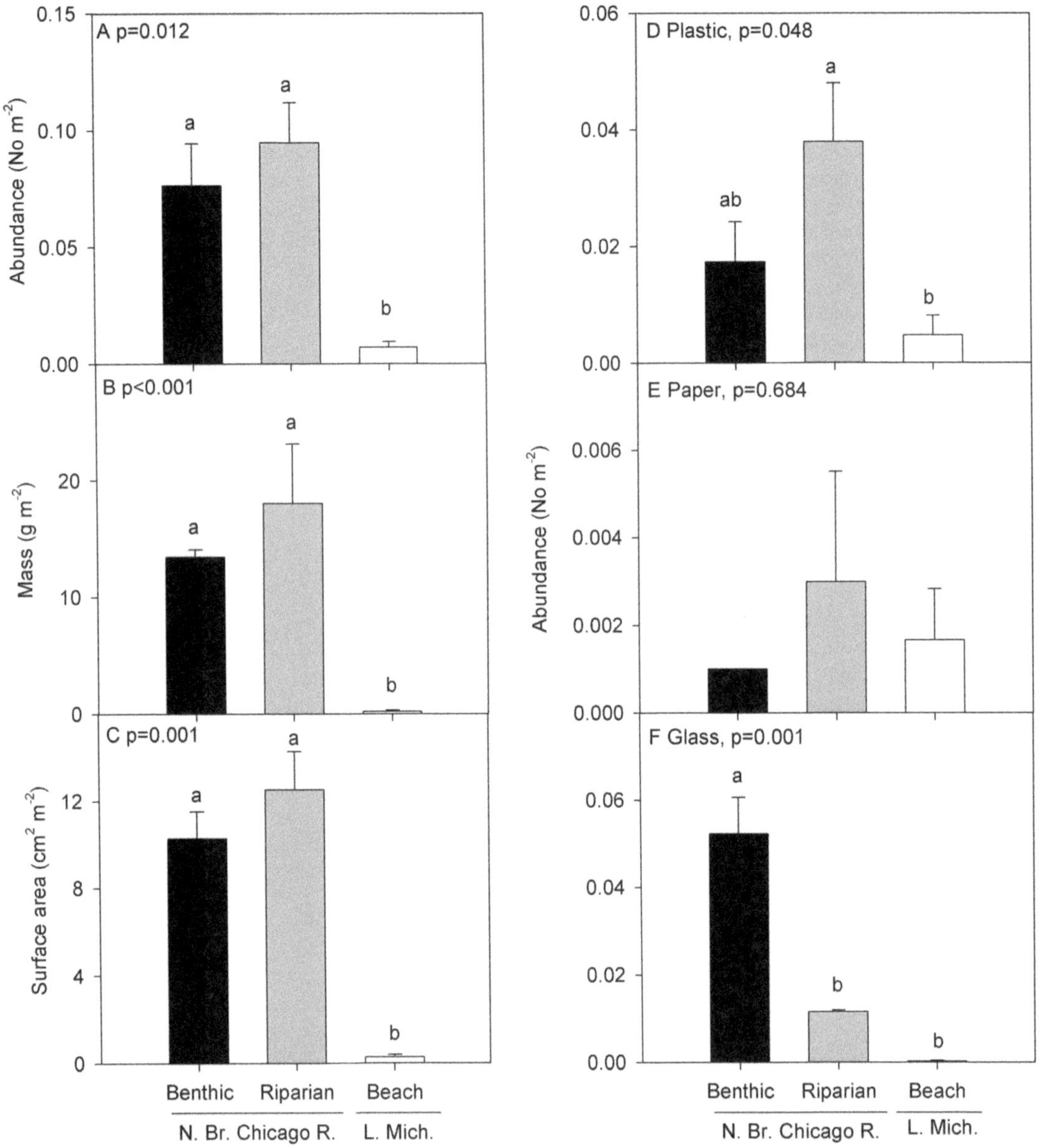

Figure 1. Mean (±SE) amounts of total anthropogenic litter (AL) in the benthic and riparian zones of the North Branch of the Chicago River and a Lake Michigan beach in Chicago, IL, USA by (A) abundance (number m^{-2}), (B) mass (g m^{-2}), (C) surface area (cm^2 m^{-2}), and by individual categories including abundance of (D) plastic, (E) paper, and (F) glass.

While plastic was more abundant in the marine beaches than freshwaters, the mass of plastic was fairly consistent across the 2 marine beaches and the Chicago River riparian and benthic zones (range = 25–55 g m^{-1}; Figure 2). The high density of glass in the Chicago River benthos was unique among habitats. Metal represented >50% of the AL mass in the riparian zone, more than other habitats.

Several clear patterns emerged when comparing relative AL abundance across terrestrial, freshwater, and marine sites. Plastic AL represented 17.6–53.5% of total AL in the terrestrial, transitional, and shallow aquatic environments, but was only 3.5% in the offshore benthic environment (50–200 m depth; Figure 3). Paper and cigarette litter was more abundant in terrestrial ecosystems, representing 65% of AL in city blocks, 28–36% of AL on lake and marine beaches, and <1% of AL in the aquatic habitats. AL consisting of fishing items was restricted to marine settings, and increased in abundance from the beach (8.4%) to shallow marine benthos (31.3%) to the offshore marine benthos (65.6%). Metal was generally found in greater abundance in aquatic environments relative to terrestrial or transitional habitats. Finally, glass represented 53.5% of AL in the river benthos, more than other sites.

Biofilm activity on AL incubated in freshwater ecosystems

AL incubated in artificial streams showed significant variation in GPP through time (RM-ANOVA $p<0.001$) and among substrates (RM-ANOVA $p<0.001$; Figure 4A). GPP on cardboard and leaves were not different from each other, but rates were lower than GPP on the hard substrates (glass, aluminum, tile and plastic). On the hard substrates, GPP was highest on glass, aluminum was

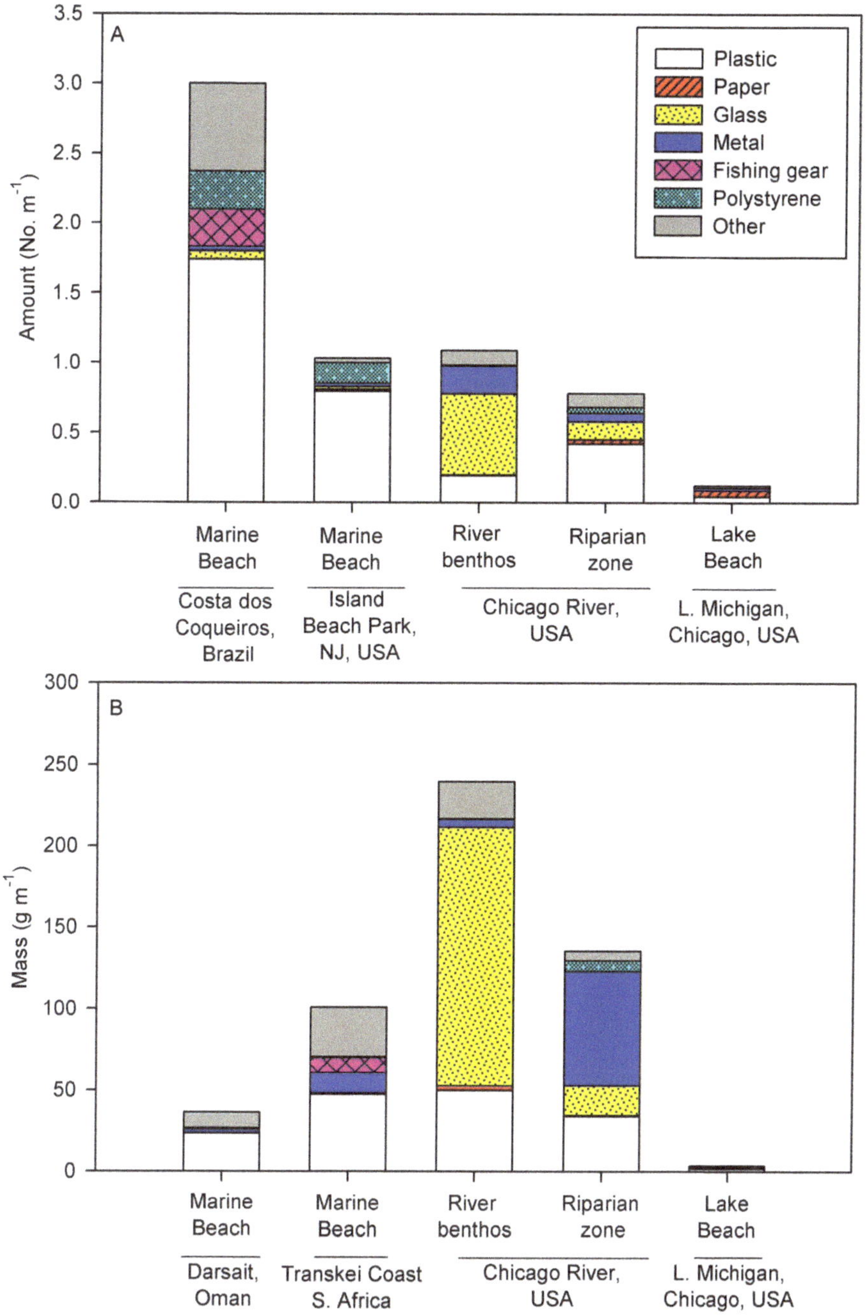

Figure 2. Density (A) and mass (B) of anthropogenic litter (AL) in 2 marine beaches and in 3 habitats from this study: benthic and riparian zones of the North Branch of the Chicago River and a Lake Michigan beach. Data for marine beaches in (A) are from [24] and [23], and data for marine beaches in (B) are from [25] and [26].

lowest, and tile and plastic were intermediate. CR was uniform among biofilms for all AL surfaces (RM ANOVA $p = 0.075$) but rates were different among sampling dates (RM ANOVA $p < 0.001$), attributed to very low rates on the first date (Figure 4B).

Biofilm activity on AL surfaces incubated *in situ* showed different patterns in the pond and river sites (Figure 5). GPP on aluminum and glass was lower than tile and plastic in the Chicago River (ANOVA $p = 0.021$; Figure 5A). Similarly, river biofilms had higher CR on tile than aluminum and plastic, while glass was intermediate (ANOVA $p = 0.006$; Figure 5B). Organic substrates were completely decomposed, ingested, or fragmented during the incubation in the river. In the pond, biofilm GPP was highest on tile, intermediate on glass, plastic, and aluminum, and lowest on cardboard and leaves (ANOVA $p = 0.037$, Figure 5C). There was no difference in biofilm CR among substrates in the pond (ANOVA $p = 0.258$; Figure 5D).

Bacterial community composition on AL incubated in freshwater ecosystems

Analysis of biofilm bacterial communities based on high-throughput sequencing of partial 16S rRNA genes demonstrated that all AL substrates were colonized by biofilms containing

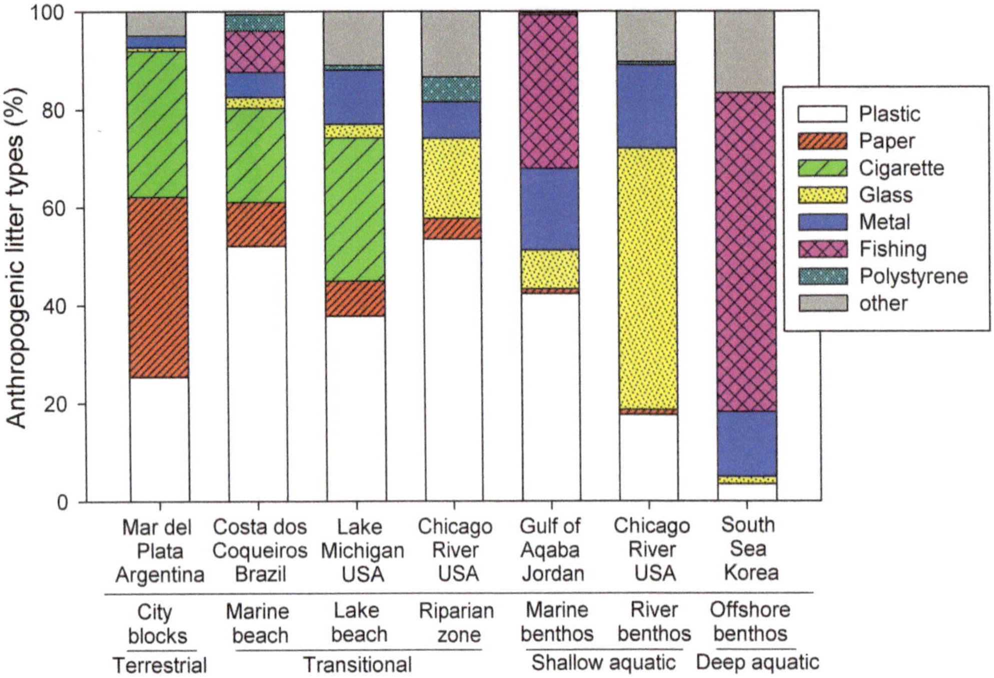

Figure 3. The relative proportion of anthropogenic litter (AL) from our study in the Chicago area (in bold; developed lake beach, riparian zone, and river benthos) relative to city blocks in Argentina [27], a developed marine beach in Brazil [23], a shallow marine benthos site in the Red Sea [1] and an offshore marine benthos site in the South Sea of Korea [29].

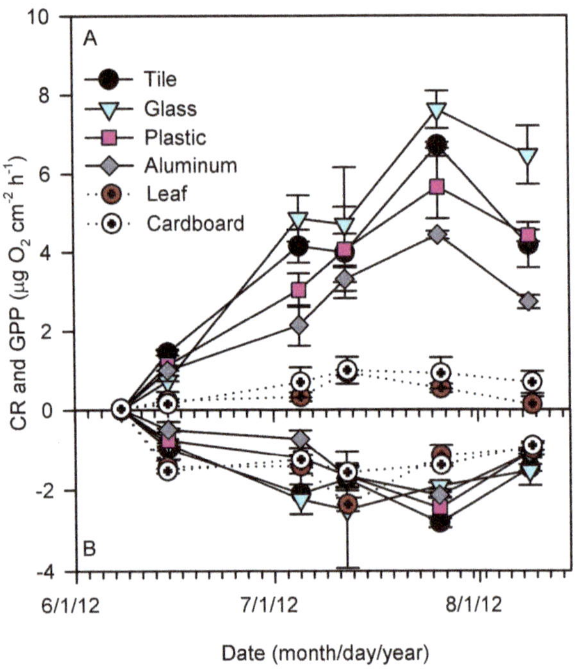

Figure 4. Mean (±SE) gross primary production and community respiration for biofilms colonizing natural substrates (unglazed ceramic tile and leaves) and anthropogenic litter substrates including glass, hard plastic, aluminum, and cardboard, incubated in artificial streams during summer 2012 (N = 18 date^{-1}).

diverse bacterial assemblages. On average, 1,552 bacterial OTUs were identified per sample. Bacterial community composition was significantly different among the 3 incubation sites. For example, there was a significant effect of site on the number of bacterial OTUs detected within the biofilms (ANOVA, $p<0.001$), with fewer OTUs detected on average in the artificial stream biofilms (958 OTUs) than in the pond (1,870 OTUs) or river (2,020 OTUs) biofilms. Bacterial communities colonizing substrates incubated in the river, pond and artificial stream were well differentiated on the nMDS ordination (with the exception of the leaf samples from the pond; Figure 6) and communities from the three habitats were significantly different from each other based on ANOSIM (Table S1).

The dominant bacterial phyla within the biofilms included Proteobacteria, Bacteroidetes, Firmicutes and Verrucomicrobia, but there were broad differences in the composition of biofilm bacterial communities among sites (Figure 7). For example, river biofilms had significantly higher relative abundances of Acidobacteria and Nitrospira and significantly lower relative abundances of Proteobacteria than the other sites, while the pond biofilms had significantly higher relative abundances of Proteobacteria, Firmicutes and Chloroflexi, and significantly lower relative abundances of Bacteroidetes than the other sites (Table S2). SIMPER analysis identified specific bacterial families that made the most significant contributions to differences in composition of the biofilms across the three sites, and several families were significantly different across sites (Table S3). Examples include significantly higher relative abundance of Burkholderiales, Nitrospiraceae, Nitrosomonadaceae and unclassified Gammaproteobacteria, as well as significantly lower relative abundance of Erythrobacteraceae in the river biofilms as compared to the other sites.

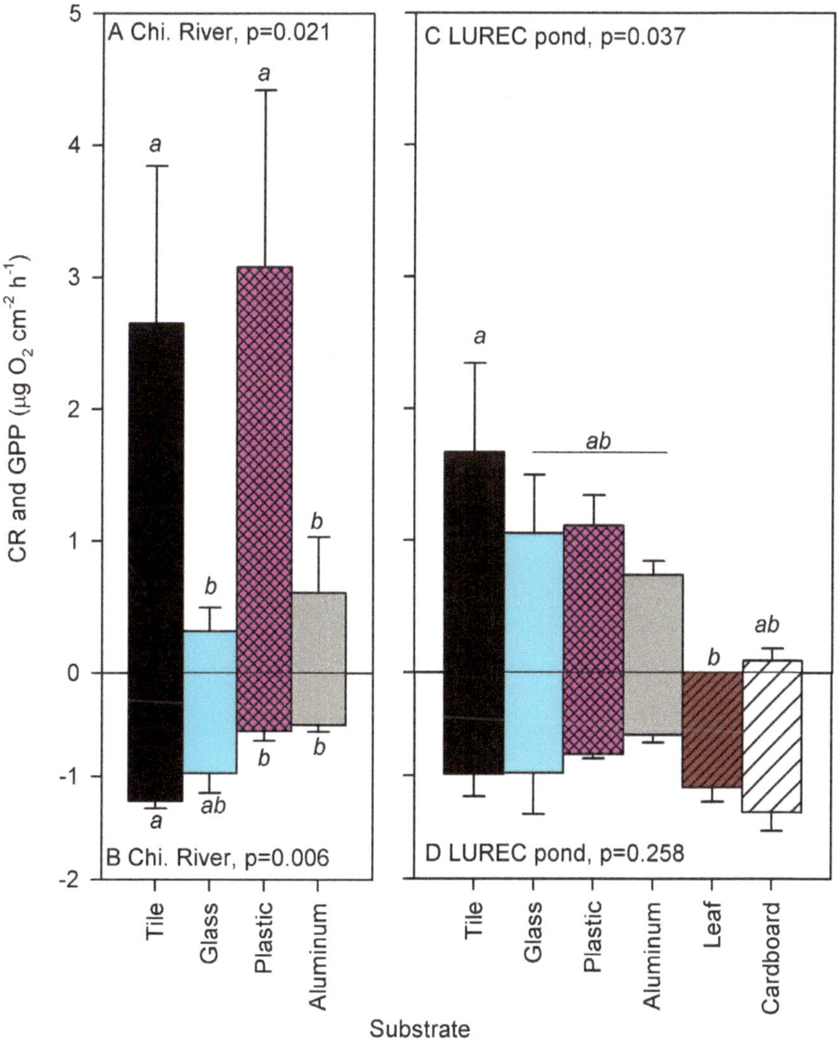

Figure 5. Mean (±SE) gross primary production (GPP) and community respiration (CR) for biofilms colonizing natural substrates (unglazed ceramic tile and leaves) and anthropogenic litter substrates including glass, hard plastic, aluminum, and cardboard incubated in (A) North Branch of the Chicago River, and (B) a pond at Loyola University Retreat and Ecology Campus (LUREC; N = 3 for each substrate at each site). Results (p-values) from 1-way ANOVA among substrates for each metric are shown in each panel, small letters indicate significant differences among substrates ($p<0.05$) as indicated by Tukey's multiple comparison test.

Substrate type also had significant effects on biofilm bacterial community composition. Specifically, biofilm bacterial communities colonizing the organic substrates (cardboard and leaves) in both the pond and artificial stream were clearly distinct from the communities colonizing the hard substrates (glass, plastic, aluminum and tile) on the nMDS ordination (Figure 6, Table S1). We were unable to analyze biofilms colonizing cardboard and leaves in the river because these substrates were no longer present at the conclusion of the incubation. For the pond, river, and artificial stream, the nMDS ordination indicated no differences in bacterial community composition among biofilms on the 4 hard substrates. In addition, there were no significant differences in community composition among the hard substrates in the 3 habitat types (Table S1). Therefore, the biofilm communities colonizing all of the hard substrates were treated as a group by ANOSIM, and were compared to the communities from the biofilms on cardboard and leaves. For both the pond and artificial stream sites, ANOSIM confirmed significant differences in bacterial community composition between the leaf and cardboard substrates and the hard substrates (Table S1). The nMDS (Figure 6) and ANOSIM (Table S1) analyses also indicated that for both the pond and artificial stream sites, the bacterial communities colonizing the cardboard were significantly different from the communities colonizing the leaves ($p<0.10$).

SIMPER analysis identified the bacterial families making the largest contributions to differences in community composition among the biofilms colonizing the cardboard, leaves and hard substrates (Tables S4 and S5). For the artificial stream, hard substrates had significantly higher relative abundances of Erythrobacteraceae, and the cardboard and leaves had significantly higher relative abundances of Caulobacteraceae (Table S4). Within the artificial stream there were also some significant differences between leaves and cardboard, including significantly higher relative abundances of Cytophagaceae and Opitutaceae on the cardboard, and significantly higher relative abundances of Caulobacteraceae, Rhizobiaceae and Xanthomonadaceae on the leaves (Table S4). For the pond, the biofilm communities colonizing the cardboard and leaves did not show any common

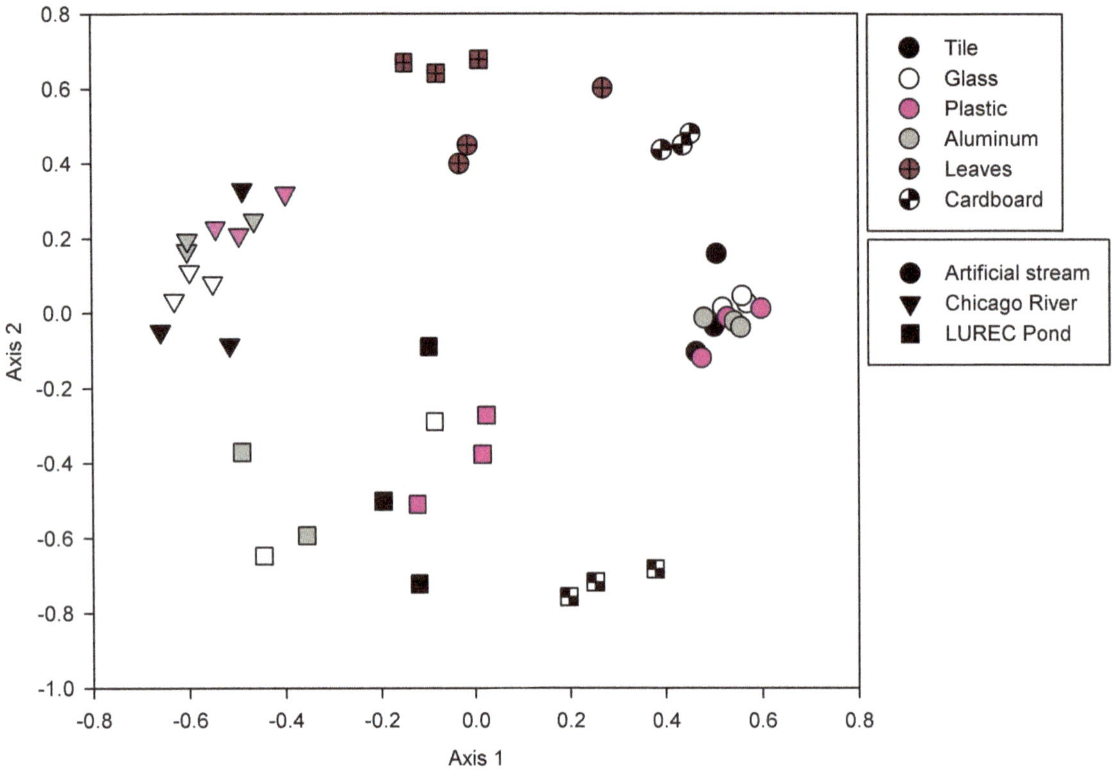

Figure 6. Non-metric multidimensional scaling (nMDS) ordination of bacterial communities based on high-throughput sequencing of bacterial 16S rRNA genes from biofilms colonizing anthropogenic litter (AL) substrates (glass, plastic, aluminum, and cardboard) and natural surfaces (leaves and tile) in artificial streams at Loyola University Chicago, the North Branch of the Chicago River, and a pond at Loyola University Retreat and Ecology Campus (LUREC).

differences when compared to the hard substrates; rather each varied from the hard substrates in distinct ways. Specifically, the cardboard had significantly higher relative abundances of Ruminococcaceae and Desulfovibrionaceae than the hard substrates, whereas leaves had higher relative abundances of Oxalobacteraceae, Enterobacteriaceae and Rhizobiaceae than the hard substrates (Table S5). Finally, there was no significant effect of substrate type on community diversity calculated as either Inverse Simpson or Shannon indices (Table S6).

Discussion

AL abundance in freshwaters

As expected, we found AL in all freshwater habitats. However, AL density in the river benthos, riparian zone, and lake beaches showed clear differences among ecosystem types, likely driven by differences in hydrology, AL movement and breakdown, and human activity. AL totals on Lake Michigan beaches were unexpectedly low, especially as the beaches studied receive many daily visitors. Low AL density may be attributed to daily beach grooming by the Chicago Park District. Anecdotally, Parks District employees suggested our results for AL density would be much higher if we measured AL collected by the maintenance equipment (C. Breitenbach, personal communication). A more accurate assessment of AL abundance for public beaches will require 1) abatement of AL clean up, or 2) collection of AL from both the beach and from the grooming machines. There are very few studies of AL density on Great Lakes beaches for comparison of our results. High density of small plastic fragments were found on Lake Huron beaches (~37 pieces m^{-2}), however, over 90% were pellets <5 mm in size and required sieving for quantification [39]. Our assessment did not account for AL particles at this scale. More measurements of AL density and composition, across size categories and habitat types, are needed to understand the distribution for AL on Great Lakes beaches.

Total density of AL in the Chicago River's North Branch benthic and riparian zones was similar, but the composition of AL in both habitats suggests that direct littering and variation in the movement and retention of different AL types drive overall abundance. For example, glass was the most abundant AL type found in the river benthos, and consisted mostly of discarded liquor and beer bottles. Our study area was in the Bunker Hill Forest Preserve District of Cook County, located in northwestern Chicago. Like other Forest Preserves, the area is popular for recreation. Glass bottles may have been transported to our study reaches from upstream during floods, but we surmise much of the glass was discarded directly on site. The glass we collected was often partially broken, colonized by microbial biofilms, and contained mud and gravel. Sinking, colonization, and benthic entrainment of glass reduce export from the benthic zone, suggesting that long term accumulation of glass bottles is likely. Local recreation and consumption are also a significant source to AL on some marine beaches [28], but there are few other analyses of AL in rivers to compare our data. A study of proportional AL abundance in a Welsh river found less metal (8%), more plastic (49%), and a large amount of sewage related AL from illegal trash disposal and malfunctioning combined sewer overflows (e.g., feminine hygiene and diapers; 23%) compared to our data [40]. Despite the low number of measurements, riverine sources are reported to contribute up to 80% of global marine AL input [41–

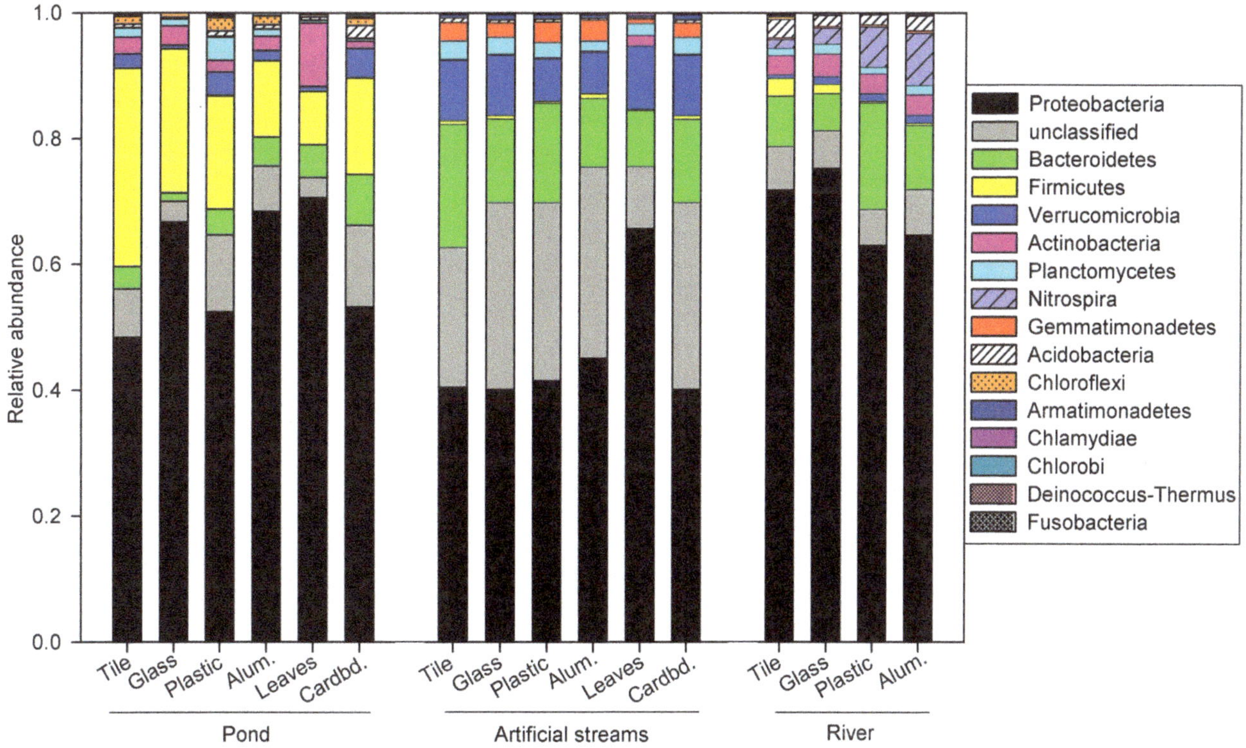

Figure 7. Relative composition of bacteria phyla based on high-throughput sequencing of bacterial 16S rRNA genes from biofilms colonizing anthropogenic litter (AL) substrates (glass, plastic, aluminum, and cardboard) and natural surfaces (leaves and tile) in artificial streams at Loyola University Chicago, the North Branch of the Chicago River, and a pond at Loyola University Retreat and Ecology Campus (LUREC).

43]. Our results suggest that selective retention of some AL types in rivers (e.g., glass and metal), combined with the influence of human activity and adjacent land-use, drive benthic AL abundance and could affect estimates for the amount of AL retained and exported in rivers worldwide.

Total AL density in the riparian zone was similar to the river benthos, but the relative composition was different, indicating that riparian AL pools were also determined by mobility and retention. In contrast to the river benthos, there was a trend of more paper and plastic in the riparian zone. Because many of these items were food related, on-site consumption and littering was also probably the major source. One notable pattern was AL accumulation in debris dams and overhanging structures such as bridges, branches, and large woody debris (Figure S1). The orientation of AL suggested that the items were deposited during high flows. This pattern has been called a "Christmas tree" effect elsewhere [40]. Given the propensity for flooding in urban watersheds [44], riparian zone AL is likely to be moved and retained in this fashion elsewhere. A potential AL reduction strategy would be to instruct volunteers to focus on these sites for efficient AL collection. In addition, the repeated wetting and drying of AL at these sites could be an important factor driving its breakdown.

Relative AL composition across multiple ecosystems

The analysis of relative AL composition of 7 sites spanning different habitats illustrates common trends in AL pools, and allows for inference of critical AL fluxes at each site (Figure 3). One unexpected result of our synthesis was the uniformity in AL composition among city blocks, a marine beach, and the Lake Michigan beach. Cigarette butts contributed a large and relatively even proportion across all 3 sites, and paper was more abundant in these ecosystems than the others in Figure 3. The low abundance of paper and cigarette butts at the aquatic sites suggests that paper and cigarettes 1) have low input rates in aquatic habitats, 2) do not move or accumulate in the same way in aquatic environments as on beaches or city blocks, or 3) rapidly decompose in aquatic habitats. Some combination of all 3 are likely important. For example, after 98 d in the Chicago River, the cardboard was absent from the bags via decomposition, ingestion, or fragmentation, and its incubation in the pond and artificial streams showed a loss of structure after ~50 days underwater (i.e. cardboard 'mush'). Cigarette butts consist of paper coverings over cellulose acetate filters, and some brands of have plasticizers added [45]. In terrestrial environments, cigarette butts have a decomposition rate of 0.265 y^{-1} or 3.7 y total (S. Haynes, unpublished data), and can have antimicrobial properties which inhibit biological degradation [45]. To our knowledge, decomposition rates for cigarette butts in aquatic habitats have not been published.

A major factor driving research on AL in the ocean (i.e. marine debris) is the amount of fishing-related garbage found on beaches, shallow marine environments, and the open ocean. Our comparison of AL among 7 sites in Figure 3 supports this pattern, as fishing-related items were found in all 3 marine locations. No fishing-related AL was documented in our freshwater sites and we witnessed no anglers during AL collection. We acknowledge our sampling area encompassed a relatively narrow geographic range. Other freshwater environments may have fishing-related AL, but none have yet been reported in the few published AL studies from rivers and the Great Lakes [40,42]. The Alliance for the Great Lakes Adopt-a-Beach program has detailed records on AL in

Great Lakes beaches since 2003, including measurements of fishing line, nets, and lures. These items typically represent a small proportion of AL on Great Lakes coastal sites, as AL is dominated by food and smoking-related items (J. Cross, personal communication).

The relative proportion of AL across the 7 sites in Figure 3 suggest that AL movement, and not just littering rate, is a critical driver of AL pools in different ecosystems. For example, all 3 benthic sites (river, shallow marine, and offshore marine) and the riparian zone had AL which was more likely to be heavy, including metal, glass, and fishing-related material. This material frequently sinks and may become entrained in the benthos and flooded riparian zones. The river benthos and offshore marine benthos were especially dominated by heavy items. In the river, glass bottles were >50% of the AL items. In the offshore benthos, the largest category was fishing gear (65.2%), which in this case included a combination of heavy materials such as fish pots, nets, octopus jars, and fishing lines [29]. Heavy items the benthic zone of the open ocean originate from direct dumping via fishing activity and not export of terrestrial or riverine sources [29].

Unlike paper items which are degraded more readily, glass and metal are long-lived [20,46], and likely to be used as habitat. For example, the glass debris we removed from the Chicago River was well-colonized with microbial biofilms, snails, and amphipods (*Gammarus* sp.). Because urban streams can have reduced benthic habitat complexity [44], relatively stable AL items such as glass or tires may represent an important habitat for stream microorganisms and macroinvertebrates. However, AL effects on macroinvertebrate communities are unknown.

A notable trend that emerges from comparison of AL among different studies is that the units used to characterize AL density have a strong impact on the patterns gleaned from the data. For example, glass and metal in the river benthos and riparian zone was abundant and heavy, so the mass of AL at those sites was higher than in the marine or lake beaches, even though the total number of AL items was intermediate (Figure 2). AL has been quantified in various units including abundance (No. m^{-1} or m^{-2}), mass (g m^{-1} or m^{-2}), and less frequently as surface area (AL cm^2 habitat m^{-2}) [1]. Other ecological metrics are reported in similar units, such as organic matter standing stock (g ash-free dry mass m^{-2}), organism density (No. m^{-2}), and abundance of large woody debris (m^2 or m^3 of wood per habitat area). For AL measurements, the choice of which units are most ecological meaningful may vary depending upon the material. An organic item such as paper may be more usefully expressed in units of mass, as it will likely decompose in similar fashion to leaf litter and represents a potential food resource. In contrast, a hard surface like glass may be more meaningfully expressed in terms of surface area, as the space available for colonization may be the most biologically meaningful aspect of its presence. Finally, the proportion of AL categories present and their relative abundance can be combined into metrics of AL diversity. Given high variation in the types, size, and abundance of AL among habitats, these metrics could synthesize AL "communities" in analogous ways to biological communities [27]. Uniting values into single numbers could also facilitate comparisons better than existing measurements of density or relative proportion.

Biofilm activity and community composition

When AL substrates were incubated in the three freshwater habitats, all AL became colonized by taxonomically diverse and metabolically active microbial biofilms. Biofilms were dominated by bacteria from the phyla Proteobacteria, Bacteroidetes and Firmicutes, all of which have been identified as common components of freshwater biofilms [47–49]. Metabolism rates ranged from −4 μg O_2 cm^{-2} h^{-1} (CR) to 8 μg O_2 cm^{-2} h^{-1} (GPP), which aligns with the range reported for the substrate-specific measurements in literature [15,16].

Site identity significantly affected the activity and composition of microbial biofilms, likely driven by differences in the physical and chemical conditions among sites. For example, biofilms in the artificial stream had higher respiration rates and lower bacterial species (OTU) richness compared to the 2 natural habitats, and mean CR on substrates in the artificial stream was higher than the river and pond. These differences were likely driven by the highly stable conditions in the artificial streams, including higher illumination (due to a lack of shading), higher temperature, and higher rates of nutrient and gas exchange (due to shallow water column and high flow rate) as compared to the field sites. These conditions may have selected for the much higher relative abundance of aerobic, heterotrophic Verrucomicrobia within the artificial streams [50]. In contrast, the higher abundance of bacteria involved in nitrification (e.g., Nitrosomonadaceae and Nitrospira) in the river biofilms was likely the result of higher water column nitrogen concentrations within the Chicago River compared to the pond and artificial stream (T. Hoellein, unpublished data). The most notable feature of the pond biofilm communities was the high relative abundance of Firmicutes, specifically the Planococcaceae, which are a family of aerobic organotrophic bacteria that are known to be halotolerant [51].

AL type affected the activity and composition of microbial biofilms, with the most significant differences observed between organic (leaves and cardboard) and hard substrates (glass, plastic, aluminum and tile). In both the pond and artificial stream, GPP was significantly lower on the organic substrates, suggesting that these communities were dominated by heterotrophic organisms. However, we note the organic substrates were situated inside the mesh bags and shading may have inhibited primary producers. There were also significant differences in bacterial community composition on organic and hard substrates which indicate use of the organic substrates for heterotrophic metabolism. Leaves had high relative abundance of several common plant-associated microbes, including Xanthomonadaceae [52], Rhizobiaceae [53] and Oxalobacteraceae [54]. Leaves in the artificial streams also had high relative abundance of Chitinophagaceae, some species of which have been shown to degrade cellulose [55]. Cardboard also showed very high abundances of bacterial groups with the ability to degrade cellulosic compounds, including Opitutaceae [56] in the artificial stream and Ruminococcaceae [57] in the pond. Cardboard in the pond had high relative abundances of two families of anaerobic bacteria, Desulfovibrionaceae and Ruminococcaceae, suggesting that the low flow conditions may have led to relatively more anaerobic niches than within the shallow, well-mixed artificial streams.

Unexpectedly, there were no significant differences in microbial community composition among tile, glass, plastic, and aluminum within each of the 3 habitats studied. The pattern suggests hard AL surfaces do not appear to affect bacterial community composition. However, patterns of GPP on hard substrates suggest variation in their algal constituents. For example, GPP on aluminum was lower than tiles in the river and artificial streams. The same pattern was observed in the pond (although not statistically significant). Reduced GPP on aluminum relative to the tiles could be attributed to microscopic differences in surface structure or chemistry which could affect algal growth (e.g., fewer attachment points or oxidation). In addition, differences in the patterns of GPP among glass, aluminum, plastic, and tile could be attributable to variation the algal community composition among

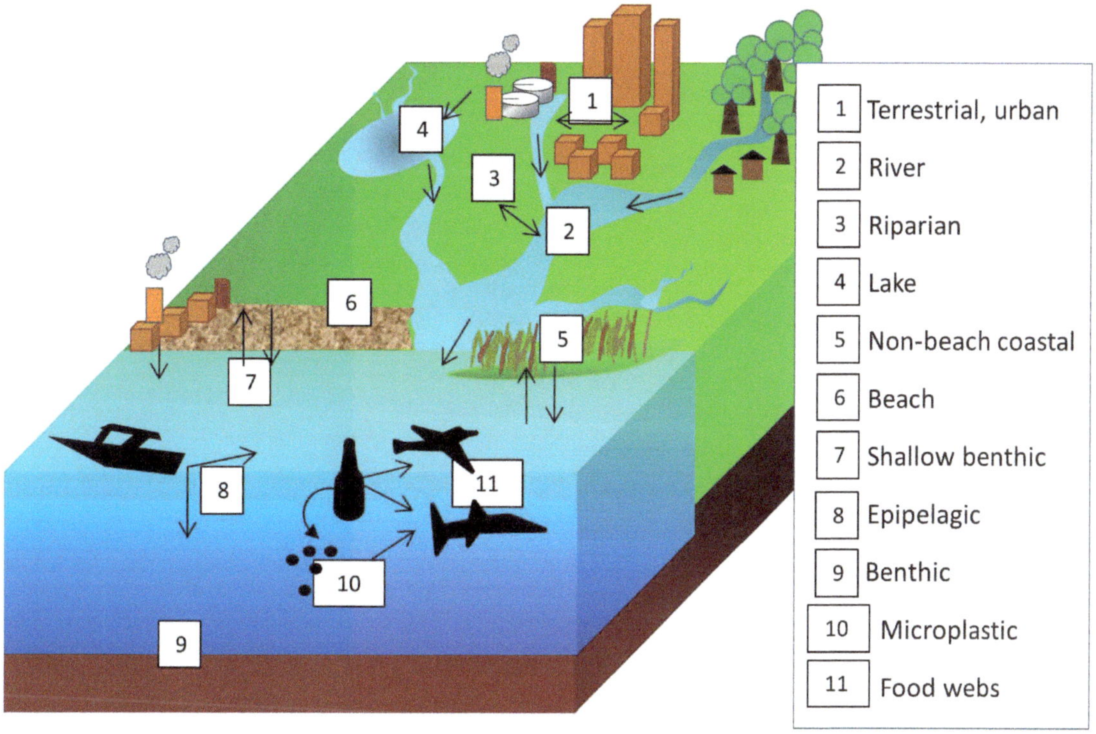

Figure 8. Conceptual diagram of major pools and fluxes of anthropogenic litter (AL) in terrestrial, freshwater, and marine ecosystems, modified from [10]. This model unites the early stages of the AL "life cycle," in terrestrial and freshwater ecosystems to stages of accumulation and transformation in coastal and pelagic environments.

the 3 ecosystems. Upon visual inspection, biofilms on all inorganic AL items in the river and pond had a brown and green appearance, with little evidence of strands of filamentous algae. In contrast, most substrates in the artificial stream had visible filamentous algae. We did not quantify AL surface texture, algal community composition, or chlorophyll *a*. However, our results suggest those analyses could be an important component of future studies which analyze the effect of AL on freshwater biofilms.

Differences in microbial community composition among habitats and AL types suggest AL movement could affect dispersal of microbial biofilm constituents among connected aquatic ecosystems. For example, biofilms colonizing AL surfaces such as plastic, which is mobile and slow degrading, may be more likely to be transported downstream intact than those on natural surfaces such as rocks (largely stationary) and leaf litter or fine sediment (which decompose). River biofilms had significantly higher Nitrosomonadaceae and Nitrospira bacterial taxa which carry out nitrification. Downstream transport of riverine AL over long distances to lake or marine ecosystems could deliver intact microbial communities through this novel mechanism, thereby affecting community structure and function in the receiving water body. This process has been demonstrated for dispersal of invasive species in the marine environment [58], but not for movement of organisms which colonize AL in freshwaters.

Revised conceptual model for AL ecology with a cross-ecosystem perspective

While AL in the marine environment is well documented [1,4,46], research on AL abundance and its role(s) in freshwater and terrestrial ecosystems lags far behind. We suggest this discrepancy may be attributed in part to an issue of nomenclature. The term "marine debris" has been used to describe AL, even when it is found in non-marine ecosystems [59]. Other studies have used disparate terms such as urban litter, riverine litter, floatables, and solid waste to refer to terrestrial or freshwater AL [22,27,40,42]. We suggest marine-focused terminology for ocean research, and lack of unity in terminology for research in other ecosystems may inhibit a more complete understanding of the sources and fate of AL at landscape and global scales. We suggest the term anthropogenic litter is most useful because 1) AL differentiates the material from natural litter or debris accumulations (e.g. leaf litter or woody debris), 2) AL describes the material independent of its collection site, and 3) the term could promote an expanded perspective on the spatial dynamics and entire 'life cycle' of AL to unify terrestrial, freshwater, and marine ecosystem research on the topic (Figure 8).

We composed a conceptual model to guide our AL research by building on the diagram for the study of marine plastics, which shows 3 pools for marine plastic including beaches, coastal waters/ sediments, and the open ocean [10]. We expanded this perspective by modifying the vocabulary (i.e. AL rather than marine debris) and adding several pools: 1) terrestrial environments, 2) rivers, 3) riparian zones, 4) lakes, and 5) marshes or other non-beach coastal habitats (Figure 8). Within the marine environment, we can separate major AL pools into 7) shallow benthos, 8) epipelagic, 9) benthic, 10) microplastic, and 11) food webs (i.e. consumption by microbes and animals).

Expanding the number of pools in the AL conceptual diagram also requires the consideration of fluxes not conventionally incorporated into models of AL movement [10]. Fluxes of AL are mediated by abiotic factors such as wind, flooding, currents, retention, and buoyancy. In addition, biotic activity (e.g., colonization, decomposition, and ingestion), and human activity affect AL flux [60]. The relative importance of these fluxes is likely to differ among terrestrial, freshwater, and marine ecosystems. For example, our data for AL pools in freshwater ecosystems suggest human activity can drive both AL input (i.e. littering), and output (i.e. beach maintenance). Some pools in Figure 8 are likely to be net sinks of AL, or sites of AL accumulation over very long time scales. Other pools are likely to represent locations where AL is only temporarily stored before it is moved, broken down, or consumed. Differentiating the net sinks from temporary storage sites requires research which incorporates measurements of biotic, abiotic, and anthropogenic fluxes along with AL pools. Building a greater understanding of the entire AL "life cycle" remains a major challenge for the science of AL ecology, but will serve as a critical tool for reducing its continued recruitment and mitigating its environmental impacts.

Supporting Information

Figure S1 Anthropogenic litter (AL) deposited on vegetation overhanging the North Branch of the Chicago River.

Table S1 ANOSIM analysis of differences in bacterial community composition by sites and across substrate types. 'Hard substrates' includes tile, glass, plastic, and aluminum.

Table S2 Relative abundances of most abundant bacterial phyla across sampling sites.

Table S3 Relative abundances of bacterial families making the largest contributions to variations in biofilm composition across sampling sites.

Table S4 Relative abundances of bacterial families within artificial stream biofilms making the largest contribution to variations between substrate types.

Table S5 Relative abundances of bacterial families within pond biofilms making the largest contribution to variations between substrate types.

Table S6 Mean (±SE) values for the Inverse Simpson and Shannon diversity (H′) indices for microbial communities colonizing substrates in the three study sites.

Acknowledgments

We thank the following students for research assistance: Owen McKenna, Steven Polaskey, Thomas Manghi, Ricardo Magallon, Michael Hassett, Simon Morgan, and Angeline David. We thank Jamie Cross at the Olga Lyandres at the Alliance for the Great Lakes Adopt a Beach program for research advice.

Author Contributions

Conceived and designed the experiments: TH JG. Performed the experiments: TH JG AP MR. Analyzed the data: TH JG AP MR JK. Contributed reagents/materials/analysis tools: TH JK. Wrote the paper: TH JG AP MR JK.

References

1. Abu-Hilal A, Al-Najjar T (2009) Marine litter in coral reef areas along the Jordan Gulf of Aqaba, Red Sea. J Environ Manage 90: 1043–1049.
2. Hammer J, Kraak MHS, Parsons JR (2012) Plastics in the Marine Environment: The Dark Side of a Modern Gift. Rev Environ Contam Toxicol 220: 1–44.
3. Law KL, Moret-Ferguson S, Maximenko NA, Proskurowski G, Peacock EE, et al. (2010) Plastic accumulation in the North Atlantic subtropical gyre. Science 329: 1185–1188.
4. Moore CJ (2008) Synthetic polymers in the marine environment: A rapidly increasing, long-term threat. Environ Res 108: 131–139.
5. Schuyler Q, Hardesty BD, Wilcox C, Townsend K (2014) Global Analysis of Anthropogenic Debris Ingestion by Sea Turtles. Conserv Biol 28: 129–139.
6. Choy CA, Drazen JC (2013) Plastic for dinner? Observations of frequent debris ingestion by pelagic predatory fishes from the central North Pacific. Mar Ecol Prog Ser 485: 155–163.
7. Phillips RA, Ridley C, Reid K, Pugh PJA, Tuck GN, et al. (2010) Ingestion of fishing gear and entanglements of seabirds: Monitoring and implications for management. Biol Conserv 143: 501–512.
8. Cole M, Lindeque P, Halsband C, Galloway TS (2011) Microplastics as contaminants in the marine environment: A review. Mar Pollut Bull 62: 2588–2597.
9. van Sebille E, England MH, Froyland G (2012) Origin, dynamics and evolution of ocean garbage patches from observed surface drifters. Environ Res Lett 7.
10. Ryan PG, Moore CJ, van Franeker JA, Moloney CL (2009) Monitoring the abundance of plastic debris in the marine environment. Philos T Roy Soc B 364: 1999–2012.
11. Browne MA, Crump P, Niven SJ, Teuten E, Tonkin A, et al. (2011) Accumulation of microplastic on shorelines woldwide: Sources and sinks. Environ Sci Technol 45: 9175–9179.
12. Davey ME, O'Toole GA (2000) Microbial biofilms: from ecology to molecular genetics. Microbiol Mol Bio R 64: 847–867.
13. Battin TJ, Kaplan LA, Newbold JD, Hansen CME (2003) Contributions of microbial biofilms to ecosystem processes in stream mesocosms. Nature 426: 439–442.
14. Hall RO, Meyer JL (1998) The trophic significance of bacteria in a detritus-based stream food web. Ecology 79: 1995–2012.
15. Hoellein TJ, Tank JL, Rosi-Marshall EJ, Entrekin SA (2009) Temporal variation in substratum-specific rates of N uptake and metabolism and their contribution at the stream-reach scale. J N Am Benthol Soc 28: 305–318.
16. Johnson LT, Tank JL, Dodds WK (2009) The influence of land use on stream biofilm nutrient limitation across eight North American biomes. Can J Fish Aquat Sci 66: 1081–1094.
17. Hoellein TJ, Tank JL, Kelly JJ, Rosi-Marshall EJ (2010) Seasonal variation in nutrient limitation of microbial biofilms colonizing organic and inorganic substrata in streams. Hydrobiologia 649: 331–345.
18. Zettler ER, Mincer TJ, Amaral-Zettler LA (2013) Life in the "plastisphere": Microbial communities on plastic marine debris. Environ Sci Technol 47: 7137–7146.
19. Fitzpatrick FA, Diebel MW, Harris MA, Arnold TL, Lutz MA, et al. (2005) Effects of urbanization on the geomorphology, habitat, hydrology, and fish index of biotic integrity of streams in the Chicago area, Illinois and Wisconsin. Am Fish Soc Symp 47: 87–115.
20. Ivar do Sul JA, Costa MF (2007) Marine debris review for Latin America and the Wider Caribbean Region: From the 1970s until now, and where do we go from here? Mar Pollut Bull 54: 1087–1104.
21. Barnes DKA, Milner P (2005) Drifting plastic and its consequences for sessile organism dispersal in the Atlantic Ocean. Mar Biol 146: 815–825.
22. Araújo MCB, Costa MF (2006) The significance of solid wastes with land-based sources for a tourist beach: Pernambuco, Brazil. Pan-Am J Aquat Sci 1: 28–34.
23. Santos IR, Friedrich AC, Ivar do Sul JA (2009) Marine debris contamination along undeveloped tropical beaches from northeast Brazil. Env Monit Assess 148: 455–462.
24. Ribic CA (1998) Use of indicator items to monitor marine debris on a New Jersey beach from 1991 to 1996. Mar Pollut Bull 36: 887–891.
25. Madzena A, Lasiak T (1997) Spatial and temporal variations in beach litter on the Transkei coast of South Africa. Mar Pollut Bull 34: 900–907.
26. Claereboudt MR (2004) Shore litter along sandy beaches of the Gulf of Oman. Mar Pollut Bull 49: 770–777.
27. Seco Pon JP, Becherucci EM (2012) Spatial and temporal variations of urban litter in Mar del Plata, the major coastal city of Argentina. Waste Manage 32: 343–348.
28. Ivar do Sul JA, Santos IR, Friedrich AC, Matthiensen A, Fillmann G (2011) Plastic pollution at a sea turtle conservation area in NE Brazil: contrasting developed and undeveloped beaches. Estuar Coasts 34: 814–823.
29. Lee D-I, Cho H-S, Jeong S-B (2006) Distribution characteristics of marine litter on the sea bed of the East China Sea and the South Sea of Korea. Estuar Coast Shelf Sci 70: 187–194.

30. Hoppe PD, Rosi-Marshall EJ, Bechtold HA (2012) The antihistamine cimetidine alters invertebrate growth and population dynamics in artificial streams. Freshwater Sci 31: 379–388.
31. Bott TL (1996) Primary productivity and community respiration. In: Hauer FR, Lamberti GA, editors. Methods in Stream Ecology. 1st Edition ed. San Diego, California: Academic Press. pp. Pages 533–556.
32. Caporaso JG, Lauber CL, Walters WA, Berg-Lyons D, Huntley J, et al. (2012) Ultra-high-throughput microbial community analysis on the Illumina HiSeq and MiSeq platforms. ISME J 6: 1621–1624.
33. Schloss PD, Westcott SL, Ryabin T, Hall JR, Hartmann M, et al. (2009) Introducing mothur: Open-Source, Platform-Independent, Community-Supported Software for Describing and Comparing Microbial Communities. Appl Environ Microbiol 75: 7537–7541.
34. Kozich JJ, Westcott SL, Baxter NT, Highlander SK, Schloss PD (2013) Development of a dual-index sequencing strategy and curation pipeline for analyzing amplicon sequence data on the MiSeq Illumina sequencing platform. Appl Environ Microbiol 79: 5112–5120.
35. Edgar RC, Haas BJ, Clemente JC, Quince C, Knight R (2011) UCHIME improves sensitivity and speed of chimera detection. Bioinformatics 27: 2194–2200.
36. Yue JC, Clayton MK (2005) A similarity measure based on species proportion. Comm Stat Theory 34: 2123–2131.
37. Clarke KR, Green RH (1988) Statistical design and analysis for a 'biological effects' study. Mar Ecol Prog Ser 46: 213–222.
38. Shannon CE (2001) A mathematical theory of communication. Mobile Comp Comm Rev 5: 3–55.
39. Zbyszewski M, Corcoran PL (2011) Distribution and degradation of fresh water plastic particles along the beaches of Lake Huron, Canada. Water Air Soil Pollut 220: 365–372.
40. Williams AT, Simmons SL (1999) Sources of riverine litter: The River Taff, South Wales, UK. Water Air Soil Pollut 112: 197–216.
41. Nollkaemper A (1994) Land-based discharges of marine debris - from local to global regulation. Mar Pollut Bull 28: 649–652.
42. Armitage N, Rooseboom A (2000) The removal of urban litter from stormwater conduits and streams: Paper 1 – The quantities involved and catchment litter management options. Water SA 26: 181–187.
43. Andrandy AL (2011) Microplastics in the marine environment. Mar Pollut Bull 62: 1596–1605.
44. Walsh CJ, Roy AH, Feminella JW, Cottingham PD, Groffman PM, et al. (2005) The urban stream syndrome: current knowledge and the search for a cure. J N Am Benthol Soc 24: 706–723.
45. Puls J, Wilson SA, Holter D (2011) Degradation of cellulose acetate-based materials: a review. J Polym Env 19: 152–165.
46. Derraik JGB (2002) The pollution of the marine environment by plastic debris: a review. Mar Pollut Bull 44: 842–852.
47. Besemer K, Singer G, Limberger R, Chlup AK, Hochedlinger G, et al. (2007) Biophysical controls on community succession in stream biofilms. Appl Environ Microbiol 73: 4966–4974.
48. Rickard AH, McBain AJ, Ledder RG, Handley PS, Gilbert P (2003) Coaggregation between freshwater bacteria within biofilm and planktonic communities. FEMS Microbiol Lett 220: 133–140.
49. McNamara CJ, Leff LG (2004) Bacterial community composition in biofilms on leaves in a northeastern Ohio stream. J N Am Benthol Soc 23: 677–685.
50. Garrity GM, Boone DR, Castenholz RW (2001) Bergey's Manual of Systematic Bacteriology, Vol. 1: The Archea and the deeply branching and phototrophic bacteria: Springer, NY.
51. DeVos P, Garrity G, Jones D, Krieg NR, Ludwig W, et al. (2009) Bergey's Manual of Systematic Bacteriology, Vol. 3: The Firmicutes. NY: Springer.
52. Mhedbi-Hajri N, Jacques MA, Koebnik R (2011) Adhesion Mechanisms of Plant-Pathogenic Xanthomonadaceae. In: Linke D, Goldman A, editors. Bacterial Adhesion: Chemistry, Biology and Physics. pp. 71–89.
53. Brenner DJ, Krieg NR, Staley JTS (2005) Bergey's Manual of Systematic Bacteriology, Vol 2: The Proteobacteria. NY: Springer.
54. Ofek M, Hadar Y, Minz D (2012) Ecology of root colonizing massilia (Oxalobacteraceae). PLoS One 7: e40117.
55. Chung EJ, Park TS, Jeon CO, Chung YR (2012) Chitinophaga oryziterrae sp nov., isolated from the rhizosphere soil of rice (*Oryza sativa* L.). Int J Syst Evol Micr 62: 3030–3035.
56. Isanapong J, Goodwin L, Bruce D, Chen A, Detter C, et al. (2012) High-quality draft genome sequence of the opitutaceae cacterium strain TAV1, a symbiont of the wood-feeding termite *Reticulitermes flavipes*. J Bateriol 194: 2744–2745.
57. Chassard C, Delmas E, Robert C, Lawson PA, Bernalier-Donadille A (2012) Ruminococcus champanellensis sp nov., a cellulose-degrading bacterium from human gut microbiota. Int J Syst Evol Micr 62: 138–143.
58. Gregory MR (2009) Environmental implications of plastic debris in marine settings-entanglement, ingestion, smothering, hangers-on, hitch-hiking and alien invasions. Philos T Roy Soc B 364: 2013–2025.
59. Sheavly SB, Register KM (2007) Marine debris & plastics: Environmental concerns, sources, impacts and solutions. J Polym Env 15: 301–305.
60. Bowman D, Manor-Samsonov N, Golik A (1998) Dynamics of litter pollution on Israeli Mediterranean beaches: A budgetary, litter flux approach. J Coast Res 14: 418–432.

7

Kernel Density Surface Modelling as a Means to Identify Significant Concentrations of Vulnerable Marine Ecosystem Indicators

Ellen Kenchington[1]*, Francisco Javier Murillo[1,2], Camille Lirette[1], Mar Sacau[3], Mariano Koen-Alonso[4], Andrew Kenny[5], Neil Ollerhead[4], Vonda Wareham[4], Lindsay Beazley[1]

1 Bedford Institute of Oceanography, Department of Fisheries and Oceans, Dartmouth, Nova Scotia, Canada, **2** Departamento de Zooloxía e Antropoloxía Física, Facultade de Bioloxía, Universidade de Santiago de Compostela, Santiago de Compostela, Spain, **3** Instituto Español de Oceanografía, Centro Oceanográfico de Vigo, Programa de Pesquerías Lejanas, Vigo, Spain, **4** Northwest Atlantic Fisheries Centre, Department of Fisheries and Oceans, St. John's, Newfoundland and Labrador, Canada, **5** Centre for Environment, Fisheries and Aquaculture Science, Lowestoft, Suffolk, United Kingdom

Abstract

The United Nations General Assembly Resolution 61/105, concerning sustainable fisheries in the marine ecosystem, calls for the protection of vulnerable marine ecosystems (VME) from destructive fishing practices. Subsequently, the Food and Agriculture Organization (FAO) produced guidelines for identification of VME indicator species/taxa to assist in the implementation of the resolution, but recommended the development of case-specific operational definitions for their application. We applied kernel density estimation (KDE) to research vessel trawl survey data from inside the fishing footprint of the Northwest Atlantic Fisheries Organization (NAFO) Regulatory Area in the high seas of the northwest Atlantic to create biomass density surfaces for four VME indicator taxa: large-sized sponges, sea pens, small and large gorgonian corals. These VME indicator taxa were identified previously by NAFO using the fragility, life history characteristics and structural complexity criteria presented by FAO, along with an evaluation of their recovery trajectories. KDE, a non-parametric neighbour-based smoothing function, has been used previously in ecology to identify hotspots, that is, areas of relatively high biomass/abundance. We present a novel approach of examining relative changes in area under polygons created from encircling successive biomass categories on the KDE surface to identify "significant concentrations" of biomass, which we equate to VMEs. This allows identification of the VMEs from the broader distribution of the species in the study area. We provide independent assessments of the VMEs so identified using underwater images, benthic sampling with other gear types (dredges, cores), and/or published species distribution models of probability of occurrence, as available. For each VME indicator taxon we provide a brief review of their ecological function which will be important in future assessments of significant adverse impact on these habitats here and elsewhere.

Editor: Simon Thrush, University of Auckland, New Zealand

Funding: Funding was received from the International Governance Strategy (IGS) fund of the Department of Fisheries and Oceans to EK. The purpose of this fund is to strengthen scientific research in support of international governance of fisheries resources, support healthy ocean ecosystems, protect Canada's economic and environmental interests and position Canada as a global leader in high-seas issues. The funders had no role in study design, data collection and analysis, decision to publish, or preparation of the manuscript.

Competing Interests: The authors have declared that no competing interests exist.

* Email: Ellen.Kenchington@dfo-mpo.gc.ca

Introduction

The 2006 United Nations General Assembly Resolution 61/105 calls upon "States to take action immediately, individually and through regional fisheries management organizations and arrangements, and consistent with the precautionary approach and ecosystem approaches, to sustainably manage fish stocks and protect vulnerable marine ecosystems, including seamounts, hydrothermal vents and cold water corals, from destructive fishing practices, recognizing the immense importance and value of deep sea ecosystems and the biodiversity they contain".

The Food and Agriculture Organization (FAO) of the United Nations International Guidelines for the Management of Deep-sea Fisheries in the High Seas [1] provide general tools and considerations for the identification of vulnerable marine ecosystems (VMEs). They include a set of criteria that should be used, individually or in combination, for the identification process. Specifically:

"i. Uniqueness or rarity – an area or ecosystem that is unique or that contains rare species whose loss could not be compensated for by similar areas or ecosystems. These include:

- habitats that contain endemic species;
- habitats of rare, threatened or endangered species that occur only in discrete areas; or
- nurseries or discrete feeding, breeding, or spawning areas.

ii. Functional significance of the habitat – discrete areas or habitats that are necessary for the survival, function, spawning/reproduction or recovery of fish stocks, particular life history stages (e.g. nursery grounds or rearing areas), or of rare, threatened or endangered marine species.

iii. Fragility – an ecosystem that is highly susceptible to degradation by anthropogenic activities.

iv. Life-history traits of component species that make recovery difficult – ecosystems that are characterized by populations or assemblages of species with one or more of the following characteristics:

- slow growth rates;
- late age of maturity;
- low or unpredictable recruitment; or
- long-lived.

v. Structural complexity – an ecosystem that is characterized by complex physical structures created by significant concentrations of biotic and abiotic features. In these ecosystems, ecological processes are usually highly dependent on these structured systems. Further, such ecosystems often have high diversity, which is dependent on the structuring organisms." [1] The Northwest Atlantic Fisheries Organization (NAFO) is the regional fisheries management organization with regulatory competence over the fishing activities that take place in international waters in the northwest Atlantic. NAFO has reviewed the invertebrate catch from depth-stratified random research vessel trawl sets, undertaken in its regulatory area, against the FAO guidelines [2–5]. They identified large-sized sponges, large and small gorgonian corals, sea pens, erect bryozoans, sea squirts, crinoids (sea lilies) and cerianthid anemones as VME indicator species from a wider list of approximately 500 taxa residing in the area [4]. All qualified under the Structural Complexity criterion (see *v* above) in combination with their Fragility and Life-history traits (see *iii* and *iv* above respectively). The guidelines do not explicitly define the distinction between a VME and a VME indicator species/taxon, although it is clear that a single occurrence does not constitute a VME, nor does the full distribution of a species/taxon [1]. However, under the Structural Complexity criterion the term "significant concentration" is used to identify the level of aggregation which is expected, even though it is given without an operational context.

In landscape ecology, such hotspots are often detected with a spatially global threshold determined through comparison of the value for a given observation with locations in the neighbourhood of the observation, in order to incorporate an explicit consideration of space. We propose a novel method to identify "significant concentrations" of benthic structure-forming VME indicators using a global threshold determined from geospatial analyses. Our approach serves to objectively separate dense aggregations that form habitats from the broader distribution of the species in question, and so serves to distinguish VMEs proper (i.e., structurally complex habitats) from single/low frequency occurrences of VME indicator species. We apply our method to four of the highly aggregating structure-forming megafaunal groups identified as VME indicator species/taxa by NAFO - specifically to large-sized sponges, sea pens, and small and large gorgonian corals. Our approach does not address questions of impairment of ecological function which ultimately are important in assessing significant adverse impacts on these habitats [1], nor does it attempt to predict biomass or occurrence in unsampled areas. It is particularly relevant to the analyses of VME-indicator catch from research vessel survey data series on continental shelves and slopes. Modeled outputs are ground-truthed using data reported from independent surveys and published species distribution analyses predicting the probability of occurrence of the VME indicator taxa based on a suite of environmental variables.

Materials and Methods

Study Area

The high seas study area in the northwest Atlantic is bounded by the Canadian exclusive economic zone (EEZ) in the west and by the 2500 m depth contour along the continental slope. This includes the Nose and Tail of Grand Bank, east of Newfoundland, Canada, and the Flemish Cap, a plateau of roughly 200 km radius at the 500 m isobath and separated from Grand Bank by the Flemish Pass which reaches depths of 1200 m (Figure 1). The study area falls within the 3LMNO management Divisions of NAFO and is referred to as the NAFO Regulatory Area (NRA).

Data

The study area is typically sampled on an annual basis by research vessel (RV) bottom trawl surveys carried out for the assessment of fish stocks by Canada and the European Union (Spain). This study did not involve human participants, specimens or tissue samples. All of the animals in this study are invertebrates and so do not require an ethics statement. Fish were caught on the same surveys but did not form a part of this study. All scientific field surveys conducted in the NAFO Regulatory area are approved by the NAFO Scientific Council on an annual basis. All of the surveys carried out for this study followed this procedure. For future surveys applicants should contact the Chair of the NAFO Scientific Council. The name and contact information of the Chair is available on the NAFO Website: http://www.nafo.int/.

All surveys follow a depth-stratified random sampling design to 1500 m optimized for the target species with vessel speeds of approximately 3 knots, however there are differences in the type and size of the gears used and the duration of the trawl sets (Table 1).

Collectively these surveys provided 2593 geo-referenced records of sponge biomass and 1478 records of coral biomass distributed as illustrated in Figure 2. Each of these records was grouped into one of the broader VME indicator species/taxa: large-size sponges, sea pens, small gorgonian corals or large gorgonian corals. Sponges were not consistently identified to species level in the surveys, however all of the large catches were comprised of massive sponges of the Families Geodiidae and Ancorinidae [6]. Sea pen catches (Cnidarian Order Pennatulacea) were dominated by two whip-like species *Anthoptilum grandiflorum* and *Halipteris finmarchica*, and the smaller fleshy *Pennatula aculeata* [7,8] though 11 species have been identified [9]. Small gorgonians coral catches (Cnidarian Order Alcyonacea) were predominately of the species *Acanella arbuscula*, although other species with similar environmental requirements, such as *Radicipes gracilis* (Cnidarian Order Alcyonacea) were included; the large gorgonian corals (Cnidarian Order Alcyonacea) comprised *Acanthogorgia* spp., *Keratoisis* spp., *Paragorgia arborea, P. johnsoni, Paramuricea* spp., and *Primnoa resedaeformis* [9].

Data Treatment

The data were drawn from three different combinations of gear type and trawling duration (Table 1). All catch weight distributions were highly right-skewed due to large numbers of small catches with few very large catches (Figure 3) as is characteristic of highly aggregating species. When spatially visualized the large

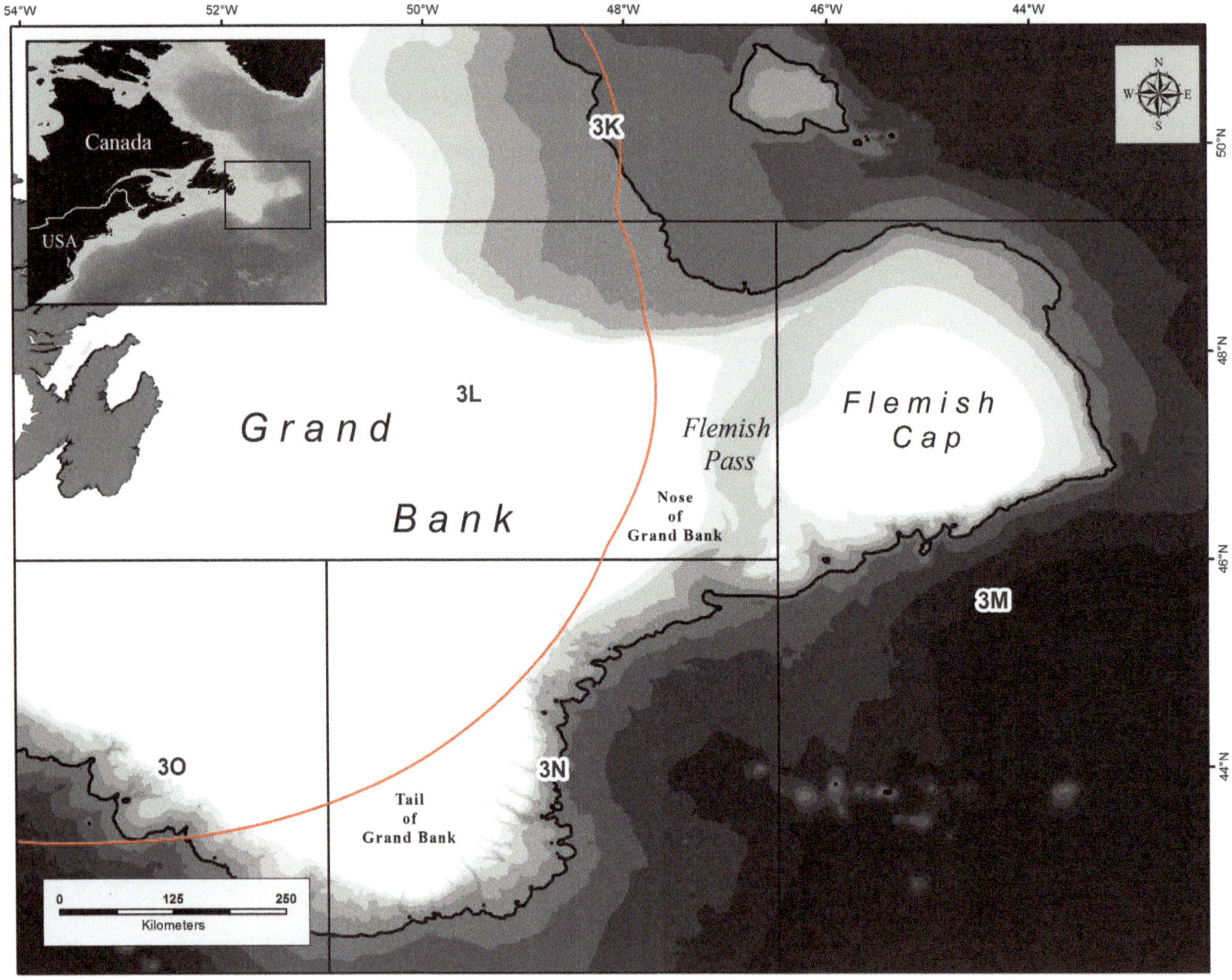

Figure 1. Location of the study area in the international waters east of Newfoundland and Labrador, Canada. Red line indicates the Canadian Exclusive Economic Zone. Heavy black line indicates the 2500 m depth contour. NAFO management divisions are indicated.

catches typically co-occurred in close proximity, whereas the smaller catches were broadly distributed.

To assess whether the different survey data should be used separately or in combination for each VME indicator taxon, we applied non-parametric statistics to the catch biomass from each of the three gear/duration data sets using all of the biomass data and only those data above arbitrary weight classes. Because each of the surveys covered different and largely non-overlapping spatial extent (Figure 2), natural differences in biomass are confounded in this approach and could enhance or mask differences between

Table 1. Details of research vessel survey data of coral and sponge biomass for the study area.

Programme	Period	NAFO Division	Trawl gear type	Mesh size in codend liner (mm)	Trawl duration (min)	Average wingspread (m)
Spanish 3NO Survey (IEO)	2002–2013	3NO	Campelen 1800	20	30	24.2–31.9
EU Flemish Cap Survey (IEO, IIM, IPIMAR)	2003–2013	3M	Lofoten	35	30	13.89
Spanish 3L Survey (IEO)	2003–2013	3L	Campelen 1800	20	30	24.2–31.9
DFO NL Multi-species Surveys (DFO)	1995–2012	3LNO	Campelen 1800	12.7	15	15–20

Notes: EU, European Union; DFO, Department of Fisheries and Oceans; NL, Newfoundland and Labrador; IEO, Instituto Español de Oceanografía; IIM, Instituto de Investigaciones Marinas; IPMA, Instituto Português do Mar e da Atmosfera.

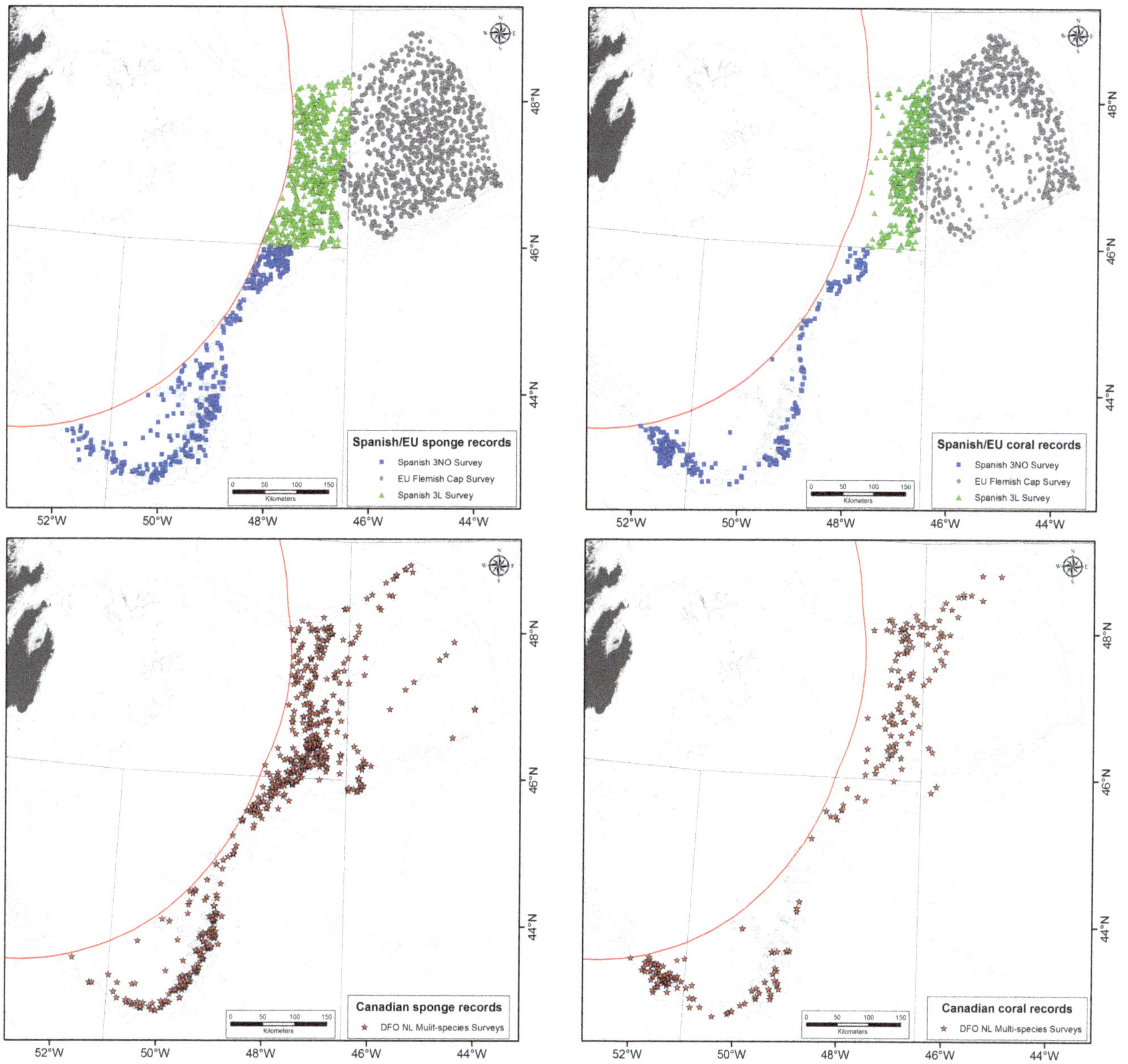

Figure 2. Location of trawl sets for the research vessel surveys in the NAFO Regulatory Area (Divs. 3LMNO) used for analyses (Table 1). Left panels: Sponges; Right Panels: Corals.

gears and/or trawl duration. This effect is most likely to affect tests involving gear comparison as the Lofoten gear was used almost exclusively on the Flemish Cap while the Campelen trawls were deployed in the Flemish Pass and on Grand Bank. Comparison of trawl duration effects between the surveys using Campelen gear cover a similar spatial extent and so are less likely to be confounded by geospatial differences in catchability (Figure 2).

Two-sample Kolmogorov-Smirnov tests using only data collected with the Campelen gear were performed to evaluate the influence of trawl duration (15 min *vs.* 30 min), a proxy of trawl length. When no significant differences were found, the data sets were combined and tested against the Lofoten gear data, to evaluate the influence of the gear type. Further, graphs of trawl length, calculated from vessel speed and duration, vs. VME indicator taxon catch weight, and of the cumulative frequency distributions were used to visualize these data.

Kernel Density Estimation and Selection of Aggregation Thresholds

The non-parametric kernel density algorithm has been increasingly used in ecology to identify home ranges from tagging data [10–12] and for identifying hotspots within a landscape [13,14] using a neighbour-based approach. For the later application, ecological hotspots (that is areas of high abundance or biomass), are defined relative to the spatial distribution of the data and are detected through the application of a spatially-defined threshold [13]. We used kernel density estimation (KDE) to create a modelled biomass surface for each VME indicator taxon using the start positions of each RV trawl set. A quadratic (Epanechnikov) kernel density function was used to fit a smooth curve over each data point in ArcGIS 10 (ESRI Canada Limited, Toronto, Ontario) using the UTM projected coordinate system North

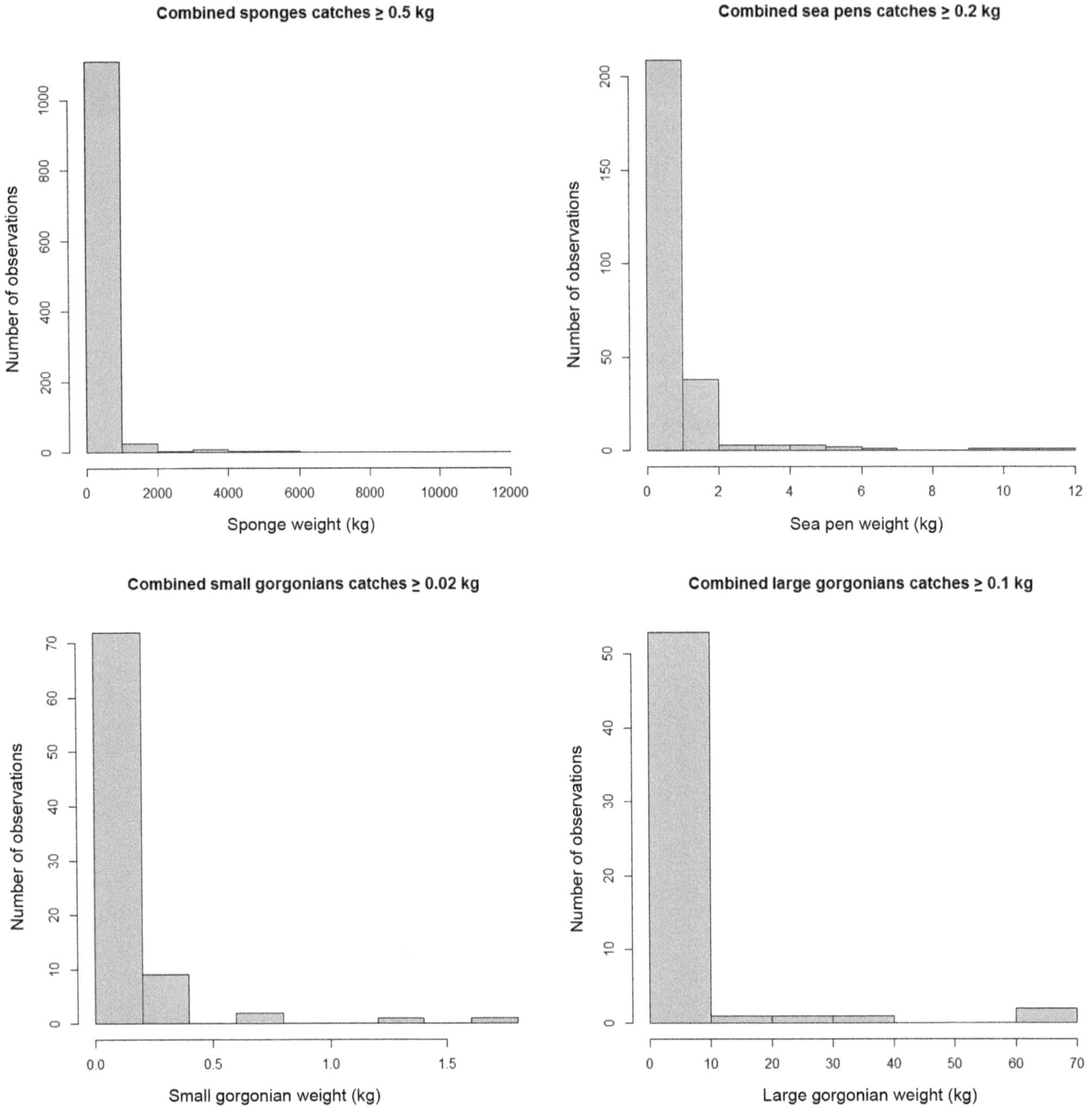

Figure 3. Catch weight distribution for all non-zero data of sponges (upper left), sea pens (upper right), small gorgonian corals (lower left) and large gorgonian corals (lower right).

American Datum 1983 Zone 23. We used an optimum search radius based on the tool's default calculation which is the shortest of the width or height of the output extent in the output spatial reference, divided by 30. In all cases the width was the shortest extent and the search radii were 22.6 km for sponges and sea pens, 22.1 km for small gorgonian corals and 16.6 km for large gorgonian corals. The surface value is highest at the location of the point and decreases outwards in all directions to reach zero at the search radius distance to define a circular neighbourhood for each point observation. Output cell size (resolution) was also based on the tool's default which is the shorter of the width or height of the output extent, divided by 250. Again using the width of the spatial extent the output cell size was ~2.7 km for sponges, sea pens and small gorgonian corals and 2.0 km for large gorgonian corals. The density at each output raster cell was calculated by adding the values of all the kernel surfaces within each cell.

Once the kernel density surface was produced, it was used to estimate the area covered by the original data points at or above selected biomass threshold levels along the kernel contours. This calculation was repeated for successively decreasing biomass threshold values in order to generate threshold-area curves; these curves allowed for identification of threshold levels associated with natural changes in the pattern of spatial aggregation in the data [7,8,15]. The polygons determining that threshold were then superimposed over the original kernel surface to differentiate the hot and non-hot spots.

The structure-forming VME indicators examined here are all aggregating taxa forming habitats commonly referred to as sponge grounds, sea pen fields and coral forests. Consequently we expected to see biomass hotspots appearing as tightly defined spatial units in the overall landscape with sharp discontinuities between neighbouring hotpots.

Typically, for these VME indicator species/taxa that form habitats through dense aggregations, the threshold-area curves initially show a slow increase in total area as the threshold values decrease. This slow increase in area reflects the fact that the arbitrary thresholds keep "mapping out" the areas that contain the dense aggregations of VME indicator species/taxa (i.e., better delineating the areas of high density, where density may decrease near their boundaries, while also by starting to incorporate smaller new aggregation areas with relatively lower densities). After this initial "phase" of slow increase in area, the threshold-area curves show a rapid and sharp increase in area as the thresholds keep decreasing; this rapid increase in area is associated with thresholds values that are beginning to capture isolated/non-aggregating individuals of the VME indicator species/taxa. Finally, as the thresholds reach its lowest value, the area covered often stabilizes again, reflecting the entire distribution of the VME indicator species/taxa in the study area.

Identification of biomass thresholds representing "significant concentrations" of VME indicator species/taxa correspond to thresholds demarcating the phase of rapid increase in area. When interpreting the catch weight defining the significant concentrations a number of criteria are simultaneously considered: 1) identification of the catch biomass which show the largest change in area after the initial establishment of the aggregations; 2) consideration of the number of data points contributing to those changes in area between successive catch thresholds; 3) examination of the spatial relationship of the polygons created by biomass thresholds greater and lesser than the potential threshold using GIS, and 4) the position of the new data points relative to previously established polygons. These two last criteria are the spatial component to criterion 2 and are necessary as polygon area can increase by the joining of two or more high density polygons. If this occurs the evidence for connecting the areas (i.e., number of points between the smaller areas) is reviewed. In this instance the threshold is considered to be valid when there is an increase in area through a reasonable number of widely spaced data points. Cases for rejecting the threshold other than insufficient data include: 1) joining of smaller polygons with little evidence for a continuous distribution within the newly formed area; 2) a gradual increase in area with every new polygon added, creating a situation where no one successive change in area is especially larger or smaller than others (this indicates that there is no aggregation); 3) an increase in area established by creation of new areas of very low density; and 4) no large increase in area.

Independent Assessments of Modelled Outputs from Published Sources

Independent data, useful for validating the modelled results, have been collected through NEREIDA (http://www.nafo.int/science/frames/nereida), a multidisciplinary research programme operating within the study area. Data from NEREIDA rock dredges, box corers and underwater imagery [16] offer an opportunity to validate modelled results. In particular, data from nine *in situ* benthic imagery transects conducted in the Sackville Spur and western Flemish Cap slope/Flemish Pass areas in 2009 and 2010 have been processed [17]. These were primarily in areas predicted to have significant concentrations of sponges identified using the above approach (i.e., sponge grounds) [16–18] and were complemented by collection of box core samples [19].

Species distribution models (SDMs) use environmental data to predict the probability of occurrence of a species or habitat. SDMs using random forest models have been applied to the presence/absence of sponge grounds, sea pens and large gorgonian corals in the study area [20,21]. These drew on the same response data sources used in our analyses and so are not fully independent; however they have, through their use of the environmental predictors, capacity to extrapolate between point observations in unsampled areas or where data are sparse.

Results

Analyses of the catch weight distributions between trawl durations (length) and gear types indicated that catch weights of sponges ≥0.5 kg, of sea pens ≥0.02 kg, of small gorgonian corals ≥0.1 kg and of large gorgonian corals ≥0.1 kg did not differ significantly between trawl duration of 15 min and of 30 min (Table 2). This is reinforced by the scatterplots presented in Figure 4 that show that above those thresholds, there is no relationship between the catch weight and length of tow. This is consistent with the aggregated distribution of the VME indicator taxa and the need for alignment between the trawl path and the maximum dimension of the aggregation to create a linear relationship. Further, except for the small gorgonian corals, the Canadian records are all above 0.01 kg catch, whereas Spanish/EU data record values to 0.001 kg; this could be due to different sampling protocols or scale precision. All of these weight thresholds for combining the data represent only a few individuals or less of each VME indicator taxon at their lowest values.

With the exception of the sea pens, none of the catch weights of the VME indicator taxa differed between gear type (Campelen and Lofoten) at the thresholds established for combining the data based on trawl duration (Table 2), justifying combining all of the data above those catch weights for further analyses. Sea pen catches ≥0.2 kg showed no significant difference between the types of gear used (Table 2), and so data records less than 0.2 kg were discarded.

For the small gorgonian corals the data reduction incurred through the above analyses (combination of data ≥0.1 kg catch weight), drastically reduced the number of point observations available for KDE analysis to less than 10% of the total small gorgonians records ($N = 36$; Table 2). In order to maximize the use of the data while respecting the differences in catch weight distribution amongst the surveys, we ran two separate KDE analyses: one model for the Flemish Cap collected with Lofoten gear ($N = 145$) and another for a subset of the spatial extent surveyed with the Campelen gear on the Tail of Grand Bank (3NO Divs.) combining survey catches for that area that were ≥ 0.02 kg ($N = 85$; $KS = 0.289$; $p = 0.058$). Similar concerns over the number of point observations for large gorgonian corals could not be addressed through this approach; however that taxon had a smaller overall spatial extent of observations.

Sponges

A KDE surface of sponge biomass was modelled from the 1154 sponge records ≥0. 5 kg (Figure 5) and the areas under polygons of decreasing weight thresholds were calculated. Between 75 kg and 50 kg polygon area increased 1.25 times from 22,439 km^2 to 28,112 km^2 through the addition of 27 new points, 13 of them in new areas outside of the polygons defined by the 75 kg threshold. Consequently, the threshold value of 75 kg meets the assessment

Table 2. Kolmogorov-Smirnov (KS) two-sample tests of the similarity of the catch weight distribution for each VME taxon testing for effects of trawl duration (15 min vs. 30 min) and gear type (Campelen vs. Lofoten trawls) (see Table 1).

VME Indicator Taxon	Comparison Groups (Samples)	Number of Records (N) for Each Test Sample for a Given Catch Biomass Threshold	KS	P
Sponges	Campelen 15 min trawl *vs.* Campelen 30 min trawl	>0 kg ($N_{Canada}=553$, $N_{EU\text{-}Spain}=1024$)	0.288	<0.001*
		≥0. 1 kg ($N_{Canada}=491$, $N_{EU\text{-}Spain}=640$)	0.142	<0.001*
		≥0. 5 kg ($N_{Canada}=391$, $N_{EU\text{-}Spain}=439$)	0.069	0.279
		≥1 kg ($N_{Canada}=339$, $N_{EU\text{-}Spain}=354$)	0.055	0.683
		≥10 kg ($N_{Canada}=137$, $N_{EU\text{-}Spain}=136$)	0.131	0.193
	Combined Campelen trawls *vs.* Lofoten trawl	≥0. 5 kg ($N_{Campelen}=830$, $N_{Lofoten}=324$)	0.065	0.284
		≥1 kg ($N_{Campelen}=693$, $N_{Lofoten}=255$)	0.059	0.5443
		≥10 kg ($N_{Campelen}=273$, $N_{Lofoten}=97$)	0.137	0.138
Sea Pens	Campelen 15 min trawl *vs.* Campelen 30 min trawl	>0 kg ($N_{Canada}=183$, $N_{EU\text{-}Spain}=489$)	0.279	<0.001*
		≥0. 01 kg ($N_{Canada}=182$, $N_{EU\text{-}Spain}=350$)	0.172	0.002*
		≥0.02 kg ($N_{Canada}=146$, $N_{EU\text{-}Spain}=299$)	0.099	0.288
		≥0.05 kg ($N_{Canada}=105$, $N_{EU\text{-}Spain}=196$)	0.094	0.583
		≥0.1 kg ($N_{Canada}=72$, $N_{EU\text{-}Spain}=118$)	0.119	0.550
		≥0.2 kg ($N_{Canada}=35$, $N_{EU\text{-}Spain}=61$)	0.207	0.299
	Combined Campelen trawls *vs.* Lofoten trawl	≥0.02 kg ($N_{Campelen}=445$, $N_{Lofoten}=436$)	0.170	<0.001*
		≥0.05 kg ($N_{Campelen}=301$, $N_{Lofoten}=324$)	0.199	<0.001*
		≥0.1 kg ($N_{Campelen}=190$, $N_{Lofoten}=239$)	0.218	<0.001*
		≥0.2 kg ($N_{Campelen}=96$, $N_{Lofoten}=166$)	0.160	0.087
		≥0.5 kg ($N_{Campelen}=32$, $N_{Lofoten}=71$)	0.134	0.822
Small Gorgonian Corals	Campelen 15 min trawl *vs.* Campelen 30 min trawl	>0 kg ($N_{Canada}=87$, $N_{EU\text{-}Spain}=172$)	0.483	<0.001*
		≥0. 01 kg ($N_{Canada}=83$, $N_{EU\text{-}Spain}=81$)	0.360	<0.001*
		≥0.02 kg ($N_{Canada}=47$, $N_{EU\text{-}Spain}=55$)	0.271	0.049*
		≥0.05 kg ($N_{Canada}=27$, $N_{EU\text{-}Spain}=25$)	0.430	0.017*
		≥0.1 kg ($N_{Canada}=19$, $N_{EU\text{-}Spain}=10$)	0.395	0.259
	Combined Campelen trawls *vs.* Lofoten trawl	≥0.1 kg ($N_{Campelen}=29$, $N_{Lofoten}=7$)	0.374	0.408
Large Gorgonian Corals	Campelen 15 min trawl *vs.* Campelen 30 min trawl	>0 kg ($N_{Canada}=42$, $N_{EU\text{-}Spain}=75$)	0.413	<0.001*
		≥0. 01 kg ($N_{Canada}=42$, $N_{EU\text{-}Spain}=44$)	0.352	0.010*
		≥0.02 kg ($N_{Canada}=29$, $N_{EU\text{-}Spain}=38$)	0.331	0.054
		≥0.05 kg ($N_{Canada}=21$, $N_{EU\text{-}Spain}=28$)	0.441	0.019*
		≥0.1 kg ($N_{Canada}=13$, $N_{EU\text{-}Spain}=27$)	0.171	0.960
	Combined Campelen trawls *vs.* Lofoten trawl	≥0.1 kg ($N_{Campelen}=40$, $N_{Lofoten}=18$)	0.292	0.242

Asterisks indicate significant differences at $\alpha = 0.05$.

criteria and can be used to approximate the limits of the sponge grounds (that is, significant concentrations of sponges, equating that density to a VME proper) (Figure 6). The 75 kg polygon area encloses 97.87% of the total sponge biomass recorded from this area and represents 19.73% of the sampled area, confirming high concentration of sponge biomass.

In situ benthic camera surveys and box core sampling have supported the occurrence of significant concentrations of sponges in Flemish Pass [18] and on Sackville Spur [16,17,19] (Figure 7) as well as on the northeast slope of Flemish Cap [17]. SDMs have recently been applied to predict the probability of occurrence of sponge grounds in the study area [20] and show excellent concordance with the location of the significant concentrations

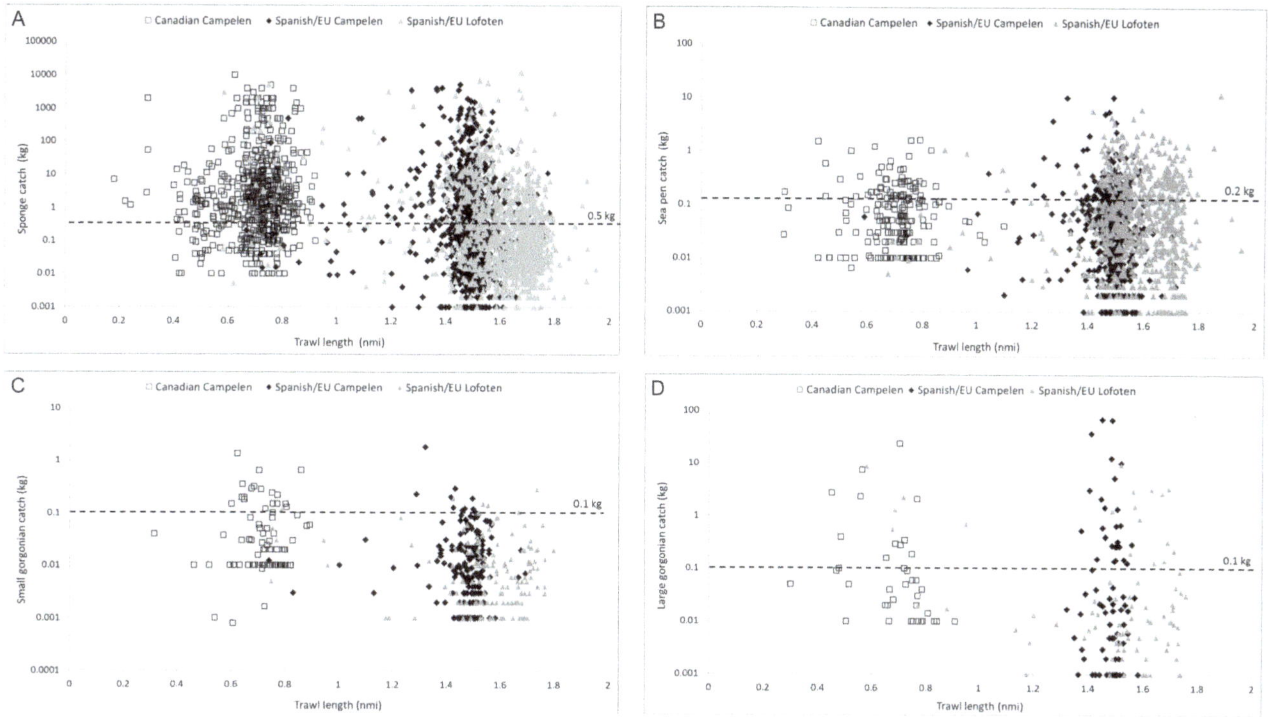

Figure 4. Catch weight (kg) of each VME indicator taxon in relation of trawl length (nmi) for each of the surveys indicated in Table 1 (Canadian surveys with Campelen gear, Spanish/EU surveys with Campelen gear and Spanish/EU surveys with Lofoten gear). The dashed line indicates the catch value above which the distributions are not significantly different (Table 2). Note that the y axis is in logarithmic scale. A) Sponges. B) Sea pens. C) Small gorgonian corals. D) Large gorgonian corals.

identified through our approach. Sponge grounds in the study area were primarily observed and predicted to exist in areas with high (>0.1 m/s) maximum bottom current [20].

Sea Pens

A KDE surface of sea pen biomass was modelled from 261 sea pen records ≥0.2 kg (Figure 8). Similar to the sponges, the area occupied by the highest catches is relatively constant, after an

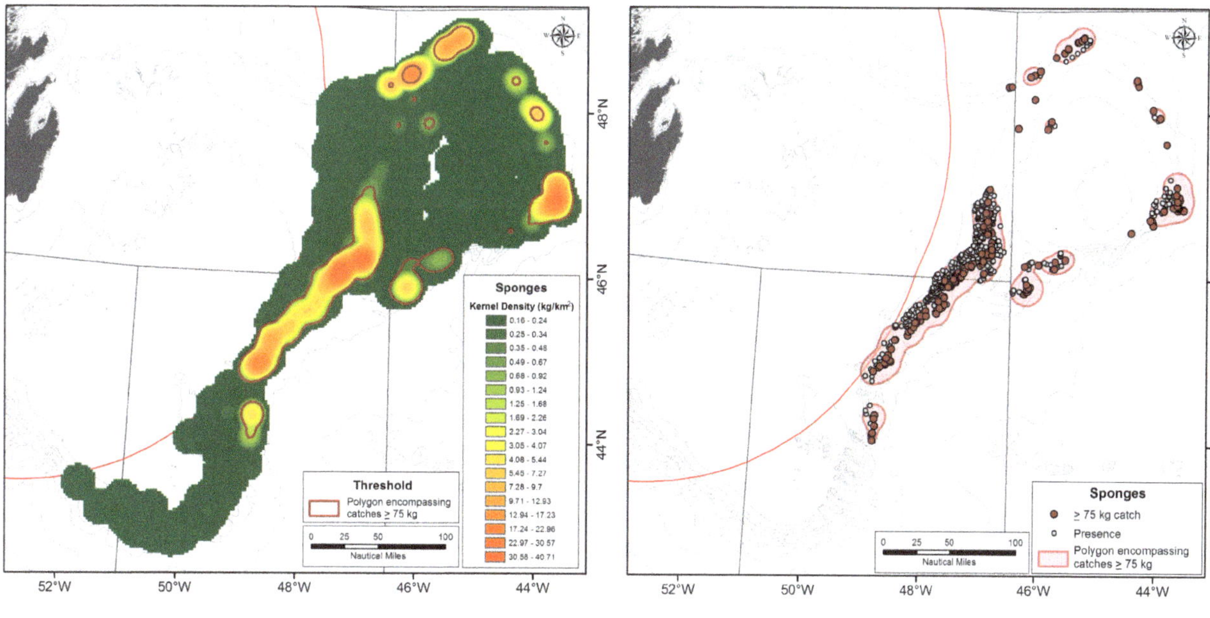

Figure 5. Left panel: Kernel density distribution of sponges in the NAFO Regulatory area with the 75 kg density polygons defining significant concentrations determined from aerial expansion thresholds superimposed in red. The green areas represent low sponge densities while the red areas indicate high sponge densities. Right panel: The location of catches greater than 75 kg (red circle) and smaller sponge catches (open circles) within the 75 kg density polygons defining the sponge ground VMEs.

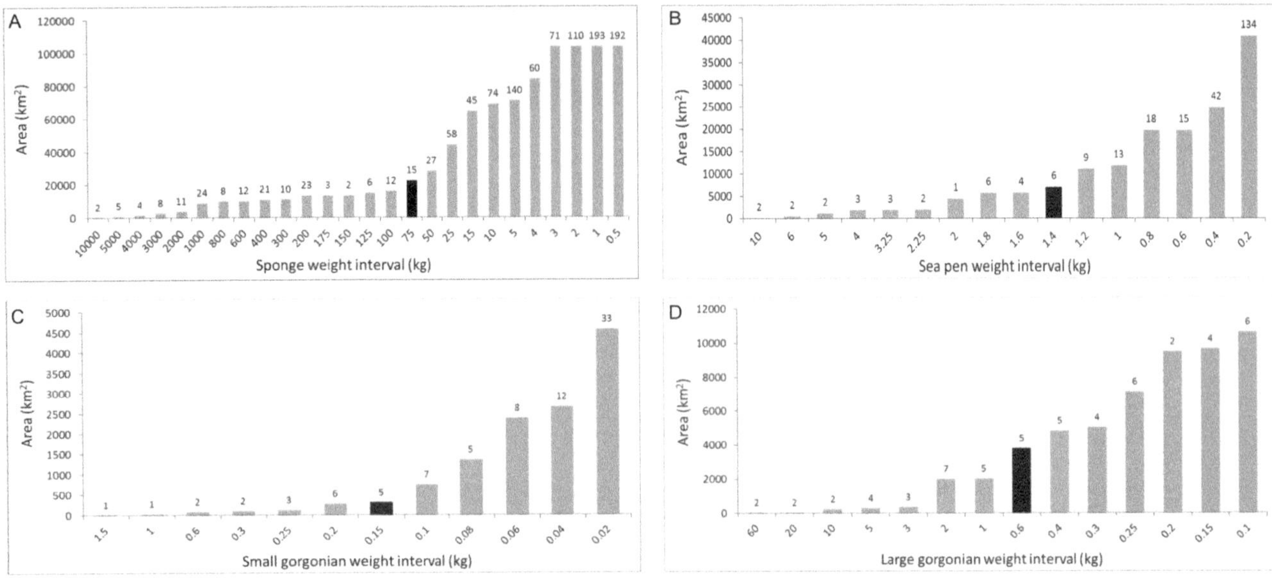

Figure 6. Histograms of the area occupied by successive weight thresholds of VME Indicator taxa. The numbers of additional point observations (over the preceding higher weight threshold) used to create each polygon are indicated above the each bar. A) Sponges. B) Sea pens. C) Small gorgonian corals. D) Large gorgonian corals.

initial increase, until the 2.25 kg interval. The area shows a first increase between 2.25 and 2 kg intervals, but this is created through the inclusion of a datum. After this, the area occupied is relatively constant until 1.4 kg. Between the 1.4 and 1.2 kg intervals the polygon area increases 1.58 times from 6,983 km^2 to 11,050 km^2 and meets the criteria outlined above for identification of a "significant concentration" threshold. The 1.4 kg polygon area encloses the 59.28% of the total sea pen biomass recorded and this area represents 5.10% of the sampled area (Figure 6).

Rock dredge data collected under the NEREIDA programme independently confirmed the presence of sea pens in the areas of significant concentrations identified above. There were no underwater imagery stations coincident with the significant concentrations of sea pens in the NRA, however, congruence between *in situ* observations of sea pen fields and locations of high research vessel trawl catches has been observed in the nearby Laurentian Channel [8] (Figure 7). The SDMs produced for sea pens in the study area show agreement with the kernel density surface. Both highlight a horse-shoe shaped distribution on Flemish Cap and significant concentrations and areas of high probability of occurrence on the Tail of Grand Bank, particularly adjacent to the Canadian EEZ in NAFO Div. 3O [21].

Small Gorgonian Corals

Two KDE surfaces for small gorgonian coral biomass were produced, one for the Flemish Cap and one for the Tail of Grand Bank as described above. For the Flemish Cap, the majority of the records were small catches (≤0.01 kg) with only 1 catch greater than 0.2 kg and 6 greater than 0.1 kg. These small catches were not highly aggregated and the analyses of the area occupied by successive density polygons supported that observation, as no clear threshold emerged with sufficient support following the criteria outlined above for identification of significant concentrations.

For the Tail of Grand Bank, the assessment criteria applied to the KDE surface identified 0.15 kg as the threshold for defining significant concentrations of small gorgonian corals (Figure 6). When superimposed on the KDE surface (Figure 9), the 0.15 kg density polygon captures all of the highest density areas (red colour on Figure 9) from the kernel analysis. Review of the data surrounding these polygons showed that three of the areas (marked with arrows on Figure 9) are based on single records with null records surrounding them. These areas require further data to determine the spatial extent of the concentrations.

Although small, the highly-branched nature of small gorgonian corals such as *Acanella arbuscula* (Figure 7), and their ability to form small but dense aggregations deems them habitat-forming, especially in areas of low topographical relief [22]. Rock dredge samples collected through the NEREIDA programme for the most part were not co-located with areas of significant concentrations of small gorgonian corals except for two locations on the Tail of Grand Bank. Both of those samples contained small gorgonian corals with high relative abundance compared to other rock dredge samples with small gorgonian coral presence. SDMs have not yet been published for this taxon.

Large Gorgonian Corals

Large gorgonian corals found in the study area (Figure 7) are very fragile and their representation in the catch is most often in the form of coral fragments rather than whole colonies. The KDE distribution identified large gorgonian coral catches in Flemish Pass, on Beothuk Knoll and on the southeastern corner of Flemish Cap (Figure 10). The 0.6 kg/RV tow density threshold emerged as defining significant concentrations (Figure 6) and when superimposed on the kernel density surface (Figure 10), it can be seen that all of the highest density areas (red colour on Figure 10) from the kernel analysis and other smaller catches are found within the defining polygons. As for the small gorgonian corals, four of the areas are based on single records. Of these four, the catch closest to the area of highest densities in the western section of the Flemish Cap, at around 500 m depth, is surrounded by other large gorgonian catches below the threshold and likely represents a significant concentration, especially considering the

A
B
C
D
E
F

Figure 7. Underwater images were taken from the locations of large research vessel trawl catches of (A) Geodid-dominated sponge grounds from the northeast Flemish Cap in the NAFO Regulatory area (NRA), (B) sea pen fields in the Laurentian Channel south of the NRA dominated by species of *Pennatula*, a smaller flesher sea pen, (C) sea pen fields in the Laurentian Channel south of the NRA dominated by the taller whip-like species *Halipteris finmarchica*, (D) large gorgonian corals in the Laurentian Channel south of the NRA (*Keratoisis* sp.), (E) large gorgonian corals from the southern wall of Flemish Pass in the NRA (*Paragorgia* sp.), (F) *Acanella arbuscula* from the Gully Marine Protected Area on the Scotian Shelf showing its dense branching architecture.

fragility of these taxa. However, the other three areas identified as significant concentrations (north Flemish Pass and slope of the Tail of the Grand Bank) are surrounded with null records and more data should be collected to resolve the spatial extent of those aggregations.

The SDM [21] predicts large gorgonian corals to be present with high probability in the area of significant concentrations of large gorgonian corals in Flemish Pass identified using the kernel density analyses. However, the area of highest probability of occurrence is along the deep, south and eastern slopes of Flemish Cap where there is little data available from the surveys. The SDM also predicts that large gorgonian corals will be present in the deeper water of the Sackville Spur area. Although underwater cameras have confirmed the presence of large gorgonian corals in the eastern portion of that area (Figure 7) they did not appear in the western transects [17]. Although the habitat may be suitable on Sackville Spur the dense sponge grounds in this area may out compete the coral.

Discussion

This study has demonstrated that kernel density analysis is useful for identifying "significant concentrations" [*sensu* 1] or hotspots from point observations of VME indicator taxon biomass from research vessel survey catches. Where independent data sources could be used to validate the presence of VMEs in the areas so identified, they consistently confirmed the modelled results.

We chose a kernel density estimation approach, over interpolation, to identify the location of biomass hotspots from a known distribution, as this method enhances habitat edges or boundaries [13] and is non-parametric. Density estimation counts the number of discrete objects in a given area and creates a field from those objects, whereas interpolation estimates the data values for points between the data. Geostatistical interpolation methods of producing biomass surfaces (e.g., kriging) require continuous and normally distributed data, while exact deterministic interpolation methods (e.g., inverse distance weighted (IDW)) dramatically underestimate areas of high biomass in right-skewed data. The highly aggregated nature of the VME indicator taxa results in a large number of zero catches and high variability in the positive catches, making it difficult to model an appropriate error term with alternative methods for producing biomass surfaces such as GLM/GAM which can give both prediction and uncertainty surfaces. Where such models have been applied to research vessel trawl-caught abundance of corals and sponges in the Pacific northwest [23], the best model fits explained between 20 and 25% of the deviance and produced low R^2 (0.04–0.21). Interestingly, in that study, the location variable (latitude, longitude) was significant in all models and accounted for ~50% of the explanatory power of most models; removing that term eroded model performance whilst including it prevented extrapolation outside of the spatial extent of the data. This dependence on location likely arises from spatial autocorrelation due to the high degree of aggregation in these species and diminishes the value that such models might have over simpler density estimation approaches. Simple plotting

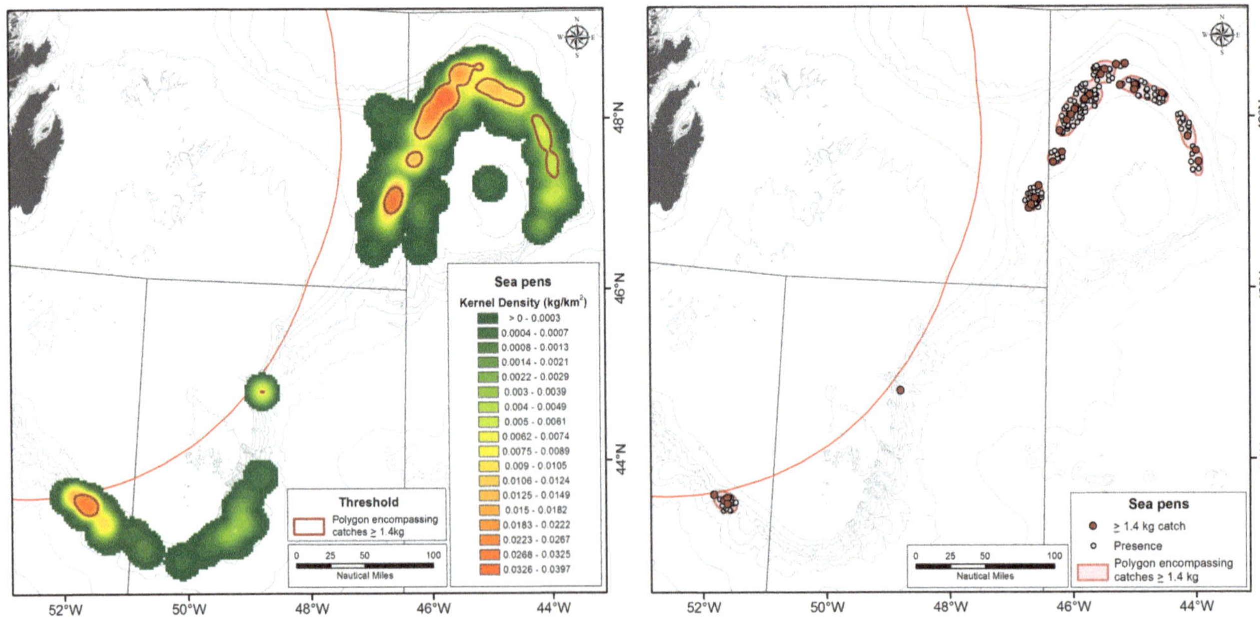

Figure 8. Left panel: Kernel density distribution of sea pens in the NAFO Regulatory area with the 1.4 kg density polygons defining the sea pen field VMEs superimposed in red. The green areas represent low sea pen densities while the red areas indicate high sea pen densities. Right panel: The location of catches greater than 1.4 kg (red circle) and smaller sea pen catches (open circles) within the 1.4 kg density polygons defining the sea pen field VMEs.

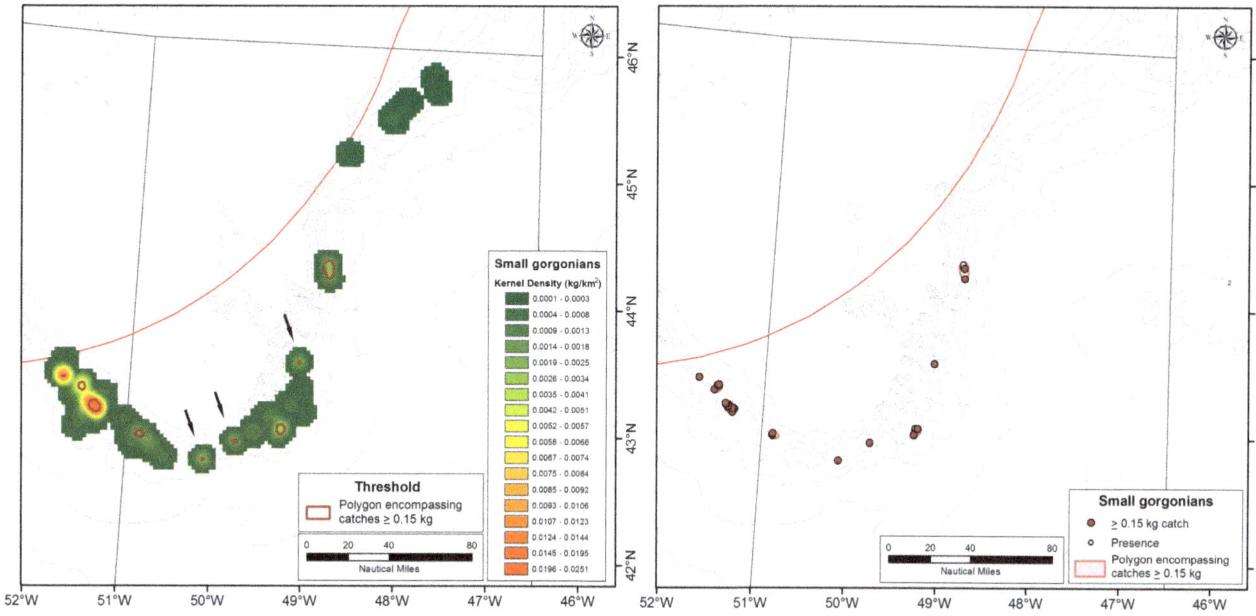

Figure 9. Left panel: Kernel density distribution of small gorgonian corals (primarily *Acanella arbuscula*) on the Tail of Grand Bank in the NAFO Regulatory area with the 0.15 kg density polygons superimposed in red. Arrows point to catches that appear to be isolated and require more data to establish their spatial extent. Right panel: The location of catches greater than 0.15 kg (red circle) and smaller small gorgonian coral catches (open circles) within the 0.15 kg density polygons defining the small gorgonian coral VMEs.

of standardized biomass point data would allow for visualizing areas of dense aggregation but any delineation of a global threshold to separate "significant concentrations" would be necessarily subjective.

The production of a smoothed biomass surface is necessary for application of our method for determining a spatially global threshold to differentiate hot and non-hot locations. Our novel use of the change in area of the polygons encircling catches of a given magnitude allows for the area of aggregation (i.e., the significant concentration or VME) to be distinguished from the broader low density distribution of the taxon. We are unaware of any other approach to arrive at a threshold density that deals with this issue from a biological basis other than the use of the point of maximum curvature [13]. However that approach is not necessarily linked to

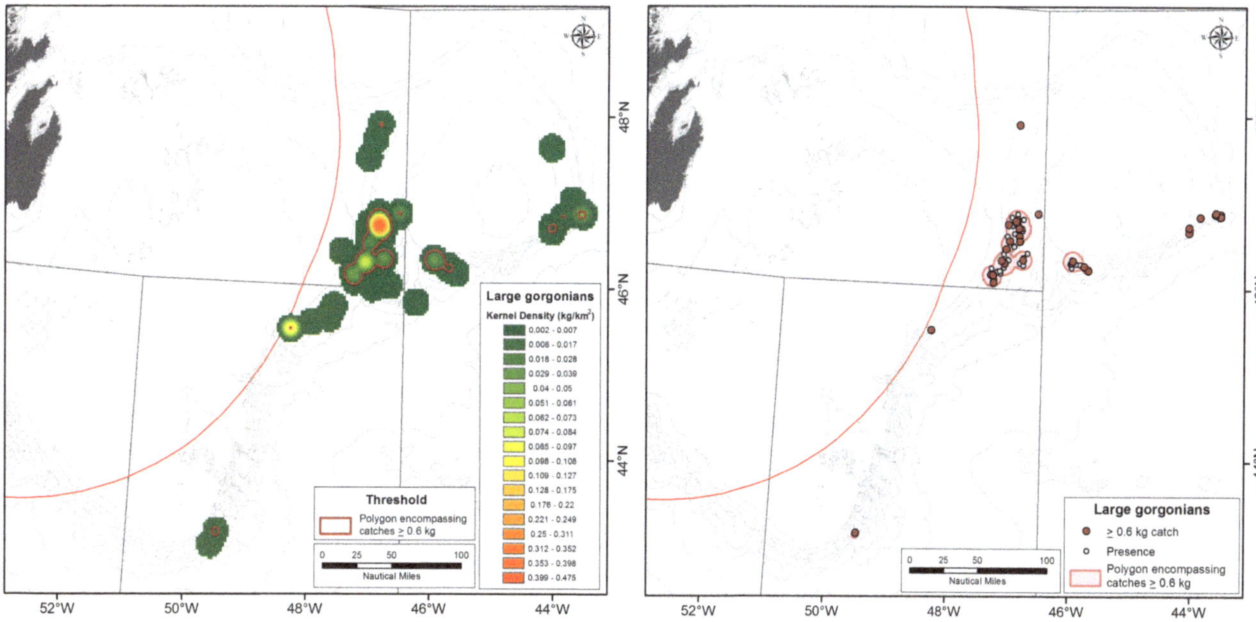

Figure 10. Left panel: Kernel density distribution of large gorgonian corals in the NAFO Regulatory area with the 0.6 kg density polygons defining significant concentrations superimposed in red. Right panel: The location of catches greater than 0.6 kg (red circle) and smaller coral catches (open circles) within the 0.6 kg density polygons defining the large gorgonian coral VMEs.

the spatial organization of the data which is critical to this application.

The outer boundaries of these VME areas are influenced by the search radius of the smoothing function and could in some instances, depending upon the distribution of the input data, extend beyond the VME habitat. For management purposes, the outer boundaries could be refined using targeted surveys, detailed surficial geology, multibeam bathymetry and/or overlay of species distribution models which incorporate environmental predictors such as depth. However, given the high vulnerability of these habitats to disturbance, the outer boundary could also be retained as a precautionary buffer zone from the significant adverse impacts of bottom trawling and dredging until further studies are made.

Although the VME areas delineated by our method utilize biomass thresholds to define boundaries for the calculation of area, the habitats are composed of catches from a wide range of biomass, including smaller catches that happen to fall amongst the larger ones (Figures 6, 8–10). These may represent areas thinned by fishing, areas of recruitment with smaller individuals dominating the catch, areas of different species composition within the VME taxon and/or minor catchability or effort differences among sets.

Of the VME indicator species/taxa in the NRA, the sponges represent the highest biomass in the research vessel catches [6] and form large-scale benthic habitats. Bell [24] divided the function of sponges in the benthic community into three different categories: 1) benthic-pelagic coupling (e.g., carbon and nitrogen cycling), 2) impacts on substrate (e.g., bio-erosion and sediment stabilization), and 3) habitat provision for other species (e.g., predation protection). The extensive sponge grounds in the NAFO NRA likely play important roles in ecosystem function, although only their role in provision of habitat has been directly looked at in any detail [18,19,25]. They are located in the deep water along the continental slopes in areas not generally targeted by the commercial fisheries, hence their persistence inside the fishing footprint.

The VME aggregations of sea pens, known as "fields", provide important structure in low-relief sand and mud habitats where there is little physical habitat complexity (Figure 7). The dominant species in the study area do not retract into the sediment and so are vulnerable to trawl gear capture [8]. These fields provide refuge for small planktonic and benthic invertebrates [26], which in turn may be preyed upon by fish [27]. In the waters adjacent to the study area sea pens have been found to be associated with redfish larvae, raising the possibility that sea pen fields are important nursery areas for commercial fish species [28]. They also alter water current flow, thereby retaining nutrients and entraining plankton near the sediment [22]. The sea pen VMES occur in shallower water than the sponges on Flemish Cap and are adjacent to commercial fishing activity, which may already have impacted their abundance [8]. The KDE method would not be overly sensitive to local depletion of biomass caused by previous fishing, as long as some of the high catches are present in the grid cell either from historical data prior to the disturbance or through the inevitable patchy nature of the fishing effort leaving areas of high biomass undisturbed. The location of the sea pen field VMEs further suggests that the populations may be connected as they form a partial ring around the cap. Knowledge of connectivity patterns among these habitats is critical for the conservation of linked populations when not all habitats are protected.

The remaining VMEs addressed here, that is, the large and small gorgonian corals, are less prominent habitats in the NRA compared with the sponge grounds and sea pen fields. The functional roles that these corals play in the deep-sea benthic ecosystem are nevertheless important, especially for structure- or reef-forming corals, and have been well documented in terms of their benefits for other species [29]. Their structural complexity creates additional microhabitat that may be utilized by other organisms as refugia from predators, as spawning and nursery grounds, and as attachment substrate for sessile invertebrates [28,30,31]. Edinger et al. [32] examined the association between groundfish and 5 classes of corals, including large gorgonians, small gorgonians, sea pens and/or cup corals, soft corals, and the total absence of corals in adjacent Newfoundland waters. They found that of all five groups, groundfish species richness was highest in sets containing small gorgonians (*Acanella arbuscula* and *Radicipes gracilis*), highlighting the potential importance of this group as fish habitat. With the identification of coral and sponge VMEs in the NRA future research can be directed towards their ecosystem function so that a full assessment of the impacts of fishing activities can be made in accordance with UNGA 61/105.

Conclusions

Large-sized sponges, sea pens, and small and large gorgonian corals have been identified previously as vulnerable marine ecosystem indicators in the study area [2]. These large structure-forming taxa can be considered ecosystem engineers [33] as they modify the physical and chemical nature of their environment and provide habit for other organisms. Kernel density estimation (KDE) surfaces applied to research vessel trawl catches of vulnerable marine ecosystem indicators can be used to identify significant concentrations of biomass, thus serving to operationalize the definition of Structure Complexity under the FAO guidelines [1] for implementation of the 2006 United Nations General Assembly Resolution 61/105. The use of the area under the polygons of successive weight categories to identify global spatial thresholds for identifying the catch levels which separate aggregations from dispersed individuals is a novel approach. It has advantages over use of the point of maximum curvature in that it is derived from a geospatial context. We contend that the significant concentrations of sponges, sea pens, small and large gorgonian corals identified herein using this method, along with the supporting literature review of their ecological functions, qualify these areas as vulnerable marine ecosystems. The boundaries of these VME polygons can be refined using underwater cameras and species distribution modeling to delineate closed areas for management.

Acknowledgments

The authors acknowledge the contributions to the development of this work made through discussions with colleagues in the Northwest Atlantic Fisheries Organization (NAFO) Working Group on Ecosystem Science and Assessment (formerly the Working Group on the Ecosystem Approach to Fisheries Management) over the years 2009–2013, and the NAFO Scientific Council over that same time period.

Author Contributions

Conceived and designed the experiments: EK. Performed the experiments: FJM CL. Analyzed the data: EK FJM CL NO MS. Contributed to the writing of the manuscript: EK FJM CL LB VW NO MK AK MS.

References

1. FAO (2009) International Guidelines for the Management of Deep-sea Fisheries in the High Seas. FAO, Rome. 73p.
2. Fuller SD, Murillo Perez FJ, Wareham V, Kenchington E (2008) Vulnerable Marine Ecosystems Dominated by Deep-Water Corals and Sponges in the NAFO Convention Area. Ser No N5524. NAFO SCR Doc 08/22, 24pp.
3. NAFO (2008) Scientific Council Meeting 22–30 October 2008. Ser No N5594. NAFO SCS Doc 08/26, 32p.
4. Murillo FJ, Kenchington E, Sacau M, Piper DJW, Wareham V, et al. (2011) New VME indicator species (excluding corals and sponges) and some potential VME elements of the NAFO Regulatory Area. Ser No N6003. NAFO SCR Doc 11/73, 20 p.
5. NAFO (2012) Scientific Council Meeting –2012. Ser No N6072. NAFO SCS Doc 12/19, 215p.
6. Murillo FJ, Duran Munoz P, Cristobo FJ, Rios P, Gonzalez C, et al. (2012) Deep-sea sponge grounds of the Flemish Cap, Flemish Pass and the Grand Banks of Newfoundland (Northwest Atlantic Ocean): distribution and species composition. Mar Biol Res 8: 842–854.
7. Murillo FJ, Kenchington E, Gonzalez C, Sacau M (2010) The use of density analyses to delineate significant concentrations of Pennatulaceans from trawl survey data. Ser No N5753. NAFO SCR Doc 10/07, 7 p.
8. Kenchington E, Murillo FJ, Cogswell A, Lirette C (2011) Development of Encounter Protocols and Assessment of Significant Adverse Impact by Bottom Trawling for Sponge Grounds and Sea Pen Fields in the NAFO Regulatory Area. Ser No 6005. NAFO SCR Doc 11/75, 53 p.
9. Murillo F J, Durán Muñoz P, Altuna A, Serrano A (2011) Distribution of deep-water corals of the Flemish Cap, Flemish Pass, and the Grand Banks of Newfoundland (Northwest Atlantic Ocean): interaction with fishing activities. ICES J mar Sci 68: 319–332.
10. Fieberg J (2007) Kernel density estimators of home range: Smoothing and the autocorrelation red herring. Ecology 88: 1059–1066.
11. Downs JA, Horner MW (2012) Analysing infrequently sampled animal tracking data by incorporating generalized movement trajectories with kernel density estimation. Comput Environ Urban Syst 36: 302–310.
12. Benhamou S, Cornélis D (2010) Incorporating movement behavior and barriers to improve kernel home range space use estimates. J. Wildl Manage 74: 1353–1360.
13. Nelson T A, Boots B (2008) Detecting spatial hot spots in landscape ecology. Ecography 31: 556–566.
14. O'Brien SH, Webb A, Brewer MJ, Reid JB (2012) Use of kernel density estimation and maximum curvature to set Marine Protected Area boundaries: Identifying a Special Protection Area for wintering red-throated divers in the UK. Biol Conserv 156: 15–21.
15. Kenchington E, Cogswell A, Lirette C, Murillo Perez FJ (2009) The Use of Density Analyses to Delineate Sponge Grounds and Other Benthic VMEs from Trawl Survey Data. Ser No N5626. NAFO SCR Doc 09/6, 18p.
16. NAFO (2011) Report of the 3rd Meeting of the NAFO Scientific Council Working Group on Ecosystem Approaches to Fisheries Management (WGEAFM). Ser No N5868. NAFO SCS Doc 10/24, 75 p.
17. Beazley LI, Anstey LJ, Kenchington ER (2013) Summary of the Location of VME Indicators on the Flemish Cap Slope Based on *in situ* Benthic Imagery. Ser No N6274. NAFO SCR Doc 13/76, 18p.
18. Beazley LI, Kenchington E, Murillo FJ, Sacau M del M (2013) Deep-sea sponge grounds enhance diversity and abundance of epibenthic megafauna in the Northwest Atlantic. ICES J Mar Sci 70: 1471–1490.
19. Barrio Froján CRS, MacIsaac KG, McMillan AK, Del Mar Sacau Cuadrado M, Large P, et al. (2012) An evaluation of benthic community structure in and around the Sackville Spur closed area (Northwest Atlantic) in relation to the protection of vulnerable marine ecosystems. ICES J Mar Sci 69: 213–222.
20. Knudby A, Kenchington E, Murillo FJ (2013) Modeling the distribution of *Geodia* sponges and sponge grounds in the northwest Atlantic Ocean. PLoS ONE 8(12): e82306. doi:10.1371/journal.pone.0082306.
21. Knudby A, Lirette C, Kenchington E, Murillo FJ (2013) Species Distribution Models of Black Corals, Large Gorgonian Corals and Sea Pens in the NAFO Regulatory Area. Ser No N6276. NAFO SCR Doc 13/78, 17p.
22. Tissot BN, Yoklavich MM, Love MS, York K, Amend M (2006). Benthic invertebrates that form habitat structures on deep banks off southern California, with special reference to deep sea corals. Fish Bull 104: 167–181.
23. Rooper CN, Zimmermann M, Prescott MM, Hermann AJ (2014) Predictive models of coral and sponge distribution, abundance and diversity in bottom trawl surveys of the Aleutian Islands, Alaska. Mar Ecol Prog Ser 503: 157–176.
24. Bell JJ (2008) The functional roles of marine sponges. Estuar Coast Shelf Sci 79: 341–353.
25. Kenchington E, Power D, Koen Alonso M (2013) Associations of demersal fish with sponge grounds on the continental slopes of the northwest Altantic. Mar Ecol Prog Ser 477: 217–230.
26. Birkeland C (1974) Interactions between a sea pen and seven of its predators. Ecol Monogr 44: 211–232.
27. Krieger K J (1993) Distribution and abundance of rockfish determined from a submersible and by bottom trawling. Fish Bull 91: 87–96.
28. Baillon S, Hamel J-F, Wareham VE, Mercier A (2012) Deep cold-water corals as nurseries for fish larvae. Front Ecol Environ 10: 351–256.
29. Buhl-Mortensen L, Vanreusal A, Gooday AJ, Levin LA, Priede IG, et al. (2010) Biological structures as a source of habitat heterogeneity and biodiversity on the deep ocean margins. Mar Ecol 31: 21–50.
30. Fosså JH, Mortensen PB, Furevik DM (2002) The deep-water coral *Lophelia pertusa* in Norwegian waters: distribution and fisheries impacts. Hydrobiologia 471: 1–12.
31. Reed JK (2002) Deep-water *Oculina* coral reefs of Florida: biology, impacts, and management. Hydrobiologia 471: 43–55.
32. Edinger EN, Wareham VE, Haedrich RL (2007) Patterns of groundfish diversity and abundance in relation to deep-sea coral distributions in Newfoundland and Labrador waters. Bull Mar Sci 81: 101–122.
33. Jones CG, Lawton JH, Shachak M (1994) Organisms as ecosystem engineers. Oikos 69: 373–386.

8

Observations and Models of Highly Intermittent Phytoplankton Distributions

Sandip Mandal, Christopher Locke, Mamoru Tanaka, Hidekatsu Yamazaki*

Department of Ocean Sciences, Tokyo University of Marine Science and Technology, Minato-ku, Tokyo, Japan

Abstract

The measurement of phytoplankton distributions in ocean ecosystems provides the basis for elucidating the influences of physical processes on plankton dynamics. Technological advances allow for measurement of phytoplankton data to greater resolution, displaying high spatial variability. In conventional mathematical models, the mean value of the measured variable is approximated to compare with the model output, which may misinterpret the reality of planktonic ecosystems, especially at the microscale level. To consider intermittency of variables, in this work, a new modelling approach to the planktonic ecosystem is applied, called the closure approach. Using this approach for a simple nutrient-phytoplankton model, we have shown how consideration of the fluctuating parts of model variables can affect system dynamics. Also, we have found a critical value of variance of overall fluctuating terms below which the conventional non-closure model and the mean value from the closure model exhibit the same result. This analysis gives an idea about the importance of the fluctuating parts of model variables and about when to use the closure approach. Comparisons of plot of mean versus standard deviation of phytoplankton at different depths, obtained using this new approach with real observations, give this approach good conformity.

Editor: Syuhei Ban, University of Shiga Prefecture, Japan

Funding: This work is supported by a fund made available from JST Crest. The funding agency has no role in study design, data collection and analysis, decision to publish, or preparation of the manuscript.

Competing Interests: The authors have declared that no competing interests exist.

* E-mail: hide@kaiyodai.ac.jp

Introduction

Seasonal variation of nutrient and dynamics of plankton in aquatic systems have been studied by many researchers during the last few decades, and mathematical models have been used to better understand these systems. Ecological models have evolved from simple two-component systems to highly complex multi-component systems [1–15]. The complexity of a model increases significantly when coupled with a physical system to make a biophysical model [16–23]. However, these models study the system dynamics in mesoscale or in larger scale but are not aimed at explaining the dynamics in microscale.

Interactions of phytoplankton, other organisms and biogenic particles in turbulent flows drive plankton dynamics and cycling of nutrients, with influences expected to be accumulative across multiple scales [24]. As technology improves and measurements of phytoplankton distributions are done with greater resolution, properly interpreting *in situ* spatial patterns becomes increasingly important for developing our knowledge of the mechanisms that create and support microscale structure and ecology of marine ecosystems [25,26]. Phytoplankton measurements of microscale distributions have recently been studied using a variety of high-resolution instruments, such as water sampling devices [27], microstructure profiling fluorometers [26,28–30], fluorescence imaging systems [31,32] and holography systems [33–35]. Various fluorescence-based measurement systems have consistently shown increasing levels of fluorescent intermittency as sample volume size is decreased [26,28,30,31], which can be attributed to the preferential detection of larger individual cells, chains and aggregates [36]. Furthermore, efforts to understand the mechanisms driving the high levels of observed intermittency using conventional (Gaussian) models, such as spectral analysis [37], appear to be limited by the discrete nature of phytoplankton distributions at small scales [26,36,38]. Highly resolved profiles of phytoplankton concentration and fluorescence would give better predictions of model parameters and improve understanding of the dynamics of phytoplankton [22]. These high-resolution instruments advance our understanding of the influence of physical processes on phytoplankton dynamics from mesoscale down to microscale. Therefore, an analysis of small-scale vertical structure in phytoplankton is required to understand the ecosystem behaviour more precisely.

Rationale for the development of a new ecosystem model

Interactions between different biotic and abiotic factors [39,40], at the microscale, generate spatial variability in the distribution of phytoplankton [31,41]. Profiles of *in vivo* chlorophyll fluorescence are commonly used to infer the distribution of phytoplankton biomass.

Figure 1A shows an observation of chlorophyll profiles (phytoplankton) collected at the mouth of Tokyo Bay, Japan, in June 2011, up to 120-m depth from the ocean surface using four sampling methods: Niskin bottle, Seapoint fluorometer, light emitting diode (LED) sensor and laser sensor (Dataset S1). Descriptions of the instruments and sampling methods are described briefly in the next section. This survey was carried out

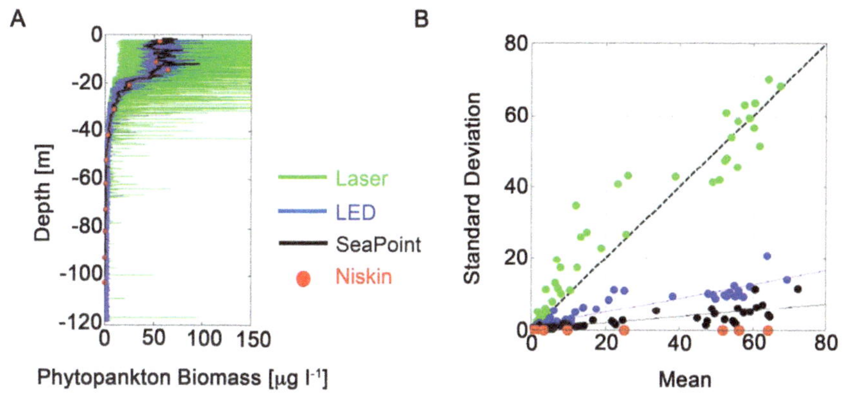

Figure 1. Examples of fluorescence microstructure measurements by Niskin bottle (red), Seapoint (black), LED sensor (blue) and laser sensor (green). (A) Plot of data at different depths from the ocean surface at the mouth of Tokyo Bay, Japan, on June 18, 2011, and (B) mean and standard deviation of chlorophyll signals by four sampling methods. Data from Seapoint, LED sensor and laser sensor were scaled to data obtained by Niskin bottle, which gives the actual concentration of phytoplankton.

with permission of the Japan Coast Guard (http://www.kaiho.mlit.go.jp/e/index_e.htm).

Comparison of fluorescence profiles using these sensors shows that the extent of phytoplankton spatial variability becomes increasingly intermittent when measured with increased resolution. However, 1-m averaged fluorescence levels (spatial mean) are consistent for all four devices. Spatial mean of the collected chlorophyll data and the corresponding standard deviations in all four methods are plotted in Figure 1B. For Niskin samples, the standard deviation of phytoplankton is zero since only a single number for average chlorophyll from the water sample is available. High-resolution data show high spatial variability of phytoplankton near the ocean surface and that variability and mean decrease with depth. In theoretical studies of phytoplankton dynamics, especially in formulating ecological models, almost all modellers focus on the large-scale dynamics of phytoplankton and the mean value of the collected data, not considering the associated small-scale spatial variability.

Our observed data suggest that considering the fluctuating part of each component of the system as a new variable is important, particularly with high-resolution data. This paper aims to study the effect of this variability on the model output and to compare this effect with real observations.

Materials and Methods

Microstructure data collection

Four sampling methods, namely Niskin bottle, Seapoint fluorometer, light emitting diode (LED) sensor and laser sensor, were used for data collection. The Niskin bottle and the Seapoint fluorometer were mounted on a CTD cage, which was tethered to a winch cable. Swaying with the motion of ship, the CTD cage generated a disturbed water column. The Niskin bottle is a discrete sampling device, which measures the "averaged" ecosystem characteristics by collecting water samples. The sampling volume of Seapoint is 0.34 mL, but the motion of the CTD cage and the configuration of the probe did not allow for sampling from undisturbed water. In contrast, the LED and the laser sensor were mounted on a free-fall microstructure profiler, TurboMAP-L [26]; thus, the sampling was made from essentially undisturbed water.

The minimum vertical resolution for the LED sensor is about 2 cm and is about 2 mm for the laser sensor [26]. Microscale spatial variability observed by the LED sensor varied by less than 2-fold across centimetre to metre scales. In the laser sensor, the microscale spatial variability varied by between 2- and 10-fold across millimetre-size peaks. The reduced sampling volume of the laser probe (32 μL), compared with that of the LED probe (4 mL), allows for measurement of the fluorescence field with increased spatial resolution.

High-resolution microstructure fluorescence data were collected in several locations in Tokyo Bay at different time periods using a laser probe mounted in a TurboMAP-L [26] (Dataset S1). TurboMAP-L is a free-falling microstructure profiler that captures physical and biological conditions. The instrument length is 2 m, the weight is 30 kg in air, the diameter is 0.12 m, and the withstanding pressure is 500 dbar (~500 m). The captured data were transmitted in real time to the shipboard PC via a 5 mm Kevlar tether attached to the end of the instrument. TurboMAP-L was deployed and freefell into the sea with profiling speeds of 0.50–0.80 ms^{-1}. After reaching the target depth, the instrument was recovered by taking up the tether using a winch exclusive to TurboMAP-L.

The sensors attached to TurboMAP-L are a laser fluorescence probe (digitized at 256 Hz), a light emitting diode (LED) fluorescence/turbidity probe (256 Hz), two turbulent shear probes (512 Hz), a FP07 fast temperature probe (512 Hz), XYZ 3-dimensional accelerometers (64–256 Hz) and CTD (Conductivity, Temperature and Depth; 64 Hz). These sensors were mounted forward of the rounded nose to ensure the measurement of undisturbed fields.

In May 2011, the experiments were conducted at three locations in Tokyo Bay: (a) the mouth of the bay (depth ~500 m), (b) inside the bay (depth ~50 m) and (c) the mouth of the Ara River (depth ~10 m). In these locations the microstructure data were collected from TurboMAP-L to a depth of ~200 m, ~50 m and ~10 m. Similarly the fluorescence data were collected at a particular place inside the bay (depth ~50 m) but in three different seasons, specifically during the months: (a) December 2006, (b) September 2007 and (c) February 2008. Normally arbitrary units of fluorescence are calibrated to give estimates of biomass in μg l^{-1}. Permission for these surveys was received from the Japan Coast Guard.

Mathematical model

Aiming to develop the methodology of a new ocean ecosystem model, at the initial stage, we concentrate only on nutrient ($\mathcal{N}$) and phytoplankton (P) as model variables and choose simple form for transfer functions [5] (Section S1 in Text S1).

Simple NP model. The time evolution equations for an NP system can be written as:

$$\frac{dP}{dt} = C\frac{N}{K+N}P - D\,P \tag{1}$$

$$\frac{dN}{dt} = -C\frac{N}{K+N}P + D\,P, \tag{2}$$

where parameters are described in Table 1. The sum of N and P variables is a conserved quantity, i.e. $N+P=A$ (Constant). This reduces the number of equations into one,

$$\frac{dP}{dt} = C\frac{(A-P)}{K+(A-P)}P - D\,P. \tag{3}$$

Studies of microscale phytoplankton distributions using high-resolution profiling fluorometers have revealed the ubiquity of fluctuating millimetre- to centimetre-sized fluorescence peaks across a range of marine systems, which also implies the local values of fluorescence are highly fluctuating [26,38] in space. Since the amount of nutrient in the system is interrelated to the biomass of phytoplankton, the nutrient would be expected to be fluctuating in space for a close system. But this spatial variability is not considered in the simple non-closure model as described in equations (1), (2) or (3).

Therefore, considering our model variables as functions of time (t) and space (r), and decomposing each variable into mean and fluctuating component, we get

$$P(r,\ t) = P_0(r,\ t) + P'(r,\ t)\ \text{and} \tag{4}$$

$$N(r,\ t) = N_0(r,\ t) + N'(r,\ t). \tag{5}$$

P_0 and N_0 are spatial mean values of the phytoplankton and nutrient respectively, and P' and N' are their fluctuating components corresponding to each mean value. Note that at a particular time point, the space-average of the fluctuating components is zero ($<P'(r)> = <N'(r)> = 0$), while temporal average can be nonzero, which also implies $<P(r)> = P_0(r)$ and $<N(r)> = N_0(r)$. This consideration is based upon our observation of Figure 1, where we see that 1-m averaged fluorescence levels (spatial mean) for all four instruments coincide. Now we are interested in seeing the temporal variation of mean and fluctuating component of P and N variables using the closure approach, which is widely used in the study of turbulence.

Closure model. Putting (4) and (5) into equations (1–2) and applying the Reynolds averaging method in space (details in Section S2 in Text S1), we get the following set of equations for the temporal variation of P_0, N_0, $<P'^2>$, $<N'^2>$ and $<N'P'>$:

$$\frac{dP_0}{dt} = C\left(\frac{N_0 P_0}{K+N_0} + \frac{K<N'\,P'>}{(K+N_0)^2} - \frac{K\,P_0<N'^2>}{(K+N_0)^3}\right) - D\,P_0 \tag{6}$$

$$\frac{dN_0}{dt} = -C\left(\frac{N_0 P_0}{K+N_0} + \frac{K<N'\,P'>}{(K+N_0)^2} - \frac{K\,P_0<N'^2>}{(K+N_0)^3}\right) + D\,P_0 \tag{7}$$

$$\frac{d<P'^2>}{dt} = 2\,C\left(\frac{N_0<P'^2>}{K+N_0} + \frac{K\,P_0<N'\,P'>}{(K+N_0)^2}\right) - 2\,D<P'^2> \tag{8}$$

$$\frac{d<N'^2>}{dt} = -2\,C\left(\frac{N_0<N'\,P'>}{K+N_0} + \frac{K\,P_0<N'^2>}{(K+N_0)^2}\right) + 2\,D<N'\,P'> \tag{9}$$

$$\frac{d<N'\,P'>}{dt} = C\left(\frac{N_0(<N'\,P'> - <P'^2>)}{K+N_0} + \frac{K\,P_0(<N'^2> - <N'\,P'>)}{(K+N_0)^2}\right) + D\,(<P'^2> - <N'\,P'>). \tag{10}$$

In the formulation of these equations we assume that N and P variables follow a joint lognormal probability distribution function, which forces the third and all higher odd order fluctuating terms to vanish. We also ignored the fourth and higher order terms in this analysis to achieve simple closure. The first two equations give the time evolution of mean terms, the next two equations give the time evolution of variance terms, and the last equation represents the evolution of the covariance term.

Note that while the sum of $N+P$remains constant, in this caseN_0+P_0 and $<N'^2> + <P'^2> + 2<N'\,P'>$, both are temporally conserved quantities. Therefore, defining $N_0+P_0 = A$ and $<N'^2> + <P'^2> + 2<N'\,P'> = B$, the above five equations can be reduced to three equations that lie in five-dimensional parameter spaces. By appropriately rescaling the equations with A and B, this dependency is reduced to three dimensionless parameters. The scaling factors and dimensionless parameters are given in Table 1, and the scaled equations can be written as follows:

$$\frac{dp_0}{d\tau} = \frac{(1-p_0)p_0}{k+(1-p_0)} + \frac{k\,\beta\,(1-x-y)}{2\,\{k+(1-p_0)\}^2} - \frac{k\,\beta\,p_0\,y}{\{k+(1-p_0)\}^3} - \varepsilon\,p_0 \tag{11}$$

$$\frac{dx}{d\tau} = 2\frac{(1-p_0)\,x}{k+(1-p_0)} + \frac{k\,p_0\,(1-x-y)}{\{k+(1-p_0)\}^2} - 2\,\varepsilon\,x \tag{12}$$

$$\frac{dy}{d\tau} = -\frac{(1-p_0)\,(1-x-y)}{k+(1-p_0)} - \frac{2\,k\,p_0\,y}{\{k+(1-p_0)\}^2} + \varepsilon\,(1-x-y) \tag{13}$$

with $n_0+p_0 = 1$ and $x+y+2\,z = 1$.

The values of the scaled variables n_0 and p_0, corresponding to the variables N_0 and P_0 respectively, lie between 0 and 1. Similarly, the values of x and y lie between 0 and 1 if the term z, which is associated with the covariance of the fluctuating components, $<N'\,P'>$, is positive. For negative covariance, the values of x and y can exceed 1. Now we define the normalized sum of variance and covariance as follows:

Table 1. Definition of different quantities used in the model and their dimensions.

Quantity	Definition	Dimension	Scaling factor	Dimensionless quantity
A	Sum of nitrate	$\mu g\ N\ l^{-1}$	-	-
B	Variance of sum of fluctuating components	$(\mu g\ N\ l^{-1})^2$	B/A^2	B
C	Maximum growth rate of phytoplankton	day^{-1}	-	-
K	Nutrient uptake half-saturation constant	$\mu g\ N\ l^{-1}$	K/A	K
D	Phytoplankton death rate	day^{-1}	D/C	E
P_0	Mean phytoplankton	$\mu g\ N\ l^{-1}$	P_0/A	p_0
$<P'^2>$	Variance for phytoplankton	$(\mu g\ N\ l^{-1})^2$	$<P'^2>/B$	X
$<N'^2>$	Variance for nutrient	$(\mu g\ N\ l^{-1})^2$	$<N'^2>/B$	Y
$<N'P'>$	Covariance	$(\mu g\ N\ l^{-1})^2$	$<N'P'>/B$	Z
t	Time	day	$t\,C$	T

$$\begin{aligned}\beta &= (<N'^2> + <P'^2> + 2<N'\,P'>)\,/A^2 \\ &= <(N'+P')^2>/A^2 = B/A^2\end{aligned} \tag{14}$$

where B is the variance of the sum of N' and P', and therefore, β value actually reflects the overall strength of the fluctuating components and modifies the model dynamics. Other dimensionless variables and parameters of the model equations (11–13) are described in Table 1.

Similarly, the non-closure equation (3) can be reduced to the following form:

$$\frac{dp}{d\tau}=\frac{p\,(1-p)}{k+(1-p)}-\varepsilon\,p \tag{15}$$

Note that for $\beta=0$, equation (11) of the closure model reduces to equation (15), which corresponds to the non-closure model.

We have found six steady state solutions (p_0*, x*, y*) for the set of closure equations (11–13), among which two are not ecologically feasible as these are complex for any realistic set of parameter values (Sections S3 and S4 in Text S1). Two other solutions lie at the boundary, and the remaining two are interior steady states. Depending on the sign of the covariance term ($<N'\,P'>$), between these two, only one interior steady state exists for the system. Also, from stability analysis, we have seen that the interior steady state is stable when the covariance term is positive (Section S5 in Text S1). Therefore, all analysis was done emphasising this stable steady state solution.

Parameter values. After reducing the equations into dimensionless variables and parameters, the influences of only three parameters on the model have to be studied. Here, the model under study is a general one and does not describe any specific ecological situation. Therefore, for numerical simulation, the assigned parameter values are adopted from previous studies [2,19,42–44]. To find the domain of parameter space for this model, we have studied a wide range of parameter values around these reported values.

Maximum growth rate of phytoplankton (C) varies with the intensity of light, although at the first stage of this work C is considered to be 2 day^{-1} [2,19], while 2 $\mu g\ N\ l^{-1}$ is chosen as the value of A [2]. Different literature shows that the death rate (D) of phytoplankton varies from 0.07 day^{-1} to 0.2 day^{-1} [2,19,42]. For the neritic zone, MacIsaac and Dugdale [43] and Eppley *et al.* [44] had considered the value for the half-saturation constant of phytoplankton nutrient uptake K to be 1 $\mu g\ N\ l^{-1}$, whereas Edwards *et al.* [19] had used 1.4 $\mu g\ N\ l^{-1}$ as this parameter value.

For our numerical simulation, we have varied the dimensionless parameter ε ($=D/C$) from 0.025 to 0.99 and k ($=K/A$) from 0.05 to 2.5, covering the reported values range. All the reported parameter values, their ranges and the corresponding dimensionless values are summarized in Table 2. The domains of parameter values, which the system of equations (11–13) shows to have stable

Table 2. Parameter values and ranges for dimensional and dimensionless systems.

Quantity	Reported value with references	Corresponding dimensionless quantity	Corresponding dimensionless value	Range of parameter values studied in this article
A	2 $\mu g\ N\ l^{-1}$ [2]	-	-	-
C	2 day^{-1} [2,19] (varies with light intensity)	-	-	-
K	1 $\mu g\ N\ l^{-1}$ [43,44], 1.4 $\mu g\ N\ l^{-1}$ [19]	k	0.5–0.7	0.05–2.5
D	0.07 day^{-1} to 0.2 day^{-1} [2,19,42]	ε	0.035-0.1 (for $C=2\ day^{-1}$); 0.35-1.0 (for $C=0.2\ day^{-1}$)	0.025–0.99

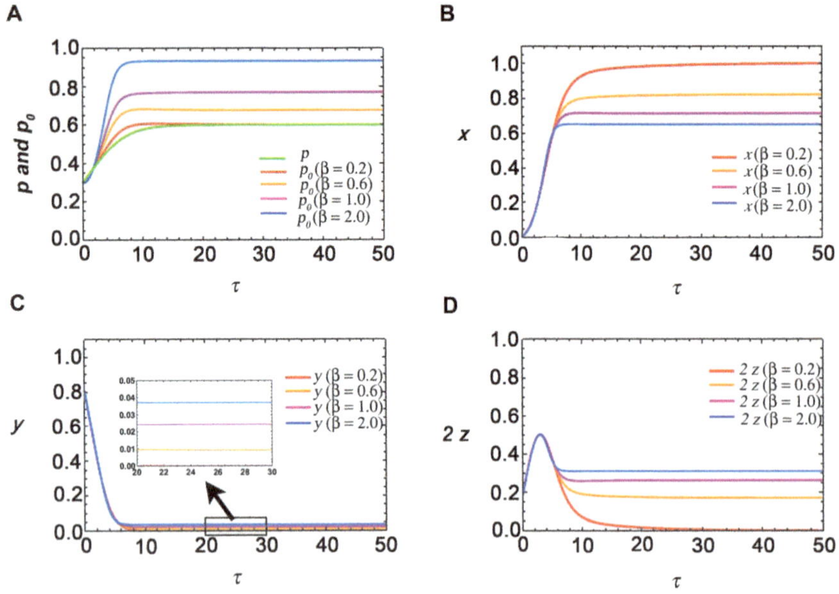

Figure 2. Time variation of variables p_0, x, y and 2z of the closure model. Plots of these variables are shown in Figures A, B, C and D respectively for β = 0.2, 0.6, 1.0 and 2.0. Variation of p of the non-closure model is simultaneously depicted in Figure A. The constant parameter values for this simulation are k = 0.6 and ε = 0.4, for which $\beta^* = 0.36$.

feasible solutions, are shown in Figures S1, S2, S3, S4, S5 and S6 (See Sections S4 and S5 in Text S1).

Results and Discussion

This article attempts to capture the spatial variability of different constituents (nutrient and phytoplankton) of an ocean ecosystem through mathematical models and to understand the contribution of the fluctuation part of model variables in the system dynamics. Here the dimensionless parameter β actually represents the ratio of overall variance of the fluctuating part to the square of total nutrient contained in the system, i.e., $\beta = (<N'^2> + <P'^2> + 2<N'\,P'>)/A^2 = <(N'+P')^2>/A^2$. We have found a critical value of β below which both closure model and non-closure model give the same steady state level.

Critical parameter value

For a certain choice of parameter values, from our studied parameter domain, Figure 2 shows the time dynamics of the variables p_0, x, y and z, which correspond to mean of phytoplankton, variance of phytoplankton, variance of nutrient and covariance of x and y. In these plots, k and ε values are kept constant at 0.6 and 0.4 respectively (chosen from the realistic parameter domain in Table 2), and for the closure model four β values are chosen, $\beta = 0.2$, 0.6, 1.0 and 2.0. Time dynamics corresponding to the variable phytoplankton for both the closure and the non-closure model are simultaneously plotted in Figure 2A. Figures 2A–D show that the values of p_0, y and z increase as the β value increases, whereas x decreases with the increase of β. According to Figures 2A and B, the variance of phytoplankton (x) can be greater or less than the mean value (p_0) depending on the value of β. Figure 2A shows that for $\beta = 0.2$ the steady state level of the closure model coincides with the steady state level of the non-closure model although their transient dynamics are different. But as β increases, beyond some critical value of $\beta = \beta*$, the steady state levels of phytoplankton in these models do not coincide, and the difference increases as β increases (also see Section S6 in Text S1, Figures S7 and S8). This gives an idea about the limit of the parameter β above which the closure approach is appropriate to use and below which the non-closure model agrees with the results from the closure model.

This critical parameter value of β can be obtained by equating the nonzero steady state value for p_0 of the closure model and p for the non-closure model (see Section S3 in Text S1). Thus, at $\beta = \beta^*$ we have the following expression for β^*:

$$\frac{2\,k}{1-\varepsilon} - \sqrt{\gamma+\delta} + \sqrt{-\gamma + 2\,\delta + \frac{2\,k\,\beta*}{(1-\varepsilon)\,\sqrt{\gamma+\delta}}} = 0 \quad (16)$$

with,

$$\delta = \frac{4\,k\,(1+k)}{3\,(1-\varepsilon)} \quad (17)$$

$$\gamma = \frac{16 \times 2^{\frac{1}{3}}\,k^2\,(1+k)^2}{3\,\eta^{\frac{1}{3}}} + \frac{\eta^{\frac{1}{3}}}{3 \times 2^{\frac{1}{3}}(1-\varepsilon)^2} \quad (18)$$

and

$$\eta = \sqrt{(128\,k^3\,(1+k)^3\,(1-\varepsilon)^3 + 27\,k^2\,\beta^2(1-\varepsilon)^4) - 16384\,k^6(1+k)^6\,(1-\varepsilon)^6} \\ + 128\,k^3(1+k)^3(1-\varepsilon)^3 + 27\,k^2\,\beta^2(1-\varepsilon)^4. \quad (19)$$

Figure 3 shows the variation of the critical parameter β^* with the variation of k and ε. For a particular value of ε and k, β^* value corresponds to the maximum value of β up to which the non-closure model and the closure model reach the same steady state value for phytoplankton. However, beyond that value, the closure model shows a different result. For example, from equation (16) we

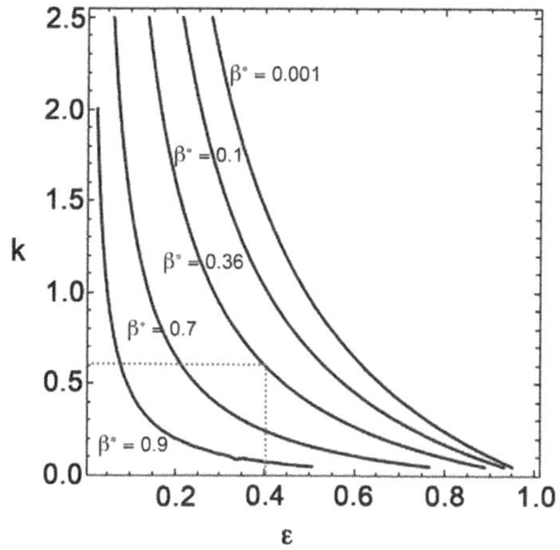

Figure 3. Plot of critical value of β (= β*), with the variation of ε for k. The red dotted line indicates the β* value corresponding to specific k and ε values.

can determine the β^* value as 0.36 for $k = 0.6$ and $\varepsilon = 0.4$, which clearly justifies our numerical results above as shown in Figure 2A: for $\beta = 0.2$ (<0.36) the closure model approaches the non-closure model, but for $\beta = 0.6$, 1.0 and 2.0 (>0.36) the closure model gives a different result from the non-closure model.

From the same figure it is also observed that as the value of the parameter ε decreases, the β^* value increases. This means that as the death rate of phytoplankton (D) decreases or as the growth rate of phytoplankton (C) increases, the critical value of β increases. Similarly, as the value of k ($= K/A$) decreases, the critical value of β increases. These phenomena imply that the closure model has a more important role where the growth rate of phytoplankton is less with respect to death rate and half-saturation constant (K) is high with respect to total nutrient content (A) of the system.

Coefficient of variation (CV) of phytoplankton in closure model

In real observations we can never measure data with full certainty. To standardize our measurement, we normally consider the mean value of several collections of that data along with its standard deviation. In this situation, a statistical parameter, which becomes important, is the coefficient of variation (CV). This is a normalized measure of variability of a distribution and is defined as the ratio of standard deviation to the mean of that variable.

Because the closure model considers the fluctuation of variables, one can expect to determine the coefficient of variation in this model. To do this, we first recall our model variables P_0, N_0, $<P'^2>$ and $<N'^2>$. P_0 and N_0 represent the mean of phytoplankton and nutrient respectively. $<P'^2>$ and $<N'^2>$ represent the variance of fluctuating component corresponding to those variables. The square root of the variance is a measure of standard deviation.

From Figure 2 we have seen that for a particular value of k and ε, all the variables reach a steady state level after some time period. This level varies as β varies. We have also seen from Figures 2A and B that the variance of the fluctuating component of phytoplankton (x) can be greater or less than the mean value of phytoplankton (p_0) depending on the value of β. Therefore, the coefficient of variation for phytoplankton can be greater or less than one, depending on the parameter values.

Microscale fluorescence measurements at different locations in Tokyo Bay and at different time periods are shown in Figure 4 (Dataset S1). Permission for these surveys was received from the Japan Coast Guard. Scattered points represent the mean and standard deviation of the fluorescence measurements over one-meter vertical intervals. Figures 4A–C show that the effective coefficient of variation at the mouth of the bay, inside the bay and at the mouth of the Ara River to varied from 1.50, 0.78 and 0.32 respectively in May 2011. These spatial structures imply that at the same period but at different locations the coefficient of variation of phytoplankton can be different. This is because of a combination of the variation of physical processes and the community structure of phytoplankton at different locations that influences the plankton ecology. Figures 4D–F demonstrate the fluorescence data at a particular place inside the bay (depth ~50 m) but in three different seasons. In December 2006, the coefficient of variation was 1.61, which becomes 0.37 (<1) in September 2007. Again, in February 2008, the value of the coefficient of variation increases to 1.36 (>1). Therefore, at a particular location the coefficient of variation changes seasonally with the seasonal change of physical dynamics as well as with the change in the community structure of phytoplankton.

Now we will see how the newly developed closure model can capture these changes of the coefficient of variation by changing different parameters. In a particular study area of ocean, when considering the abundance of a particular species of phytoplankton, we can assume that the average death rate (D) and half-saturation constant (K) of phytoplankton are constant. As an example, let us consider the values of D and K as 0.135 day^{-1} and 1.2 μg N l^{-1} respectively, which are actually the average values of the reported cases (Table 2). We also assume the total nutrient of the system to be constant, and during the first step we set its value to be 2 μg N l^{-1} (Table 2). Therefore, among three main parameters (C, K and D) of the original model, two (K and D) are constant over time and space (here space actually represents the depth). The maximum growth rate, C, depends on the intensity of light and varies with depth. As the intensity of light decreases with the depth from the sea surface, we assume that the maximum growth rate of phytoplankton decreases with depth.

To study our system of equations (11–13) we must look after the values of the dimensionless parameters. The value of k is 0.6 as $k = K/A$, and the value of $\varepsilon = D/C = 0.135/C$. Considering the maximum value of C at the surface as 2 day^{-1} gives the minimum value of $\varepsilon = 0.067$, and according to our model formulation, the maximum value of ε is <1. Therefore, the value of ε, which is the ratio of phytoplankton death rate to the maximum growth rate of phytoplankton, at a specific depth, is the minimum at the ocean surface, and its value increases as depth increases. Figure 5 shows the domain of parameters in the ε-β parameter space for p_0 to lie between zero and one. The red line indicates the value of β^* below which the mean value of phytoplankton for the non-closure and the closure model coincide. The dashed curve in the grey shaded region divides the entire parameter domain of the closure model into two sections: on one side the coefficient of variation (CV) for phytoplankton is less than one and on the other, is greater than one. We have also observed that as the total nutrient (A) of the system decreases, the domain of parameter in ε-β space for CV<1 decreases and the parameter domain for CV>1 increases (see Section S7 in Text S1, Figure S9). Now we will see the variation of mean and standard deviation for phytoplankton at different depths within this described parameter domain.

Figure 6 shows the normalized mean (P_0/A) and standard deviation ($\sqrt{<P'^2>}/A$) corresponding to phytoplankton for different β values. This is obtained by giving back the dimensions of the variables of the system of equations (11–13). Each point in this plot represents the steady state value of mean and standard

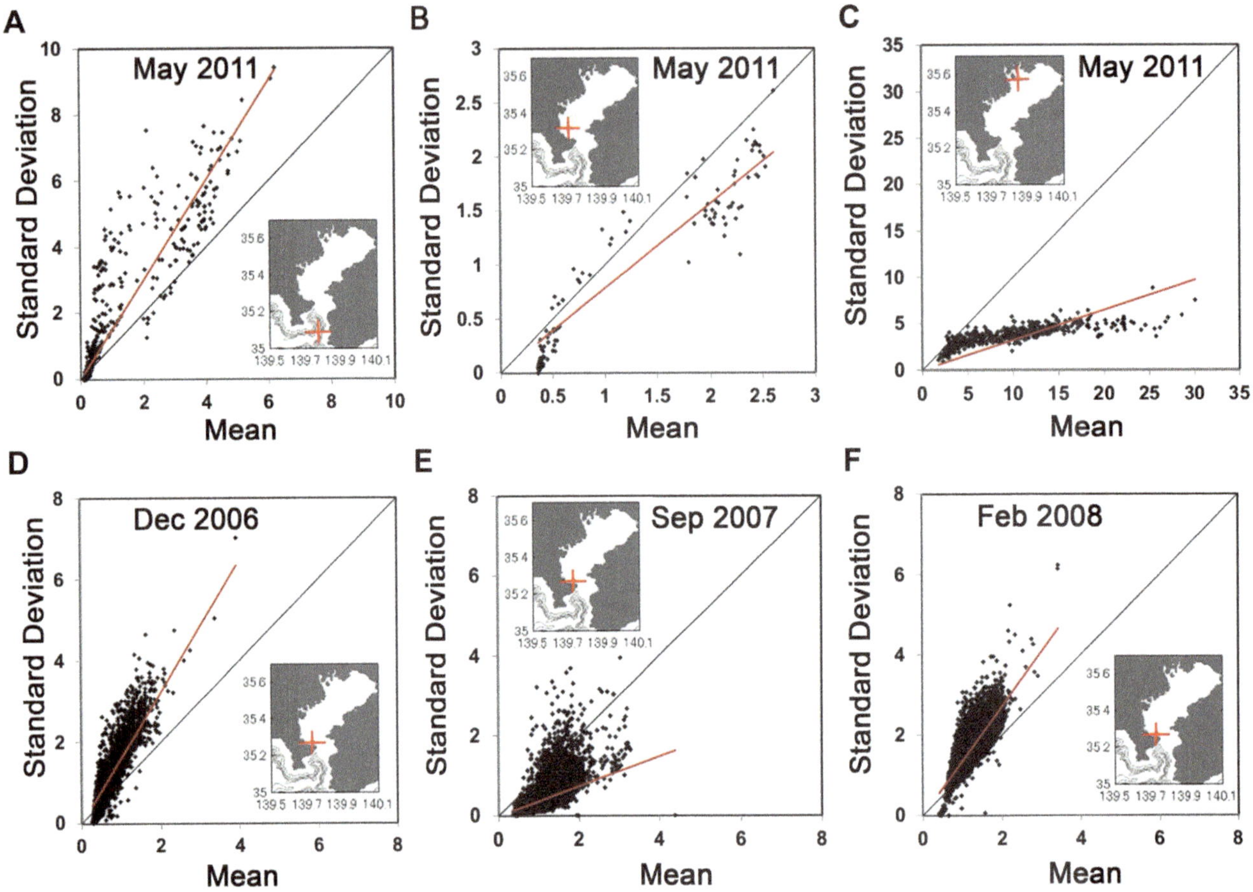

Figure 4. Observations of phytoplankton data in Tokyo Bay. Figures (A, B, C) show mean and standard deviation of fluorescence microstructure data obtained from different locations in Tokyo Bay in May 2011. (D, E, F) show data obtained from the same location in Tokyo Bay but at different time periods. Slope of the trend line passing through the origin of these data sets (red line) gives the effective value of the coefficient of variation at that location. Inset figures show the data collection locations.

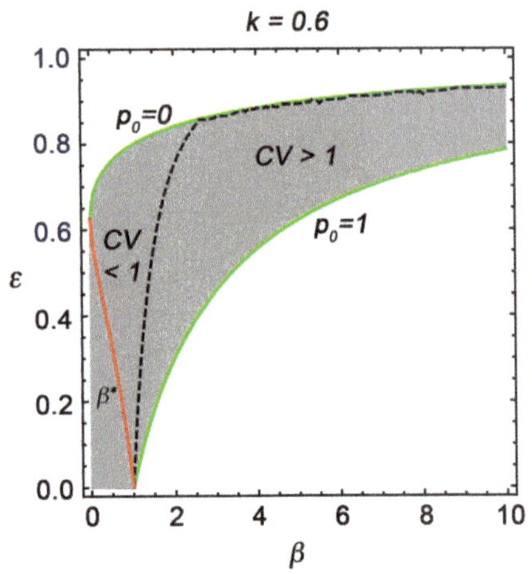

Figure 5. Domain of parameters for the value of coefficient of variation (CV) of phytoplankton. The grey region is the domain of parameters in the (ε-β) parameter space for p_0 to lie between 0 and 1. Green lines are the boundary values corresponding to the mean of phytoplankton (p_0). The dashed line divides the parameter domain for CV<1 and CV>1.

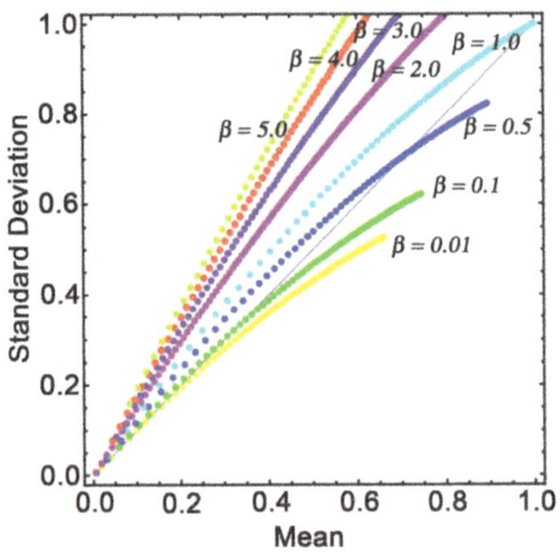

Figure 6. Plot of mean versus standard deviation (SD) of phytoplankton at different depths for eight different β values. Each point corresponds to the mean and SD of phytoplankton at a particular depth. Both axes are normalized by the value of A (=2 μg N l^{-1}). Here the depth profile is obtained by changing the parameter value C, which decreases as depth increases, and the other two parameters K and D are kept constant at 1.2 μg N l^{-1} and 0.135 day^{-1}.

deviation at a specific value of ε ($=D/C$). The value of ε increases with depth as the growth rate of phytoplankton (C) decreases with depth due to light attenuation, even if the average death rate of phytoplankton (D) remains constant. Therefore, for a particular β value, the changes of ε value give us the depth profile of phytoplankton from our model simulation. Eight β values ($\beta = 0.01$, 0.1, 0.5, 1.0, 2.0, 3.0, 4.0 and 5.0) are chosen to show these results, and for a particular β value, the ranges of ε values are determined from the parameter domain of Figure 5. As an example, for $\beta = 3.0$, the value of ε can vary from 0.46 to 0.86. From these plots, we observe that as the depth increases, the mean value of phytoplankton as well as its standard deviation decreases, which implies that the mean of phytoplankton and its spatial variability decrease with depth. This model result is justifiable, as we see similar behaviour in small-scale observation of phytoplankton data, depicted in Figure 4.

In real observations we see the coefficient of variation vary from less than one to greater than one (Figure 4). In Figure 6, the diagonal line characterizes the points where the mean and standard deviation are equal, indicating that the coefficient of variation is equal to one along this line. From our model results we also see the coefficient of variation vary on both sides of the diagonal line depending on the value of the parameter β. This figure also shows that as β increases, the coefficient of variation increases. If the total nutrient of the system, A, is conserved, then the change of β implies the change of fluctuating components of the system. This is because $\beta = B/A^2$, and B represents the variance of overall fluctuating components. Therefore, according to our analysis, as the variance of the overall fluctuating components of the system increases, the coefficient of variation increases.

We have also observed that the parameter β is more sensitive to the coefficient of variation when the total nutrient (A) of the system is high. These mathematical observations imply that in a particular area of ocean with high total nutrient, spatial variation will be low for low β value and therefore, the coefficient of variation will be low. This may be the reason for low coefficient of variation at the mouth of the Ara River (Figure 4C). Low values of CV may imply that phytoplankton species with less sticky material may dominate the community structure. As the seasons change, the ocean dynamics change, which causes the spatial variability of the phytoplankton species to vary. In the closure model this change is observed by changing the parameter β. Therefore, Figures 4D, E, and F can be explained by stating that the β value changes seasonally and causes the coefficient of variation to change seasonally.

Estimation of parameter values

In our closure model the phytoplankton death rate D is a measurable quantity and was estimated by Lehman *et al.* [42], Franks *et al.* [2], Edwards *et al.* [19] and many other researchers. Similarly, the value of half-saturation constant, K, the maximum growth rate of phytoplankton, C, and the total nutrient of the system, A, were also measured by Franks *et al.* [2], Edwards *et al.* [19], MacIsaac and Dugdale [43] and Eppley *et al.* [44]. On the other hand, in our model, B is a conserved quantity that essentially measures the sum of variance and covariance terms (or equivalently the variance of total fluctuating components) and is not easily measurable. One can guess the value of this quantity by estimating the parameter β, as $B = \beta \times A^2$. This parameter can be estimated by using β-value, at which the coefficient of variations (= Standard deviation/Mean) of the observation agrees with the coefficient of variation of the model result (Figure 6). Therefore, the closure model can also be used to estimate this unknown parameter of the system.

Conclusions

Mathematical modelling is a pillar of theoretical ecology research. Aimed at addressing various questions, several models have been developed over the decades and range from very simple to the highly complex. Predictive models are always validated or compared with measured data sets. However, sometimes a problem occurs where the measured variables are highly intermittent. Technological advances allow for measurement of phytoplankton data to greater resolution in oceanic ecosystems. High-resolution data, which displays high spatial variability, are useful for understanding the influences of physical processes on phytoplankton dynamics. In conventional methods, the mean value of the measured variable is approximated to compare with the model output. This approximation may misrepresent the reality of planktonic ecosystems, especially at microscale level. To consider spatial variability of variables, in this article, we have applied a new modelling approach, the closure approach, to the planktonic ecosystem. The closure approach has been widely used for numerical models in turbulence research. In this approach, the variables are separated into mean and fluctuating parts, and the fluctuating parts are considered new variables of the system. By taking the spatial average of the governing equations, we get the closure equations, which are then solved to see the effect of fluctuation on the mean variables.

To justify this closure approach, as a first step we have considered the simple two-component NP system. After dividing each variable into mean and fluctuating parts, using the closure approach technique we have obtained five evolution equations among which two belong to the mean value of the variables, two correspond to the variance terms, and the last corresponds to the covariance term. Analysing this set of equations, first we found the domain of parameters and conditions for stability of the system (see Sections S4 and S5 in Text S1). We have observed that the increase of variance of overall spatial fluctuation (β-value) increases the stability zone of the system (Figure S5). Then we determined the critical value of fluctuation below which the non-closure modelling approach is sufficient to describe the dynamics of the system but above which the system attains a new steady state, which shows why the closure approach is necessary above this critical limit.

As this modelling approach handles variance and covariance terms, simulated results of this model are easily comparable with observed coefficients of variation. Depending on the variance of overall fluctuation (β-value), the system may have coefficients of variation greater or less than one. We have observed from our simulation that for high fluctuation, the coefficient of variation is higher, and we have also found a parameter domain (ε-β space) where the coefficient of variation is greater or less than one. By comparing model results with observed data sets, physically unmeasured parameter values, such as variance of overall fluctuation, can also be estimated.

Although the whole analysis is performed on a simple NP model, the results of this new approach show some beautiful features that we also observe in the real oceanic ecosystem. This novel technique for planktonic ecosystem modelling can be used for a system where the observed data sets are highly intermittent and normally difficult to compare with conventional non-closure models. Also, this method gives a general idea about the contribution of fluctuating components of the variables, which are generated by various biotic and abiotic factors of the system. In the next step of our work, we plan to incorporate physical dynamics, extend model variables in a more realistic way using this methodology and observe the effect of intermittency on the variables. How these fluctuating terms will change the results of global climate models when the time integration is extended over a few decades or more is not clear, but the results would be different from current predictions using these methods. The future of the climate is still unknown.

Supporting Information

Figure S1 Domain of parameter values for which the equilibrium point E_2 exists. Grey shaded regions indicate the domain for (A) $\beta=0.1$, (B) $\beta=0.5$, (C) $\beta=1.0$, (D) $\beta=1.5$, (E) $\beta=4.0$ and (F) $\beta=10.0$. Green, blue and red lines are the boundary values corresponding to the mean of phytoplankton (p_0), variance of fluctuating component associated with phytoplankton (x) and nutrient (y) respectively.

Figure S2 Domain of parameter values for which the equilibrium point E_3 exists. Grey shaded regions indicate the domain for (A) $\beta=0.1$, (B) $\beta=0.5$, (C) $\beta=1.0$, (D) $\beta=1.5$, (E) $\beta=4.0$ and (F) $\beta=10.0$. Green, blue and red lines are the boundary values corresponding to the mean of phytoplankton (p_0), variance of fluctuating component associated to phytoplankton (x) and nutrient (y) respectively.

Figure S3 Stability zone of the equilibrium point E_0. Two-parameter bifurcation diagram to show the stability zone for (A) $\beta=0.1$, (B) $\beta=0.5$, (C) $\beta=1.0$, (D) $\beta=1.5$, (E) $\beta=4.0$ and (F) $\beta=10.0$.

Figure S4 Stability zone of the equilibrium point E_1. Two-parameter bifurcation diagram to show the stability zone for (A) $\beta=0.1$, (B) $\beta=0.5$, (C) $\beta=1.0$, (D) $\beta=1.5$, (E) $\beta=4.0$ and (F) $\beta=10.0$.

Figure S5 Stability zone of the inner equilibrium point E_2. Two-parameter bifurcation diagram to show the stability zone for (A) $\beta=0.1$, (B) $\beta=0.5$, (C) $\beta=1.0$, (D) $\beta=1.5$, (E) $\beta=4.0$ and (F) $\beta=10.0$.

Figure S6 Stability zone of the inner equilibrium point E_3. Two-parameter bifurcation diagram to show the stability zone for (A) $\beta=0.1$, (B) $\beta=0.5$, (C) $\beta=1.0$, (D) $\beta=1.5$, (E) $\beta=4.0$ and (F) $\beta=10.0$.

Figure S7 Change of steady state level due to parameter variation for the closure model. For (A) $\beta=0.1$, (B) $\beta=1.0$, (C) $\beta=4.0$, (D) $\beta=10.0$. Initial value of $(p_0, x, y)=(0.3, 0.01, 0.8)$.

Figure S8 Change of steady state level due to change of parameter values for the non-closure model. Initial value of $p=0.3$.

Figure S9 Change of domain of CV for phytoplankton, for the changes of total nutrient A of the system. Here half-saturation constant K is kept constant at 1.2 μg N l^{-1}. Green lines are the boundary values corresponding to the mean of phytoplankton (p_0). The dashed line divides the parameter domain for CV<1 and CV>1.

Text S1 Supplementary text. This supplementary text contains seven sections (Section S1–S7) explaining model development and analysis referred in the main article.

Dataset S1 Experimental dataset. Phytoplankton data as collected by four sampling methods, namely Niskin bottle, Seapoint fluorometer, light emitting diode (LED) sensor and laser sensor, are given in this file.

Acknowledgments

We are grateful to the anonymous reviewers for their valuable comments and suggestions for improving the content of this article. We thank Dr. Dennis McGillicuddy of Woods Hole Oceanographic Institution for suggesting a new, more appropriate title for this manuscript. We thank Ms. Lynn Allmon for careful reading of the manuscript. We would also like to thank the captain and the crew of the T/R/V Seiyo Maru (Tokyo University of Marine Science and Technology) for their assistance with the micro-structure measurements presented in this paper.

Author Contributions

Conceived and designed the experiments: SM CL HY. Performed the experiments: SM CL. Analyzed the data: SM MT. Contributed reagents/materials/analysis tools: SM MT. Wrote the paper: SM HY.

References

1. Steele JH, Henderson EW (1981) A simple plankton model. Am Nat 117: 676–691. (DOI: 10.1086/283752).
2. Franks PJS, Wroblewski JS, Flierl GR (1986) Behavior of a simple plankton model with food-level acclimation by herbivores. Marine Biol 91: 121–129. (DOI: 10.1007/BF00397577).
3. Fasham MJR (1993) Modelling the marine biota. In: Heimann M, editor. The global carbon cycle. Berlin: Springer-Verlag. pp. 457–504.
4. Clark DR, Flynn KJ (2000) The relationship between the dissolved inorganic carbon concentration and growth rate in marine phytoplankton. Proc R Soc B 267: 953–959. (DOI: 10.1098/rspb.2000.1096).
5. Edwards AM (2001) Adding Detritus to a Nutrient–Phytoplankton–Zooplankton Model: A Dynamical-Systems Approach. J Plankton Res 23: 389–413. (DOI: 10.1093/plankt/23.4.389).
6. Loreau M (2001) Microbial diversity, producer-decomposer interactions and ecosystem processes: a theoretical model. Proc R Soc B 268: 303–309. (DOI: 10.1098/rspb.2000.1366).
7. Franks PJS (2002) NPZ models of plankton dynamics: their construction, coupling to physics, and application. J Oceanogr 58: 379–387. (DOI: 10.1023/A:1015874028196).
8. Armstrong RA (2003) A hybrid spectral representation of phytoplankton growth and zooplankton response: The "control rod" model of plankton interaction. Deep-Sea Res II 50: 2895–2916. (DOI: 10.1016/j.dsr2.2003.07.003).
9. Flynn KJ (2003) Modelling multi-nutrient interactions in phytoplankton; balancing simplicity and realism. Prog Oceanogr 56: 249–279. (DOI: 10.1016/S0079-6611(03)00006-5).
10. Huisman J, Thi NNP, Karl DM, Sommeijer B (2006) Reduced mixing generates oscillations and chaos in the oceanic deep chlorophyll maximum. Nature 439: 322–325. (DOI: 10.1038/nature04245).
11. Mandal S, Ray S, Roy S, Jørgensen SE (2006) Order to chaos and vice versa in an aquatic ecosystem. Ecol Model 197: 498–504. (DOI: 10.1016/j.ecolmodel.2006.03.020).
12. Gentleman WC, Neuheimer AB (2008) Functional responses and ecosystem dynamics: how clearance rates explain the influence of satiation, food-limitation and acclimation. J Plankton Res 30(11): 1215–1231. (DOI: 10.1093/plankt/fbn078).
13. Andersen KH, Pedersen M (2010) Damped trophic cascades driven by fishing in model marine ecosystems. Proc R Soc B 277: 795–802. (DOI: 10.1098/rspb.2009.1512).

14. Poulin FJ, Franks PJS (2010) Size-structured planktonic ecosystems: constraints, controls and assembly instructions. J Plankton Res 32(8): 1121–1130. (DOI: 10.1093/plankt/fbp145).
15. Segura AM, Kruk C, Calliari D, Garcia-Rodriguez F, Conde D, et al. (2013) Competition drives clumpy species coexistence in estuarine phytoplankton. Sci Rep 3: 1037. (DOI: 10.1038/srep01037).
16. Doney SC, Glover DM, Najjar RG (1996) A new coupled, one-dimensional biological-physical model for the upper ocean: Applications to the JGOFS Bermuda Atlantic time-series study (BATS) site. Deep-Sea Res 43: 591–624. (DOI: 10.1016/0967-0645(95)00104-2).
17. Franks PJS (1997) New models for the exploration of biological processes at fronts. ICES J Mar Sci 54: 161–167. (DOI: 10.1006/jmsc.1996.0191).
18. Denman KL, Peña MA (1999) A coupled 1-D biological/physical model of the northeast subarctic Pacific Ocean with iron limitation. Deep-Sea Res II 46: 2877–2908. (DOI: 10.1016/S0967-0645(99)00087-9).
19. Edwards CA, Batchelder HP, Powell TM (2000) Modeling microzooplankton and macrozooplankton dynamics within a coastal upwelling system. J Plankton Res 22 (9): 1619–1648. (DOI: 10.1093/plankt/22.9.1619).
20. Ji RB, Davis CS, Chen C, Beardsley RC (2009) Life history traits and spatiotemporal distributional patterns of copepod populations in the Gulf of Marine-Georges Bank region. Mar Ecol Progr Ser 384: 187–205. (DOI: 10.3354/meps08032).
21. Kishi MJ, Ito S-i, Megrey BA, Rose KA, Werner FE (2011) A review of the NEMURO and NEMURO.FISH models and their application to marine ecosystem investigations. J Oceanogr 67: 3–16. (DOI: 10.1007/s10872-011-0009-4).
22. Prairie JC, Franks PJS, Jaffe JS, Doubell MJ, Yamazaki H (2011) Physical and biological controls of vertical gradients in phytoplankton. Limnol Oceanogr Fluids Environ 1: 75–90. (DOI: 10.1215/21573698-1267403).
23. Stegert C, Ji R, Li N, Davis CS (2012) Processes controlling seasonality and spatial distribution of *Centropages typicus*: a modeling study in the Gulf of Maine/Georges Bank region. J Plankton Res 34: 18–35. (DOI: 10.1093/plankt/fbr084).
24. Azam F, Worden AZ (2004) Microbes, molecules, and marine ecosystems. Science 303: 1622–1624. (DOI: 10.1126/science.1093892).
25. Levin SA (1992) The problem of pattern and scale in ecology. Ecology 73: 1943–1967. (DOI: 10.2307/1941447).
26. Doubell MJ, Yamazaki H, Kokubu Y (2009) An advanced laser-based fluorescence microstructure profiler (TurboMAP-L) for measuring bio-physical coupling in aquatic systems. J Plankton Res 31(12): 1441–1452. (DOI: 10.1093/plankt/fbp092).
27. Waters RL, Mitchell JG, Seymour J (2003) Geostatistical characterisation of centimetre-scale spatial structure of *in vivo* fluorescence. Mar Ecol Prog Ser 251: 49–58. (DOI: 10.3354/meps251049).
28. Desiderio RA, Moore C, Lantz C, Cowles TJ (1997) Multiple excitation fluorometer for *in situ* oceanographic applications. App Optics 36: 1289–1296. (DOI: 10.1364/AO.36.001289).
29. Wolk F, Seuront L, Yamazaki H (2001) Spatial resolution of a new micro-optical probe for chlorophyll and turbidity. J Tokyo Univ Fish 87: 13–21.
30. Doubell MJ, Seuront L, Seymour JR, Patten NL, Mitchell JG (2006) High-resolution fluorometer for mapping microscale phytoplankton distributions. Appl Environ Microbiol 72: 4475–4478. (DOI: 10.1128/AEM.02959-05).
31. Franks PJS, Jaffe JS (2001) Microscale distributions of phytoplankton: initial results from two-dimensional imaging fluorometer, OSST. Mar Ecol Prog Ser 220: 59–72. (DOI: 10.3354/meps220059).
32. Prairie JC, Franks PJS, Jaffe JS (2010) Cryptic peaks: Invisible vertical structure in fluorescent particles revealed using a planar laser imaging fluorometer. Limnol Oceanogr 55(5): 1943–1958. (DOI: 10.4319/lo.2010.55.5.1943).
33. Katz J, Donaghay PL, Zhang J, King S, Russell K (1999) Submersible holocamera for detection of particle characteristics and motions in the ocean. Deep-Sea Res I 46: 1455–1481. (DOI: 10.1016/S0967-0637(99)00011-4).
34. Davis CS, Thwaites FT, Gallager SM, Hu Q (2005) A three-axis fast-tow digital Video Plankton Recorder for rapid surveys of plankton taxa and hydrography. Limnol Oceanogr 3: 59–74. (DOI: 10.4319/lom.2005.3.59).
35. Malkiel E, Abras JN, Widder EA, Katz J (2006) On the spatial distribution and nearest neighbor distance between particles in the water column determined from *in situ* holographic measurements. J Plankton Res 28: 149–170. (DOI: 10.1093/plankt/fbi107).
36. Franks PJS (2005) Plankton patchiness, turbulent transport and spatial spectra. Mar Ecol Prog Ser 294: 295–309. (DOI: 10.3354/meps294295).
37. Denman KL, Gargett AE (1995) Biological-physical interactions in the upper ocean: the role of vertical and small-scale transport processes. Ann Rev Fluid Mech 27: 225–255. (DOI: 10.1146/annurev.fl.27.010195.001301).
38. Doubell MJ, Prairie JC, Yamazaki H (2014) Millimeter scale profiles of chlorophyll fluorescence: Deciphering the microscale spatial structure of phytoplankton. Deep-Sea Res II 101: 207–215. (DOI: 10.1016/j.dsr2.2012.12.009).
39. Blackburn N, Fenchel T, Mitchell J (1998) Microscale nutrient patches in planktonic habitats shown by chemotactic bacteria. Science 282: 2254–2256. (DOI: 10.1126/science.282.5397.2254).
40. Young WR, Roberts AJ, Stuhne G (2001) Reproductive pair correlations and the clustering of organisms. Nature 412: 328–331. (DOI: 10.1038/35085561).
41. Waters RL, Mitchell JG (2002) Centimetre-scale spatial structure of estuarine *in vivo* fluorescence profiles. Mar Ecol Prog Ser 237: 51–63. (DOI: 10.3354/meps237051).
42. Lehman JT, Botkin DB, Likens GE (1975) The assumptions and rationales of a computer model of phytoplankton production dynamics. Limnol Oceanogr 20: 343–364.
43. MacIsaac JJ, Dugdale RC (1969) The kinetics of nitrate and ammonia uptake by natural populations of marine phytoplankton. Deep-Sea Res 16: 45–57. (DOI: 10.1016/0011-7471(69)90049-7).
44. Eppley RW, Rogers JN, McCarthy JJ (1969) Half-saturation constants for uptake of nitrate and ammonium by marine phytoplankton. Limnol Oceanogr 14: 912–920.

9

Integration at the Round Table: *Marine Spatial Planning in Multi-Stakeholder Settings*

Erik Olsen[1]*, David Fluharty[2], Alf Håkon Hoel[3], Kristian Hostens[4], Frank Maes[5], Ellen Pecceu[4]

1 Institute of Marine Research, Bergen, Norway, **2** School of Marine and Environmental Affairs, University of Washington, Seattle, Washington, United States of America, **3** Institute of Marine Research, Tromsø, Norway, **4** Animal Sciences, Aquatic Environment and Quality, Bio-environmental Research, Institute for Agricultural and Fisheries Research (ILVO), Oostende, Belgium, **5** Maritime Institute, Ghent University, Ghent, Belgium

Abstract

Marine spatial planning (MSP) is often considered as a pragmatic approach to implement an ecosystem based management in order to manage marine space in a sustainable way. This requires the involvement of multiple actors and stakeholders at various governmental and societal levels. Several factors affect how well the integrated management of marine waters will be achieved, such as different governance settings (division of power between central and local governments), economic activities (and related priorities), external drivers, spatial scales, incentives and objectives, varying approaches to legislation and political will. We compared MSP in Belgium, Norway and the US to illustrate how the integration of stakeholders and governmental levels differs among these countries along the factors mentioned above. Horizontal integration (between sectors) is successful in all three countries, achieved through the use of neutral 'round-table' meeting places for all actors. Vertical integration between government levels varies, with Belgium and Norway having achieved full integration while the US lacks integration of the legislature due to sharp disagreements among stakeholders and unsuccessful partisan leadership. Success factors include political will and leadership, process transparency and stakeholder participation, and should be considered in all MSP development processes.

Editor: Judi Hewitt, University of Waikato (National Institute of Water and Atmospheric Research), New Zealand

Funding: The project was jointly funded by the EU FP-7 research project "Monitoring and evaluation of spatially managed areas (MESMA)" www.mesma.org, and by internal funding from the authors institutions: the Institute of Marine Research, University of Washington, Ghent University and the Institute for Agricultural and Fisheries Research (ILVO). The funders had no role in study design, data collection and analysis, decision to publish, or preparation of the manuscript.

Competing Interests: The authors have declared that no competing interests exist.

* Email: eriko@imr.no

Introduction

The management of marine ecosystems and their human activities underwent dramatic changes since the 1992 Rio Declaration on Environment and Development. Today sustainable development is still a fundamental principle, but following the 2002 Johannesburg Declaration, sustainable management has been expanded to encompass not only single species or sectors, but the whole ecosystems through the "ecosystem approach to management" (EBM) [1,2]. EBM is a powerful concept based on analysing and managing the ecosystems from a holistic approach, taking into account all components, pressures and impacts [3,4].

Because ecosystems are spatially explicit, area-based management approaches offer a suitable and efficient way of implementing EBM into practice [5,6]. One such approach is marine spatial planning (MSP), which offers an effective perspective to deal with the challenging issues of multiple use and multiple (cumulative) impacts in EBM [7]. Even though a spatial perspective has been successful in terms of leading to novel and forward-looking ecosystem-based management plans, like in Norway and Australia [8–11], the development of marine spatial plans is a complex process at the borders between science, management and politics. Several approaches, like the UNESCO 10-step approach [12], have been developed to lead the practitioners safely from start to finish during planning. The establishment of effective governance of the planning, implementation and review processes, is a fundamental step to develop sound MSP [12,13].

Because MSP by definition is multi-sectoral, a potentially large number of managers, stakeholders and policy-makers are involved, each accustomed to operate on his own (*i.e.* within specific sectors). Successful MSP means getting all these actors to communicate and work together in an integrated way. Therefore, integration means crossing boundaries at professional, physical, institutional or administrative level [14]. To develop the appropriate measures in an integrated (ecosystem) MSP setting, the integration of concerns and interests across sectors (horizontal integration) and between governmental levels or between government and stakeholders (vertical integration) is required. Integration is fundamental to MSP and especially important to pro-actively resolve spatial conflicts [14].

In this paper we explore how such integration has been dealt with in three markedly different MSP processes: the Belgian MSP covering the whole Belgian EEZ in the North Sea, the Norwegian Integrated Management plans – three plans covering the Norwegian EEZ, and the current US National Ocean Policy (Figure 1) regional planning process now being implemented. We illuminate how the integration of concerns and interests varies with

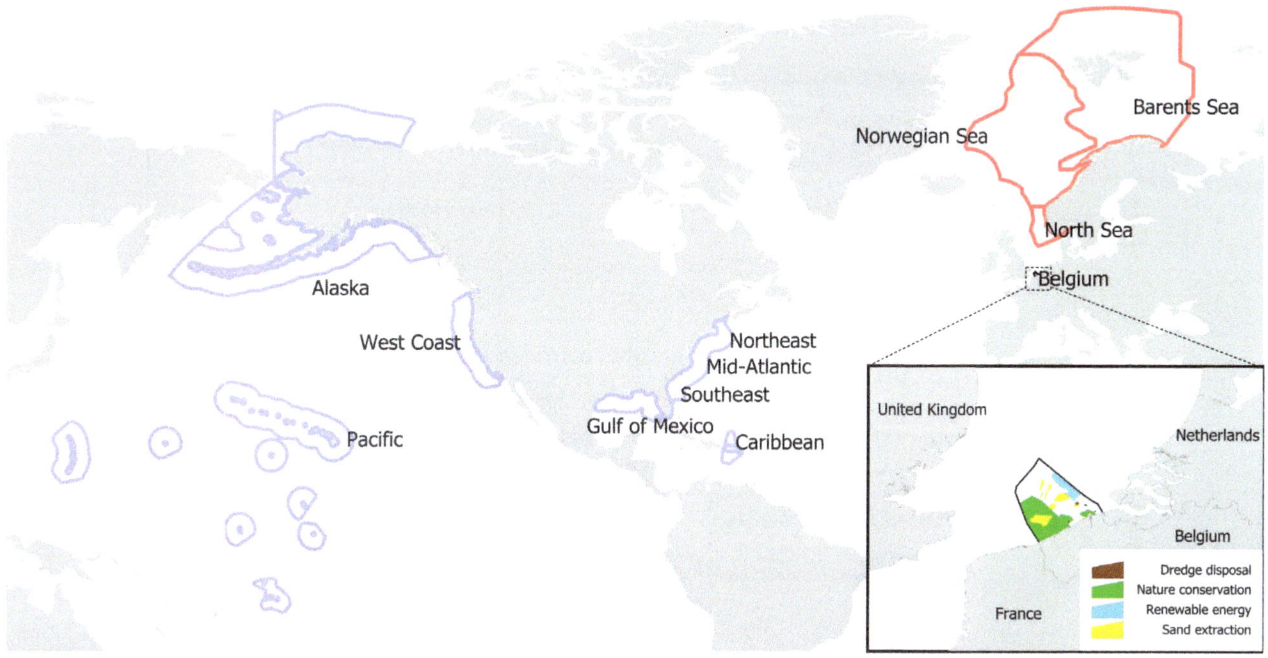

Figure 1. Marine spatial planning (MSP) study areas around the North Atlantic. United States (blue), Norway (red) and Belgium (Green, detail in inset). The US ocean policy area is divided into 9 planning areas (named in figure) that span the EEZ, but the borders between each of these areas have not been officially defined.

the context and commonalities that allow for a successful integration of different viewpoints and hence successful governance of MSP.

The Three Case Studies

Belgium: from Masterplan to MSP

In March 2014, Belgium approved a legally binding Marine Spatial Plan (Royal Decree of 20 March 2014 on the adoption of the MSP) for the Belgian part of the North Sea (BPNS), a small area of ca. 3500 km^2 covering the territorial sea, the continental shelf and the entire Belgian Exclusive Economic Zone (EEZ). MSP in Belgium is the result of a long process, going back to initial attempts to implement the 1999 Marine Environmental Protection Act (MEPA), followed by the so-called "Masterplan for the BPNS" in 2003–2005. Since 2012 an explicit legal basis for MSP in Belgium is embedded in the MEPA, which clearly indicates the environmental roots of MSP. Therefore, governments and users of the sea in Belgium have to take into account several legal principles, namely the principle of prevention, the precautionary principle, the polluter pays principle and the restoration principle (art. 5 MEPA).

Belgium functions as a multi-level government with authorities divided among local, regional, federal (and European) levels. The marine competences are divided between the federal state and the Flemish region, and within each level, competences are fragmented over several departments [15]. No less than 17 governmental institutions (both Flemish and federal) have some form of competence at sea. Both governments are legally equal in adopting their legislation within their fields of competence (Figure 2). Consultation between the two governments can be held at ministerial level, between their cabinets or between their administrations. Matters in which both governments have different exclusive competences can be jointly dealt with in formal cooperation agreements.

It took until 2003, with the appointment of a federal Minister for the North Sea, to set a major step forward in the Belgian MSP process. This Minister was mandated to spatially co-ordinate all activities and competences at sea, except for fisheries which is a regional (Flemish) competence. He initiated a more strategic approach to the (potential) conflicting claims between users of the BPNS which emerged between 1999 and 2003. He developed the BPNS Masterplan, which actually combined several political decisions concerning marine matters that were taken by the federal Council of Ministers and implemented by a number of Royal decrees. The plan was further adopted in two phases: in 2003 with an agreement on the delimitation of several zones for two economic activities- aggregate extraction and offshore renewable energy - and later in 2005 with the delimitation of nature conservation areas for birds and habitat protection. However, it remained essentially a zoning plan, without a legal basis for a planning process or an integrated policy approach, nor a clear and transparent process for stakeholder involvement and public participation [16].

In Belgium, there is strong vertical integration between the highest levels (Ministers) and their public administrations. Moreover, each governmental level has its own advisory bodies (Figure 2), representing the major stakeholders, e.g. the Federal Council for Sustainable Development (FRDO), the Flemish Council for Environment and Nature (MiNa Council) and the Flemish Strategic Advisory body for Agriculture and Fisheries (SALV). It is important to recognize that despite the existence of sectoral cooperation agreements and advisory bodies, the concerned parties only have the mandate and responsibility for their own competences.

Stakeholders have been involved in the MSP process since the Masterplan was presented in 2003, but this happened only on a sectoral basis and mostly through bilateral consultations behind closed doors [17]. This was in line with an old Belgium practice of informal consultations within and between the governments and

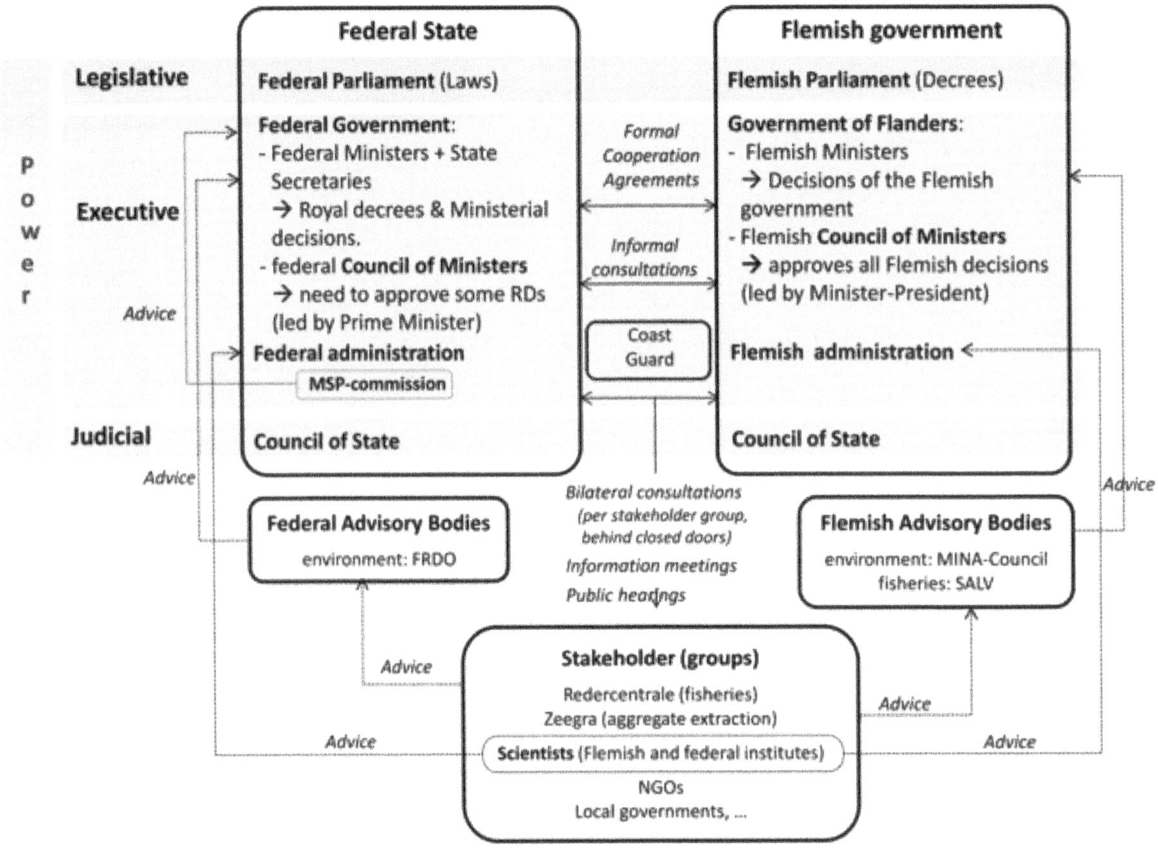

Figure 2. Institutional and stakeholder integration in Belgium. Institutional bodies, stakeholder participation and integration, related to marine and maritime governance in the Belgian part of the North Sea.

administrations, as well as between the governments and stakeholders. As such the Masterplan was drawn up on an '*ad hoc*' basis, taking into account the individual and selective demands of sectoral organizations and individual stakeholders. As a consequence, consultation with other sea users than the ones in the focus of the Masterplan was not deemed necessary. After 2006, the Masterplan has been adjusted a few times, in particular to take into account shipping interests and to designate a new marine protected area (2008–2010). However, these developments were once again dealt with at sectoral level.

In 2012, the newly elected federal government re-appointed the same Minister for the North Sea, and a second phase of MSP started. The new marine spatial plan departed from the BPNS Masterplan, but with a clear aim to establish a 'legally binding' marine spatial plan and process. The draft spatial plan, which was again based on informal and formal consultations with stakeholders, was subjected to a series of administrative consultations and advice from a new Advisory Commission on MSP. The latter was formally established by a Royal Decree in November 2012 and involves all federal authorities with competences at sea, next to 'invited' Flemish authorities. As such, an integration between all federal governmental departments concerning MSP became formally embedded in the new MSP legislation.

Next to the MSP Advisory Commission, experts were consulted, stakeholder participation was organized and a strategic environmental assessment (SEA) of the spatial plan was conducted. In July 2013, an updated plan was approved by the Council of Ministers, followed by consultation with the public and neighboring countries and advice from advisory bodies and regional authorities. The Ministry of Environment took into account 140 remarks and finally the Council of Ministers approved the MSP by Royal Decree on 20 March 2014.

The new Belgian marine spatial plan comprises the coordinates of all delimitated zones for the activities that are allowed and limited or prohibited in the BPNS, including some new zones for future activities (energy atolls, sustainable aquaculture, etc.). The Royal Decree also contains several annexes outlining the spatial context and explaining the policy choices that have been made. The long term vision to implement MSP in Belgium foresees concrete objectives for the period 2014–2020 and a revision of the spatial plan every six years, although in between adaptation of the plan is not excluded.

A major shortcoming of the Belgian MSP might be that fisheries is only partly included in the MSP, as this is a not a competence of the federal government (Flemish competence). Nevertheless, consultations took place at the highest political (ministerial) level, and certain limitations for fisheries are included in the new MSP, such as the prohibition of fisheries in the offshore renewable energy concession zones, and a limitation for certain fishing techniques (mainly classic bottom disturbing trawling techniques) in specific parts of the Natura 2000 area 'Vlaamse Banken' in order to achieve a good environmental status for the BPNS.

Norway: Integrated Management Plans

As a response to the international [1] and regional drivers (North Sea ministerial conference in 1997) for EBM as well as national pressures from the petroleum industry to get access to new areas further north and along the coast, the new government heralded marine Integrated Management Plans (IMPs) in its coming-to-power declaration in 2001 [18]. The development of these IMPs started in 2002, with the first plan, for the Lofoten–Barents Sea area implemented in 2006 [8,18], the Norwegian Sea plan in 2009 [9], and the last plan for the North Sea and Skagerrak area in 2013 [19]. These IMPs are regional plans integrating all human uses of the area in a spatial context through sectoral zoning and geographical analysis. The Lofoten-Barents Sea plan has been revised once in 2011, the Norwegian Sea plan is due for revision in 2014.

An inter-ministerial steering group led by the Ministry of Environment was set up to coordinate the development and implementation of these IMPs. The steering group had members from all relevant sector-ministries (e.g., Foreign Affairs, Fisheries and Coastal Affairs, Petroleum and Energy). This strong political (top-down) steering group tasked the different institutions and directorates under each ministry to contribute to the development and implementation of the plan. Each plan was customized for the particular ecosystem, but with a similar overarching strategic objective: "*to allow for sustainable use while ensuring continued ecosystem health*" [20]. The duality in this main strategic objective reflects the underlying conflict to integrate two widely different objectives, namely the push for increased industrial (petroleum) developments and the protection of the ecosystems (to ensure the health and survival of species and habitats) [21].

Integration of different interests and concerns is central to the Norwegian MSP process, and already commenced at the start of the planning process in 2002. The various sectors and levels (of government) have been integrated by jointly developing the knowledge base and management measures across sectors and between levels of government (Figure 3). Joint groups, forums and meeting places across sectors have been set up to achieve horizontal – cross-sectoral integration. Hearings, open public meetings and sectoral meetings with the government help to foster cooperation between the various governmental levels and with stakeholders in particular.

The Norwegian planning process is centralized, where decisions are made by the government and subjected to parliamentary approval. As such, all IMPs have been passed with broad political approval in the Parliament. Involvement of local government levels (e.g. counties) has been poor, mainly because ocean management is a national issue in Norway, with only limited local control over marine resources or management of marine space.

The overarching conflict of industrial use versus nature protection has been a central issue for debate at all levels, in particular in the management and risk forums (Figure 3). In these multi-sector groups the different views allowed for cross-sectors insight and discussion, and although this has not resolved the conflict, it has increased mutual understanding and respect, based on the personal, professional and institutional relationships that have been established during the planning and implementation process. The networking has increased trust among the participants, which made it possible to discuss difficult and challenging issues.

Imposing consensus between the sectoral government institutions was instrumental to develop and implement the Lofoten-Barents Sea plan in a 4-year time frame. This first IMP spearheaded the development of the other plans and created much of the structure and methods to allow the two later plans to be created in 3-year processes. However, seeking consensus could also lead to suppression of substantial and value-based differences. It can be questioned whether the decision-makers would be better served if the differing views and options in the central conflict (industrial use vs. nature conservation) would have been clearly presented as different choices rather than hidden in a compromise. A consensus approach limits the options for the decision-makers, who normally prefer a range of options which allows them more room to govern.

The existing sectoral legislation and management structures have been the main barrier for integration. Using soft-law in processes like the development of IMP's, is a tradition in Norway, and is fast, pragmatic and efficient. Also, by implementing through existing legislation and institutional structures, the policy is rapidly and effectively translated into practice without spending much time on developing new legislation or governance structures. However, this approach does little to integrate concerns across the existing sectoral legislation. Nevertheless, many new sectoral laws passed since 2002, took the integrated management approach into consideration and specifically required integration between sectors.

US: Ocean Policy Executive Order

Influenced by international MSP developments and strongly promoted by major environmental organizations, the US started a process towards developing an Ocean Policy [22] including MSP. The first steps were already taken by the Bush administration, but development efforts increased in 2008 with the new Obama administration. After 18 months of intensive planning, the National Ocean Policy was effectuated through President Obama's Executive Order 13547 (July 19, 2010). This Executive Order established the National Ocean Council, consisting of 27 federal agencies, departments and offices. Additionally, the US marine and Great Lakes areas are divided into nine regions, which can form regional planning bodies on a voluntary basis to develop the planning process in their respective regions (Figure 4). The National Ocean Council published an Implementation Plan [23], that further gives guidance to the administration and partner institutions on how the Ocean Policy should be implemented [24]. The Obama Ocean Policy approach relies on existing legal mechanisms [25].

The Ocean Policy can be seen as a MSP process, although the term 'marine spatial planning' is not used in the most recent documents as it has become politicized to the point that the US Congress in 2012 and 2013 refused to fund any MSP related activities. The majority in the House of Representatives see MSP as an unnecessary layer of government, hampering the development of business activities at sea. Also industrial (oil, fishermen, etc.) lobby groups (e.g., National Ocean Policy Coalition, see http://oceanpolicy.com) advocated against MSP. Therefore, MSP implementation has ended up in the middle of the partisan battlefield, even though the first steps were taken by the Bush Administration with bipartisan support. As a result, the federal support for the Ocean Policy has decreased compared to what was anticipated in 2010. Currently, the Ocean Policy implementation is totally dependent on support from the State and regional levels, who are leading the way while the federal level lags behind. Substantial financial support has however been provided by private donors to MSP efforts in the Northeast and Mid-Atlantic regions. It is critically important that top-down and bottom-up processes are linked in the implementation of the US Ocean Policy [26], however the situation today is charitably defined as experimental and otherwise defined as chaotic when looking across all nine regions.

Some States have taken a leadership positions in MSP for their coastal environments, e.g., Rhode Island, Massachusetts, Oregon, New York, Hawaii and Florida, with Washington State not far behind [27]. Also, California has taken major steps to develop

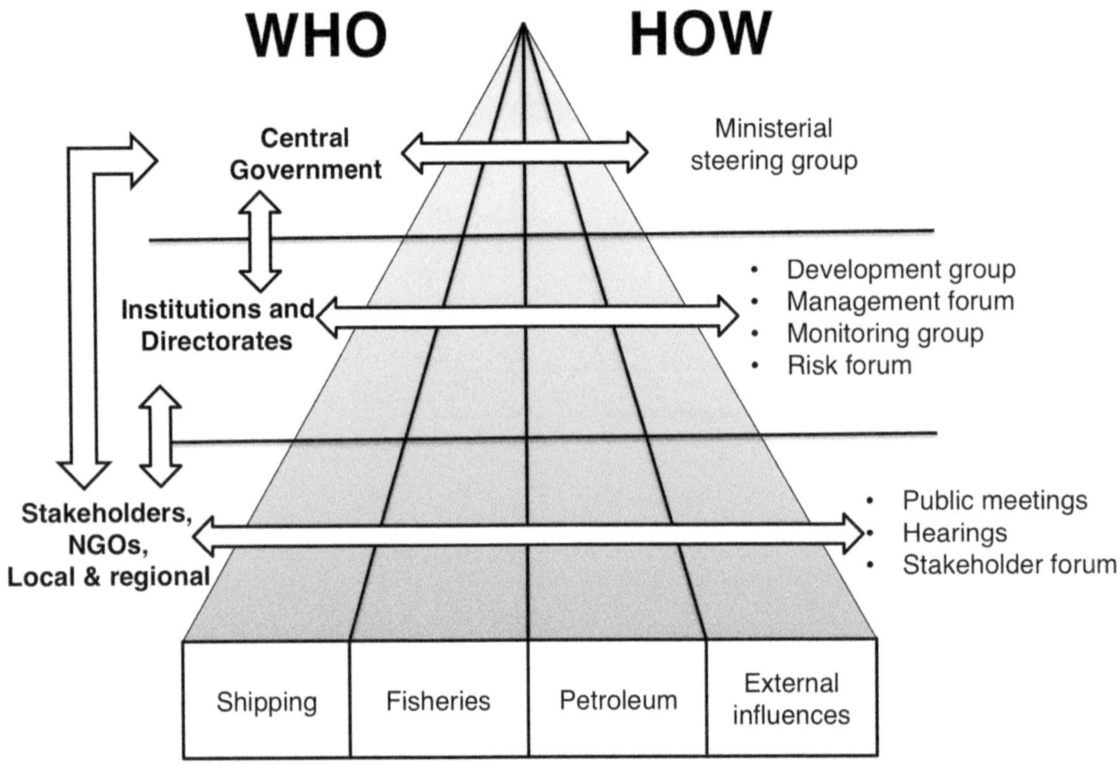

Figure 3. Institutional, government, and stakeholder integration in Norway. Institutional bodies, stakeholder participation and integration, related the development and implementation of the Integrated Management plan for the Lofoten – Barents sea area.

marine spatial planning through a marine reserve network in state waters. In addition, Maine, New Hampshire, Massachusetts, Rhode Island and Connecticut together with federal partners formed the Northeast Regional Planning body (RPB) as envisioned by the Ocean Policy (Figure 4). The Northeast RPB constituted itself, meets regularly, and is on its way to develop a regional plan in 2015. Also, the Mid-Atlantic and Southeast Atlantic have formed RPBs, with the Mid-Atlantic having started a planning process in 2014, while the Southeast is a bit delayed by the limited participation from Florida. These RPBs are expected to produce regional MSPs by 2015 for review and approval by the National Ocean Council but this timeline is ambitious.

Other regions, except Alaska, have taken up on the Ocean Policy and have or are in the process of forming RPBs. Hawaii (part of the Pacific region) had a state plan under development when the Ocean Policy was being put forward. The greatest challenge in the Pacific region is distance, as it includes Hawaii and all US Pacific Islands, spanning seven time-zones and the international date-line. Finding time, place and funding to bring regional representatives together, is logistically very demanding. The Pacific RPB may ultimately adopt a more archipelagic approach as has been done for fisheries management [28].

The National Ocean Policy envisions a regional approach to develop MSPs for large management regions, *in casu* ecosystems, based on the regional administrative capacity. However, it is recognized that the desire for standard approaches across the nation may not be accommodated at the regional level. Therefore, a flexible structure of Federal agency engagement with States and local governments is required. All current RPBs have wide participation of federal, state and tribal governmental institutions that have formal mandates for ocean management (Figure 4). The RPB work and meetings are open to the public and public interaction is actively encouraged through specific sessions where the public is allowed to ask questions or make comments to the full RPB. These occasions are actively used by self-selected and motivated NGOs, interest groups and the public to voice their perspectives.

Discussion - Comparing Integration in the Three Cases

Spatial scale of the managed area

Belgium, Norway and the US represent three different approaches of implementing MSP, stemming from both geography and governance [26]. Whereas the Belgium plan spans the whole EEZ and Belgian national waters, an area of a few thousand square kilometres, both the Norwegian and US plans span millions of square kilometres under different climate and biogeographic regions, making them much more complex than the Belgium plan area. To better handle the varied climatological and geographic patterns, both the US and Norway chose to divide their marine areas into (eco)regions and are developing separate plans for each area. Regardless of the size of the spatially managed area, it remains a complex issue to manage all types of human activities under one umbrella, requiring integration between different sectors, stakeholders and governmental levels. Reducing the size of the planning area will reduce complexities, but only to a certain level, as the multi-sectoral cross-political aspects remain, irrespective of spatial scale.

Legislation, political will and leadership

The three countries differ markedly in terms of governance: Belgium is an EU member and marine governance has to adhere to EU policies (e.g. Common Fisheries Policy) and directives (e.g.,

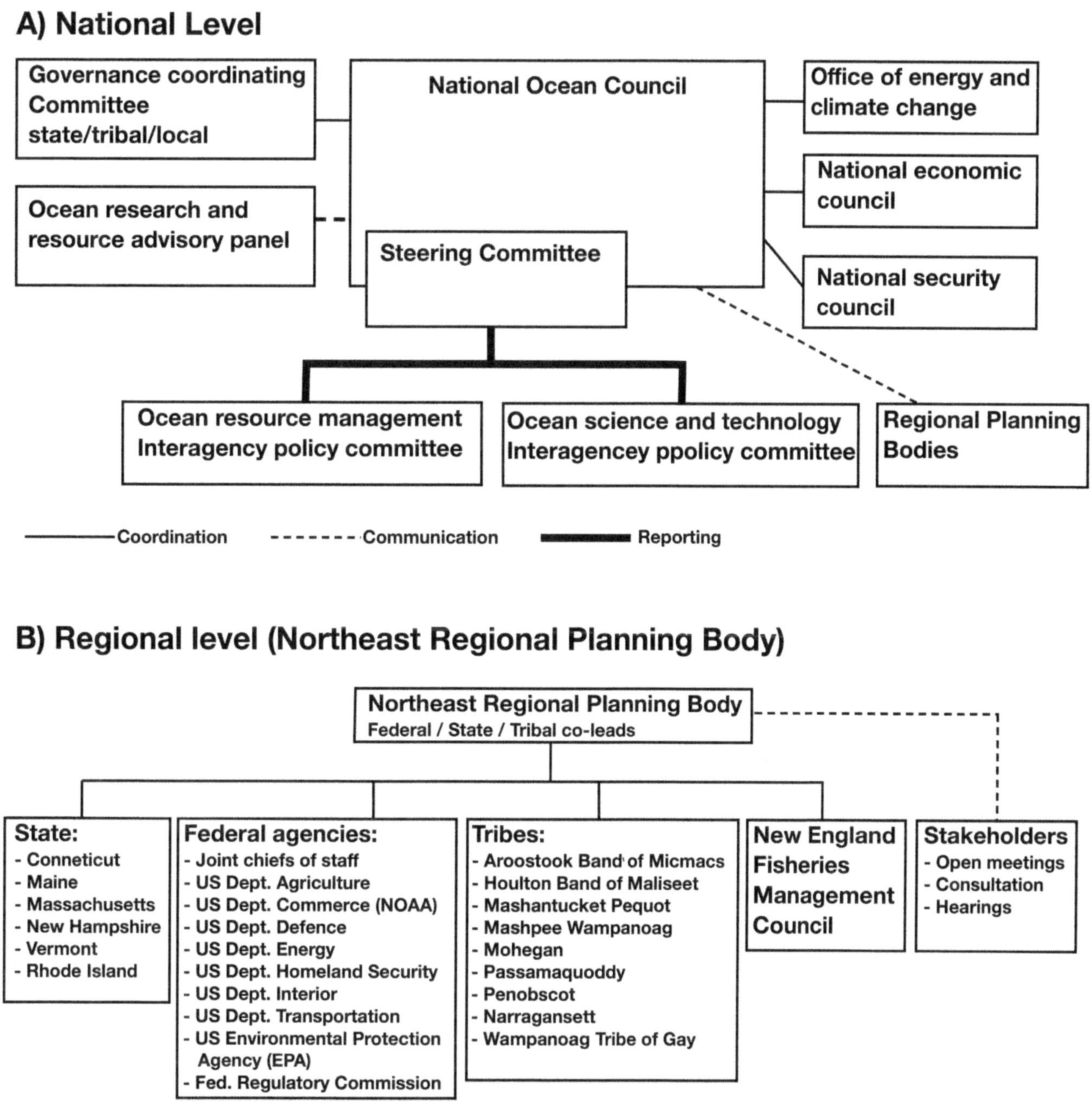

Figure 4. Governance structure and integration in the US Ocean Policy development. A) At a national level leading the Ocean Policy development. B) Regional implementation in regional planning bodies (RPB), exemplified by the Northeast RPB.

Habitats and Bird, Marine Strategy Framework, Water Framework). Belgium has a strong formal division of power between the federal and regional levels, although integration between both governmental levels is assured through formal agreements and informal consultations. This is also reflected in the protracted Belgian MSP process: after more than 10 years, the MSP plan and process are firmly embedded in environmentally driven legislation.

Norway is not a member of the EU and has an ethnically homogenous population. The government is well integrated across sectors both at central and regional levels. The Norwegian Integrated Management Plans (IMPs) were passed through the Parliament, but are not embedded in Norwegian legislation. They are rather government reports with MSP objectives and a vision on how they will achieve these objectives through a range of management options, including spatial zoning.

In the US the independency of the states led to a geographic division in marine management at the border between federal and state waters, mostly three nautical miles offshore. The US Ocean Policy process, based on a Presidential Executive Order, does not have backing in the Congress, which led to funding restrictions to develop or continue the MSP process. The US Ocean Policy became heavily politicized in 2010 and has since been drawn into a struggle among stakeholders largely along commercial users of ocean space and biodiversity conservation, spilling over into the partisan debates in Congress. Even though embedding MSP in legislation creates a more stable solution in heterogeneous societies like Belgium and the US, achieving overall trust and buy-in is instrumental for a successful implementation (and funding) of the MSP process and plan.

In contrast to Norway and Belgium, broad political support for implementation of MSP has not been reached in the US, due to the lack of strong stakeholder support for MSP by commercial and non-commercial users. Strong political leadership was instrumental to achieve that bipartisan support for MSP in both Belgium and Norway. In Belgium the personal charisma, tenacity and clout of the Minister of the North Sea was pivotal to lead the planning process and persuade the different parties of the necessity of MSP. In Norway it was the tenacity and clout of the Ministry of Environment, strongly supported by the Ministries of Fisheries and Foreign Affairs that pushed the MSP process. Moreover, achieving broad political buy-in was relatively easy in Norway, as the MSP process was partly driven by Norway's foreign policy. Foreign priorities are very stable in Norway and do not change with changing governments. In the US leadership has certainly been shown at the Executive level but it is not sufficient to carry through to the legislative leadership.

Drivers and incentives

Leaving aside political and ethnic differences between the countries, socio-economics, i.e. the importance of certain marine industries in the national economy of each country, help to explain some of the differences in their governance approach to MSP. In Belgium, maritime infrastructure (shipping and ports) is very important, next to aggregate extraction and to a lesser extent fisheries. Since 2003 offshore renewable energy (wind-farms) developments have been the main driver for MSP in Belgium. However, adherence to EU environmental directives (Habitats, Bird, Marine Strategy Framework Directive) has created a second driver in MSP in order to identify areas of high biological value (Natura2000 sites) and to achieve a "good environmental status" of the ecosystem. In Norway, the marine natural resources (oil and fish) form the basis for the country's welfare, and sea-based industries (oil/gas, fisheries and aquaculture) are of fundamental socio-economic importance. It is a high priority for all governmental levels to create (sustainable) marine value in Norway. The health of the marine ecosystem is understood as an important prerequisite for sustainable value creation, but this is not seen as an absolute hindrance to human uses. Therefore, protecting areas by excluding economic activities is not common in Norway.

In the US, sea-based industries, shipping and fishing are of great socio-economic importance, but the country is not as dependent on the exploitation of marine resources (oil/gas and fish) as Norway. The importance and political power of the different sectors vary regionally, e.g., in the Gulf of Mexico the petroleum industry is given priority, while in the Northeast region offshore renewable wind energy has been a driver for MSP. The US has a strong terrestrial tradition of using protection (national parks and forests) as a management measure. This has been transferred to the marine environment with the establishment of several MPAs with a high protection level like the Papahānaumokuākea Marine National Monument (MNM) in the north-western Hawaiian islands, the Marianas Trench MNM and the Pacific Remote Islands MNM.

The economic incentives have led to different objectives in the respective MSP processes. Although all three countries try to combine ecosystem health with economic use, they differ in their approach. While Norway has a clearly stated dual goal of sustainable use while ensuring a healthy ecosystem, Belgium is focusing on protection of valuable habitats and ecosystem components while finding room for new uses and minimizing conflicts. The US objectives are complex and not prioritized in terms of trade-offs between economic use and nature conservation.

Stakeholder participation and transparency

Integrating stakeholders in the planning process has been important, but done in different ways in the three countries. In Norway and Belgium, stakeholders were brought early on board, which led to acceptance of the MSP process, although with much debate about specific details of the plan. In the US, stakeholders were invited to comment during the 18 month process leading up to the federal Ocean Policy Executive Order. This process built on many years of responding to the US Commission on Ocean Policy and the non-governmental Pew Oceans Commission processes. However, there was considerable surprise that MSP was the most tangible action required by the Executive Order. This shift in approach resulted in several industrial stakeholder groups forming lobby-groups against MSP implementation, which blocked necessary incremental funding of the MSP process. The resultant stakeholder skepticism conveniently fed into the partisan political battle in Washington D.C. In the meantime, stakeholder involvement and transparency is much more effective at the state and regional level in the US. Threats and opportunities are tackled depending on one's perspective from existing spatial conflicts, and through new drivers primarily associated with both non-renewable (import/export) and renewable (offshore) energy. A credible threat to established interest adds to the acceptance and continuation of the MSP process in most US regions. Full transparency, open meetings, public documents and active solicitation of stakeholder opinions, make the US stand out compared to Belgium and Norway, where transparency was more limited and controlled.

Conclusion – Integration at the MSP Round Table

Ecosystem-based management in general and MSP in particular is a multifaceted management approach that combines the management of multiple sectors and goals under one roof through integration. Integration can be measured both in relation to geographic scale and governance scope [14] according to which all three cases (Belgian, Norwegian and the US) can be classified as "highly integrated". However, there are some marked differences between the cases in their horizontal and vertical integration (Table 1).

Horizontally, all three processes are similarly well-integrated, but it has been achieved through different governance approaches. In Belgium a "Minister for the North Sea" was appointed to lead the process, Norway set one ministry (Environment) in charge to lead a multi-ministerial steering group in which the existing ministries were tasked to cooperate, while the US set down an Executive (Ocean Policy) Task Force, directly reporting to the Council on Environmental Quality in the White House and operating through government agencies on a (eco)regional basis. Both the Belgian and Norwegian processes are integrated on a national level, while the US proposed to integrate on a regional level (which is sensible as US regions are much larger than most European states). Horizontal integration across sectors relies strongly on each sector perceiving it to be treated in a fair and equitable manner. Therefore, meeting places and forums should be neutral – amalgamating the Arthurian 'round table' of Camelot that created peace among the noble but imperious knights closest to the king (in our examples the relevant ministers or president).

Vertically, Belgium and Norway are well-integrated from parliament (through the executive government levels) to stakeholders, while the US is well integrated from the Executive level

Table 1. Horizontal, vertical and stakeholder integration in the MSP processes in Belgium, Norway and the US.

	Horizontal integration	Vertical integration	Stakeholder integration
Belgium	Good integration. Minister for the North Sea ensured a high government focus and top-level integration. Important federal and partly regional integration achieved	All levels of government well integrated	Consultation and formal roles in planning group
Norway	Good integration both at top government levels and at government institutional levels	All levels of government integrated	Consultation and participation in some meetings
US	Good integration between sectors at state and regional levels where direct threats or opportunities are perceived. Poor integration at national level due to lack of prioritization of objectives and failure to produce adequate resources to support on-going efforts	Lacking integration with lawmakers (congress).	Open public meetings of regional planning bodies. Consultation at national level. Strong stakeholder lobbying

downwards, but lacks integration (acceptance) from the lawmakers (Congress).

Next to informal consultations with sectoral stakeholder groups, stakeholders also have a formalized role in some planning forums in Belgium. In Norway and the US, stakeholder participation remains informal and voluntary or self-selected (through consultations), although in the US stakeholder involvement is more comprehensive, reflected in the strong lobbying work of stakeholder groups to the government and the ability to file lawsuits based on administrative procedures and environmental compliance. Planning meetings are open to the public in the US, in contrast to Norway where meetings in governmental steering groups are held behind closed doors. The latter was also true for Belgium in the first MSP (2003–2005), but changed with the adoption process of the second MSP (2013–2014) when stakeholder participation and public consultation were included in the process. Although the current political situations may be regarded as challenging, the three cases illustrate the benefit of having a wide political backing to efficiently implement broad management processes like MSP. Where the MSP process tends to be top-down, formally defined processes and leadership are deemed necessary, while also bottom-up processes are highly valuable when it comes to integrating across vertical and horizontal scales.

In Belgium and Norway the process has led to effective and practical implementation of MSP, while in the US implementation is still pending. Our analysis shows that all three planning efforts can be regarded as well-integrated in intent, and that integration is important for a successful MSP. In addition to seeking and supporting political leadership, planners should keep in mind that processes and structures that facilitate effective integration are needed. Transparency and openness of access to stakeholders to make their perspectives be heard and analysed in the total process is a large part of what is required to ensure fair and equitable treatment, in relation to both horizontal and vertical integration.

Acknowledgments

This work was carried out as part of the EU FP-7 project "Managing and evaluation of spatially managed areas" (MESMA).

Author Contributions

Conceived and designed the experiments: EO DLF AHH FM EP. Performed the experiments: EO DLF AHH KH FM EP. Analyzed the data: EO DLF AHH KH FM EP. Contributed reagents/materials/analysis tools: EO DLF AHH KH FM EP. Contributed to the writing of the manuscript: EO DLF AHH KH FM EP.

References

1. United Nations (2002) Report of the World Summit on Sustainable Development. Johannesburg. 173pp.
2. Barange M (2003) Ecosystem science and the sustainable management of marine resources: from Rio to Johannesburg. Frontiers in Ecology and the Environment 1: 190–196.
3. Link JS, Browman HI (2014) Integrating what? Levels of marine ecosystem-based assessment and management. ICES Journal of Marine Science. doi:10.1093/icesjms/fsu026.
4. Browman HI, Stergiou KI (2004) Perspectives on ecosystem-based approaches to the management of marine resources. Marine Ecology Progress Serie 274: 269–303.
5. Ogden JC (2010) Marine spatial planning (MSP): A first step to ecosystem-based management (EBM) in the Wider Caribbean. Revista de biología tropical 58(3): 71–79.
6. Douvere F (2008) The importance of marine spatial planning in advancing ecosystem-based sea use management. Marine Policy 32: 762–771.
7. Katsanevakis S, Stelzenmüller V, South A, Sørensen TK, Jones PJS, et al. (2011) Ecosystem-based marine spatial management: Review of concepts, policies, tools, and critical issues. Ocean & Coastal Management 54: 807–820.
8. Olsen E, Gjøsæter H, Røttingen I, Dommasnes A, Fossum P, et al. (2007) The Norwegian ecosystem-based management plan for the Barents Sea. ICES Journal of Marine Science 64: 599–602.
9. Ottersen G, Olsen E, van der Meeren GI, Dommasnes A, Loeng H (2011) The Norwegian plan for integrated ecosystem-based management of the marine environment in the Norwegian Sea. Marine Policy 35: 389–398.
10. Day V, Paxinos R, Emmett J, Wright A, Goecker M (2008) The marine planning framework for South Australia: A new ecosystem-based zoning policy for marine management. Marine Policy 32: 535–543.
11. Kenchington RA, Day JC (2011) Zoning, a fundamental cornerstone of effective Marine Spatial Planning: lessons learnt from the Great Barrier Reef, Australia. J Coast Conserv 15: 271–278.
12. Ehler C, Douvere F (2009) Marine Spatial Planning: a step-by-step approach toward ecosystem-based management. Paris: Intergovernmental Oceanographic Commission and Man and the Biosphere Programme. IOC Manual and Guides No. 53, ICAM Dossier No. 6. 99pp.
13. Qiu W, Jones PJS (2013) The emerging policy landscape for marine spatial planning in Europe. Marine Policy 39: 182–190.
14. Portman ME (2011) Marine spatial planning: achieving and evaluating integration. ICES Journal of Marine Science 68: 2191–2200.
15. Pecceu E, Hostens K, Maes F (2013) The evolution of marine protected areas in the Belgian Part of the North Sea. A case study report for Work Package 6 of the MESMA project (www.mesma.org). MESMA report. 49pp.
16. Maes F, Vanhulle A, Lescrauwaet AK (2013) Marine spatial planning. In: Lescrauwaet, AK, Pirlet, H, Vereleye, T, Mees, J, Herman, R, editors. Compendium for Coast and Sea: Integrating Knowledge on the Socio-economic, Environmental and Institutional Aspects of the Coast and Sea in Flanders and Belgium. Ostend. pp. 261–273.
17. Bogaert D, Cliquet A, De Waen D, Maes F (2008) The designation of marine protected areas in Belgium. An analysis of the decision making process. In: Bogaert D, Maes F, editors. Who rules the coast?. Antwerp. pp. 59–105.
18. Anon (2001) Politisk grunnlag for en samarbeidsregjering. Sem-erklæringen. Sem. 52pp. [in Norwegian]

19. Hoel AH, Olsen E (2012) Integrated Ocean Management as a Strategy to Meet Rapid Climate Change: The Norwegian Case. AMBIO 41: 85–95.
20. Anon (2006) Helhetlig forvaltning av det marine miljø i Barentshavet og havområdene utenfor Lofoten (forvaltningsplan). Oslo: Det kongelige miljøverndepartement. Oslo. 141pp. [in Norwegian]
21. Misund OA, Olsen E (2013) Lofoten–Vesterålen: for cod and cod fisheries, but not for oil? ICES Journal of Marine Science 70: 722–725.
22. Lubchenco J, Sutley N (2010) Proposed U.S. Policy for Ocean, Coast, and Great Lakes Stewardship. Science 328: 1485–1486.
23. Anon (2013) National Ocean Policy Implementation Plan. Executive Office of the President - National Ocean Council. Washington DC. 36pp.
24. National Ocean Council (2013) Marine planning handbook. Executive Office of the President. Washington DC. 22pp.
25. Environmental Law Institute (2009) Marine Spatial Planning in US Waters: An Assessment of Existing Legal Mechanisms, Anticipated Barriers, and Future Opportunities. Washington DC. 94pp.
26. Sievanen L, Leslie HM, Wondolleck JM, Yaffee SL, McLeod KL, et al. (2011) Linking top-down and bottom-up processes through the new U.S. National Ocean Policy. Conservation Letters 4: 298–303.
27. Collie JS, Vic Adamowicz WL, Beck MW, Craig B, Essington TE, et al. (2013) Marine spatial planning in practice. Estuarine, Coastal and Shelf Science 117: 1–11.
28. Glazier E (2011) Ecosystem Based Fisheries Management in the Western Pacific. Blackwell-Wiley Publishers, Hoboken, New Jersey. 400pp.

10

A Spatially Distinct History of the Development of California Groundfish Fisheries

Rebecca R. Miller[1,2]*, John C. Field[2], Jarrod A. Santora[3], Isaac D. Schroeder[4], David D. Huff[1], Meisha Key[5], Don E. Pearson[2], Alec D. MacCall[2]

1 Institute of Marine Sciences, University of California Santa Cruz, Santa Cruz, California, United States of America, **2** Fisheries Ecology Division, Southwest Fisheries Science Center, National Marine Fisheries Service, National Oceanic and Atmospheric Administration, Santa Cruz, California, United States of America, **3** Center for Stock Assessment Research, University of California Santa Cruz, Santa Cruz, California, United States of America, **4** Environmental Research Division, Southwest Fisheries Science Center, National Marine Fisheries Service, National Oceanic and Atmospheric Administration, Pacific Grove, California, United States of America, **5** California Department of Fish and Wildlife, c/o, National Marine Fisheries Service, Southwest Fisheries Science Center, Santa Cruz, California, United States of America

Abstract

During the past century, commercial fisheries have expanded from small vessels fishing in shallow, coastal habitats to a broad suite of vessels and gears that fish virtually every marine habitat on the globe. Understanding how fisheries have developed in space and time is critical for interpreting and managing the response of ecosystems to the effects of fishing, however time series of spatially explicit data are typically rare. Recently, the 1933–1968 portion of the commercial catch dataset from the California Department of Fish and Wildlife was recovered and digitized, completing the full historical series for both commercial and recreational datasets from 1933–2010. These unique datasets include landing estimates at a coarse 10 by 10 minute "grid-block" spatial resolution and extends the entire length of coastal California up to 180 kilometers from shore. In this study, we focus on the catch history of groundfish which were mapped for each grid-block using the year at 50% cumulative catch and total historical catch per habitat area. We then constructed generalized linear models to quantify the relationship between spatiotemporal trends in groundfish catches, distance from ports, depth, percentage of days with wind speed over 15 knots, SST and ocean productivity. Our results indicate that over the history of these fisheries, catches have taken place in increasingly deeper habitat, at a greater distance from ports, and in increasingly inclement weather conditions. Understanding spatial development of groundfish fisheries and catches in California are critical for improving population models and for evaluating whether implicit stock assessment model assumptions of relative homogeneity of fisheries removals over time and space are reasonable. This newly reconstructed catch dataset and analysis provides a comprehensive appreciation for the development of groundfish fisheries with respect to commonly assumed trends of global fisheries patterns that are typically constrained by a lack of long-term spatial datasets.

Editor: David Hyrenbach, Hawaii Pacific University, United States of America

Funding: NOAA Climate Data Modernization Project (CDMP), the NMFS Stock Assessment Improvement Plan, and the SWFSC California Current Integrated Ecosystem Assessment (IEA). The funders had no role in study design, data collection and analysis, decision to publish, or preparation of the manuscript.

Competing Interests: The authors have declared that no competing interests exist.

* Email: Rebecca.Miller@noaa.gov

Introduction

Historical fisheries data provides insights into the development of fisheries activities and can inform efforts to interpret long-term trends, changes in species catch composition and the location of productive fishing habitats over time. Many fish populations worldwide undergo extraction activities where individual fishermen decide which species to pursue based upon market conditions and available gear technology [1], [2]. Selection of fishing grounds is determined from a variety of factors such as habitat, ocean productivity, distance to port and weather conditions [3]. Optimal fishing habitats are selected first where the expectation of catch is high and economic risks are low [4]. When catches become less reliable at previous fishing locations, harvest activities often shift to new habitats, leading to the geographic expansion of fisheries and fisheries impacts. Waters fished in the early years of the fishery may be associated with serial depletion in catch rates and species abundance [5], [6]. Both historical data and meta-analyses of global catch data suggest a large scale geographic expansion and serial depletion of fisheries across multiple stocks and regions [7–9], an increase in the spatial extent of the global fishing "footprint" [10], and a trend towards fishing greater depths over time [11], [12]. A number of empirical case studies have also documented patterns of fisheries development and consequent serial depletion of resources at regional scales for marine invertebrate populations [13], [14], [5], temperate water reef fishes [15], [16] shelf finfish species [17], [18], [6], coastal pelagic species [19], and deepwater seamount populations [20], [21]. Importantly however, geographic expansion is not necessarily synonymous with serial depletion and the effects of fisheries expansion are expected to vary by species, fishery and ecosystem [22].

The potential for geographic expansion and serial depletion are particularly problematic for evaluating fished populations in stock assessments, as it can lead to a perception of stability in catch size-structure and rates which could mask the true status of exploited populations [23], [24]. Most single-species stock assessments assume homogeneity in demographic structure and fishing mortality across space, and are typically incapable of explicitly addressing complex spatiotemporal patterns of fisheries development and/or spatial fisheries management [23–25]. While spatial assumptions in stock assessments are not likely to pose problems for stocks with high mixing or migratory behavior, other assessments of stocks with limited mobility, heterogeneity in population structure and/or distribution could be substantially flawed if incorrect assumptions regarding the behavior of the fishery are made, leading to bias in assessment results [25], [26]. Concerns over geographic expansion in California have been raised for several U.S. west coast groundfish stock assessments [27–29] and anecdotal accounts of serial depletion for some groundfish species provide evidence for its effect in California. For example, in a review of historical southern California fisheries, interviewed fishermen stated that historically vessels fishing with vertical line gear would "completely decimate" local rockfish *(Sebastes spp.)* concentrations in some habitats, facilitating a perpetual shift to new fishing locations that tended to be further offshore and at greater depths [30]. Regional-scale attempts to demonstrate geographic expansion found a similar transition in catch composition from shallow to deeper recreational rockfish species in Monterey Bay, California [31].

The goal of this study was to examine the spatiotemporal patterns and geographic expansion of commercial and recreational groundfish fisheries in California. We mapped the distribution and magnitude of groundfish catch by spatially summarizing the total catch per habitat area (CPA). For commercial groundfish, we map the year in which a given grid-block experienced catches that represented half of the cumulative total (Yr50) for species by preferred habitat depth stratum. We then used Generalized Linear Models (GLMs) to model Yr50 using variables that describe temporal patterns of historical fisheries development and oceanographic conditions over space including 1) distance from port, 2) depth, 3) the percent of days with wind speeds exceeding 15 knots (wind), 4) yearly mean chlorophyll-a (CHL) and 5) annual mean sea surface temperature (SST). The modeled geographic expansion for the entire state of California is the first of its kind and will provide a full picture of geographic expansion that will be useful in future stock assessments, catch reconstruction efforts, and marine spatial planning.

California Groundfish Development

Rockfish, flatfish and other groundfish have been harvested commercially, recreationally and for subsistence in California waters for well over a century, with the earliest fisheries targeting nearshore and inland habitats, while later harvest expanded to more distant habitats. Native Americans first harvested groundfish for subsistence fisheries in nearshore reefs, collecting rockfish and other marine species when land-based food was in short supply [32]. Commercial fisheries activities in San Francisco Bay accelerated in the 1850s when the Gold Rush lured immigrants from around the globe. The immigrants brought new harvest techniques, such as the Italian paranzella (two-boat trawl), first introduced in 1876. By 1892 the port of San Francisco contributed 93% of reported catch for the entire state of California [33], with much of the fishing activities taking place in the Bay and Sacramento River. However, depletion concerns over inshore groundfish stocks led area closures of trawl nets for some species in San Francisco Bay in 1906 [34] and waters off southern California in 1913 [35] forcing vessels to pursue fish habitats further from ports. In order to keep track of fish harvest and sales a monthly fish ticket system, also known as landing receipts, was enacted in 1911 by California Department of Fish and Wildlife (CDFW, previously California Department of Fish and Game). Each ticket recorded the species or market category, pounds caught, date, and price. Further ticket refinements in 1933 included a spatial-catch "grid-block" system which was implemented out of necessity because distances traveled to fish no longer reflected the locality of landing port [36]. In order to travel to more distant harvest grounds, fishing vessels switched from gasoline motors to diesel engines and increased capacity with ice and refrigeration. The need for inexpensive protein and vitamin A from shark livers to support the war effort led to a sharp increase in trawl effort and landings, particularly in northern California waters after 1941 [37]. After a slight post-WWII decline in landings the groundfish fishery diversified and included a wide range of rockfish, flatfish and roundfish species. Many of these species were previously harvested only by hook and line, due to greater depths and rough habitats, but by 1940s and 1950s were targeted by otter trawl gear. The 1950s ushered in a new generation eager for recreational fisheries leisure; fiberglass rod and reels allowed anglers to easily fish deep-water rockfishes, with some ports reporting a 400–500% increase over an 8-year span [38]. Many recreational fishers either used their own personal watercraft or paid a fare to board a chartered commercial passenger fishing vessel (CPFV) also known as "Party Boats." Recreational fisheries continue to be a significant proportion of total catch throughout the state, representing the majority of total catch for many nearshore and shelf species, particularly throughout the Southern California Bight (SCB) [39]. Globalization of the 1960s–1970s lured foreign factory trawlers to California where they harvested rockfish and other groundfish populations offshore, outside of what was then considered to be a three nautical-mile territorial sea. Foreign vessel rockfish catches were substantial, with approximately 50,000 tons harvested in California [40], nearly 35% of the domestic rockfish catch during this period. To ensure that coastal resources were harvested domestically rather than by foreign vessels, national policies encouraged rapid growth of the US West Coast groundfish fleet and banned foreign vessels to the 200 mile limit off the US coastline [41], [42]. During this period ships became evermore hardy and the ability to fish and travel in rough weather was less of a constraint. In the 1980s development of midwater species, such as widow rockfish *(S. entomelas)*, also led to new fishing opportunities, but yet again, rapid depletion of the newly developed resources often followed [43]. Increasing vulnerability concerns of the fishery resource relative to fishing power constrained harvest levels and landings quotas in the 1990s as stock assessments began to recognize the low productivity, longevity, and vulnerability of rockfish populations to overexploitation [44–46]. In the early 2000s, seven rockfish species were declared to be overfished, and efforts to rebuild populations resulted in tremendous reductions in allowable catches of most rockfish species, and other co-occurring groundfish species, for both commercial and recreational fisheries [44], [47]. Efforts to rebuild populations led to the implementation of a substantial network of spatial areas closed to groundfish fishing [48], [49], the largest being the nearly 11,000 km^2 Cowcod Conservation Area (CCA) in the SCB. The boundary of the CCA was established based on observation of higher relative abundance in this area when the stock was declared overfished in 2001 [50]. Consequently, the future management system for West Coast groundfish is likely to require spatially explicit management for some time.

Methods

California Groundfish Spatial Datasets

A unique set of spatially referenced, long-term commercial groundfish catch data from 1933–1968 were recently recovered

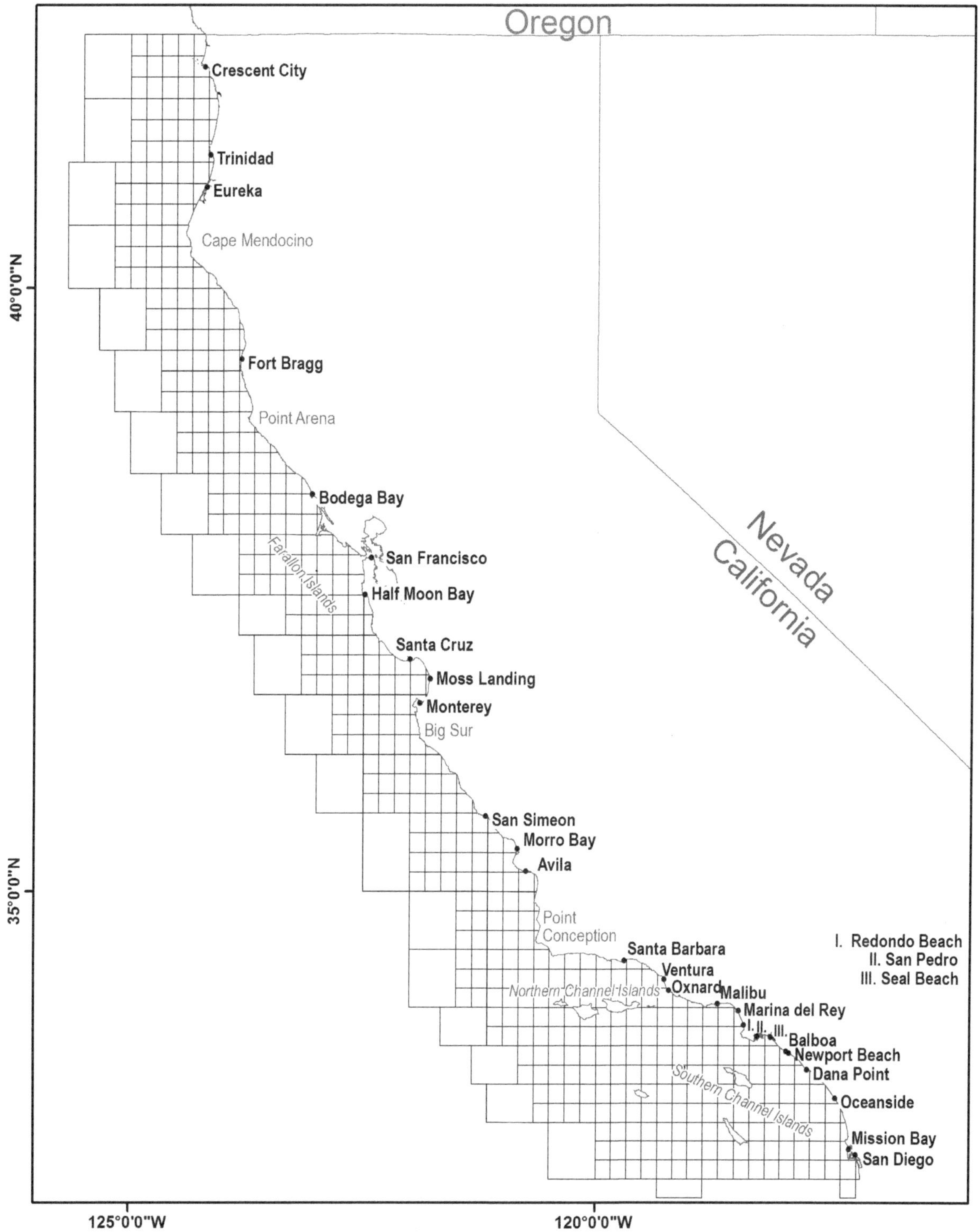

Figure 1. California Department of Fish and Game spatial blocks established in the 1930s. The commercial and recreational ports accounting for 95% of total catch are labeled in black. Geographic features are labeled gray.

and digitized from microfiche records as part of an ongoing catch reconstruction effort between the National Marine Fisheries Service (NMFS) and CDFW [49]. In 1933 the CDFW established a spatially distinct "grid-block" data collection system where gridded cells representing ocean habitat were partitioned into 554 blocks at a 10×10 minute grid resolution [51], [52] (Figure 1). When commercial fishermen bring fish to port a landing receipt is produced. This receipt holds detailed information on species type, pounds caught, date, price and spatial block [53]. The record keeping in general is detailed, however information such as gear type and effort were not consistently available, therefore gear-specific and effort analyses are not possible for the early period of

Table 1. Evaluated commercial groundfish market categories and the fraction of the total catch from 1935–2000.

Common name	Scientific name	% of total
0 to 200 meters depth		
English sole	*Parophrys vetulus*	7.14%
Sanddab, Pacific, Speckled, other	*Citharichthys sordidus, C. stigmaeus*	1.91%
California Halibut	*Paralichthys californicus*	1.77%
Starry Flounder	*Platichthys stellatus*	1.06%
Skate, Big, California, other	*Raja binoculata, R. inornata*	0.73%
Sole, Rock, Butter, other	*Lepidopsetta bilineata, Isopsetta isolepis*	0.40%
Sand sole	*Psettichthys melanostictus*	0.26%
California Scorpionfish	*Scorpaena guttata*	0.17%
Leopard shark	*Triakis semifasciata*	0.11%
Turbot, Curlfin, Hornyhead, other	*Pleuronichthys decurrens, P. verticalis*	0.11%
Cabezon	*Scorpaenichthys marmoratus*	0.10%
Kelp Greenling	*Hexagrammos decagrammus*	0.00%
Sculpin, Staghorn, Yellowchin, other	*Leptocottus armatus, Icelinus quadriseriatus*	0.00%
0 to 400 meters depth		
Pacific Whiting	*Merluccius productus*	5.54%
Petrale Sole	*Eopsetta jordani*	4.72%
Lingcod	*Ophiodon elongatus*	3.07%
Arrowtooth Flounder	*Atheresthes stomias*	0.25%
Pacific Halibut	*Hippoglossus stenolepis*	0.07%
Pacific Cod	*Gadus macrocephalus*	0.00%
Pacific Tomcod	*Microgadus proximus*	0.00%
Slender Sole	*Lyopsetta exilis*	0.00%
0 to 600 meters depth		
Unspecified rockfish	*Sebastes spp.*	21.61%
Bocaccio rockfish	*Sebastes paucispinis*	4.56%
Widow rockfish	*Sebastes entomelas*	3.04%
Chilipepper rockfish	*Sebastes goodei*	1.46%
Splitnose rockfish	*Sebastes diploproa*	0.50%
Yellowtail rockfish	*Sebastes flavidus*	0.44%
100 to 1200 meters depth		
Dover sole	*Microstomus pacificus*	20.03%
Sablefish	*Anoplopoma fimbria*	11.73%
Thornyheads, Longspine, Shortspine	*Sebastolobus alascanus, S. altivelis*	5.72%
Rex sole	*Glyptocephalus zachirus*	2.73%
Spiny Dogfish shark	*Squalus acanthias*	0.60%
Grenadiers, Giant, Pacific, other	*Coryphaenoides acrolepis, Albatrossia pectoralis*	0.19%

Species are grouped into habitat depth strata that best approximated their greatest population densities.

the fishery. Landing receipts from 1969–2010 were queried from the California Cooperative Survey (CalCOM) database [54] using both trawl and non-trawl gear types. As the data used here reflect the landings data from fish tickets, they were analyzed with respect to market categories, rather than species-specific estimates. Table 1 lists the species and market categories that accounted for virtually all groundfish landings analyzed, including the fraction of the total catch accounted for by each market category over the time period evaluated. As most groundfish spend the majority of their adult life stage near the ocean bottom, species and market categories were assembled into representative depth strata that best approximated their greatest population densities, as a proxy for habitat. The depth strata included were 0–200 m, 0–400 m, 0–600 m or 100–1200 m. For most species, the market categories are relatively "pure" to the species level, however the rockfish market categories in particular typically reflect a broad (and non-stationary) grouping of species that required considerable analysis to "reconstruct" back to species-level catches [51]. With the exception of Pacific halibut (*Hippoglossus stenolepis*), managed by the International Pacific Halibut Commission, and California halibut (*Paralichthys californicus*), managed by the CDFW, virtually all groundfish species in this study are managed under the Pacific Fishery Management Council's Groundfish Fishery Management Plan [50], adopted in 1982. The recreational rockfish block summary data from CPFV's were archived in a CDFW dataset beginning in 1936 in southern California and 1957 in central and northern California, and is based on self-reported numbers of fish landed (i.e. vessel operator logbooks), as documented in Hill and Schneider [55]. Rockfish were seldom reported to the species level before the 1980s, and only sporadically thereafter, so we lumped all recreational rockfish into a single *Sebastes spp.* category. As with the commercial fishery, over 50 species of rockfish are either commonly or infrequently encountered in recreational fisheries. The most commonly encountered species include blue (*Sebastes mystinus*), black (*S. melanops*), vermillion (*S. miniatus*), bocaccio (*S. paucispinis*), olive (*S. serranoides*), chilipepper (*S. goodei*), yellowtail (*S. flavidus*) and canary (*S. pinniger*) rockfish [51]. Catches are reported as number of fish kept.

Historical reported catches were reviewed and adjusted to address gaps in spatial block reporting. Although spatial block reporting to CDFW was technically mandatory, not all of the catches included block-specific information. The first several years of this program were associated with very low reporting rates, however an increase in reporting rates occurred by the mid–1930s and by the 1950s–1975 approximately 80–100% of the commercial catch was associated with spatial information (Figure 2a) (see [51] for details). This was the basis for the decision to begin our analysis of commercial data in 1935, when approximately 50% of catches had plausible spatially-explicit catch information. Reporting requirements on landing receipts were relaxed in the 1970s and 1980s and the rate of block reporting declined, however landings associated at the regional scale and trawl logbook data collection remained mandatory. Spatial data for recreational rockfish were not archived in all California regions until 1957, therefore for consistency purposes we chose a start date of 1957. While recreational catches prior to 1957 are not trivial, catches were modest relative to those from 1957–2000. An end date of 2000 was used for both commercial and recreational fisheries because spatial management measures enacted at that time had substantial impacts on both magnitude of catches and ability of fishermen to select habitats [49]. The catch for some grid-blocks were considered implausible because the blocks include habitat beyond the depth range of the reported species (e.g. rockfish species reported in depths greater than 800 meters, at which very few individuals in this genus are encountered [56], [57]). This is likely a consequence of at least two factors; intentional misreporting in which fishermen deliberately reported the wrong statistical area of the catch, and inadvertent misreporting in which fishermen may have fished multiple species in different locations on a single trip but only included a single block on their landing receipt. To address these concerns, we removed blocks in which catches were highly unlikely to occur and scaled the catches within the remaining blocks such that the sum of the year-specific landings equaled the commercial landings estimates by groundfish species category; rockfish, flatfish, roundfish, and elasmobranchs (Figure 2b and Figure 2c). Historical catches from foreign fishing vessels were excluded, due to lack of spatial information and the fact that catches from these at-sea fisheries operations would not be related to proximity to ports. Consequently, this analysis focuses on the total magnitude of the catch and the temporal pattern of fisheries development, rather than an analysis of spatially explicit catch rates.

Mapping Groundfish Yr50 and CPA Values

For each grid-block, we spatially mapped the values of geographic expansion and total historical catch for groundfish species by preferred habitat depths. The spatiotemporal development of California groundfish catch history is described as the year at 50% cumulative historical catch (Yr50). The Yr50 is the year in which half of the total reported catch within a given block was reported, representing the temporal element of geographic expansion for these fisheries. By necessity, the previously described depth strata for each species or market category were simplified to reflect the majority distribution of a given species where fishing occurs, despite the fact that some species may occur infrequently over shallower or greater depths. A 1200 meter cutoff was based on the Pacific Fishery Management Council's (PFMC) Essential Fish Habitat (EFH) analysis which found historical fishing effort rarely extended past 1200 meters and usage of trawl gear is currently prohibited in depths greater than 1280 meters [58]. Recreational rockfish were assembled into a single 0–200 m depth stratum, as most recreational fishermen rarely fish deeper than 200 m. Blocks were stratified by depth based upon the U.S. Coastal Relief Model bathymetry data, which is the best available seamless bathymetry dataset containing all blocks at the highest resolution, 3-arc seconds (90m) [59]. The total catch per area (CPA) was calculated as metric tons per block habitat (km^2) for all evaluated years (respective to each fishery type). For commercial groundfish we map CPA for each depth stratum and spatially summarize. Blocks containing less than 1 km^2 habitat were removed in order to reduce potential CPA over-inflation biases associated with small areas. CPA provides the relative magnitude of cumulative removals relative to total available habitat and represents a combined consequence of exploitation and productivity of a given depth-defined habitat. Yr50 and CPA values were calculated for each fishery for all depth strata and mapped using an equal quantile distribution, where the 10 classification breaks represents each subsequent 10th percentile.

Modeling Geographic Expansion of Groundfish

GLM covariates were developed for each block, representing oceanographic and habitat conditions that fishermen might consider before selecting fishing habitats (i.e. blocks). Habitat covariates were calculated for each depth stratified block, where a centroid was considered a representative location since it is unknown where within each block groundfish species were caught. The minimum Euclidian distance between habitat centroids and ports was calculated. Port landings for groundfish are unavailable prior to 1969, so to be consistent we used ports which received 95% of the cumulative rockfish catch during the period 1969–2000. Since only a small number of ports met this criterion in northern California, we included the top 4 ranked ports in this region, as the port of Eureka comprised a large proportion of the groundfish catch. Mean bathymetry per block was calculated using the Zonal Statistics tool (Spatial Analyst, ESRI ArcMAP v.10 [60].

For each block we developed an annual mean of SST, CHL and % days with high winds using satellite data over 10 years (1999–2009). SST, CHL, and wind speed were derived from Pathfinder, SeaWiFS, and QuikSCAT satellites data products, respectively [61–63] and all data products were downloaded from NOAA CoastWatch (http://coastwatch.pfeg.noaa.gov/erddap/). Spatial subsets and monthly means of SST (day and night) and CHL were used in the creation of the annual means to improve spatial coverage due to missing data from cloud cover [64]. We selected percentage of days during 1999–2009 with wind magnitudes greater than 15 knots as an indicator of how likely a given habitat may have been fishable, under the assumption that winds greater than 15 knots are likely to be constraining to effective fishing operations. Bilinear interpolation was used to place the satellite annual means onto the CDFW grid. QuikSCAT is unable to measure nearshore winds resulting in only 381 of the 554 blocks with wind estimates. These missing blocks were filled by using

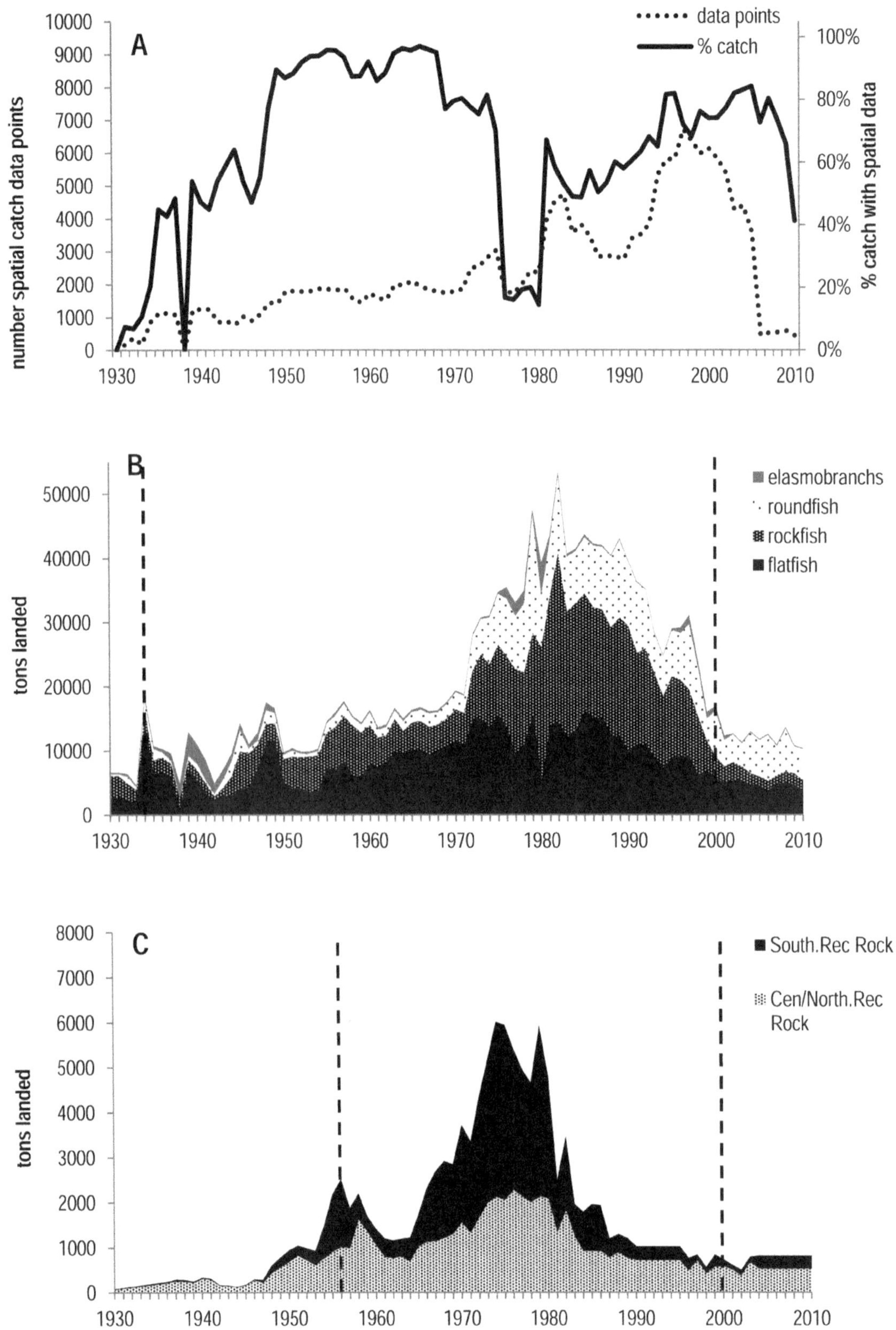

Figure 2. California groundfish landing receipts from 1930–2010. Figure 2a shows the total number of observations for commercial fisheries and the percentage of the estimated total catch associated with spatially distinct block information. Figure 2b shows the total California commercial groundfish landings in metric tons for the 1930–2010 period. Dashed line indicates begin and end year selected for analysis. Figure 2c shows the total estimated metric tons of recreational rockfish landed for the 1930–2010 period (from [51]). Dashed line indicates begin and end year selected for analysis.

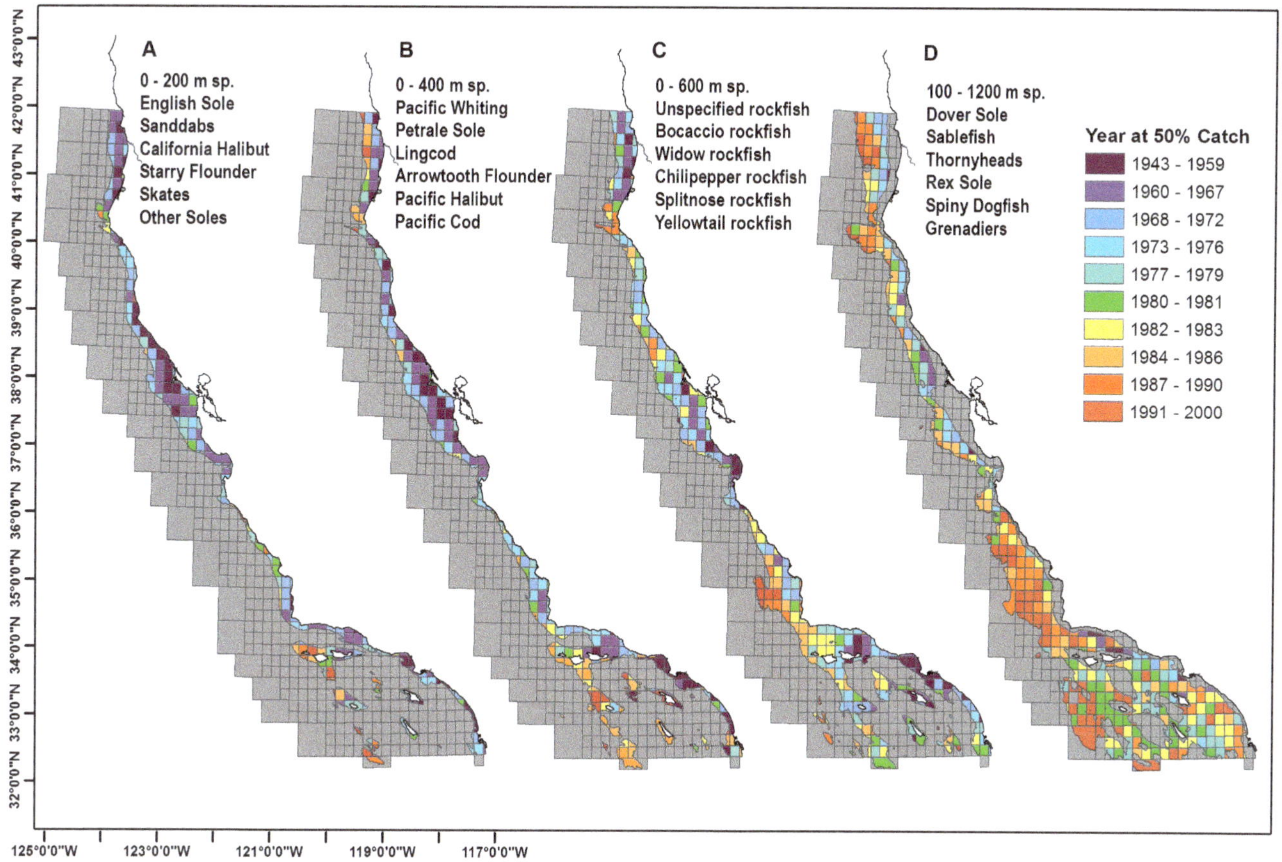

Figure 3. Year at 50% cumulative catch (Yr50) values by block for commercial groundfish. Spatial blocks are stratified by depth strata representing habitat of each groundfish species and include Fisheries Management Plan (FMP) species and a subset of state-managed species that are generally landed or encountered with FMP species. Named species are top 6 by landing weight. 3a. Yr50 values for 0–200 m species. 3b. Yr50 values for 0–400 m species. 3c. Yr50 values for 0–600 m species. 3d. Yr50 values for 100–1200 m species. Yr50 classifications are based on an equal quantile distribution.

nearest neighbor interpolation or merged with nearshore buoy data (NDBC bouys and M1 mooring in Monterey Bay). To maintain data normality, we arcsin square-root transformed wind percentages and log-transformed chlorophyll-a values for analysis purposes.

Analysis

variable Yr50 to identify which combinations of explanatory covariates best explain commercial groundfish and recreational rockfish geographic expansion. In order to reduce collinearity we calculated a Pearson's correlation coefficient for all potential explanatory covariate pairs, where a high correlation was considered >0.6. We found that SST was highly correlated to wind ($r = 0.73$ commercial and $r = -0.69$ recreational) and logCHL was highly correlated to distance from port ($r = -0.62$ commercial). Therefore, we removed SST and logCHL from the list of potential commercial fisheries explanatory variables and SST from the list of potential recreational variables in order to avoid distortion in model estimation [65]. Each candidate model was fitted using R software [66]. For model selection, we chose a single model with the lowest Akaike Information Criterion (AIC) score [67] if it was allocated most of the AIC weight (MuMln package [68]). We plotted the response of each model covariate across the range of data values while holding the other model covariates at their median values [69]. We mapped the model predictions based on sample covariates from each geographic block (MGET [70]).

Results

Mapping Groundfish Yr50 and CPA Values

The blocks for commercial species, bounded by depth strata representing habitat, demonstrate shallower species were caught earlier than deep water species (Figure 3a–d). For some blocks the Yr50 occurred as early as 1943, while the latest occurred in the year 2000. The 200 m and 400 m species had an earlier Yr50 catch near San Francisco and Eureka, and later Yr50 catch near the offshore Channel Islands. The 600 m species, all rockfish, have earlier catches near Eureka, Monterey, San Pedro and the nearshore Channel Islands, while the later catches occurred near Cape Mendocino and the Santa Lucia Bank, west of Avila. The 100–1200 m species have many blocks where Yr50 occurred after 1976, and blocks concentrated west of Avila, Cape Mendocino and Crescent City have a Yr50 during the 1980s–2000s.

The historical total CPA for the shallow water commercial (0–200 m) species, including California Halibut, Greenlings (Hexagrammidae) and Sanddabs (*Citharichthys spp.)*, tended to be located near Crescent City and Eureka, and shallow sandy areas outside of San Francisco Bay (Figure 4a). The 0–400 m species, including Pacific Whiting *(Merluccius productus)* and Petrale Sole (*Eopsetta jordani)*, had the largest catches near the shelf breaks just west of

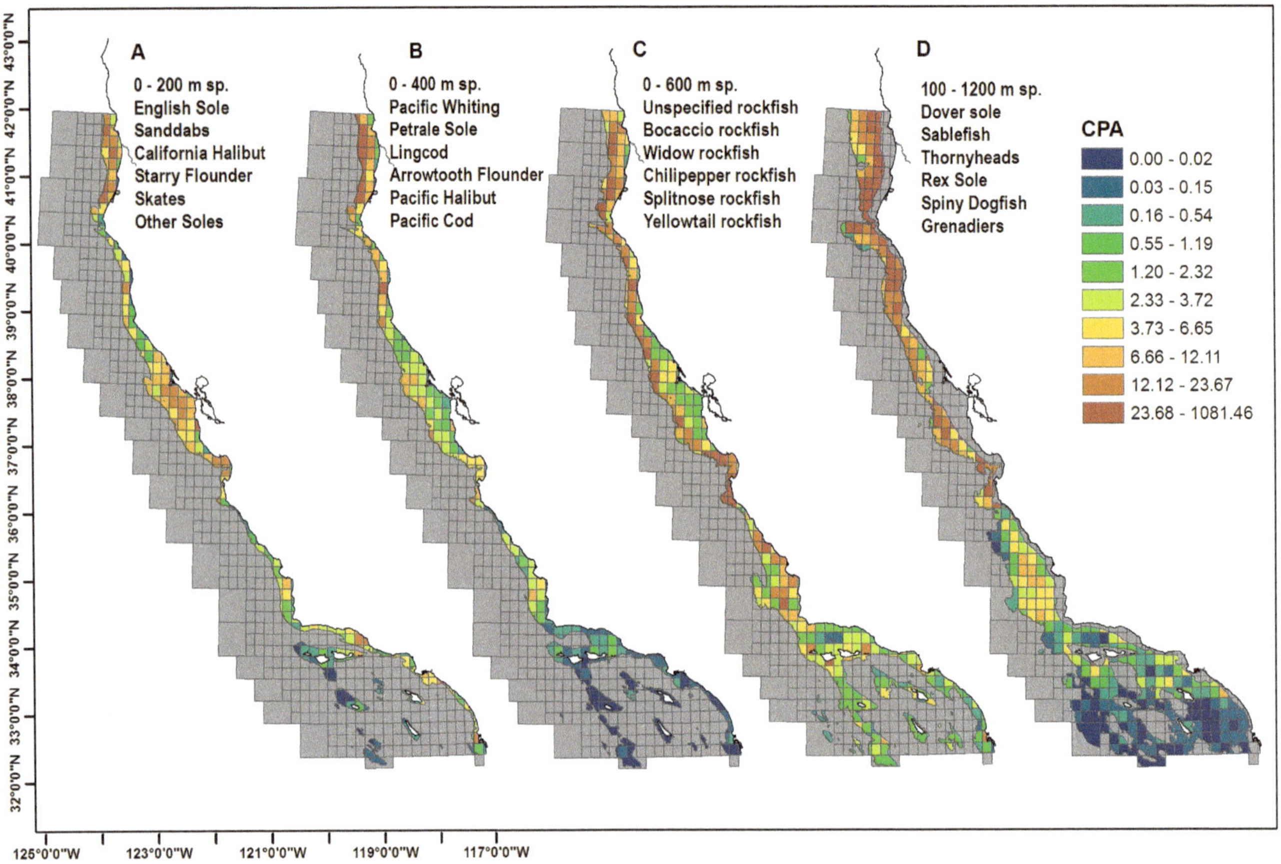

Figure 4. Commercial groundfish total Catch per Area (CPA) in metric tons per block habitat (km^2) for the years 1935–2000. Spatial blocks are stratified by depth strata representing habitat of each groundfish species and include Fisheries Management Plan (FMP) species and a subset of state-managed species that are generally landed or encountered with FMP species. Named species are top 6 by landing weight. Total Catch is reported in tons. 4a. CPA for 0–200 m species. 4b. CPA for 0–400 m species. 4c. CPA for 0–600 m species. 4d. CPA for 100–1200 m species. Yr50 classifications are based on an equal quantile distribution.

Crescent City and Eureka (Figure 4b). The 0–600 m species, all rockfish spp., were concentrated on the shelf breaks from northern California to Morro Bay (Figure 4c). The deepwater species (100–1200 m) largely sablefish (*Anoplopoma fimbria)*, dover sole (*Microstomus pacificus)* and thornyheads (*Sebastolobus spp.*), were concentrated throughout Northern California and Monterey Bay, while a moderate amount was caught west of Avila (Figure 4d).

The spatially summarized CPA for commercial groundfish had the largest catches near the continental shelf breaks in northern and central California while CPA for recreational fisheries was located at inshore blocks near the SCB, Morro Bay and Monterey (Figure 5 and 6). The summarized CPA values for all commercial groundfish were concentrated in the Northern and Central California continental shelf breaks and were much greater than in the SCB. Recreational rockfish CPA values show large catches for inshore blocks in Southern California near Redondo Beach, Terminal Island and at many of the off-shore islands of Santa Catalina, Santa Cruz, Santa Barbara and San Clemente. In the central coast, large recreational catches occurred near Morro Bay, Monterey and Half Moon Bay, along with the offshore sites of Cordell Bank and Farallon Islands (Figure 6). Blocks along the northern California coast had a smaller recreational rockfish CPA with respect to southern and central California.

Modeling Geographic Expansion (Yr50) of Groundfish

The environmental covariates for the California seascape demonstrate regional off-shore and latitudinal patterns (Figure 7a–e). Distance (km) from port is the greatest in the SCB, with the most distant blocks occurring up to 180 km from port. The ten-year average SST corresponds to a latitudinal gradient of colder water in the northern blocks and warmer waters in the southern blocks. The percent of days exceeding 15 knots is patchy, with calm winds in nearshore areas of the SCB, Half Moon Bay, between Eureka and Crescent City and within the Monterey Bay. High winds are located near Cape Mendocino, Point Arena, Big Sur, Point Conception and 'outside' the Channel Islands. Chlorophyll-a demonstrated a regional gradient with northern blocks having relatively high values of Chlorophyll-a, whereas southern blocks have medium to low Chlorophyll-a, with nearshore blocks having a higher value of Chlorophyll-a than offshore blocks. The mean depth of each block shows a general pattern of shallower water nearshore and deeper water further offshore.

The GLMs demonstrated that geographic expansion occurred for both commercial groundfish and recreational rockfish fisheries with a shift to deeper, more offshore habitat and more inclement weather conditions over time. The selected model for commercial groundfish Yr50 catch (Table 2) and associated plots show an increase in depth, distance from port and wind, demonstrating

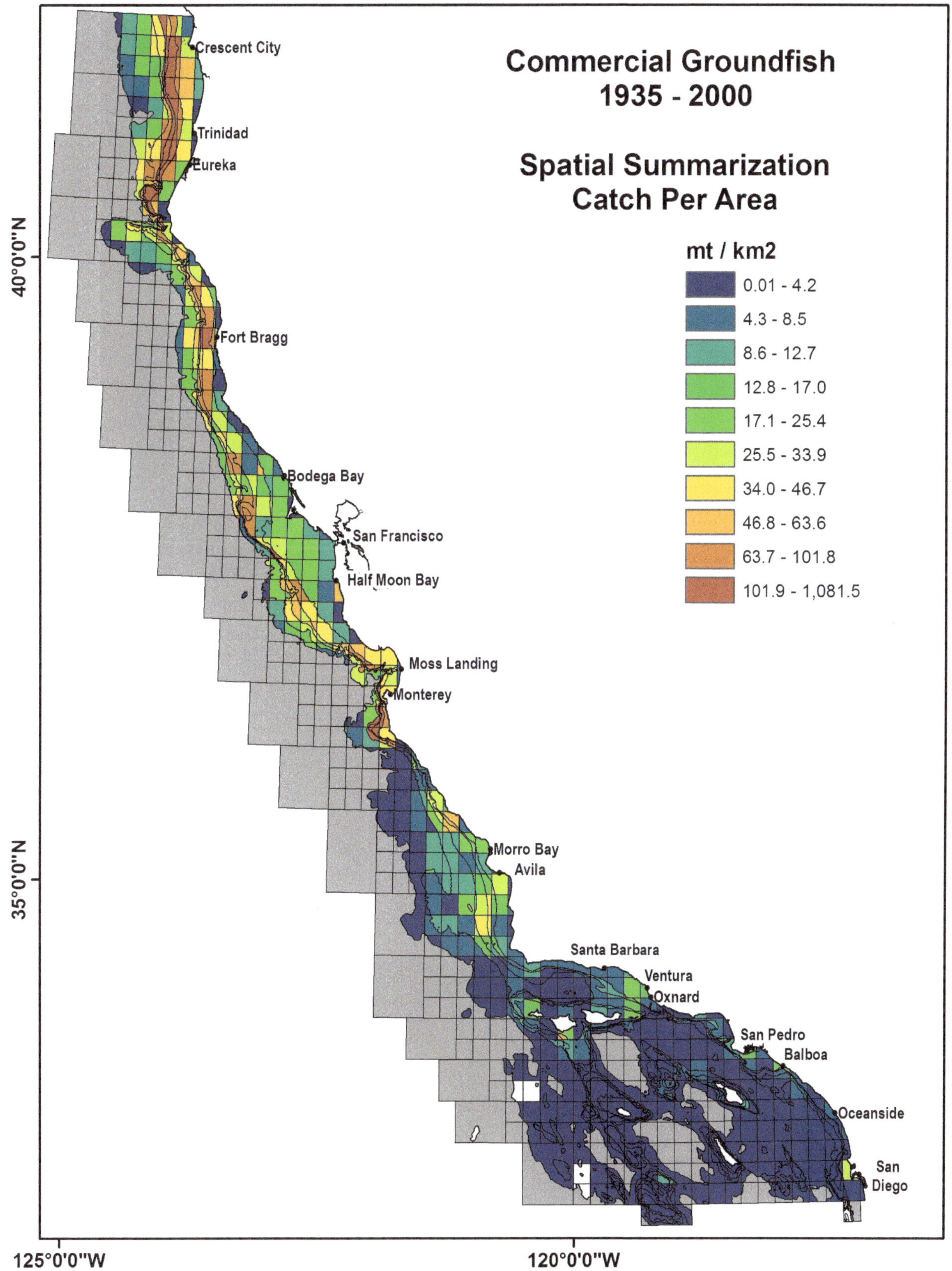

Figure 5. Spatial summarization of commercial groundfish total Catch per Area (CPA) in metric tons per block habitat (km^2) for the years 1935–2000. Gridded overlay represents catch blocks stratified by depth strata representing habitat of each groundfish species, 200 m, 400 m, 600 m, 1200 m. Ports shown are identified as the top 95% commercial ports by total catch.

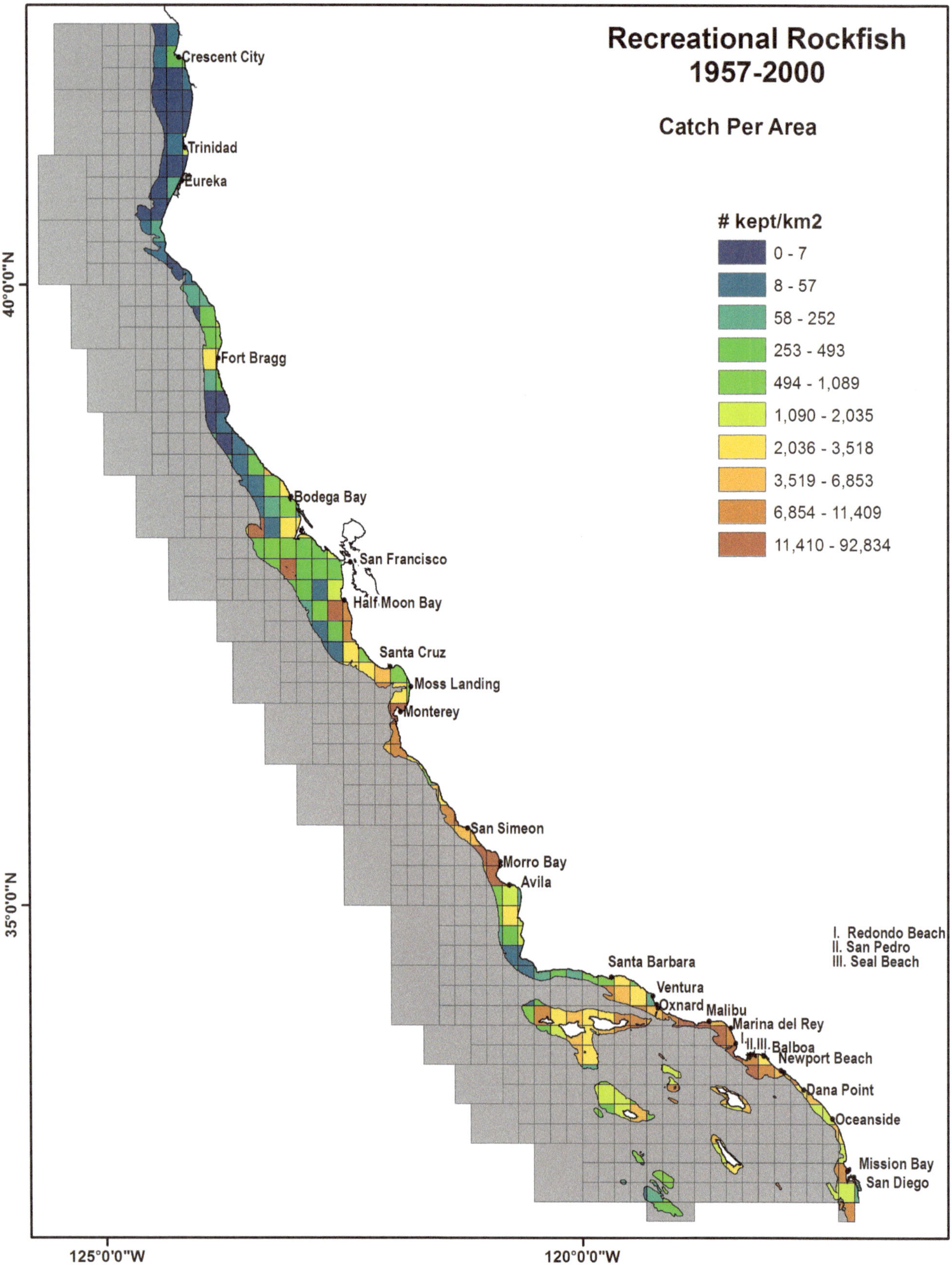

Figure 6. Spatial summarization of recreational rockfish total Catch per Area (CPA) in number of fish kept per block habitat (km²) for the years 1957–2000. Gridded overlay represents catch blocks stratified by 200 m depth. Ports shown are identified as the top 95% recreational ports by total catch.

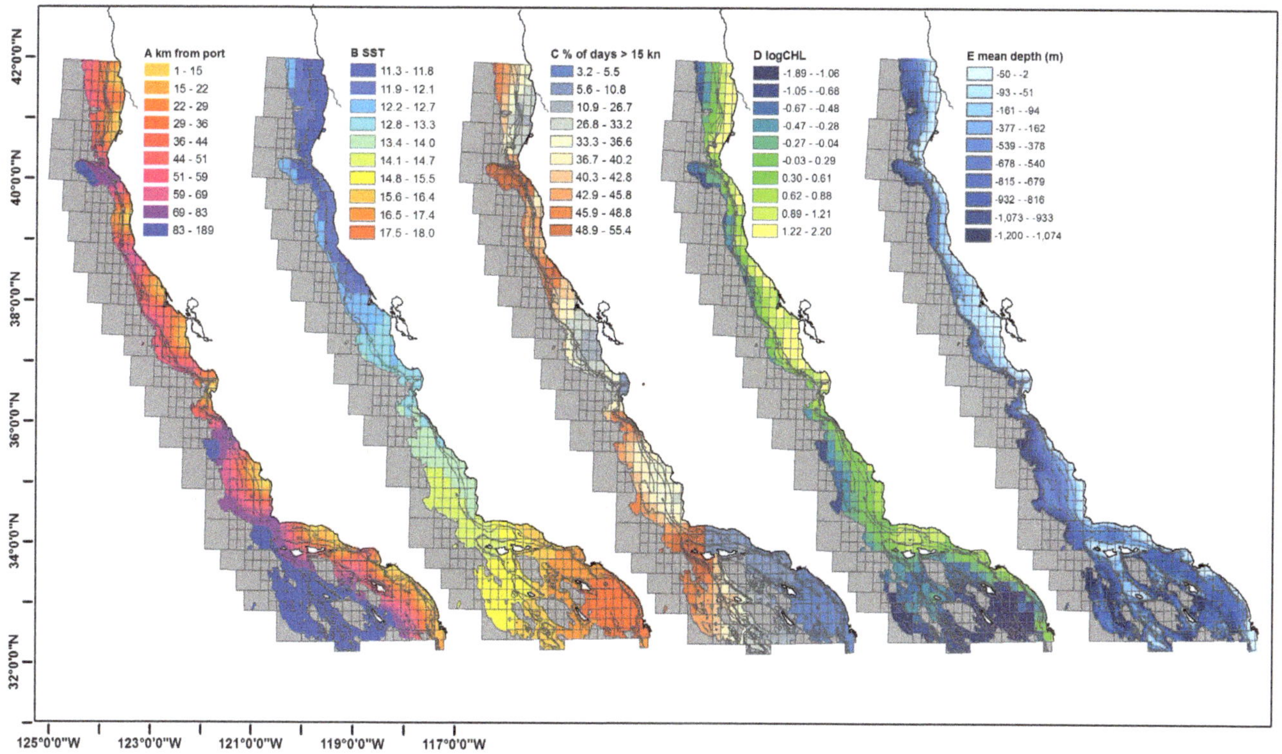

Figure 7. Potential explanatory covariates for GLMs by blocks. From left to right, 8a. distance (in kilometers) of the centroid of a given block to the nearest port. 8b. mean sea surface temperature (for the 1999–2009 period), 8c. percentage of days in which average winds in a given block exceed 15 kn/hour (for the 1999–2009 period), 8d. mean chlorophyll a levels (for the 1999–2009 period) and 8e. the mean depth of the habitat by block. Shown are the 100 m, 200 m, 400 m, 600 m, and 1200 m depth contours. Note that recreational fisheries only include 0–200 m blocks.

historical catches generally occurred earlier in shallower, nearshore habitat, closer to port, and in regions characterized by more favorable weather (less wind) (Figure 8a–c). For recreational rockfish Yr50, the selected model (Table 2) indicated wind, distance from port and logCHL had a steep positive slope, while depth demonstrated a moderate positive response with Yr50 (Figure 9a–d).

The modeled predictions for Yr50 catch for commercial groundfish demonstrated a nearshore/offshore gradient while recreational rockfish indicated a patchy regional latitudinal gradient (Figures 10 and 11). The commercial fisheries Yr50 modeled values ranged from 1960–1996 and occurred earlier near islands, historically active ports of Eureka, San Francisco, Monterey and the nearshore SCB. Yr50 occurred later for offshore, distant reefs such as Tanner/Cortez Bank in the SCB, the Santa Lucia Banks west of Avila and in the region near Cape Mendocino. Catches occurred later at deeper blocks, including deep blocks located just west of San Diego. The recreational fisheries Yr50 modeled values ranged from 1973–1989 and clearly show catch occurred earlier in the SCB (mid-1970s) and the nearshore Channel Islands (Anacapa, Santa Barbara and Catalina). Catch for Northern California recreational rockfish fisheries occurred in the 1980s with blocks just north and south of Fort Brag have Yr50 in the late 1980s.

Discussion

Geographic expansion within the coastal waters of California was demonstrated for both commercial groundfish and recreational rockfish fisheries. The findings provide historical trends for fishing conditions experienced over the time period evaluated and the shift to other species and habitat blocks as fishery resources were exploited. The selection and movement to different blocks over the history of the fishery provides a proxy for the relative costs and benefits of fishing, where economic forces drive the expansion of fishing activity to more optimal habitats when other accessible habitat resources are fully utilized or depleted [1]. Our results show an increase in fishing depth during the modeled period for California groundfish. Consistent with our general observation is a historical comparison by Karpov et al. [71] who found that recreational rockfish catch shifted from typically shallow to deepwater species during the 1980s with an accompanying decline in average weight of the nearshore species. Movement towards more productive waters (Chlorophyll-a) increased over time for recreational rockfish, indicating a potential harvest shift to habitats where species experience more rapid life histories. In a comparison study in the SCB, rockfish occurring in more productive waters grew larger relative to age and had an earlier age at recruitment [72]. Wind and distance had a steep positive slope in our model response plots indicating that as time progresses, many fishermen have increased their willingness to fish more inclement weather and travel farther to available habitats in order to potentially catch slightly larger or greater numbers of rockfish. A former CPFV operator working out of Santa Barbara found similar long-term patterns using his own high resolution logbook data with respect to rockfish catch rates and the distance from port, depth, and climate covariates (M. McCrea, University of California Santa Barbara, unpublished data).

We used a novel approach combining historical satellite data and CDFW habitat grid-blocks to best represent a suite of factors that fishermen may consider before selecting areas to fish, however

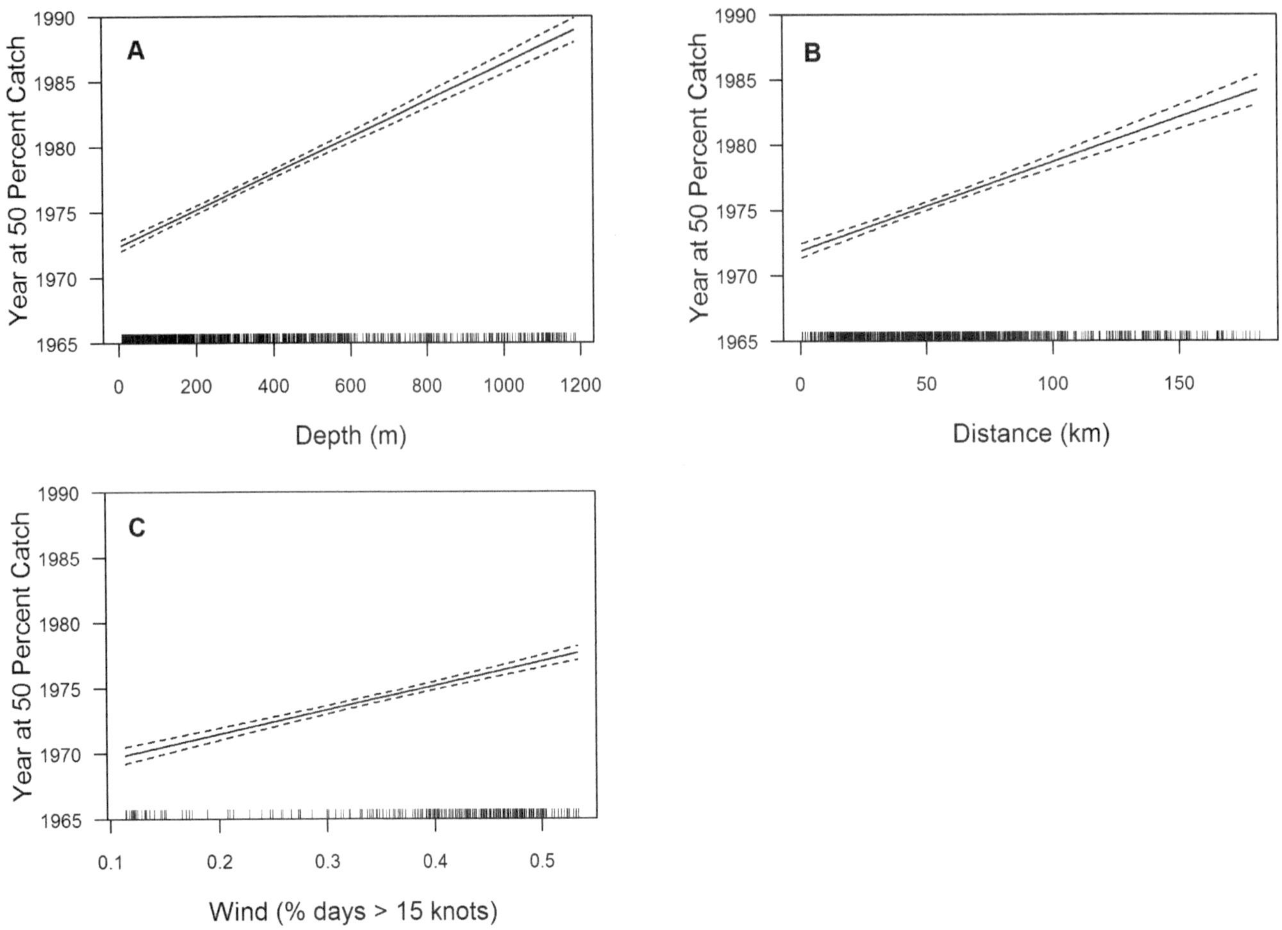

Figure 8. Generalized linear model fits for commercial groundfish. Where the y-axis is Year at 50% Catch (Yr50) and the x-axis is model variable, 8.a depth, 8.b distance from port, and 8.c wind. The dashed line indicates 95% confidence interval. Tick marks across the x-axis represent the distribution of observations.

the some of the variables were correlated and we had to estimate the distance from port. Our Pearson's analysis showed that the percentage of windy days was highly correlated to SST (commercial and recreational) and CHL was highly correlated to distance from to port (commercial). The strong couplings are expected because wind stress mixes the top layers of seawater thus influencing temperature while CHL is most abundant at narrow alongshore upwelling zones where the fishing ports are located [73]. For the GLMs, we included the percentage of windy days and distance from port because these factors would be more important considerations in day-to-day fishing operations than SST or CHL [74]. The study assigned the minimum Euclidian distance from the reported catch block to ports, due to the limited port information during the early years of the fishery, however it is possible that the real distances travelled to fishing grounds may be much greater than evaluated here as many fishermen would travel to and from their home port, rather than the nearest port.

Many of the species in this study were historically overexploited, and the maps generated suggest serial depletion may have occurred in some areas. It is evident that the commercial groundfish fishery selected the shallow 200 m species habitat blocks earlier and have more recently concentrated selection for deeper habitats and species. These deep water blocks were typically less intensively exploited for the time period evaluated. The Yr50 and CPA maps taken together may further provide evidence for serial depletion. For example, in areas where groundfish were caught early (Yr50 = purple or blue) and total cumulative catch is high (CPA = brown) (see Monterey Bay and nearshore northern California for commercial fisheries and nearshore SCB for recreational fisheries) it may be that more recent levels of effort in these habitat blocks is lower, potentially due to limited abundance of groundfish when compared to historical conditions. Although existing data cannot resolve within-species shifts in abundance and distribution, these patterns are consistent with many of the available stock assessment results for the species discussed here. For example, many shelf species of commercial significance, such as English Sole (*Parophrys vetulus)* and Petrale Sole, experienced high or peak landings, followed by significant declines in abundance by the 1940s and 1950s [75], [76], while more broadly but also more deeply distributed species, such as sablefish, Dover sole and thornyhead rockfishes have long exploitation histories but tended to experience peak landings and stock declines considerably later, as fisheries developed into deeper habitats [77–79].

The summarized CPA and plotted covariate response indicates that there may be spatial disaggregation between commercial groundfish and recreational rockfish resource extraction. Even though the historical recreational and commercial fisheries data did not allow for direct comparison of CPA (metric tons vs. # fish kept), a relative magnitude of catch can be compared. For

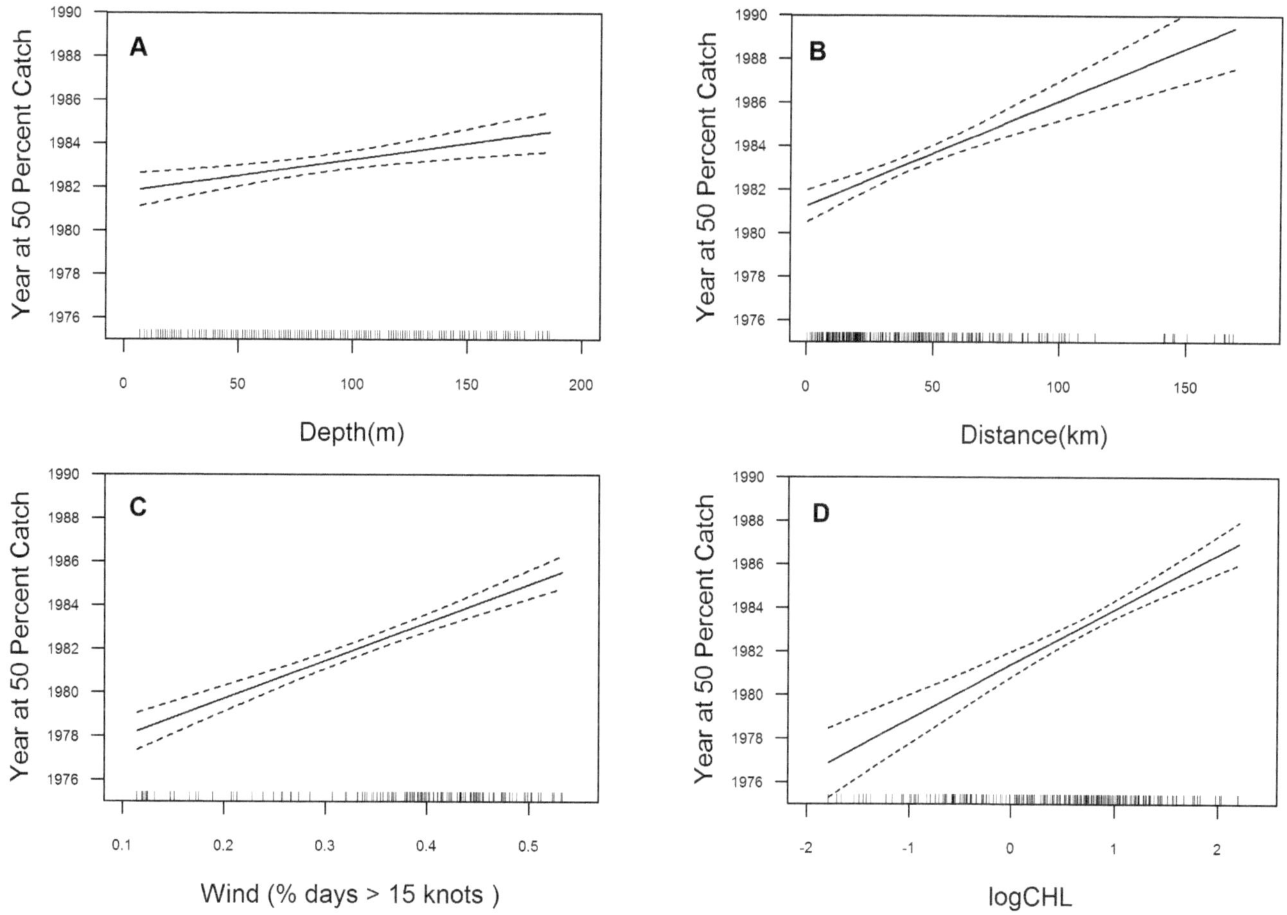

Figure 9. Generalized linear model fits for recreational rockfish. Where the y-axis is Year at 50% Catch (Yr50) and the x-axis is model variable, 9.a depth, 9.b distance from port, 9.c wind, and 9.d Chlorophyll-a. The dashed line indicates 95% GLM confidence intervals. Tick marks across the x-axis represent the distribution of observations.

recreational rockfish, a larger CPA value was demonstrated at inshore blocks located near ports, especially in central and southern California waters. A greater number of recreational ports with 95% of catch were geographically located in southern and central California, which agrees with the majority of the human population. In other regions of the world, fishing pressure increases with increasing population size [80]. In contrast, the CPA for commercial groundfish is further offshore and primarily located in central and northern California, where fewer ports and population is located. While the habitat selection over time increased in distance, depth and wind for both commercial and recreational fisheries, the plotted covariates for depth had a steeper slope in commercial fisheries than recreational fisheries, perhaps indicating that technological advances and new gear types to catch fish in newly available deeper habitats were important.

The legacy effects of geographic expansion may be evident in contemporary groundfish population abundance and size structures. Our maps demonstrate that SCB habitat within the Cowcod Conservation Area (CCA) had a different exploitation history than those habitats closer to ports. In the CCA, catch history (Yr50) from our modeled surfaces suggest fishing activity is more recent and the magnitude of fishing impacts are less (CPA), this is consistent with demographic studies which found larger, older, and more diverse fish occur in these offshore areas [27], [81]. In

Table 2. Candidate generalized linear models representing Year at 50% Cumulative Catch (Yr50) for California commercial groundfish and recreational rockfish.

Fishery	Model	AICc	R2	weight
Commercial	**Depth, Wind, Distance**	**9434.4**	**0.31**	**1**
Recreational	**Wind, Distance, Depth, logCHL**	**1506.2**	**0.31**	**0.62**
Recreational	Wind, Distance, logCHL	1514.9	0.30	0.36

Modeling considered depth, distance from port, wind and chlorophyll-a (recreational only) as potential covariates. Presented are the candidate models where the weight was >0.00. The final selected model is shown in bold face.

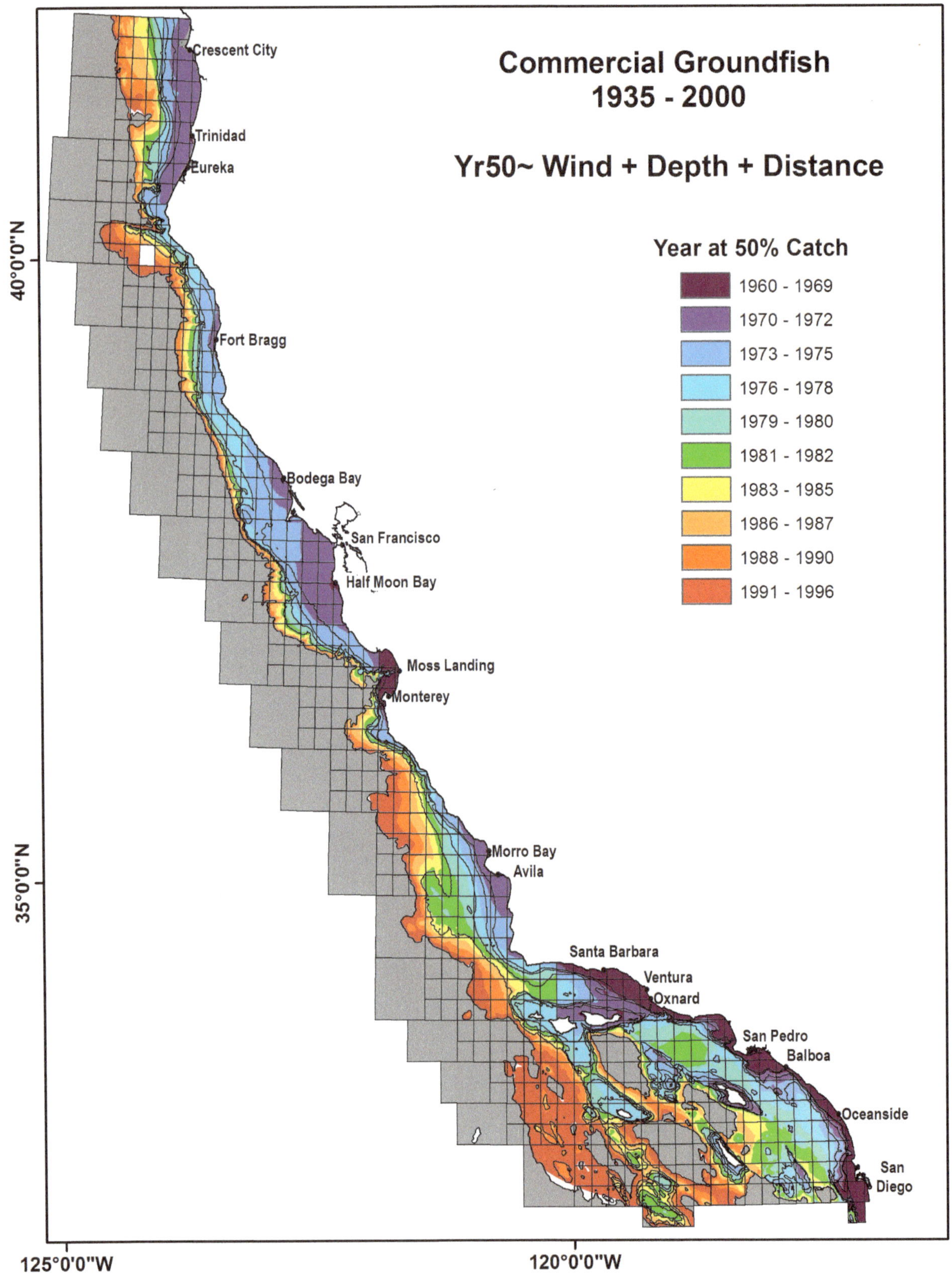

Figure 10. Modeled GLM for Year at 50% cumulative catch (Yr50) for commercial groundfish species. Modeled Yr50 was between the years 1960–1996. Classifications are based on an equal quantile distribution. Gridded overlay represents catch blocks stratified by depth strata representing habitat of each groundfish species, shown are 200 m, 400 m, 600 m, 1200 m. Ports shown are identified as the top 95% commercial ports by total catch.

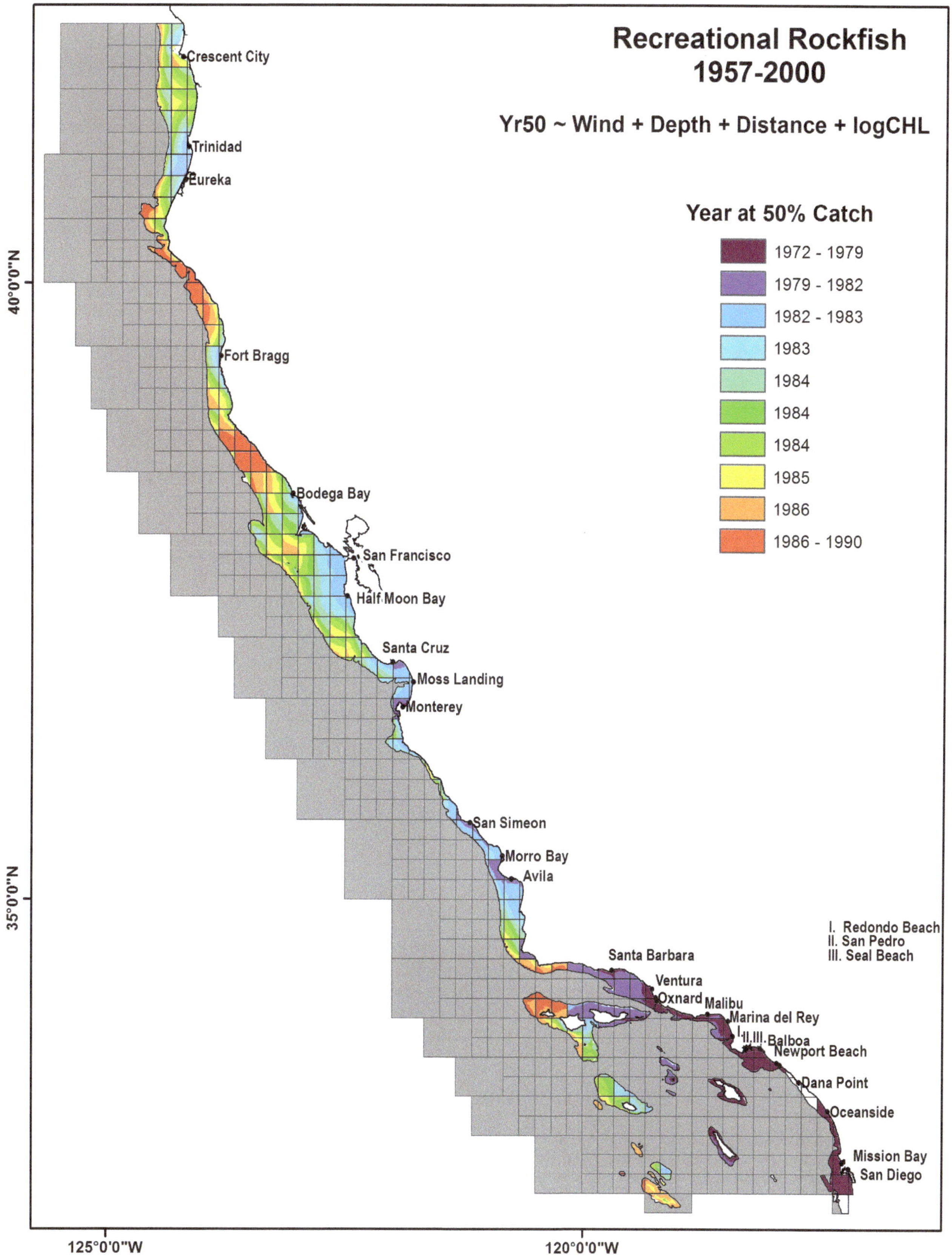

Figure 11. Modeled GLM for Year at 50% cumulative catch (Yr50) for recreational groundfish species. Modeled Yr50 was between the years 1973–1990. Classifications are based on an equal quantile distribution. Gridded overlay represents catch blocks stratified by 200 m representing fished habitat. Ports shown are identified as the top 95% recreational ports by total catch.

contrast, for SCB habitats closer to ports, the modeled historical catches (Yr50) occurred much earlier and fishing magnitude is relatively greater (CPA), such nearshore areas have been associated with an increase in the smaller, dwarf rockfish species complex [82]. Fishing legacy, represented here by Yr50 and CPA, may be an important variable for many fisheries demographic patterns and may be much more common than presently documented in other regions due to the lack of historical data.

This is the first effort to model geographic expansion of groundfish in California and will provide a better representation of the spatiotemporal catch reconstruction in stock assessments. Geographic expansion models are of great importance for species-of-concern stock assessments, such as the rockfish complex. Over 60 species in the rockfish genus occur in California waters, however the catch reconstruction dataset prior to 1969 lumped rockfish into 13 market categories, and individual species were not adequately recorded until the late 1970s and early 1980s. This inability to separate out rockfish species led to considerable challenges in estimating historical catches at the species level to support assessment and management activities [51]. Consequently, in the most recent catch reconstruction efforts, there was a potential bias towards overestimating historical catches of deep water species in the early years of the fishery, because deep water habitats were more likely to be targeted in the 1970s and 1980s than in the period from the 1930s to 1960s. Understanding the spatial patterns in which fisheries developed will improve species resolution of historical catches. The modeled historical catch will also improve the spatiotemporal accuracy of fish distributions for stock assessments, as well as surveys and habitat assessments which are presently based upon contemporary fish distribution patterns.

The geographic expansion model and recovered historical dataset provides resource managers and stakeholders an invaluable window to the historical context of California fisheries and interaction between users and the environment. Our results indicate that over the history of these fisheries, catches have taken place in areas of increasingly deeper habitat, with an increasing distance between catch locations and ports, and in more inclement weather conditions. The growing recognition of serial depletion and geographic expansion detected in the California Current ecosystem will contribute to efforts related to marine spatial planning, habitat impact assessments, quantification of ecosystem services, and ultimately to ecosystem-based approaches to management of marine fisheries and other resources [83–85]. The future of fisheries management will benefit from the appreciation of both historical and contemporary patterns of fisheries effort and catch.

Acknowledgments

This analysis would not have been possible without the dedication of fishermen, port samplers, state and federal biologists, who for nearly a century dutifully reported and collected the fisheries catch dataset. We would also like to thank Rosemary Kosaka, Merit McCrea, Milton Love and Cindy Thomson for comments on earlier drafts of this manuscript. We also thank Shannon Lyday, Michael Bell and David Hyrenbach for their thoughtful comments and improvements to the manuscript. M1 mooring wind data was provided by MBARI's Biological Oceanography Group lead by Francisco P. Chavez. All satellite data and NDBC buoy data was obtained through the West Coast node of NOAA's CoastWatch program.

Author Contributions

Conceived and designed the experiments: RRM JCF JAS IDS DDH MK DEP ADM. Performed the experiments: RRM JCF IDS DDH MK DEP. Analyzed the data: RRM JCF IDS DDH. Contributed reagents/materials/analysis tools: RRM JCF JAS IDS DHH MK DEP ADM. Wrote the paper: RRM JCF JAS IDS DDH MK.

References

1. Sethi S, Branch T, Watson R (2010) Global fishery development patterns are driven by profit but not trophic level. Proceedings of the National Academy of Sciences of the United States of America 107: 12163–12167.
2. Kasperski S, Holland D (2013) Income diversification and risk for fishermen. Proceedings of the National Academy of Sciences of the United States of America 110: 2076–2081.
3. Wilen J, Smith M, Lockwood D, Botsford L (2002) Avoiding surprises: incorporating fisherman behavior into management models. Bulletin of Marine Science 70: 553–575.
4. Dreyfus-Leon M (1999) Individual-based modeling of fishermen search behaviour with neural networks and reinforcement learning. Ecological Modeling 120: 287–297.
5. Huitric M (2005) Lobster and conch fisheries of Belize: a history of sequential exploitation. Ecology and Society 10: 21.
6. Cardinale M, Bartolino V, Llope M, Maiorano L, Sköld M, et al. (2011) Historical spatial baselines in conservation and management of marine resources. Fish and Fisheries 12: 289–298.
7. Kirby M (2004) Fishing down the coast: Historical expansion and collapse of oyster fisheries along continental margins. Proceedings of the National Academy of Sciences in the United States of America 101: 13096–13099.
8. Berkes T, Hughes T, Stenenk R, Wilson J, Bellwood D, et al. (2006) Globalization, roving bandits and marine resource. Science 331: 1557–1558.
9. Anderson S, Flemming J, Watson R, Lotze H (2001) Serial exploitation of global sea cucumber fisheries. Fish and Fisheries 12: 317–339.
10. Swartz W, Sala E, Tracey S, Watson R, Pauly D (2010) The spatial expansion and ecological footprint of fisheries (1950 to Present). Public Library of Science ONE.
11. Morato T, Watson R, Pitcher T, Pauly D (2006) Fishing down the deep. Fish and Fisheries 7: 24–34.
12. Watson R, Morato T (2013) Fishing down the deep: Accounting for within-species changes in depth of fishing. Fisheries Research 140: 63–65.
13. Karpov K, Haaker P, Taniguchi I, Rogers-Bennett L, editors (2000) Serial depletion and the collapse of the California abalone (Haliotis spp.) fishery. Ottawa: NRC Research Press.
14. Orensanz J, Armstrong J, Armstrong D, Hilborn R (1998) Crustacean resources are vulnerable to serial depletion-The multifaceted decline of crab and shrimp fisheries in the Greater Gulf of Alaska. Reviews in Fish Biology and Fisheries 8: 117–176.
15. Epperly S, Dodrill J (1995) Catch rates of snowy grouper, Epinephelus niveatus, on the deep reefs of Onslow Bay, Southeastern U.S.A. Bulletin of Marine Science 56: 450–461.
16. Rudershausen P, Williams E, Buckel J, Potts J, Manooch C (2008) Comparison of reef fish catch per unit effort and total mortality between the 1970s and 2005–2006 in Onslow Bay, North Carolina. Transactions of the American Fisheries Society 137: 1389–1405.
17. Cadigan S, Hutchings J (2001) Nineteenth-century expansion of the Newfoundland fishery for Atlantic cod: an exploration of underlying causes. In: Holm P, editor. The exploited seas: new directions for marine environmental history. St. John: International Maritime Economic History Association. 31–65.
18. Cardinale M, Linder M, Bartolino V, Maiorano L, Casini M (2009) Conservation value of historical data: reconstructing stock dynamics of turbot during the last century in the Kattegat-Skagerrak. Marine Ecology Progress Series 386: 197–206.
19. Cardinale M, Nugroho D, Jonson P (2011) Serial depletion of fishing grounds in an unregulated, open access fishery. Fisheries Research 108: 106–111.
20. Clark M (1999) Fisheries for orange roughy (Hoplostethus atlanticus) on seamounts in New Zealand. Oceanologica Acta 22: 593–602.
21. Koslow J, Boehlert G, Gordon J, Haedrich R, Lorance P, et al. (2000) Continental slope and deep-sea fisheries: implications for a fragile ecosystem. ICES Journal of Marine Science 57: 548–557.
22. Worm B, Hilborn R, Baum J, Branch T, Collie J, et al. (2009) Rebuilding global fisheries. Science 325: 578–585.
23. Field J, Punt A, Methot R, Thomson C (2006) Does MPA mean 'Major Problem for Assessments'? Considering the consequences of place-based management systems. Fish and Fisheries 7: 284–302.
24. Punt A, Methot R (2004) Effects of marine protected areas on the assessment of marine fisheries. In: Shipley J, editor. Aquatic Protected Areas as Fisheries Management Tools (Proceedings of the American Fisheries Society Symposium 42, Quebec, Canada, 11 August–12 August 2003). Bethesda: American Fisheries Society. 133–154.
25. Ralston S, O'Farrell M (2008) Spatial variation in fishing intensity and its effect on yield. Canadian Journal of Fisheries and Aquatic Sciences 65: 588–599.

26. Cope J, Punt A (2011) Reconciling stock assessment and management scales under conditions of spatially varying catch histories. Fisheries Research 107: 22–38.
27. Dick E, Ralston S, Pearson D (2007) Status of cowcod, Sebastes levis, in the Southern California Bight. Portland: Pacific Fishery Management Council Stock Assessment and Fishery Evaluation. 90 p.
28. Field J, Pearson D (2011) Status of the blackgill rockfish, Sebastes melanostomus, in the Conception and Monterey INPFC areas for 2011. Portland: Pacific Fishery Management Council. 311 p.
29. Ralston S, MacFarlane B (2010) Population estimation of bocaccio (Sebastes paucispinis) based on larval production. Canadian Journal of Fisheries and Aquatic Sciences 67: 1005–1020.
30. Kronman M (1999) Santa Barbara hook and line fishery. Santa Barbara Maritime Museum. 37 p.
31. Mason J (1998) Declining rockfish length in the Monterey Bay, Califonria, Recreational Fishery, 1959–94. Marine Fisheries Review 60: 15–28.
32. McEvoy A (1986) The Fisherman's Problem: Ecology and law in the California Fisheries 1850–1980. New York: Cambridge University Press. 368 p.
33. Skinner J (1962) A historical review of the fish and wildlife resources of the San Francisco Bay area. California Department of Fish and Game 226 p.
34. Leipzig P (2001) Pacific groundfish buy-back proposal and the final summary and analysis. Eureka: Fisherman's Marketing Association. 23 p.
35. Barsky K (1990) History of the Commercial California halibut industry. In: The California Halibut, Paralicthys californicus, resource and fisheries: Fish Bulletin No. 174. Department of Fish and Game. 174–475 p.
36. Scofield W (1954) California Fishing Ports, Fish Bulletin No. 96. California Division of Fish and Game. 163 p.
37. Scofield W (1948) Trawling gear in California, Fish Bulletin No. 72. California Division of Fish and Game. 60 p.
38. Love M (2006) Subsistence, Commercial and Recreational Fisheries. In: Allen LG PD, Horn MH, editor. The ecology of marine fishes: California and adjacent waters. Berkeley: University of California Press. 567–594.
39. Schroder D, Love M (2002) Recreational fishing and marine fish populations in California. California Cooperative Oceanic Fisheries Investigation Reports. 182–190 p.
40. Rogers J (2003) Species allocation of Sebastes and Sebastolobus sp. caught by foreign countries from 1965 through 1976 off Washington, Oregon, and California, USA. NOAA Tech. Memo. NMFS-NWFSC-57 NOAA 117 p.
41. Magnuson W (1968) The opportunity is waiting... make the most of it. In: Gilbert D, editor. The Future of the Fishing Industry of the United States: University of Washington Publications in Fisheries. 7–9.
42. Mansfield B (2001) Property regime or development policy? Explaining growth in the US Pacific groundfish fishery. The Professional Geographer 53: 384–397.
43. Gunderson D (1984) Great widow rockfish hunt of 1980–1982. North American Journal of Fisheries Management 4: 465–468.
44. Berkeley S, Hixon M, Larson RJ, Love M (2004) Fisheries sustainability via protection of age structure and spatial distribution of fish populations. Fisheries 29: 23–32.
45. Parker S, Berkeley S, Golden J, Gunderson D, Heifetz J, et al. (2000) Management of Pacific rockfish. Fisheries 25: 22–30.
46. Ralston S (2002) West Coast groundfish harvest policy. North American Journal of Fisheries Management 22: 249–250.
47. Hilborn R, Stewart I, Branch T, Jensen O (2012) Defining trade-offs among conservation, profitability, and food security in the California Current bottom-trawl fishery. Conservation Biology 26: 257–268.
48. Bellman M, Heppell S, Goldfinger C (2005) Evaluation of a US west coast groundfish habitat conservation regulation via analysis of spatial and temporal patterns of trawl fishing effort. Canadian Journal of Fisheries and Aquatic Sciences 62: 2886–2900.
49. Mason J, Kosaka R, Mamula A, Speir C (2012) Effort changes around a marine reserve: The case of the California Rockfish Conservation Area. Marine Policy 36: 1054–1063.
50. Pacific Fishery Management Council (2011) Pacific Coast Groundfish Fishery Management Plan for the California, Oregon, and Washington Groundfish Fishery. Portland. 158 p.
51. Ralston S, Pearson D, Field J, Key M (2010) Documentation of the California catch reconstruction project. NOAA-TM-NMFS-SWFSC-461. 80 p.
52. California Department of Fish and Game (1935) The commercial fish catch of California for the years 1930–1934, inclusive. Fish Bulletin 44. Bureau of Commercial Fisheries. 124 p.
53. California Department of Fish and Game (1952) The commercial fish catch of California for the year 1950 with a description of methods used in collecting and compiling the statistics. Fish Bulletin 86. Bureau of Marine Fisheries. 120 p.
54. Pearson D, Erwin B (1997) Documentation of California's commercial market sampling data entry and expansion programs. NOAA-TM-NMFS-SWFSC-240. 62 p.
55. Hill K, Schneider N (1999) Historical logbook databases from California's commercial passenger fishing vessel (partyboat) fishery, 1936–1997. SIO Reference Series No. 99–19 SIO Reference Series No. 99–19.
56. Keller A, Wallace J, Horness B, Hamel O, Stewart I (2012) Variations in eastern North Pacific demersal fish biomass based on the US west coast groundfish bottom trawl survey (2003–2010). Fishery Bulletin 110: 205–222.
57. Vetter R, Lynn E (1997) Bathymetric demography, enzyme activity patterns, and bioenergetics of deep-living scorpaenid fishes (genera Sebastes and Sebastolobus): Paradigms revisited. Marine Ecology Progress Series 155: 173–188.
58. Pacific Fishery Management Council (2005) Amendment 19 Essential Fish Habitat to the Pacific Coast Groundfish Fishery Management Plan for the California, Oregon and Washington Groundfish Fishery. Portland. 88 p.
59. NOAA National Geophysical Data Center (2011) Coastal Relief Model. http://www.ngdc.noaa.gov/mgg/coastal/crm.html (accessed 1 August 2011).
60. ESRI (2010) ArcGIS 10.0 Environmental Systems Research Institute. Redlands.
61. Kilpatrick K, Podesta G, Evans R (2001) Overview of the NOAA/NASA Advanced Very High Resolution Radiometer Pathfinder algorithm for sea surface temperature and associated matchup database. Journal of Geophysical Research: Oceans 106: 9179–9197.
62. Hooker S, McClain C (2000) The Calibration and Validation of SeaWiFS Data. Progress in Oceanography 45: 427–465.
63. Pickett M, Wenqing T, Rosenfeld L, Wash C (2003) QuikSCAT Satellite Comparisons with Nearshore Buoy Wind Data off the U.S. West Coast. Journal of Atmospheric and Oceanic Technology 20: 1869–1879.
64. Thomas A, Mendelssohn R, Weatherbee R (2013) Background trends in California Current surface chlorophyll concentrations: A state-space view. Journal of Geophysical Research: Oceans 118: 5296–5311.
65. Dormann CF, Elith J, Bacher S, Buchmann C, Carl G, et al. (2012) Collinearity: a review of methods to deal with it and a simulation study evaluating their performance. Ecography 36: 27–46.
66. R Core Team (2014) R: A language and environment for statistical computing. R Foundation for Statistical Computing, Vienna, Austria URL http://wwwR-projectorg/.
67. Akaike H (1974) A new look at the statistical model identification. Institute of Electrical and Electronics Engineers Transactions on Automatic Control 19: 716–723.
68. Bartoń K (2011) MuMl package: Multi-model inference. Model selection and model averaging based on information criteria (AICc and alike). CRAN.
69. Ihaka R, Gentleman R (1996) R: a language for data analysis and graphics. Journal of Computational and Graphical Statistics 5: 239–314.
70. Roberts J, Best B, Dunn D, Treml E, Halpin P (2010) Marine Geospatial Ecology Tools: An integrated framework for ecological geoprocessing with ArcGIS, Python, R, MATLAB, and C++. Environmental Modelling & Software 25: 1197–1207.
71. Karpov K, Albin D, Buskirk WV (1995) The marine recreational fishery in northern California and central California: a historical comparison (1958–86), status of stocks (1980–1986), and effects of changes in the California Current. California Department of Fish Game. Fish Bulletin 176 Fish Bulletin 176. 192 p.
72. Wilson J, Kay M, Colgate J, Qi R, Lenihan H (2012) Small-Scale Spatial Variation in Population Dynamics and Fishermen Response in a Coastal Marine Fishery. Public Library of Sciences ONE.
73. Huyer A (1983) Coastal upwelling in the California current system. Progress in Oceanography 12: 259–284.
74. Christensen A, Raakjaer J (2006) Fishermen's tactical and strategic decisions A case study of Danish demersal fisheries. Fisheries Research 81: 258–267.
75. Stewart I (2007) Updated US English sole stock assessment: Status of the resource in 2007. Portland: Pacific Fisheries Management Council.
76. Haltuch M, Ono K, Valero J (2013) Status of the U.S. petrale sole resource in 2012. Portland: Pacific Fishery Management Council.
77. Stewart I, Thorson J, Wetzel C (2011) Status of the US sablefish resource in 2011. Portland: Pacific Fishery Management Council.
78. Hicks A, Wetzel C (2011) The Status of Dover Sole (Microstomus pacificus) along the U.S. West Coast in 2011. Portland: Pacific Fishery Management Council.
79. Taylor I, Stephens A (2013) Stock Assessment of Shortspine Thornyhead in 2013. Portland: Pacific Fishery Management Council.
80. Stewart K, Lewison R, Dunn D, Bjorkland R, Kelez S, et al. (2010) Characterizing fishing effort and spatial extent of coastal fisheries. Public Library of Sciences ONE.
81. Yoklavich M, Love M, Forney K (2007) A fishery-independent assessment of an overfished rockfish stock, cowcod (Sebastes levis), using direct observations from an occupied submersible. Canadian Journal of Fisheries and Aquatic Sciences 64: 1795–1804.
82. Love M, Caselle J, Buskirk WV (1998) A severe decline in the commercial passenger fishing vessel rockfish (sebastes spp.) catch in the Southern California Bight, 1980–1996. California Cooperative Oceanic Fisheries Investigation Report. 180–195 p.
83. Jackson J, Kirby M, Berger W, Bjorndal K, Botsford L, et al. (2001) Historical overfishing and the recent collapse of coastal ecosystems. Science 27: 629–637.
84. Halpern B, Walbridge S, Selkoe K, Kappel C, Micheli F, et al. (2008) A global map of human impact on marine ecosystems. Science 319: 948–952.
85. Halpern B, Kappel C, Selkoe K, Micheli F, Ebert C, et al. (2009) Mapping cumulative human impacts to California Current marine ecosystems. Conservation Letters 2: 138–148.

11

Ecohydrodynamics of Cold-Water Coral Reefs: A Case Study of the Mingulay Reef Complex (Western Scotland)

Juan Moreno Navas[1]*, Peter L. Miller[2], Lea-Anne Henry[1], Sebastian J. Hennige[1], J. Murray Roberts[1,3,4]*

1 Centre for Marine Biodiversity and Biotechnology, Heriot-Watt University, Edinburgh, United Kingdom, 2 Remote Sensing Group, Plymouth Marine Laboratory, Plymouth, United Kingdom, 3 Center for Marine Science, University of North Carolina, Wilmington, North Carolina, United States of America, 4 Scottish Association for Marine Science, Scottish Marine Institute, Oban, United Kingdom

Abstract

Ecohydrodynamics investigates the hydrodynamic constraints on ecosystems across different temporal and spatial scales. Ecohydrodynamics play a pivotal role in the structure and functioning of marine ecosystems, however the lack of integrated complex flow models for deep-water ecosystems beyond the coastal zone prevents further synthesis in these settings. We present a hydrodynamic model for one of Earth's most biologically diverse deep-water ecosystems, cold-water coral reefs. The Mingulay Reef Complex (western Scotland) is an inshore seascape of cold-water coral reefs formed by the scleractinian coral *Lophelia pertusa*. We applied single-image edge detection and composite front maps using satellite remote sensing, to detect oceanographic fronts and peaks of chlorophyll *a* values that likely affect food supply to corals and other suspension-feeding fauna. We also present a high resolution 3D ocean model to incorporate salient aspects of the regional and local oceanography. Model validation using *in situ* current speed, direction and sea elevation data confirmed the model's realistic representation of spatial and temporal aspects of circulation at the reef complex including a tidally driven current regime, eddies, and downwelling phenomena. This novel combination of 3D hydrodynamic modelling and remote sensing in deep-water ecosystems improves our understanding of the temporal and spatial scales of ecological processes occurring in marine systems. The modelled information has been integrated into a 3D GIS, providing a user interface for visualization and interrogation of results that allows wider ecological application of the model and that can provide valuable input for marine biodiversity and conservation applications.

Editor: Judi Hewitt, University of Waikato (National Institute of Water and Atmospheric Research), New Zealand

Funding: This paper is a contribution towards the European Commission FP7 projects HERMIONE (grant 226354) and EPOCA (grant 211384) and the UK Ocean Acidification programme (NERC grant NE/H017305/1). JMN and JMR acknowledge support from Heriot-Watt University's Environment & Climate Change theme and Javier Campuzano from the Technical University of Lisbon. The funders had no role in study design, data collection and analysis, decision to publish, or preparation of the manuscript.

Competing Interests: The authors have declared that no competing interests exist.

* E-mail: J.Moreno_Navas@hw.ac.uk (JMN); j.m.roberts@hw.ac.uk (JMR)

Introduction

In a time of unprecedented global climatic change and human pressures in the ocean, there is an urgent need to understand marine ecosystems, their dynamics and the mechanisms that underpin their structure and functioning. Hydrodynamics are an important but inadequately considered constraint on marine ecosystems, despite much of the fauna and flora depending on adequate water flow for nutrient and food provision, oxygen supply and vital biological processes such as larval dispersal. Turbulent mixing, transport by waves and small scale circulation cells tend to increase ecosystem efficiency by augmenting particle encounter rates and giving species a better chance of finding a favorable environment [1]. Bottom layer turbulence and resuspension of eroded sediments help disperse food for benthic and demersal communities. Hydrodynamic processes operating on ecological timescales also affect biogeochemical interactions that in turn constrain chemical and biological properties of ecosystems [2],[3].

Ecohydrodynamic approaches investigate the hydrodynamics constraints on ecosystems across different temporal and spatial scales [2] encompassing the physical, chemical and biological characteristics of the ocean layers that sustain marine ecosystems. Hydrodynamics around coral colonies and entire coral reefs span a range of spatial scales in which fluid motions operate, starting with large-scale flow phenomena such as eddies produced by reef wakes, to smaller scale turbulent features created by reef topology and to finer spatial scales of flow occurring around single coral colonies and polyps [4], [5]. Lateral and vertical advection of particles also plays a significant role in the functioning of coral ecosystems, and reef geometry plays a key part in determining patterns and rates of coral reef fish larval transport, reef connectivity and transport of other particles [6],[7].

Cold-water reef framework-forming scleractinian corals such as the globally distributed *Lophelia pertusa* (Linnaeus 1758) lack symbiotic algae, feeding solely on particles advected by water motion [8],[9]. Hydrodynamics thus play an important role in the transport of food to some of the most biologically diverse ecosystems on the deep seafloor, and are themselves mediated by processes including internal waves and tidal currents, all of which can locally enhance food supply. For example, in continental margin settings, cold-water coral reefs occur where geostrophic currents and internal waves steepen and break along the continental slope [10]. Seasonal changes in current, temper-

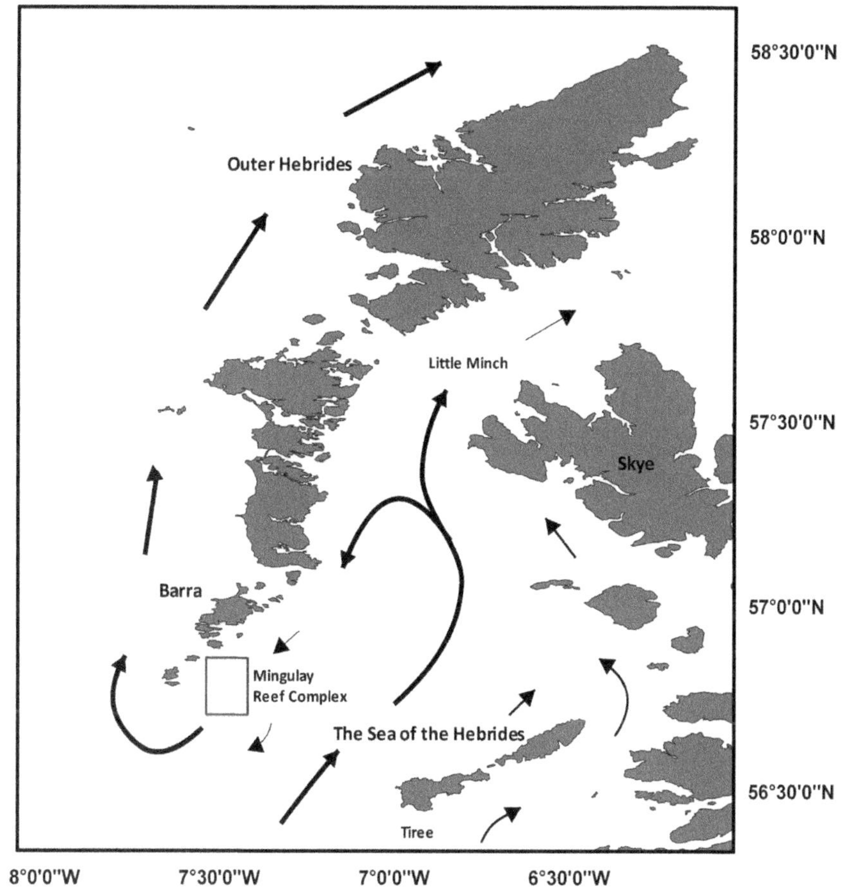

Figure 1. General current circulation. Surface circulation of the Hebridean island chain (adapted from ([29]).

ature and salinity patterns also occur in areas due to seasonal changes in the slope current [11],[12]. Near-seabed currents are significantly influenced by local topography and the patchy growth of the corals is often related to the varied speed and direction of local currents [11]. Active sediment transport by strong local near-bed currents helps prevent coral burial and also supplies sediment that becomes baffled in the dead coral framework contributing to the vertical growth of mounds [11],[13],[14].

Recent studies in these ecosystems suggest a coupling between the reef fauna and the surface productivity due to the local hydrography and sedimentary dynamics [10],[15],[16]. *Lophelia pertusa* is frequently associated with areas where surface waters show above average primary productivity [17]. In some continental shelf areas, domes of dense nutrient-rich waters can be generated over the banks, enhancing surface productivity that can be delivered down to the corals via several oceanographic dynamics [14]. Surface production and copepod species appear to be important food sources due to the fatty acid compositions of these animals and phytodetritus from the surface [11],[18]. The annual cycle in primary flux as well as the seasonal variability in near-bed hydrodynamic might also control coral growth and reproduction [19]. The ecohydrodynamic conditions around cold-water coral reefs therefore determine both the supply of food particles and recruits to the reef.

Remote sensing provides frequent views of a wide range of physical and biological processes occurring at the sea surface that may influence the ecology of corals. These processes include fronts, mesoscale eddies, currents, upwelling, phytoplankton and algal blooms. Three-dimensional (3D) hydrodynamic modelling can provide spatially explicit information on the key variables governing the oceanographic dynamics between surface phenomena and deep marine ecosystems with enough temporal resolution to give valuable information about turbulence, current regime and large scale oceanographic processes.

Our study is the first to apply these principles to ecosystems engineered by reef framework-forming cold-water corals. Our main goal is to characterize the oceanographic conditions using remote sensing techniques and the development of a 3D hydrodynamic model in order to assess the ecological constrains in a cold coral reef at medium temporal and spatial scales. The Mingulay Reef Complex, located 13 km to the east of the island of Mingulay in the Sea of the Hebrides (western Scotland, Figure 1), was chosen as a case study since the area has been intensively studied since its discovery in 2003 [20],[21].

Regional hydrodynamic setting

The Scottish western continental shelf has three water sources: Atlantic water propagating northward from the west of Ireland, Irish Sea water passing northward through the North Channel, and coastal water created by the high volume of river runoff from the Scottish mainland [22–27]. The overall circulation pattern comprises a net northward transport along the Scottish west coast, giving a northward flux of about 11×10^4 m^3 s^{-1} passing through the Little Minch [26] with variability superimposed by meteorological events [28]. Drifting buoys revealed a bifurcation of the northward coastal current in the Sea of the Hebrides: a water mass

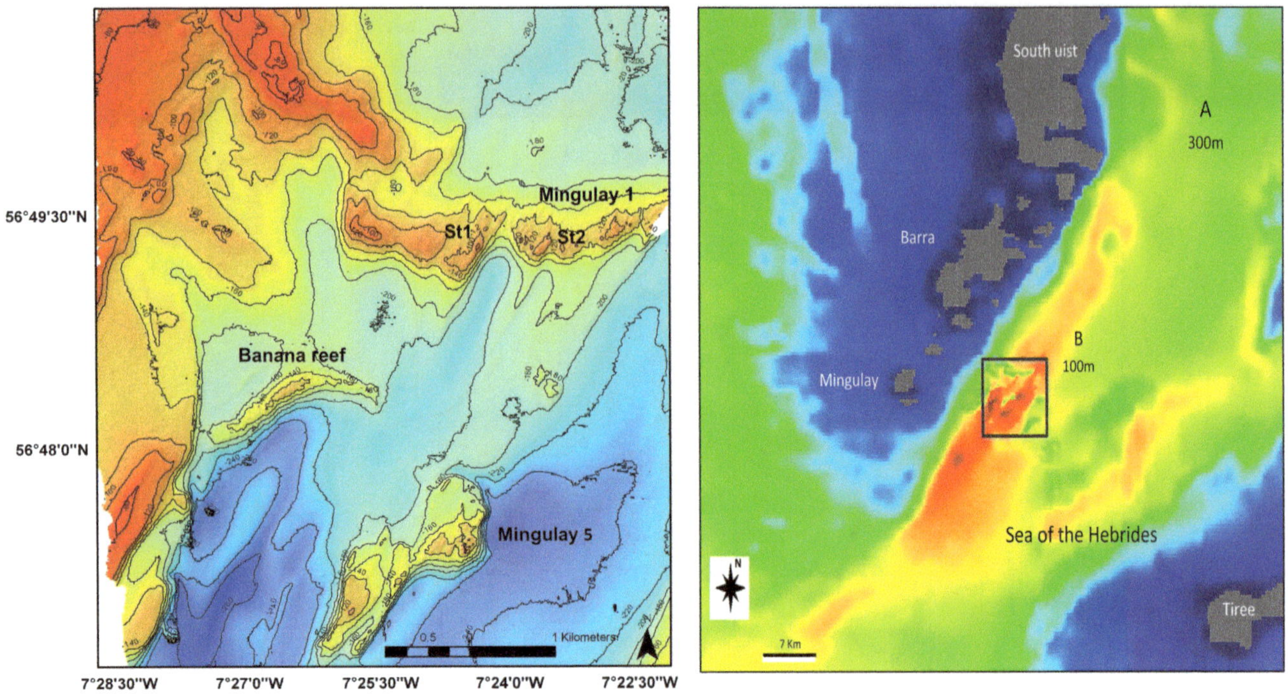

Figure 2. Detail maps of the study area. Bathymetry of Mingulay reef complex (left), and the bathymetry grid resolution used in the hydrodynamic model (right).

passes through the Little Minch and a recirculating southward current towards Barra Head [29] (Figure 1). The recirculated current includes two barotropic processes, tidal rectification and topographic steering of large scale flows, and a third mechanism whereby a baroclinic pressure gradient is associated with the observed dome of dense saline Atlantic water which extends into the centre of the channel from the south [30].

The regional oceanography is characterized by an intrusion of high salinity, high nutrient Atlantic water into the Sea of the Hebrides, and a weaker intrusion in the North Minch that produces a cyclonic circulation system [27],[29]. During summer months surface heating over topographic depression regions with weak tidal stirring, can lead to the formation of domes of cold dense bottom water trapped beneath the thermocline. The dense water dome is isolated horizontally from adjacent water by bottom fronts, which drive a cyclonic, near-surface circulation around the dome [29].

The seasonal water column characteristics and the cycles of physical and chemical properties of the whole Hebridean island chain are broadly repeated each year [27]. Sea surface temperature varies between 7–8 °C in spring with the warmest temperatures in autumn (13–14 °C). Deeper water temperatures show an intrusion of Atlantic Water into the southwest corner of the Sea of the Hebrides [22–27]. Salinity and nutrient levels

Table 1. Parameters used in the model calculations.

Physical parameter	Numerical value
Time step:	10 s
Grid mesh:	300,100 m
Horizontal cells (x,y):	(193,244) (114,126)
Vertical coordinate:	Sigma
Vertical layers:	8
Horizontal Eddy Viscosity:	4.3810 $m^2 \cdot s^{-1}$
Vertical Eddy Viscosity:	Derived from GOTM
Drag coeffcient:	0.0025
River discharges;	No
Salinity:	35 psu
Temperature:	11°C
Forced:	Tide from FES2004

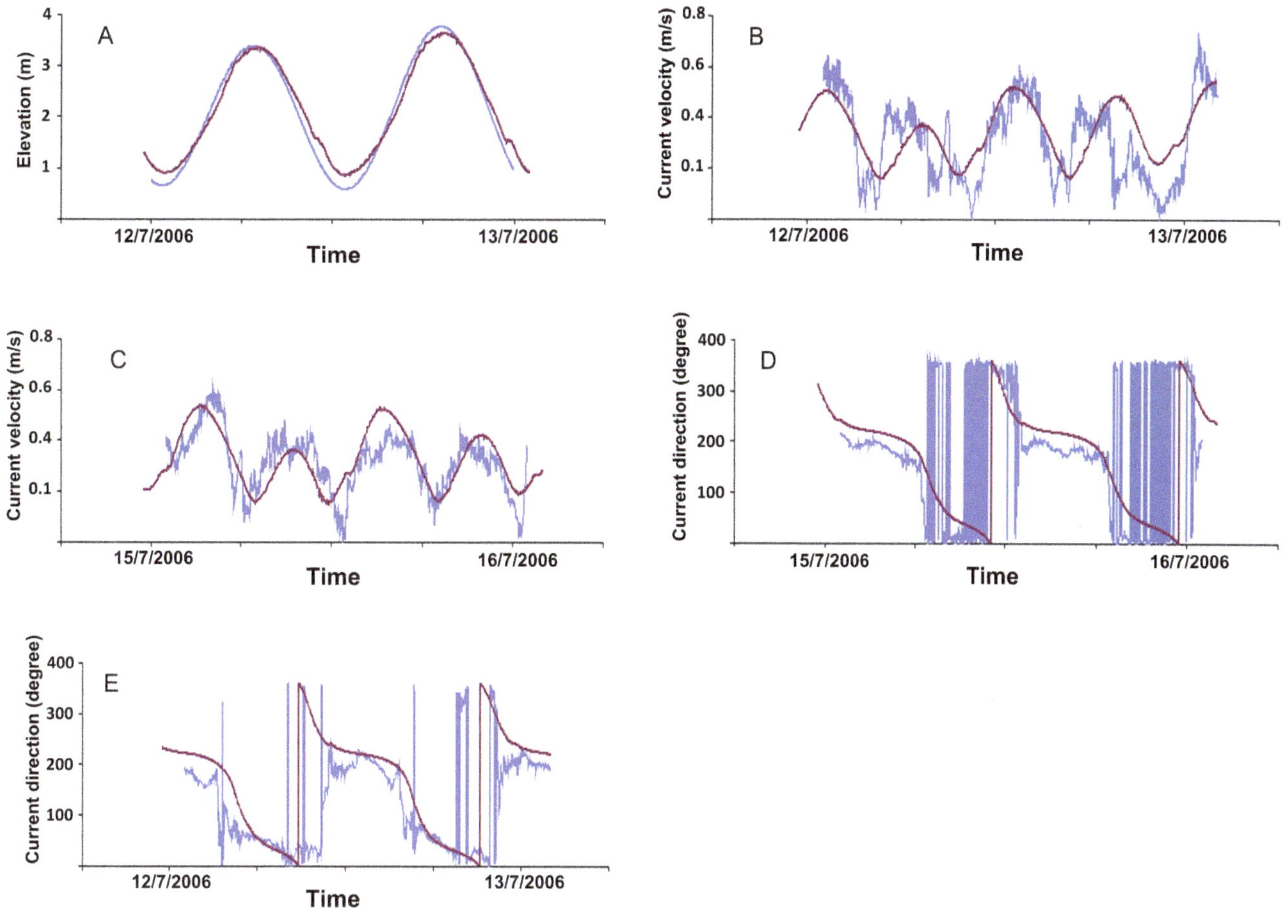

Figure 3. Model validation. The modelled hydrographic parameters were derived for two sampling stations within the reef complex and were recorded for model runs 15 days in duration (6/07/06 to 21/07/06). Time series of the intensity and current direction (blue line) and modelled (dark line) of one day at two stations St1 (B,E) and St2 (C,D) at the bottom and elevation measurements (A).

followed a positive east to west gradient, with higher values in the west mainly caused by freshwater discharge from the main Scottish landmass. Elevated chlorophyll-*a* levels in surface coastal waters were found during spring, suggesting a phytoplankton bloom that begins near the Scottish mainland, Little Minch and the Sea of the Hebrides [27],[30],[31].

The distribution of physical and chemical parameters in the Sea of the Hebrides reveal a weak thermal stratification in spring and a strong stratification during summer and autumn, the column water being well mixed during winter [27]. The salinity was always >35 ppt throughout the year, reaching the maximum of 35.3 ppt during spring, summer and autumn. The density distribution confirms a pool of dense Atlantic water throughout spring and autumn, with surface nutrient depletion of nitrate and silicate during summer and autumn. High nitrate phosphate and silicate concentrations correlated positively with high values of salinity except ammonia and chlorophyll-*a*, which were inversely correlated [27].

Geological history and present-day biodiversity

The bedrock of the Mingulay reef complex comprises Mesozoic sediments and a Paleocene intrusion that have been eroded by the recent glaciation [32]. This created a seascape that contains hollows >250 m deep and rocky ridges that rise more than 100 m above the surrounding seafloor. The interaction between water currents and irregular topography has exposed bedrock and boulders in elevated positions with bare rock predominating on topographic highs. The sediments in the area are dominated by muddy sand with broken shells and variable gravel content, with patches of coarse sediments and gravel in the north and west of the Mingulay Reef Complex [20],[21].

Fifteen distinct habitats including *L. pertusa* reef and coral rubble were identified from the first acoustic backscatter, video and grab surveys of the Mingulay reefs [20] with further mapping in 2006 allowing a total area of coral reef and rubble habitat in the complex to be estimated at 5.4 km^2 [33]. Over 500 macroinvertebrate species inhabit the reef complex, which form benthic communities that change across space due to interactions between changes in seabed terrain and current speed [34],[35]. Ecohydrodynamic phenomena are thought to partially explain why oviparous egg-laying sharks are using specific areas on the reef complex for spawning [36].

The area has been bathymetrically mapped using multibeam echosounders, revealing several reefs formed by the cold-water coral *Lophelia pertusa* [20],[21],[37]. The Mingulay reefs have developed intermittently during the Holocene with the oldest coral material so far dated from 7.7 ka [38]. A number of cases of apparently recent established coral colonies have been noted suggesting recruitment is still occurring [24]. Previous studies have shown that the Mingulay reefs have two dominant food supply

Table 2. Values of several descriptive statistics.

		Station st1			Station st2		
		obs	mod300	mod100	obs	mod300	mod100
Current Intensity (m/s)	Mean (m/s)	0.2926	0.3422	0.3377	0.3263	0.3381	0.3023
	Minimum (m/s)	0.0072	0.1241	0.1154	0.0048	0.1367	0.078
	Maximum (m/s)	0.6448	0.5413	0.5881	0.6027	0.5330	0.5279
	Variance (m/s)	0.0245	0.0117	0.0187	0.0117	0.0123	0.0198
	Sta dev (m/s)	0.1564	0.1121	0.1367	0.1084	0.1108	0.1408
	IoAd		0.75	0.67		0.83	0.64
	RMAE		0.16	0.15		0.04	0.07
Elevation (m)	IoAd	0.987					
	RMAE	0.03					

The IoAD and RMAE of water current intensity and elevation measured (obs) and modelled (mod) at the seabed at two stations (st1, st2).

mechanisms; a regular rapid downwelling of surface water delivering pulses of warm water and a periodic advection of high turbidity bottom waters [16]. In the Mingulay tidal regime, the hydraulic control of the flow over a bank can lead to a depression of the density structure downstream of it (hydraulic jump). When the tidal flow over the reef weakens and reverses, this depression propagates as an internal wave over the summit in the previously upstream direction, driving superficial waters downward and over the reef [16]. This sequence occurs on both tidal phases but is strongest on the flood to ebb because of the asymmetric profile of the tidal current. The higher turbidity in the advected water is most likely due to the fine grain sediment surrounding the reef area that becomes entrained in the bottom water because high frictional turbulence and accelerated tidal currents on the reef top enhance the delivery of particles suspended in deeper water [16].

The three-dimensional (3D) water modelling system, MOHID

Three dimensional hydrodynamic modelling has been used extensively to understand the flow dynamics in both shallow and oceanic waters. There have been many detailed studies of the hydrodynamics of tropical coral reefs [4],[5] but only two previous studies used hydrodynamic models [10],[15] to examine the ecology of cold-water coral reefs.

The hydrodynamic model MOHID (Modelo Hidrodinámico) used in this study was developed by the Marine and Environmental Technology Research Centre (MARETEC) at Instituto Superior Técnico, Technical University of Lisbon. This model has been previously used in ocean circulation, transport and mixing simulations along the continental shelf edge, in oil spill management, operational oceanography, nutrient load and residence time models in several places around the world (see http://www.mohid.com/wiki/index.php?title=Mohid_Bibliography#2013).

The model solves the equations of a three-dimensional flow for incompressible fluids and an equation of state relating density to salinity and temperature [39–41].

The Cartesian coordinate framework [41] equations are as follows:

$$\frac{\partial u_i}{\partial x_i}=0 \tag{1}$$

$$\frac{\partial u_1}{\partial t}+\frac{\partial(u_j u_1)}{\partial x_j}=fu_2-g\frac{\rho_\eta}{\rho_0}\frac{\partial \eta}{\partial x_1}-\frac{1}{\rho_0}\frac{\partial p_S}{\partial x_1}-\frac{g}{\rho_0}\int_z^\eta \frac{\partial \rho'}{\partial x_1}dx_3+\frac{\partial}{\partial x_j}\left(A_j\frac{\partial u_1}{\partial x_j}\right) \tag{2}$$

$$\frac{\partial u_2}{\partial t}+\frac{\partial(u_j u_2)}{\partial x_j}=-fu_1-g\frac{\rho_\eta}{\rho_0}\frac{\partial \eta}{\partial x_2}-\frac{1}{\rho_0}\frac{\partial p_S}{\partial x_2}-\frac{g}{\rho_0}\int_z^\eta \frac{\partial \rho'}{\partial x_2}dx_3+\frac{\partial}{\partial x_j}\left(A_j\frac{\partial u_2}{\partial x_j}\right) \tag{3}$$

$$\frac{\partial p}{\partial x_3}=-\rho g \tag{4}$$

Where (u_i) is the velocity componentx_i, η is the free surface elevation, A_j is the turbulent viscosity coefficient, f is the Coriolis parameter, p_s is the atmospheric pressure, g is the gravitational acceleration, ρ is the density and ρ' represents the density anomaly as the depth mean density minus the density at particular height. The density is calculated as a function of salinity and temperature using the equation of state [42].

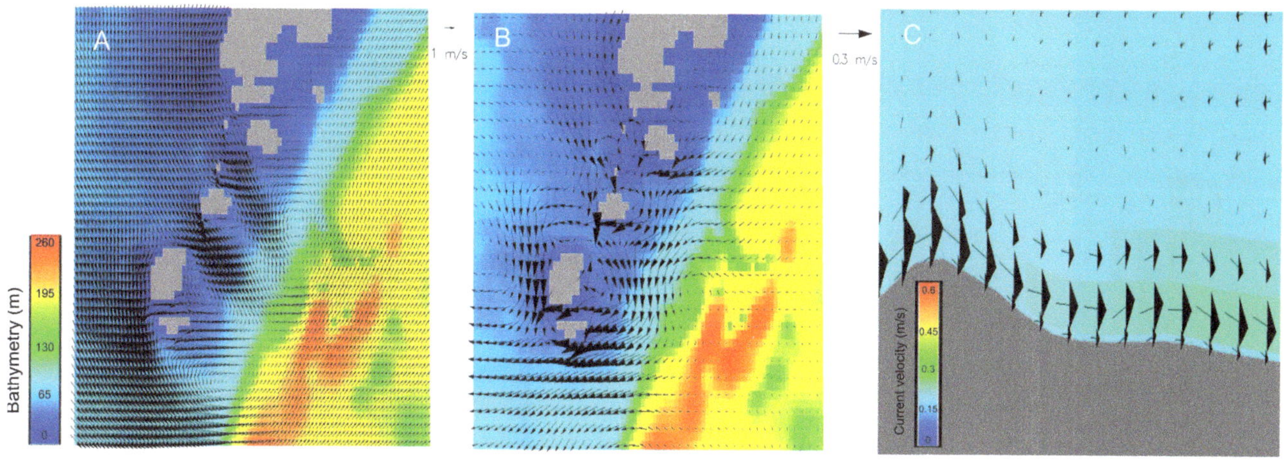

Figure 4. Horizontal water circulation in the Mingulay Reef Complex. (a) Snapshot of the seabed current from the model simulation (11/07/06). Arrows illustrate the direction and speed of the current. A very clear circulation structure can be indentified with three anticlockwise eddies in different areas in the islands. (b) The residual circulation in the seabed layer during the period modelled. (c) the downwelling at Mingulay Area 1 reproduced using this high resolution 3D hydrodynamic model (vertical exaggeration, 4x).

The residual current u_e was estimated by averaging the horizontal flow u over the modelled time period T:

$$u_e = \frac{1}{T}\int_0^T u dt \quad (5)$$

The nested system used in this model consists of two sub-components: a coarse-resolution outer model covering part of the Sea of the Hebrides with a horizontal resolution of roughly 300 m, and a fine-resolution inner model covering the Mingulay Reef Complex with a horizontal resolution of roughly 100 m. In order to ensure model stability and a smooth transition between modelled areas certain constrains are imposed following the suggested protocol for nesting a hydrodynamics model [43].

A bathymetry model was obtained from GEBCO and from two multibeam echosounder surveys carried out across the reef complex. The first survey led to the discovery of the Mingulay Reef Complex in 2003 and was carried out using a Kongsberg EM2000 multibeam echosounder [20]. The second survey in 2006 extended this area using a Kongsberg EM300 echosounder leading to the discovery of the Banana Reef [21],[37],[44]. These data were used to produce an initial GIS vector data layer followed by interpolation to 300 m and 100 m grid resolutions (Figure 2). An Arakawa C grid was used for spatial discretization [45] and in this study a sigma coordinate was chosen with eight vertical layers.

The modelled hydrographic parameters were derived for two sampling stations within the reef complex and were recorded over model runs of 15 days (6/07/06 to 21/07/06) covering a lunar tidal cycle and coinciding with the period during which measurements were available [16]. Two hydrologic time series over two days were selected (11-13/07/06 at Station 1 and 14-15/

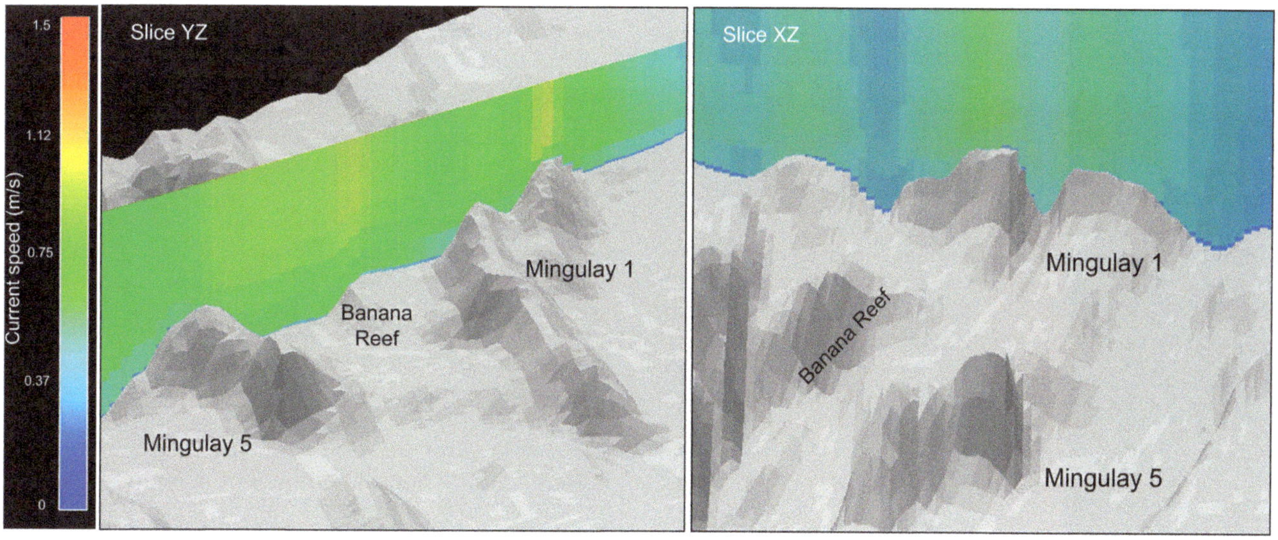

Figure 5. Vertical water circulation in the Mingulay Reef Complex. Vertical current velocity snapshots covering the whole Reef (Slice YZ) and in detail at Mingulay 1 (Slice XZ) with a 3x vertical exaggeration.

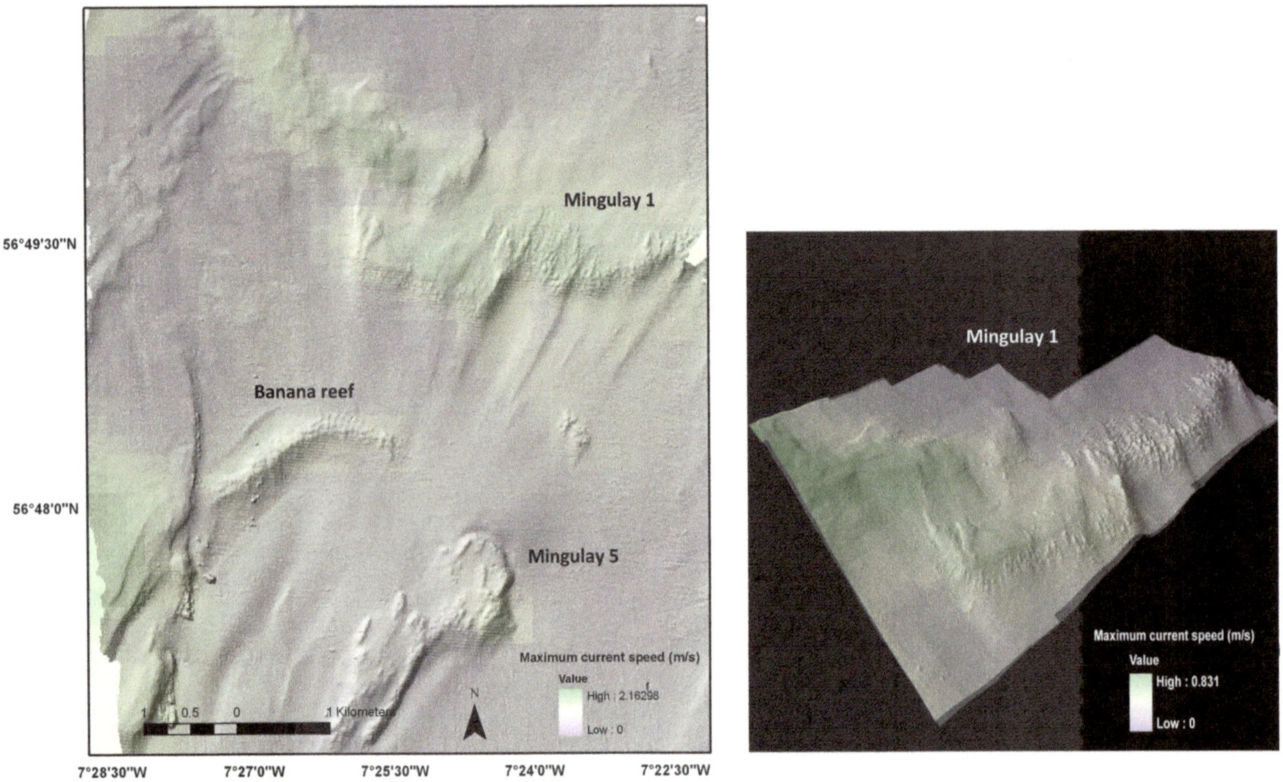

Figure 6. Example of integration between the hydrodynamic model in 3D GIS. The shading illustrates the maximum current velocity at the seabed across the Mingulay reef complex (plan view, left) and across Reef Area 1 (3-D view with vertical exaggeration (2x), right).

07/06 at Station 2); two data cross-sections of one day were separated for use in the calibration and validation processes, and independently for cross-validation. Vertical eddy viscosity/diffusivity was determined with a turbulence closure model selected from those available in the General Ocean Turbulence Model (GOTM) [46].

In this application a 3D model forced with tide was implemented. Water level imposed at the boundary was taken from the FES2004 global tide solution [47], which gave heights of tidal constituents as it was not possible to use local tide gauge information. The model incorporates the islands of South Uist, Barra, Mingulay and Tiree (see Figure 2) covering an area 100 km north-south and 80 km east-west.

We compared the tidal water elevation calculated from the tidal constituents as per FES2004 with modelled values of tidal elevations from MOHID. The parameters used in the model calculations are summarized in Table 1.

Rates of residual flow were computed because this helps to understand long-term water exchange in the area. The residual current was estimated by averaging the horizontal flow over a modelled time period of 15 days (lunar tidal cycle).

For the validation process several descriptive statistics (mean, variance and standard deviation), the relative mean absolute error (RMAE) [48] and the Index of Agreement (IoAd) [49],[50] were calculated.

Ocean colour

Local oceanic fronts were studied using both thermal and colour data from satellite observations. For thermal fronts, 10 years of Advanced Very High Resolution Radiometer (AVHRR) sea-surface temperature data from December 1998 to November 2008 were processed, comprising several passes per day over the UK at a maximum 1.1 km resolution. Chlorophyll-*a* fronts were processed using data from the Moderate Resolution Imaging Spectroradiometer (MODIS) sensor on the NASA Aqua satellite, from 2009 to 2011 at 1 km resolution. The composite front map technique was used to combine the location, gradient, persistence and proximity of all fronts observed over each month into a single map [51]. The monthly front maps were then aggregated into seasonal front climatologies to identify frequently occurring features [52],[53].

Results

Figure 3 shows a time series of the water current intensity of one day used for validation. The direction and elevation measured (blue line) and modelled (dark line) for two stations St1 and St2 at the bottom using the lowest sigma layer with around 3 m of thickness. For sea surface elevation, the calculated values of IoAd were close to 1, and low values of RMAE revealed a good agreement between the prediction of the model [48] and the observations (table 2). A nested model with high resolution (mod100) was applied in detail at the Mingulay Reef Complex and produced similar current characteristic as given by the coarse resolution model (mod300). The water current intensity IoAd values varied from 0.64 and 0.84, suggesting that our model produced acceptable results. The values of mean, variance and standard deviation of the observed and modelled current intensity at the two stations were similar. The RMAE values were very low, suggesting that the model produced good average values (Table 2).

Our model reproduced the tidally-driven current regime and showed several eddies due to tidal current jets entering the area

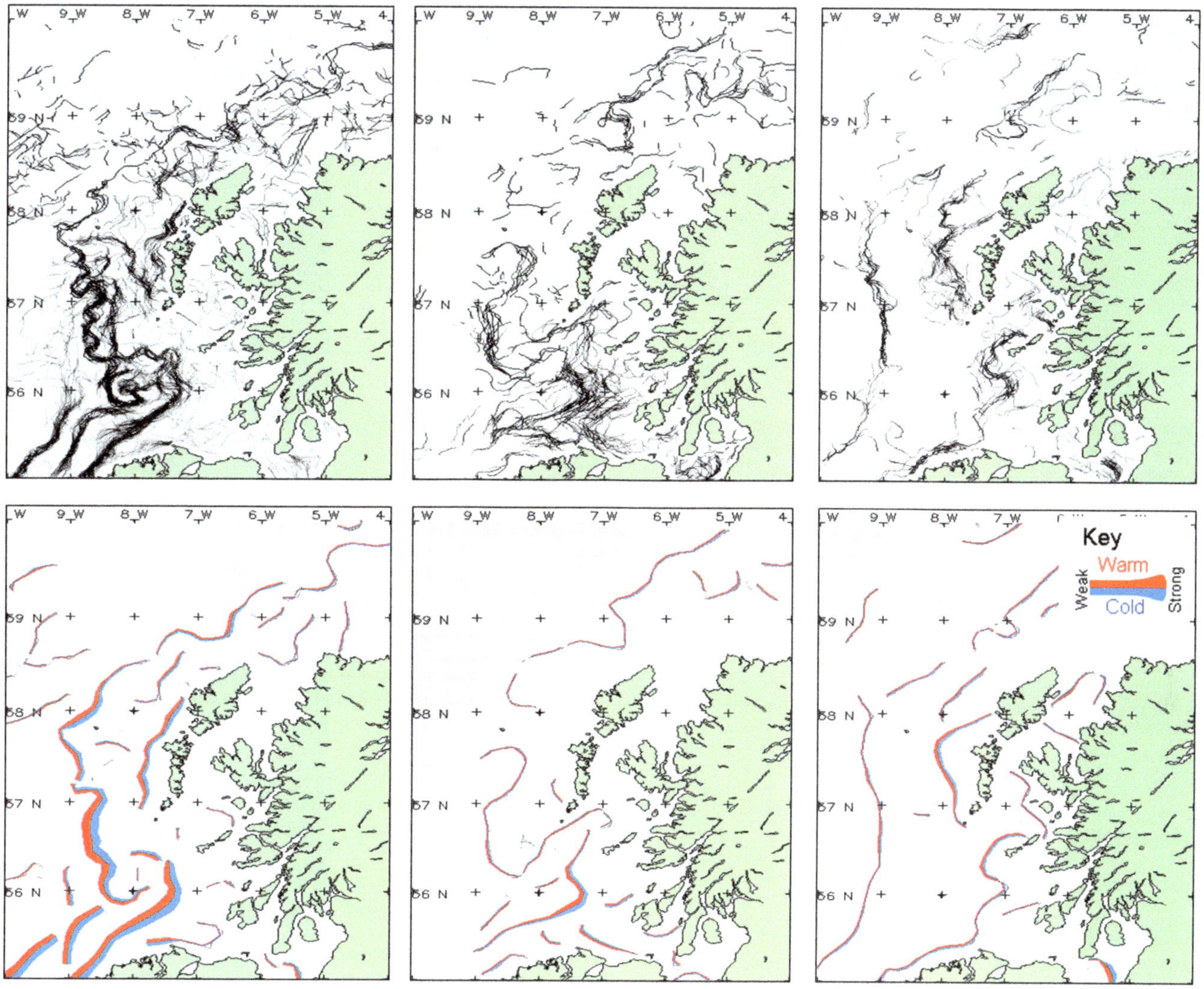

Figure 7. Remote Sensing I. Thermal fronts in several months in 2010. Thermal fronts observed during selected months in 2010. Left column: March; center column: May; right column: December. The upper maps show the composite front map derived from all cloud-free observations during that month; whereas the lower maps depicts a simplified version of the fronts, coloured red-blue to indicate the warm-cold side of the front, and the thickness of the line indicates the strength of the front.

through the channels between the islands of Barra and Mingulay (Figure 4a). These jet currents influence the internal circulation structure. A circular current regime can be observed, with current direction flowing south then turning to the west and back again through the channels at the end of the Hebridean islands in the Mingulay area). A drifter could confirm this modeled circular structure around the Mingulay island (see www. http://martech.sams.ac.uk/fastnet/map_all.php, drifter number 25). In the east of the study area, the residual current calculated during a full tidal cycle is predominantly northward whereas in the westernmost portions, closer to the Hebridean island chain, the residual currents flow in a southward direction (Figure 4b). The hydrodynamic model successfully recreated the current regime, as well as downwelling phenomena [16] observed *in situ* at Mingulay Area 1 (Figure 4c). The model also provided evidence of downwellings in Mingulay 5 and Banana Reef with the same characteristics as describe in Mingulay 1 (Figure 4c). These several downwelling phenomena occurs with ~30–60 minutes delays, but is difficult to ascertain the timings with high precision.

We can observe areas of high-speed current velocity in the whole water column in several areas in the Mingulay Reef Complex (Figure 5, slice YZ) creating high turbulence and enhancing the downwelling phenomena. Differences can also be noticed in the current speed at different water depths and lateral areas in Mingulay 1 (Figure 5,slice XZ).

The mean, maximum and the standard deviation values of the modelled current velocities were incorporated into GIS to provide an adaptable graphical user interface for 2D, 3D and temporal visualization (see Figure 6). The results show a change in current velocity from west to east with higher current speeds in the western areas closer to the island chain. High current speed areas were found in Mingulay 1, Mingulay 5 and Banana Reef.

Figure 7 shows thermal fronts observed over a wide region during selected months of 2010, in order to illustrate seasonal differences. In March the fronts had a strong thermal signature in the Irish Sea and on the Hebrides section of the continental shelf. In May, a weak thermal front was close to Mingulay Reef

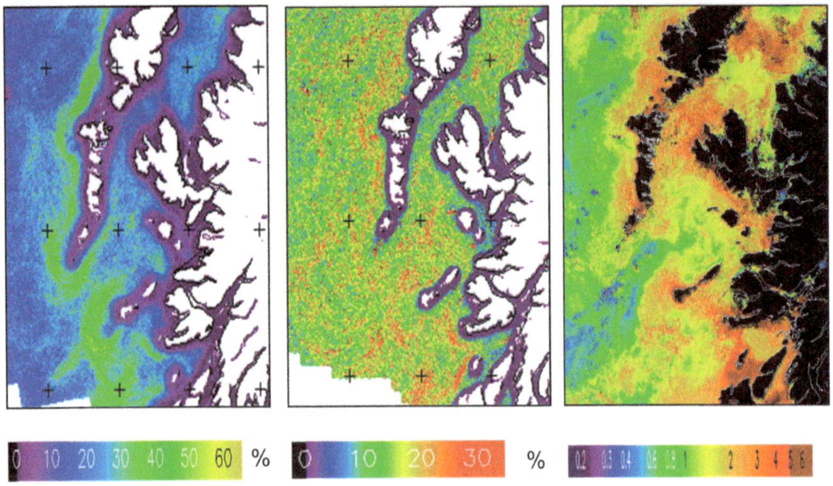

Figure 8. Remote Sensing II. Mean all-season thermal and Chl-*a* fronts. Mean all-seasons front frequency for the Outer Hebrides and Tiree: (a) thermal fronts; (b) Chl-a fronts. (c) MERIS Full-resolution (300 m) resolution chl-a composite for 23–30 Apr. 2011 showing chl-a fronts in similar location s to frequent thermal fronts.

Complex, and in December several fronts were observed but were generally weak in different ocean areas.

The ocean fronts observed in this region by satellite were primarily thermal, concentrated in the areas west of the Hebridean islands and the isle of Tiree (Figure 8a). There was a slightly increased frequency in the Aqua-MODIS 1 km Chl-*a* fronts in areas affected by thermal fronts (Figure 8b).

Discussion

Ecohydrodynamics

Relevant temporal scales vary from less than a second, dominated by diffusion and turbulence, to months over which tidal regimes and major oceanographic phenomena come into play. Relevant spatial scales vary from centimeters where flow within and across corals affects local biogeochemical fluxes and skeletal morphologies up to kilometers across entire reef complex-es.

Ecohydrodynamics are widely accepted as key drivers of marine ecosystem structure and functioning in deep water settings such as seamounts, ridges and canyons, but empirical data and models of complex flow are still largely lacking.

Over this broad region, the existence of fronts (Figure 7 and 8 a,b) had been verified by several studies [23],[24],[26],[54–56].A thermal front is usually present on the Scottish shelf during spring and summer, marking the boundary between the cooler tidally-mixed water near to the coast, and the warmer stratified surface water further offshore. In December there are also fronts visible along the shelf-break, probably related to the slope current. However, the distribution of chl-a fronts is not as clear. This could be explained by the position of chl-*a* fronts being more variable than the physical features, or a low surface chl-*a* concentration; there may be a deep chl-*a* maximum that cannot be detected using EO data. However, ocean colour fronts may be observed occasionally, as in this example from April 2011 (Figure 8c), where there was a sharp contrast between the nutrient-rich shelf water and less productive oceanic water during the spring bloom. The front southwest of Tiree is related to the shallow inner shelf that extends southwest from the island and causes tidal mixing [56].

We observed thermal fronts and peaks of chl *a* values that may influence the coral reef areas in the Hebrides Sea. The elevated chl-*a* distribution in surface coastal water found during spring suggests the phytoplankton bloom began near the Scottish mainland, Little Minch and the Sea of the Hebrides. These high productivity areas and the majority of the thermal fronts are outside or very close to the model boundaries, and for these reasons linking information between the remote sensing features and the model outputs are problematic. Our modelled time scale is ~ one tidal cycle (15 days) whereas the oceanographic features could range from hours to months. Additionally, the intrusion of Atlantic water into the Sea of the Hebrides is not recreated by the model. However we can assume that the currents, the down-welling, and the residual current regime in a southward direction (as indicated by the model) suggests that the reef areas can be feed by phytoplankton and particles from the Little Minch and Hebrides coastal areas. Due to the proximity of the model boundary to Tiree Island, our model cannot provide specific information about the thermal fronts in this area.

Previously only two studies used hydrodynamics to model ecological dynamics on cold-water coral reefs [15],[57] with coarse and reasonable high spatial resolution. We have developed and applied the first high resolution (100 m) temporally dynamic 3D ocean model to improve our understanding of how regional ecohydrodynamics affect cold-water coral reef ecosystems. The hydrodynamics of our study area had been previous modeled [58],[59] with different spatial resolutions provided information about the influence of tide, wind and upstream inflow. The tidal residual flow modelled in our study matched the patterns and directions of previous study [59].

Although numerical modelling was challenging due to the scarcity of information for 3D model parameterization, several open boundary extensions were tested, and all available hydro-graphic data was used for validation (assuming barotropic flow is a good characteriser of hydrodynamic regime). Subsequent com-parison between predicted and ground-truthed measurements of what is a hydrodynamically complicated area showed good agreement. However, it is important to remember the importance of taking density-driven dynamics into consideration. The next steps will be to advance the state-of-the-art by developing a fully three-dimensional baroclinic hydrodynamic model as suggested by

Hill et al. [29] across a longer time period to better characterise the residual currents and extend the open boundaries for a better understanding of oceanographic features such as thermal and chlorophyll fronts.

Some studies [27],[29] provided evidence of an intrusion of Atlantic water into the Sea of the Hebrides and found that during summer each year surface heating over the isolated regions with weak tidal stirring such as the topographic depressions found in the area could lead to the formation of domes of cold, dense bottom water trapped beneath the thermocline [29]. Our model suggests that the Mingulay Reef Complex has a very strong tidal current regime and that water exchange is too rapid to actually account for such a cold lens.

The ecohydrodynamic temporal scale at Mingulay Reef complex range is about hours to minutes dominated by the tidal regime that affect the coral. The downwellings every 6 hours and a high current tidal regime are the key oceanographic features. The hydrodynamic model recreated a realistic tidal current regime and the downwelling phenomenas [16], which now provides us with the basis to model and predictively map areas with dense coral growth that also have high coral-associated biodiversity of sessile suspension feeders at the complex [34], [35].

Recently a study considered that a downwelling and mixing play an important role in the nutrition of cold-water corals on a fjordic sill in the Norwegian Skagerrak as well as reef geomorphology [60]. The role of downwelling of surface matter was also confirmed at Mingulay, where *Lophelia* feeds on herbivorous copepod crustaceans [18] and surface algal matter [37]. The current regime in the Sea of the Hebrides transports zooplankton from the surrounding areas to the coral reef mainly from the Little Minch, and advects particles from high productivity areas in the west coast of the Hebrides island chain. When the inshore-offshore distribution of plankton species on the west coast of Scotland were compared, it was found that a large oceanic copepod species was more abundant with increasing distance from the mainland [61]. In addition, current velocity affects food uptake by *Lophelia pertusa*, with several studies suggesting velocities values between 2.5 cm/s to 5 cm/s as an optimal range for prey capture, although the range could exceed 40 cm/s [37],[62],[63]. Our results suggest that the current regime with high velocities, created high turbidity that can play an important role advecting food particles to the corals, and that the area has an active sediment transport with strong local near-bed currents preventing the coral to be buried.

Drifting buoys drogued at 15 m depth revealed that seawater density plays an important role in the Scottish coastal circulation with a significant contribution to the northward transport and recirculation patterns [29]. These features were predicted to have great significance for larval transport including for commercially important species such as the Norway lobster (*Nephrops norvegicus*), and may also be relevant to the distribution of *Lophelia* off Mingulay. In the case of cold-water corals, a seawater density envelope of sigma-theta between 27.35 and 27.65 kg m^{-3} could act as a physical boundary condition for coral growth and distribution depending on larval buoyancy [64]. Little information exists about the timing of gamete release, larval competency and recruitment in deep water scleractinians [65], however reproductive particles could behave as passive tracers with water density playing an important role [29]. A fully baroclinic 3D hydrodynamic model coupled with a particle tracking model could help to explain larval transport and the distribution of these species.

Ecosystem-scale relevance

In addition to close coupling between hydrodynamic regimes and the growth and distribution of cold-water corals, ecohydrodynamics shape the structure and function of biological communities inhabiting the reef complex. Species diversity and composition of these communities change in response to interactions between hydrography and seabed terrain [35]. Current speed was particularly important to community assembly, with few fauna being able to withstand an exposed high current speed environmental niche. These interactions are also important for local sharks, which seem to spawn in areas not directly exposed to strong currents but with adequate water motion to ventilate egg capsules deposited by pregnant females [36]. We believe that our hydrodynamic model has already provided valuable additional input to ecological models of marine biodiversity [35] and could be used in future studies with conservation applications, for example in models that predict coral larval dispersal trajectories and reproductive output.

Conclusions and future perspectives

Once more satellite data has been collected under future research programmes, connectivity between satellite and modelled data can be improved. The results from the hydrodynamic model here have been incorporated into GIS to provide an adaptable graphical user interface for 2D, 3D and temporal visualization for interrogation of results and as an input to other spatial models or decision support systems. This integration also offers the possibility of combining the data with layers of spatial information about economic and social aspects, communications and security of the study area to develop an integrated socio-ecological approach for the selection of marine protected areas. The process of 3D circulation modelling is challenging; however the benefits of such a decision support tool or modelling future climate change scenarios are also considerable.

The oceans have entered a period of rapid change with rates of warming and acidification forecasted to reach unprecedented levels. In almost all marine ecosystems we lack a rigorous modelled framework to assess the ecological implications of predicted change. It is recognised that at present rates of carbon dioxide release the depth of the Atlantic's aragonite saturation horizon will shoal rapidly in the coming century [17],[66]. Recent work shows that the downwelling at the Mingulay Reef Complex brings more alkaline surface waters potentially helping mitigate the overall trend of declining saturation states [67]. Future models need to account for this natural variability and attempt to incorporate carbonate chemistry and other key parameters if we are to achieve a properly integrated picture of the ramifications of global ocean change. Similarly, ecohydrodynamic approaches will be important in fully understanding the implications of marine renewable energy and related offshore installations.

Acknowledgments

JMN and JMR acknowledge support from Heriot-Watt University's Environment & Climate Change theme and Javier Campuzano from the Technical University of Lisbon. Mike Heath is thanked for constructive comments on the manuscript.

Author Contributions

Conceived and designed the experiments: JMN. Performed the experiments: JMN PM. Analyzed the data: JMN PM. Contributed reagents/ materials/analysis tools: JMN JMR LH SH PM. Wrote the paper: JMN JMR LH SH PM.

References

1. Nihoul JCJ, Djenidi S (1991) Hierarchy and scales in marine ecohydrodynamics. Earth Science Reviews 26: 163–189.
2. Nihoul JCJ (1981) Marine hydrodynamics at ecological scale. In: Ecohydrodynamics: Proceedings of the 12th International Liège Colloquium On Ocean Hydrodynamics. Amsterdam: Elsevier Scientific Pub. Co. pp 1–12.
3. Warwick RM, Uncles RJ (1980) Distribution of benthic macrofauna associations in the Bristol Channel in relation to tidal stress. Marine Ecology Progress Series 3: 97–103.
4. Hearn CJ (2011) Geomorphic zonation. In: Hopley D, editor. Encyclopedia of modern coral reefs: Structure, form and process. Springer-Verlag Earth Science Series. Part 8, pp 563–573.
5. Monismith SG (2007) Hydrodynamics of Coral Reefs. Annual Review of Fluid Mechanics 39: 37–55.
6. Black KP (1993) The relative importance of local retention and inter-reef dispersal of neutrally buoyant material on coral reefs. Coral Reefs 12: 44–53.
7. Cowen RK, Lwiza KMN, Sponaugle S, Paris CB, Olson DB (2000) Connectivity of marine populations: open or closed?. Science 287: 857–59.
8. Roberts JM, Wheeler AJ, Freiwald A (2006) Reefs of the deep: the biology and geology of cold-water coral ecosystems. Science 312: 543–547.
9. Roberts JM, Wheeler A, Freiwald A, Cairns SD (2009) Cold-water Corals: The Biology and Geology of Deep-sea Coral Habitats. Cambridge University Press. 334 pp.
10. White M, Mohn C, De Stigter HC, Mottram G (2005) Deep-water coral development as a function of hydrodynamics and surface productivity around the submarine banks of the Rockall Trough, NE Atlantic. In: Freiwald A, Roberts JM (Eds.) Cold-Water Corals and Ecosystems. Springer, Berlin. pp 503–514.
11. Mienis F, de Stigter HC, White M, Duineveld G, de Haas H, et al. (2007) Hydrodynamic controls on cold-water coral growth and carbonate-mound development at the SW and SE Rockall Trough Margin, NE Atlantic Ocean. Deep-Sea Research I: Oceanographic Research Papers 54: 1655–1674.
12. Pingree RD, Le Cann B (1989) Celtic and Armorican slope and shelf residual currents. Progress in Oceanography 23: 303–338.
13. Dorschel B, Hebbeln D, Foubert A, White M, Wheeler AJ (2007) Hydrodynamics and cold-water coral facies distribution related to recent sedimentary processes at Galway Mound west of Ireland. Marine Geology 244: 184–195.
14. Wheeler AJ, Kozachenko M, Beyer A, Foubert A, Huvenne V, et al (2005) Sedimentary processes and carbonate mounds in the Belgica Mound province, Porcupine Seabight, NE Atlantic. In: Freiwald, A.,Roberts, M.J. (Eds.), Cold-Water Corals and Ecosystems. Springer, Berlin. pp 503–514.
15. Thiem O, Rvagnan E, FossåJH, Berntsen J (2006) Food supply mechanisms for coldwater corals along a continental shelf edge. Journal of Marine Systems 60: 207–219.
16. Davies AJ, Duineveld GCA, Lavaleye MSS, Bergman MJN, van Haren H, et al. (2009) Downwelling and deep-water bottom currents as food supply mechanisms to the cold-water coral Lophelia pertusa (Scleractinia) at the Mingulay Reef complex. Limnology and Oceanography 54: 620–629.
17. Guinotte JM, Orr JC, Cairns SD, Freiwald A, Morgan L, et al (2006) Will human-induced changes in seawater chemistry alter the distribution of deep-sea scleractinian corals?. Frontiers In Ecology And The Environment 4: 141–146.
18. Dodds LA, Black KD, Orr H, Roberts JM (2009) Lipid biomarkers reveal geographical differences in food supply to the cold-water coral *Lophelia pertusa* (Scleractinia). Marine Ecology Progress Series 397: 113–124
19. Mienis F, De Stigter HC, De Haas H, Van Weering TCE (2009) Near-bed particle deposition and resuspension in a cold-water coral mound area at the Southwest Rockall Trough margin, NE Atlantic. Deep Sea Research Part I: Oceanographic Research Papers 56: 1026–1038.
20. Roberts JM, Brown CJ, Long D, Bates CR (2005) Acoustic mapping using a multibeam echosounder reveals cold-water coral reefs and surrounding habitats. Coral Reefs 24: 654–669.
21. Roberts JM, Davies AJ, Henry LA, Duineveld GCA (2009) Mingulay reef complex, northeast Atlantic: an interdisciplinary study of cold-water coral habitat, hydrography and biodiversity. Marine Ecology Progress Series 397: 139–151.
22. Craig RE (1959) Hydrography of Scottish coastal waters. Marine Research 1958:30
23. Ellett DJ (1979) Some oceanographic features of Hebridean waters. Proceedings of the Royal Society of Edinburgh 77B:61–74
24. Ellett DJ, Edwards A (1983) Oceanography and inshore hydrography of the Inner Hebrides. Proceedings of the Royal Society of Edinburgh 83B: 143–160
25. McKay WA, Baxter MS, Ellett DJ, Meldrum DT (1986) Radiocaesium and circulation patterns west of Scotland. Journal of Environmental Radioactivity 4: 205–232.
26. McKinley IG, Baxter MS, Ellett DJ, Jack W (1981) Tracer applications of radiocaesium in the Sea of the Hebrides. Estuarine, Coastal and Shelf Science 13: 69–82.
27. Gillibrand PA, Sammes PJ, Slesser G, Adams RD (2003) Seasonal Water Column Characteristics in the Little and North Minches and the Sea of the Hebrides. I. Physical and Chemical Parameters. Fisheries Research Services Internal Report No 08/03. FRS Marine Laboratory. Aberdeen. 40pp
28. Simpson JH, Hill AE (1986) The Scottish coastal current. In The Role of Freshwater Outflow in Coastal Marine Ecosystems (Skreslet, S., ed.) NATO ASI Series. Springer-Verlag, Berlin, Heidelberg. G7: 295–308.
29. Hill AE, Horsburgh KJ, Garvine RW, Gillibrand PA, Slesser G, et al (1997) Observations of a density-driven recirculation of the Scottish coastal current in the Minch. Estuarine Coastal and Shelf Science 45: 473–484.
30. Savidge G, Lennon HJ (1987) Hydrography and phytoplankton distribution in north-west Scottish waters. Continental Shelf Research 7: 45–66.
31. Fehling J, Davidson K, Bolch CJS, Brand TD, Narayanaswamy BE (2012) The Relationship between Phytoplankton Distribution and Water Column Characteristics in North West European Shelf Sea Waters. PLoS ONE 7(3): e34098. doi:10.1371/journal.pone.0034098
32. Long D, Wilson CK (2003) Geological background to coldwater coral occurrences in the Minch. British Geological Survey Internal Report, IR/03/149
33. Watmough T (2008) Mapping the Mingulay Reef Complex and developing video analysis techniques for deep-sea research. Heriot-Watt University.
34. Henry LA, Davies AJ, Roberts JM (2010) Beta diversity of cold-water coral reef communities off western Scotland. Coral Reefs 29: 427–436.
35. Henry LA, Moreno-Navas J, Roberts JM (2013) Multi-scale interactions between local hydrography, seabed topography, and community assembly on cold-water coral reefs. Biogeosciences 10: 2737–2746.
36. Henry LA, Moreno-Navas J, Hennige SJ, Wicks LC, Vad J, et al (2013) Cold-water coral reef habitats benefit recreationally valuable sharks. Biological Conservation 161: 67–70.
37. Duineveld GCA, Jeffreys RM, Lavaleye MSS, Davies AJ, Bergman MJN, et al (2012) Spatial and tidal variation in food supply to shallow cold-water coral reefs of the Mingulay Reef complex (Outer Hebrides, Scotland). Marine Ecology Progress Series 444: 97–115.
38. Douarin M, Elliot M, Noble SR, Sinclair D, Henry L-A, et al (2013) Growth of north-east Atlantic cold-water coral structures during the Holocene: a high resolution U series and ^{14}C chronology. Earth & Planetary Science Letters 375:176–187
39. Martins F, Leitao P C, Silva A, Neves R J(2001) 3D modelling in the Sado estuary using a new generic vertical discretization approach. Oceanologica Acta 24: 51–62.
40. Martins FA, Neves RJ, Leitão PC (1998) A three-dimensional hydrodynamic model with generic vertical coordinate. In: Babovic V, Larsen LC (Eds.), Proceedings of Hidroinformatics 98, 2, Balkerna/Rotterdam, Copenhague, Denmark, August 1998. pp.1403–1410
41. Santos AP (1995). Modelo hidrodinámico de circulacáo oceaánica e estuarina (in portuguese). PhD Thesis. IST. Lisbon.
42. Leendertsse J, Liu S (1978) A three-dimensional turbulent energy model for non-homogeneous estuaries and coastal systems. In: Hydrodynamics of Estuaries and Fjords, J. C. J. Nihoul. Elsevier, Amsterdam, pp. 387–405.
43. Danish Hydraulic Institute DHI (1995) MIKE 21 HD (Hydrodynamic module), Release 2.5. User guide and reference manual. Danish hydraulic Institute. 90 p.
44. Maier C (2006) Biology and ecosystem functioning of cold-water coral bioherms at Mingulay (Hebrides), NE Atlantic Cruise Report BIOSYS 2006 Cruise 64PE250 on R/V Pelagia, Oban-Oban, 7–23 July 2006. Royal Netherlands Institute for Sea Research, Texel.
45. Arakawa A, Lamb V (1977) Computational Design of the basic dynamical processes of the UCLA General Circulation Model. Methods of Computational Physics 17: 174–264.
46. Burchard H, Bolding K, Villarreal MR (1999) GOTM, a General Ocean Turbulence Model. Theory, implementation and test cases. European commission, EUR 18745 EN.
47. Lyard F, Lefevre F, Letellier T, Francis O (2006) Modelling the global ocean tides: modern insights from FES2004. Ocean Dynamics 56: 394–415
48. Walstra DJR, van Rijn LC, Blogg H, van Ormondt M (2001) Evaluation of a hydrodynamic area model based on the Coast3D Data at Teignmouth 1999. Proceedings of coastal dynamics 2001 conference, Lund, pp D4.1–D4.4.
49. Dawson CW, Abrahart RJ, See LM (2007) HydroTest: a web-based toolbox of evaluation metrics for the standardised assessment of hydrological forecasts. Environmental Modelling and Software 22: 1034–1054.
50. Wilmott CJ (1981) On the validation of models. Physical Geography 2: 184–194.
51. Miller PI (2009) Composite front maps for improved visibility of dynamic sea-surface features on cloudy SeaWiFS and AVHRR data. Journal of Marine Systems 78(3): 327–336.
52. Miller PI, Christodoulou S (2014) Frequent locations of oceanic fronts as an indicator of pelagic diversity: application to marine protected areas and renewables. Marine Policy 45: 318–329.
53. Miller PI, Xu W, Lonsdale P (2014) Seasonal shelf-sea front mapping using satellite ocean colour to support development of the Scottish MPA network. Scottish Natural Heritage Commissioned Report No. 538. pp.33
54. Pingree RD, Holligan PM, Mardell GT(1978) The effects of vertical stability on phytoplankton distributions in the summer on the northeast European shelf. Deep Sea Research 25: 1011–1028.
55. Fearnhead PG (1975) On the formation of fronts by tidal mixing around the British Isles. Deep Sea Research 22: 311–321.

56. Simpson JH, Edlestein DJ, Edwards A, Morris NCG, Tett PB (1979) The Islay Front: physical structure and phytoplankton distribution. Estuarine, Coastal and Shelf Science 9: 713–726.
57. Mohn C, Rengstorf A, White M, Duineveld G, Mienis F, et al (2014) Linking benthic hydrodynamics and cold-water coral occurrences: A high-resolution model study at three cold-water coral provinces in the NE Atlantic. Progress in Oceanography. In press, http://dx.doi.org/10.1016/j.pocean.2013.12.003
58. Proctor R, Davies AM (1996)A three dimensional hydrodynamic model of tides off the north-west coast of Scotland. Journal of Marine Systems 7:43–66.
59. Xing J, Davies AM (1996) A numerical model of the long term flow along the Malin-Hebrides shelf. Journal of Marine Systems 8: 191–218.
60. Wagner H, Purser AT, Laurenz CC, Lundälv T(2011) Particulate organic matter fluxes and hydrodynamics at the Tisler cold-water coral reef.Journal of Marine Systems 85: 19–29.
61. Mauchline J (1987) Zooplankton in the Oban area and especially in Upper Loch Etive. Internal Report 144. Scottish Marine Biological Association.
62. Mortensen PB (2001) Aquarium observations on the deepwater coral *Lophelia pertusa* (L., 1758) (Scleractinia) and selected associated invertebrates. Ophelia 54: 83–104.
63. Purser A, Larsson AI, Thomsen L, van Oevelen D (2010) The influence of flow velocity and food concentration on Lophelia pertusa (Scleractinia) zooplankton capture rates Journal of Experimental Marine Biology and Ecology 395: 55–62.
64. Dullo WC, Flogel S, Ruggeberg A (2008)Cold-water coral growth in relation to the hydrography of the Celtic and Nordic European continental margin. Marine Ecology Progress Series 371:165–176.
65. Waller RG (2005) Deep-water Scleractina (Cnidaria: Anthozoa). In: Freiwald, A., Roberts, J.M. (Eds.). Cold-Water Corals and Ecosystems. Springer, Berlin, pp. 691–700.
66. Orr JC, Fabry VJ, Aumont O (2005) Anthropogenic ocean acidification over the twenty-first century and its impact on calcifying organisms. Nature 437: 681–686.
67. Findlay HS, Wicks L, Moreno Navas J, Hennige S, Huvenne V, et al (2013) Tidal downwelling and implications of future ocean acidification and warming on cold-water coral reefs. Global Change Biology 19: 2708–2719.

12

Local Ecological Knowledge and Scientific Data Reveal Overexploitation by Multigear Artisanal Fisheries in the Southwestern Atlantic

Mariana G. Bender[1], Gustavo R. Machado[2], Paulo José de Azevedo Silva[3], Sergio R. Floeter[1], Cassiano Monteiro-Netto[2], Osmar J. Luiz[4], Carlos E. L. Ferreira[2]*

1 Departamento de Ecologia e Zoologia, Universidade Federal de Santa Catarina, Florianópolis, SC, Brazil, **2** Departamento de Biologia Marinha, Universidade Federal Fluminense, Niteroi, RJ, Brazil, **3** Fundação Instituto de Pesca, Arraial do Cabo, RJ, Brazil, **4** Department of Biological Sciences, Macquarie University, Sydney, NSW, Australia

Abstract

In the last decades, a number of studies based on historical records revealed the diversity loss in the oceans and human-induced changes to marine ecosystems. These studies have improved our understanding of the human impacts in the oceans. They also drew attention to the shifting baseline syndrome and the importance of assessing appropriate sources of data in order to build the most reliable environmental baseline. Here we amassed information from artisanal fishermen's local ecological knowledge, fisheries landing data and underwater visual census to assess the decline of fish species in Southeastern Brazil. Interviews with 214 fishermen from line, beach seine and spearfishing revealed a sharp decline in abundance of the bluefish *Pomatomus saltatrix*, the groupers *Epinephelus marginatus, Mycteroperca acutirostris*, *M. bonaci* and *M. microlepis*, and large parrotfishes in the past six decades. Fisheries landing data from a 16-year period support the decline of bluefish as pointed by fishermen's local knowledge, while underwater visual census campaigns show reductions in groupers' abundance and a sharp population decline of the Brazilian endemic parrotfish *Scarus trispinosus*. Despite the marked decline of these fisheries, younger and less experienced fishermen recognized fewer species as overexploited and fishing sites as depleted than older and more experienced fishermen, indicating the occurrence of the shifting baseline syndrome. Here we show both the decline of multigear fisheries catches – combining anecdotal and scientific data – as well as changes in environmental perceptions over generations of fishermen. Managing ocean resources requires looking into the past, and into traditional knowledge, bringing historical baselines to the present and improving public awareness.

Editor: Konstantinos I. Stergiou, Aristotle University of Thessaloniki, Greece

Funding: Conselho Nacional de Desenvolvimento Científico e Tecnológico (CNPq - www.cnpq.org) provided individual grants to CELF, SRF and CMN. CNPq, Fundação do Amparo à Pesquisa do Estado do Rio de Janeiro (FAPERJ - www.faperj.br) and EcoHub (www.ecohub.com.br) supported field sampling. The funders had no role in study design, data collection and analysis, decision to publish, or preparation of the manuscript.

Competing Interests: The authors have declared that no competing interests exist.

* Email: carlosferreira@id.uff.br

Introduction

Fishing is the most ancient form of exploitation of coastal resources, preceding all other human disturbances to marine ecosystems, such as pollution, eutrophication, habitat loss, disease outbreaks, human induced climate change and species invasions [1–4]. Fishing has caused the worldwide depletion of large predatory fishes, including shark populations [5,6], and driven several species across different ecosystems to ecological extinction [1]. Moreover, fishing is reported to impact fish populations since prehistoric times [2,7–10], affecting species diversity and size [11,12]. Today, fishing remains as the major source of impact upon marine and coastal environments, contributing to global biodiversity loss [13–15]. The continuity of unsustainable historical fishing has left few truly pristine ecosystems in the marine realm [1,5]. In the Southwestern Atlantic, patterns of marine resources' overexploitation are similar to what have been reported worldwide [5,14,16], with widespread population declines and collapsed stocks [17–19]. It has been estimated that 23% of all Brazilian marine fish stocks are fully exploited and 33% are overexploited, including species from low trophic levels [20–22]. Overfishing has been changing the density and the size structure of reef fish top predators occurring in Brazil [17,23], consequently elevating their threatened status [24,25].

Environmental changes and lack of baselines for pristine marine ecosystems have profound implications in our perceptions of what is a natural environment [4,26]. The compromised ability of people in perceiving environmental modifications and past ecological conditions is called the shifting baseline syndrome (SBS) [27]. This phenomenon was first noted among fisheries scientists, who perceived as a 'natural' baseline for stock size and composition the condition they observed at the start of their careers, while failing to incorporate past and historical data [27–29]. As generations change, environmental baselines become increasingly shifted, misinforming fisheries management. In order to adjust shifted environmental perceptions, old travel diaries, naturalists' observations, historical data and fishermen' anecdotes are important sources of information [28–32], as well archaeological and paleontological data [1,4]. Baselines constructed from historical data are critical to better gauge and interpret long-term changes and to set appropriate targets for management and restoration [1,4,33].

The Local Ecological Knowledge (LEK) held by traditional fishing communities is considered an important tool for the assessment and management of tropical fisheries [29,30,37]. Local Ecological Knowledge is a set of perceptions and experiences of traditional communities regarding its surrounding natural environment [34,35], this knowledge being handed-down through generations by cultural transmission [36]. This knowledge includes perceptions of fishing resources – fisheries' composition and abundance patterns, as well as fish species biological and ecological aspects [38,39] – all of which are important to conservation and management strategies [40]. Because humans have modified natural systems, local perceptions of the status of species and ecosystem resources are unlikely to remain constant over time [41]. Thus, the SBS could influence the validity of LEK derived data, whose use for species and ecosystems assessments is becoming more common [42,43]. However, specific conditions are required for SBS to occur, which include a combination of environmental events and observer perceptions [44]. Here we combine multiple sources of information to understand the status of local fisheries of Arraial do Cabo, Rio de Janeiro state, Southeastern Brazil (Fig. 1). Specifically, we aim to (i) investigate the occurrence of SBS among fishermen communities of Arraial do Cabo; (ii) utilize LEK to obtain past estimates of abundance and population trends for exploited fish species; and (iii) demonstrate the importance of anecdotal data to adjust perceptions regarding marine resources in the subtropical coast of Brazil.

Methods

Study site

Arraial do Cabo (Rio de Janeiro state, Brazil) constitutes a traditional fishing village with approximately 20,000 residents and 1,340 active fishermen. The area was declared a Marine Extractive Reserve (MER) in 1997 – the first of its kind in Brazil – a collaboratively managed marine protected area where only local fishermen are allowed to exploit resources. In Arraial do Cabo, while there are some fishing regulations, enforcement is poor or non-existent, resulting in a typical overexploitation scenario [17]. Among fishing gears, hook and line, gillnet, beach seine and spearfishing are the most used. The beach seine fishery is the most traditional practice, using large canoes and seine nets that encircle passing schools of bluefish (*Pomatomus saltatrix*), mullets (Mugilidae), jacks and trevallies (Carangidae), and tunas (Scombridae). Spearfishing commenced *c.* 50 years ago in Brazil, being today widespread along the coast. The region is formed by an isthmus and two islands dominated by rocky shores and sand beaches. Coastal morphology associated to prevailing winds trigger small-scale upwelling events, favoring a rich marine environment (Fig. 1).

Data collection

Ethical considerations. The Marine Biology Graduate Program at Universidade Federal Fluminense (PPGBM-UFF), Niterói, Rio de Janeiro state, Brazil, approved the data collection of our study that comprised interviews with local fishermen communities. Prior to each interview, fishermen were informed on research purpose and we obtained verbal consent from participants. This methodology was chosen because some interviewees would lack reading and writing skills. Once the participant agreed on being interviewed, we recorded participant consent by writing the participant's name and interview date on questionnaire sheets. This procedure was considered adequate by the Graduate Program review board. We also informed participants that all information provided in interviews would be anonymized.

Interviews. To assess fishermen perceptions regarding the status of overexploited fish species, we conducted individual interviews with 214 fishermen – from August 2007 to August 2008 – in Arraial do Cabo, Southeastern Brazilian coast (Fig.1). Interviews with fishermen included three different gears: hook and line, beach seine and spearfishing. As line and beach seine are traditional fishing gears in the region and have been practiced for centuries, this enabled a large number of experienced fishermen to be surveyed. Making possible the unveiling of a rich ecological

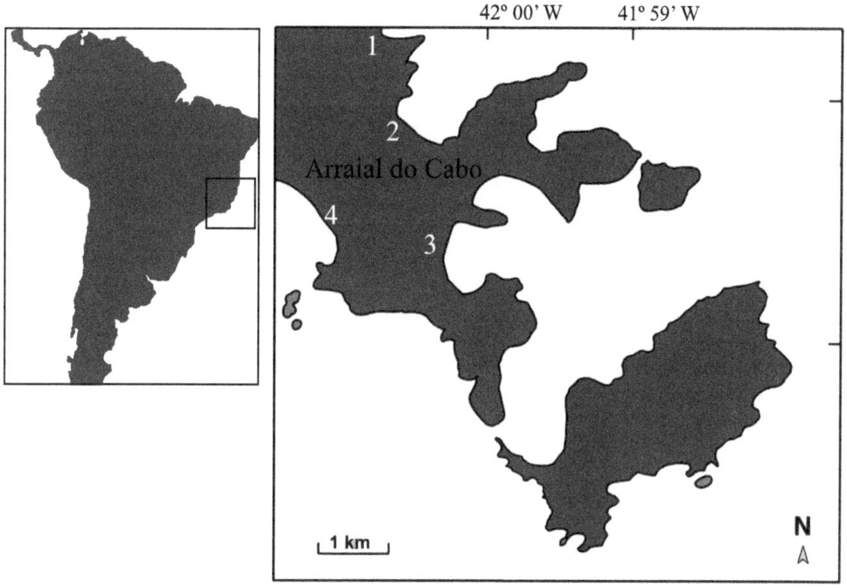

Figure 1. Locations of interview sites at Arraial do Cabo.

knowledge related to these gears and a robust historical baseline for local fisheries. Respondents were categorized into three groups according to fishing experience. For line fishing and beach seine fishing the categories were: beginners (<16 years of practice, n = 45), intermediate (16–30 years of practice, n = 50) and experienced (>30 years of practice, n = 43). However, since spearfishing is a recent fishing practice when compared to beach seine and line fishing, experience categories for spearfishermen were adapted to a smaller period: beginners (<8 years of practice, n = 25), intermediate (8–15 years of practice, n = 28) and experienced (>15 years of practice, n = 23).

The survey started a few weeks following arrival to the fishing village. Fishermen were randomly approached at the beach – as they were encountered either going fishing or making repairs on nets and fishing gears – and inquired on the availability to participate on the research. Following this approach, fishermen in different groups (experience category per fishing gear) were selected through snowball sampling [37,45], where one or more peers indicate the respondent. This method was applied so that we could achieve the target sample size in each experience category and fishing gear (line, beach seine and spearfishing). This method also led us to interview some elder and retired informants at their homes. Respondents were always interviewed separately in order to avoid interfering from others. Our questionnaire constitued of structured and open-ended questions [29] (see File S1 for questionnaire).

The bluefish, *Pomatomus saltatrix* was chosen among various species based on its importance as a resource in the region [45,47]. We asked fishermen about the best bluefish catch they ever remembered landing and the year in which this catch was made [47]. For the spearfishing practice interviews, we selected four Epinephelidae (groupers) species [48] that are important targets in the region (*Mycteroperca acutirostris*, *M. bonaci*, *M. microlepis* and *Epinephelus marginatus*) [17]. To explore the possible fishing impacts at lower trophic levels, we included in questionnaires four endemic large herbivorous-detritivorous species of the subfamily Scarinae (parrotfishes), namely, *Scarus trispinosus, Sparisoma axillare, Sparisoma frondosum* and *Sparisoma amplum*. These fishes became targets for spearfishing in the last 20 years [17,20]. For each species we asked divers about the best catch they remembered spearing (number of fishes/individuals), the largest individual they ever caught (in kilograms), and the year in which these catches were made [29]. When fishermen were asked about the largest fishes they had ever caught, the majority of respondents provided information relative to species weight (in kilograms). However, when fishermen referred to fish body length, species length–weight relationships were used to convert to weight [49]. Finally, we assessed shifts in environmental perceptions among generations of fishermen. We inquired each respondent to cite the number of overexploited species and sites in the region, as well as the number of species that were formerly discarded but are now fishing targets. Given the richness of common names for Brazilian reef fish [46], we used photographs to clarify species identifications during the interviews [37].

Bluefish landing data. We have included fisheries landing data in our analysis in order to improve the overall understanding of the status of fish stocks in the region. Bluefish (*Pomatomus saltatrix*) catch data (total catch – kg and effort – fishing hours) was recorded daily and pooled over the year during 16 years (1992 to 2008) at the local landing point (Marina dos Pescadores, Porto do Forno) in Arraial do Cabo.

Reef fish abundance. The abundance of groupers (Epinephelidae) and parrotfishes (Labridae, Scarinae) was assessed through an underwater visual census (UVC) program performed in Arraial do Cabo region across different years (1992, 2012). The program is led by one of the authors (C.E.L.F.), and includes UVC and other sampling techniques applied to studies/monitoring of the reef community at Arraial do Cabo. Fish density was estimated using 20×2 m strip transects [50]. In each transect, all fishes within the sample area were identified, counted and categorized into five size classes. Data across all sampled sites were pooled together to show temporal changes for target species in Arraial do Cabo region.

Data analysis

Interviews. From the data obtained through interviews with local fishermen we measured the decline in fish species catches, as well as the differences in the set of environmental perceptions across fishermen generations. With the purpose of measuring the decline in fish species abundance and body-size patterns across the years, we applied regressions and plotted (i) the best day catch reported by each respondent, versus the year in which the catch was made; and (ii) the larger individual ever caught by the respondent (kg) versus the year in which a given species was caught. We tested distinct regression types (exponential, polynomial and linear regressions) and r^2 values were compared to assess the quality of the regression fit to data distribution. Fish catches measured for time intervals are presented in the results section as mean ± standard deviation.

To investigate the potential differences in respondents perceptions, we tested differences in the mean number of overexploited species, overexploited sites, and the mean number of target species (formerly discarded) mentioned by fishermen in different categories (beginner, intermediate, experienced) through one-way ANOVA ($\alpha = 0.05$; Student-Newman-Keuls post-hoc test) [51]. All data was tested for normality (Kolmogorov-Smirnov test) and transformed when necessary (square root or log transformed) before conducting analysis.

Bluefish landing data. In order to measure *P. saltatrix* catch trends across the years (1992 to 2008) from landing data we have applied regression analysis, plotting the catch per unit effort (tons/hour) throughout the years. Again, we tested for distinct regression types (exponential, polynomial and linear regressions) and r^2 values were compared to assess the quality of the regression fit to data distribution.

Reef fish abundance. In order to examine the population decline trends within targeted reef fish families (groupers and parrotfishes) in Arraial do Cabo we tested the Spearman's rank correlation between the mean density (of groupers and parrotfishes) per transect and year of sampling. Only the most abundant and large sized parrotfish species (*Scarus trispinosus, Sparisoma axillare, S. frondosum, S.amplum*) were used for analysis [52].

Results

Multiple sources reveal fish species declines

The pattern for bluefish catches across the years reveals significant reductions in *P. saltatrix* stocks in Arraial do Cabo region over time (Fig. 2a). Captures (mean ±standard deviation) of 1,200 kg (±186) were considered a good catch during the 1960s among line fishermen. This number decreased to an average of 970 kg (±260) in the 1980s. Today, 370 kg (±220) of bluefish is considered a great catch (Fig. 2a). Results reveal that in fifty years (1960–2010) the amount of bluefish caught on a good day catch was reduced by approximately 70%.

Experienced beach seine fishermen also reported great bluefish catches in Arraial do Cabo decades ago, and approximately 80% decline in the past six decades, a pattern similar to that identified

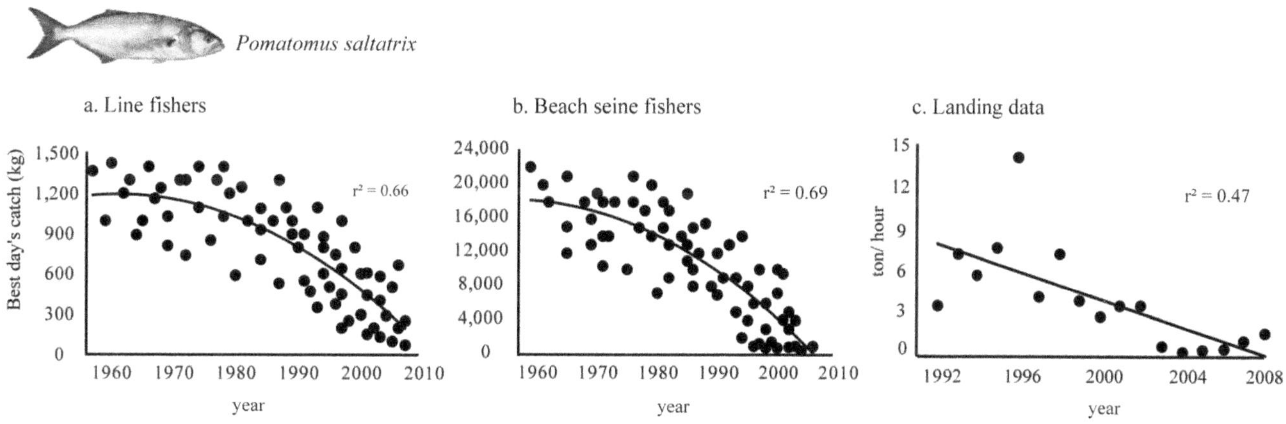

Figure 2. Bluefish *(Pomatomus saltatrix)* catches according to line fishermen (a) and beach seine fishermen (b), both are second order polynomial regressions; and catch per unit effort ($t \times h^{-1}$) (c) from 1992 to 2008 in Arraial do Cabo, Brazil.

by line fishermen (Fig. 2b). During the 1960s, large catches were in the order of 18,000 kg (±3,517) on average. In the 1980s, this number has fallen to13,570 kg (±3,753), and today *P. saltatrix* catches greater than 3,700 kg (±3,340) are rare. Moreover, landing data from *P. saltatrix* catches in the region reveal the same pattern found by accessing the knowledge of local fishermen (Fig. 2c), with a declining trend in bluefish fisheries along 16 years (1992–2008). CPUE varied between 14.2 tons/hour in 1996 and 0.3 tons/hour in 2004, with an annual average of 4 tons hour (Fig. 2c).

Fish species exploited by spearfishing also exhibit declining trends. There was a decrease in the abundance of the assessed Epinephelidae species (Fig. 3). The relationships considering the largest catch and year again indicate that large individuals of *E. marginatus*, *M. bonaci*, *M. acutirostris* and *M. microlepis* have declined in the region, and that the big ones were frequently caught not so far back, in the 1970s (Fig. 3). For the endemic parrotfishes *Sparisoma axillare, S. frondosum* and *S. amplum* there were no significant differences neither in the abundance nor in the size of individuals caught along the years (Fig. S1 in File S1). However, for *Scarus trispinosus*, we have identified reductions on the greatest individuals captured (kg) and on the best catches made (Fig. 3).

Data from visual census campaigns conducted in the region from 1992 to 2012 – revealed a declining trend in the local densities of *Scarus trispinosus*, in fact, individuals of the species are not recorded on transects conducted in the region since 1995 (Fig. 4). The decline in *S. trispinosus* is followed by an overall decline of Scarinae (Labridae) species. For Epinephelidae species, the situation is similar, with decreasing densities in local reef fish communities from 1997 onwards (Fig. 4).

Shifting baselines across multigear fisheries

Out of 214 interviewed fishermen, ~83% (n = 177) recognized reductions in the abundance of exploited fish species in the region of Arraial do Cabo. Less experienced fishermen are less aware of overexploitation, indicating the occurrence of the SBS among local fishermen. Significant age-related differences were evident in assessed fishing communities in terms of respondent perception regarding fisheries declines. Experienced line, beach seine and spearfishermen mentioned a greater number of overexploited species when compared to less experienced ones (Table 1).

We detected significant experience-related differences in terms of respondent perception of shifts in target species (Table 1). Not only for line and beach seine fishermen (line fishing: experienced = 6.1, intermediate = 3.1 and beginners = 1.6; beach seine: experienced = 5.5, intermediate = 3.4 and beginners = 1.4), but also for spearfishing there were differences between the mean number of species cited by each category (spearfishing: experienced = 7.4, intermediate = 5.1 and beginners = 1.8) (see Table 1). For instance, surgeonfishes (e.g. *Acanthurus* spp.) were identified as targets exclusively by beginner spearfishermen. The mean number of sites mentioned as depleted by different fishermen categories also suggests the occurrence of the SBS. The more experienced and older fishermen referred to a greater number of overexploited sites when compared to younger, less experienced fishermen. For instance, experienced line fishermen cited approximately 11 sites, while intermediate and beginners referred to 4 and 2 sites, respectively (see Table 1).

Discussion

From local to global: the impacts of changes in fishermen's perceptions

Since Daniel Pauly reported the global phenomenon that he termed 'the shifting baseline syndrome' (SBS) [27], much supporting evidence have been reported across marine [29,53], freshwater [41,54] and terrestrial ecosystems [43,55]. Fishermen gather their environmental references based on personal experiences, which may lead to perception differences among generations. In Arraial do Cabo, experienced line and beach seine fishermen often mentioned the old days when top predators such as the sharks *Carcharhinus milberti*, *Squatina argentina*, *Isurus oxyrinchus* and the shovelnose ray *Rhinobatos percellens* were abundant in their catches, being witnesses to the decline of those species. Even though most of the older beach seine fishermen (73%) recognized the shark *Carcharhinus maculipinnis* as overexploited, beginners believe that the species is naturally rare in the region. Also, contrary to older fishermen, young spearfishermen did not mention the snappers *Lutjanus analis, Lutjanus cyanopterus*, and the groupers *Mycteroperca microlepis* and *Epinephelus morio* as depleted. The decline of these fish species forced fishermen to change their target species. Those species that were not important to fisheries in the past – or were only utilized as bait – turned out to be today's main targets. This replacement process is occurring slowly and gradually over the years, making it more difficult for future generations of fishermen to perceive these changes. Experienced and older fishermen have seen the effects of

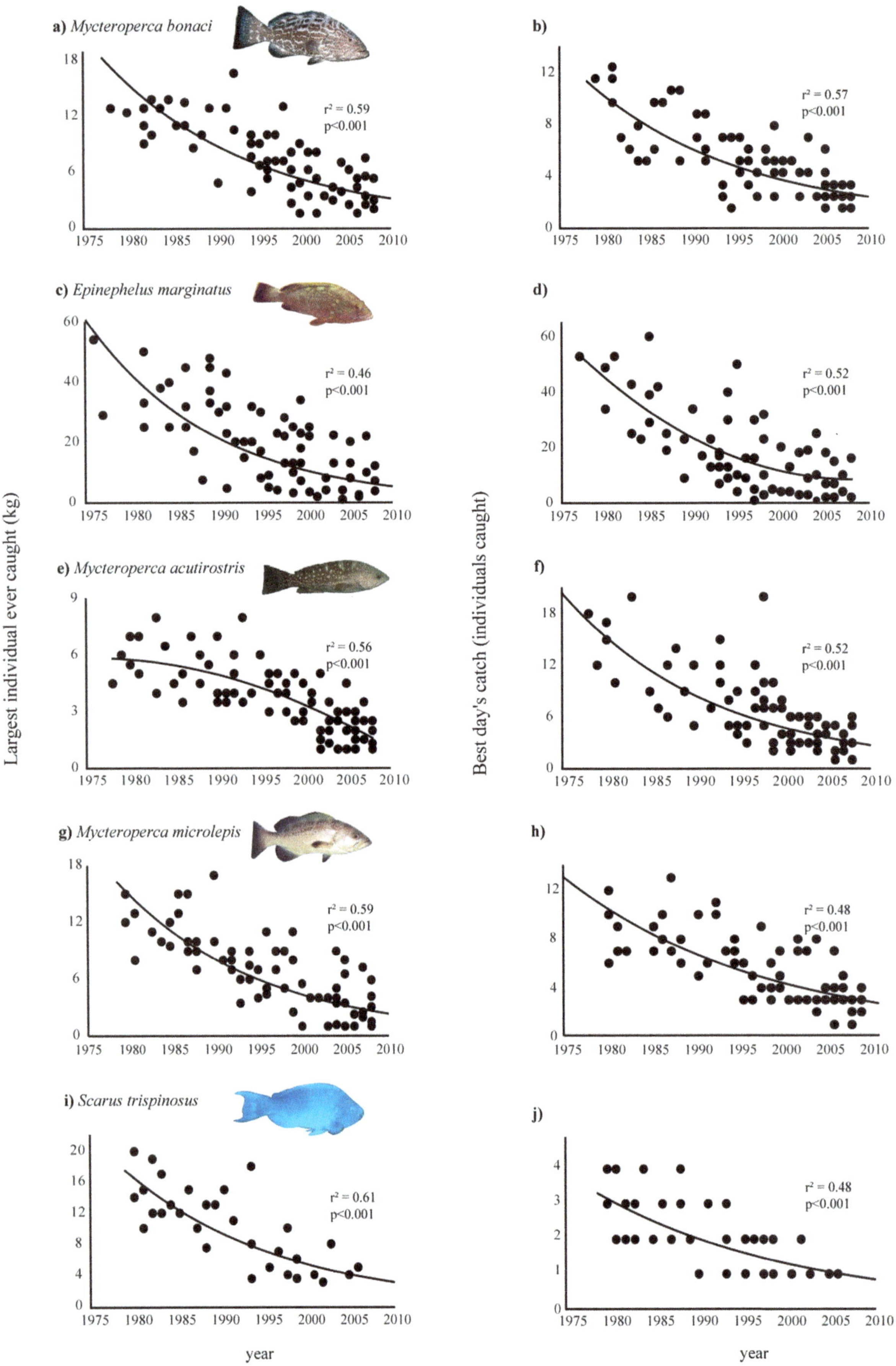
a) Mycteroperca bonaci
r² = 0.59
p<0.001
b)
r² = 0.57
p<0.001
c) Epinephelus marginatus
r² = 0.46
p<0.001
d)
r² = 0.52
p<0.001
e) Mycteroperca acutirostris
r² = 0.56
p<0.001
f)
r² = 0.52
p<0.001
g) Mycteroperca microlepis
r² = 0.59
p<0.001
h)
r² = 0.48
p<0.001
i) Scarus trispinosus
r² = 0.61
p<0.001
j)
r² = 0.48
p<0.001
Largest individual ever caught (kg)
Best day's catch (individuals caught)
1975 1980 1985 1990 1995 2000 2005 2010
year
year

Figure 3. Largest individual (in kilograms) and greatest catch (number of individuals caught) that spearfishermen remembered landing. a) and b) *Mycteroperca bonaci*, second order polynomial regression; c) and d) *Epinephelus marginatus*, exponential regression; e) and f) *Mycteroperca acutirostris*, second order polynomial regression; g) and h) *Mycteroperca microlepis*, second order polynomial regression; i) and j) *Scarus trispinosus*, exponential regression.

overexploitation upon marine ecosystems for many more years, whereas young fishermen begun fishing when stocks were either reduced or on the path for depletion. This led young fisherman to assume that sharks and large groupers were naturally rare in Arraial do Cabo, and that overexploited fishing grounds were still very productive. Thus perceptions are subjected to the effects of time and possibly compromised knowledge transmission between community members [30].

Baseline shifts regarding the past conditions of the world's oceans might lead to a collective amnesia, making people more tolerant to population declines and biodiversity losses [56]. Despite the fact that fishermen are usually in contact with nature on a daily-basis, exhibiting greater awareness of environmental changes, baselines shift rapidly even among them [29,41]. In Arraial do Cabo, sixty years of fisheries exploitation and impacts upon marine ecosystems have affected the environmental baselines of three consecutive generations. Such changes in the perceptions of fishermen regarding the status of marine ecosystems are of special concern given the importance of LEK to adjust and improve our understanding of the oceans' past conditions.

A possible short-term solution to slow the reference loss process is to recreate old conditions of marine life through historical data [4] such as films, photos [57], booklets and reports [31,50] of early explorers. Such kind of data can provide important detail on former species size and catch composition [30], potentially improving our knowledge on the effects of overfishing. Recreating past environmental baselines is a unique opportunity to assess the historical changes that ecosystems have undergone, as well as for improving social awareness on the status of biodiversity. Recent studies have pointed that underestimating empirical data is a mistake [27,29,45,56,58,59], especially in places like Arraial do Cabo, where written records on species' former size and abundance are rare.

When assessing fishing impacts, few studies make such parallel use of LEK and species population status, mostly because LEK is considered anecdotal, non-scientifically sampled, and of limited application among scientists and managers [40,60]. Our work is among the first to combine landing, underwater visual census and LEK data to assess the temporal patterns of exploited fish populations [58,61]. Combining data from different sources renders more robust and reliable assessments, which are especially important in times where decision makers require scientific certainty. More importantly, local management and conservation efforts will certainly gain more commitment from fishermen when their own experience is taken into account on the establishment of management strategies.

Fishing collapse in the Southwestern Atlantic

The combined methodologies utilized to assess multigear fisheries reveal the decline of different fish species in the region of Arraial do Cabo. The decline of the bluefish (*P. saltatrix*) fishery revealed through information provided by fishermen was supported by evidence from landing data. The declining trend for Epinephelidae species (*E. marginatus, Mycteroperca bonaci, M. acutirostris* and *M. microlepis*) in the subtropical reefs of Arraial do Cabo has been demonstrated through information provided by fishermen, and through data from underwater visual census campaigns conducted in the region. Another study for the region underscores recent declines in the abundance of groupers using the underwater visual census technique [17]. Not only the abundance of fish is diminishing, but also the larger individuals are consistently more difficult to capture and sight in the field [17].

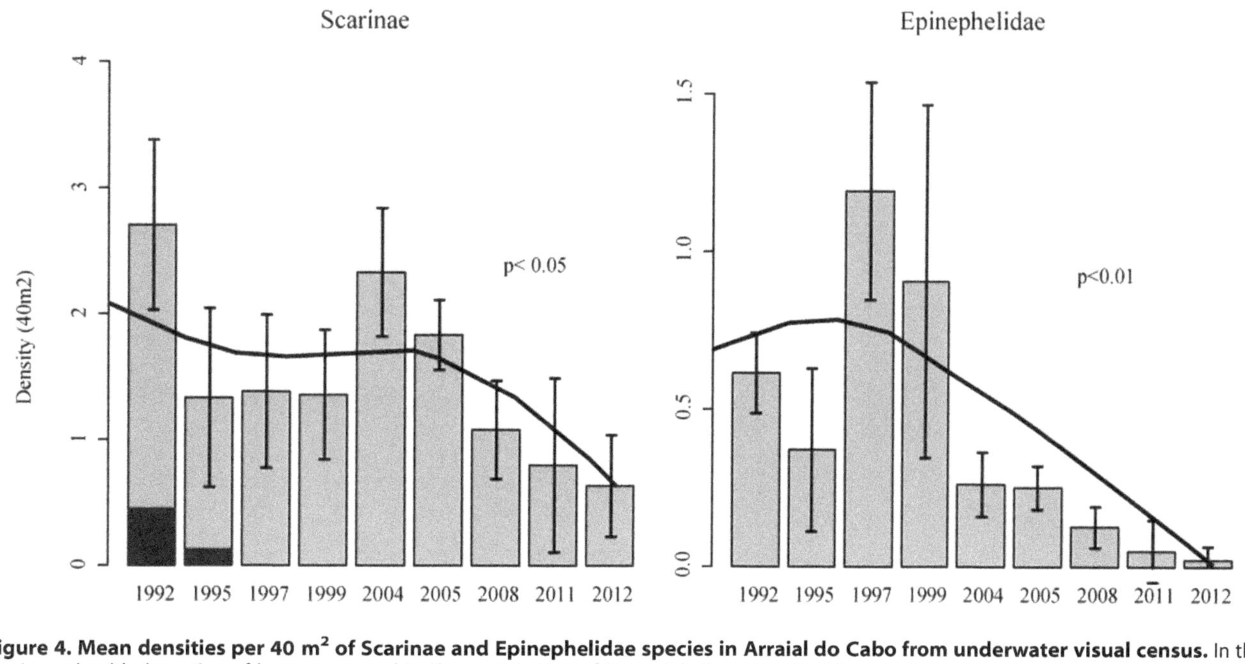

Figure 4. Mean densities per 40 m^2 of Scarinae and Epinephelidae species in Arraial do Cabo from underwater visual census. In the Scarinae plot, black portion of bars correspond to the contribution of *Scarus trispinosus* to densities.

Table 1. Mean number of overexploited fish species, new target species and depleted sites mentioned by fishermen from different fishing gears and according to experience categories.

Fishing modality	Fishers experience category (n)	Overexploited species (mean ±SD)	ANOVA*	New target species (mean ±SD)	ANOVA*	Depleted sites (mean ±SD)	ANOVA*
Line	Beginners (22)	2.0±1.0	*f*= 139.3	1.6±1.1	*f*= 35.7	2.0±1.0	*f*= 135
	Intermediate (25)	5.5±1.8	*p*<0.001	3.1±1.5	*p*<0.001	4.5±1.9	*p*<0.001
	Experienced (21)	10.5±1.7		6.0±1.0		11.3±2.3	
Beach seine	Beginners (23)	2.3±1.5	*f*= 76.1	1.4±1.5	*f*= 60.8	2.4±2.1	*f*= 63.6
	Intermediate (26)	6.0±2.2	*p*<0.001	3.4±1.1	*p*<0.001	5.5±2.0	*p*<0.001
	Experienced (21)	9.0±1.5		5.5 ± 1.9		10.5±3.1	
Spearfishing	Beginners (25)	1.4±0.9	*f*= 187.7	1.8±0.7	*f*= 107.2	1.8±1.4	*f*= 133.5
	Intermediate (28)	5.8±1.5	*p*<0.001	5.1±1.7	*p*<0.001	6.3±1.7	*p*<0.001
	Experienced (23)	8.9±1.7		7.4±1.9		11.4±2.9	

**f* and *p*-values of each ANOVA (one-way) test.
All values associated to different age categories were significantly different according to ANOVA (one-way) and Student-Newman-Keuls comparison test ($p<0.05$).

Overfishing is affecting the size of fish individuals, removing the largest ones at an unsustainable rate, and preventing the smallest ones from reaching large sizes. This can seriously compromise fish population rebound since larger and older females produce greater quantity and better quality offspring when compared to younger fish [62]. Most of these grouper species are important fishing targets elsewhere in the Brazilian coast, and the overfished state of populations has been demonstrated by different studies [17,20]. Moreover, they are highly vulnerable to overexploitation given its critical combination of biological attributes [25].

As populations of top predators collapse, other large-bodied species at lower trophic level become new targets, as is the case of the endemic greenbeak parrotfish, *Scarus trispinosus*. Being the largest of all Brazilian parrotfishes, the greenbeak was the first herbivorous-detritivorous reef fish targeted by spearfishermen, but other large-bodied parrotfishes of the genus *Sparisoma* also became targeted in the last few years. Different species of parrotfishes perform specific functional roles, some reported as critical in maintaining the resistance and resilience of reef ecosystems [63,64]. Along the Brazilian coast, populations of *S. trispinosus* have strongly declined due to spearfishing [17], and the species is considered ecologically extinct in Arraial do Cabo [50]. The species was recently listed by the IUCN [65] as Endangered (EN) yet there is no regulation or national listing to guarantee its protection [25].

Here we have shown the decline of reef fisheries in the Southwestern Atlantic from multiple sources of information: local ecological knowledge from multigear fisheries (line, beach seine and spearfishing), fisheries landing data and even underwater visual census data. The pattern identified for six different target species follows a similar pattern of global deterioration of marine ecosystems reported elsewhere [1,66]. In the Southwestern Atlantic, as elsewhere in the world, fishing has moved from upper to lower trophic levels: groupers used to be fisheries' major target, but today, parrotfishes and even surgeonfishes are targeted. If going back to the early days of the 1960's tell us a very different story on fish abundance and size in the region of Arraial do Cabo, one could only imagine what a 150 or 200-year-old baseline would inform us on those fish species. We emphasize the importance of combining LEK with scientific data in order to improve our understanding of the status of marine habitats and its associated biodiversity, but the occurrence of changes in environmental baselines must be taken into account. Only through a clear picture of the status of natural resources in the past can we help people improve their awareness of the importance of preserving natural resources and ecosystem services.

Supporting Information

File S1 File contains Questionnaire and Figure S1. Questionnaire applied in interviews with fishermen in Arraial do Cabo. Figure S1. Catch trends for *Sparisoma* (Scarinae) species reported by spearfishermen in Arraial do Cabo. Linear regressions of the largest individual caught (kg) and best day's catch (number of individuals) of the parrotfishes *Sparisoma axillare*, *S. amplum* and *S. frondosum* that spearfishers remembered landing (from top to bottom).

Author Contributions

Conceived and designed the experiments: CELF CMN GRM PJAS. Performed the experiments: CELF CMN GRM PJAS. Analyzed the data: CELF CMN GRM PJAS MGB OJL SRF. Contributed reagents/materials/analysis tools: CELF CMN GRM PJAS MGB OJL SRF. Wrote the paper: CELF CMN GRM PJAS MGB OJL SRF.

References

1. Jackson JBC, Kirby MX, Berger WH, Bjorndal KA, Botsford LW, et al. (2001) Historical Overfishing and the Recent Collapse of Coastal Ecosystems. Science 293: 629–637.
2. Erlandson JM, Rick TC, Vellanoweth RL (2004) Human impacts on ancient environments: a case study from California's Northern Channel Islands. Voyages of discovery: The archaeology of islands: 51–83.
3. Jackson JB (1997) Reefs since Columbus. Coral reefs 16: S23–S32.
4. Lotze HK, Worm B (2009) Historical baselines for large marine animals. Trends in ecology & evolution 24: 254–262.
5. Myers RA, Worm B (2003) Rapid worldwide depletion of predatory fish communities. Nature 423: 280–283.
6. Worm B, Davis B, Kettemer L, Ward-Paige CA, Chapman D, et al. (2013) Global catches, exploitation rates, and rebuilding options for sharks. Marine Policy 40: 194–204.
7. Barrett JH, Locker AM, Roberts CM (2004) The origins of intensive marine fishing in medieval Europe: the English evidence. Proceedings of the Royal Society of London Series B: Biological Sciences 271: 2417–2421.
8. Barrett JH, Nicholson RA, Cerón-Carrasco R (1999) Archaeo-ichthyological evidence for long-term socioeconomic trends in northern Scotland: 3500 BC to AD 1500. Journal of Archaeological Science 26: 353–388.
9. Erlandson JM, Rick TC (2010) Archaeology Meets Marine Ecology: The Antiquity of Maritime Cultures and Human Impacts on Marine Fisheries and Ecosystems. Annual Review of Marine Science 2: 231–251.
10. Smith I (2005) Retreat and resilience: fur seals and human settlement in New Zealand. The exploitation and cultural importance of sea mammals: 6–18.
11. Morales A, Rosello E, Canas J (1994) Cueva de Nerja (prov. Malaga): a close look at a twelve thousand year ichthyofaunal sequence from southern Spain [Paleolithic, Neolithic, Chalcolithic].
12. Wing S, Wing E (2001) Prehistoric fisheries in the Caribbean. Coral reefs 20: 1–8.
13. Dulvy N, Polunin NV, Mill A, Graham NA (2004) Size structural change in lightly exploited coral reef fish communities: evidence for weak indirect effects. Canadian Journal of Fisheries and Aquatic Sciences 61: 466–475.
14. Jackson JB (2008) Ecological extinction and evolution in the brave new ocean. Proceedings of the National Academy of Sciences 105: 11458–11465.
15. Roberts C, Hawkins JP, Campaign WES (2000) Fully-protected marine reserves: a guide: WWF Endangered seas campaign Washington, DC.
16. Halpern BS, Walbridge S, Selkoe KA, Kappel CV, Micheli F, et al. (2008) A global map of human impact on marine ecosystems. Science 319: 948–952.
17. Floeter S, Halpern BS, Ferreira C (2006) Effects of fishing and protection on Brazilian reef fishes. Biological Conservation 128: 391–402.
18. Frédou T (2004) The fishing activity on coral reefs and adjacent ecosystems: a case study of the Northeast of Brazil. Cybium 28: 274.
19. Rezende SdM, Ferreira BP (2004) Age, growth and mortality of dog snapper Lutjanus jocu (Bloch & Schneider, 1801) in the northeast coast of Brazil. Brazilian Journal of Oceanography 52: 107–121.
20. Ferreira C, Gonçalves J (1999) The unique Abrolhos reef formation (Brazil): need for specific management strategies. Coral reefs 18: 352–352.
21. Freire KM, Pauly D (2010) Fishing down Brazilian marine food webs, with emphasis on the east Brazil large marine ecosystem. Fisheries Research 105: 57–62.
22. Pauly D, Christensen V, Dalsgaard J, Froese R, Torres F (1998) Fishing down marine food webs. Science 279: 860–863.
23. Luiz OJ, Edwards AJ (2011) Extinction of a shark population in the Archipelago of Saint Paul's Rocks (equatorial Atlantic) inferred from the historical record. Biological Conservation 144: 2873–2881.
24. Bender M, Floeter S, Hanazaki N (2013) Do traditional fishers recognise reef fish species declines? Shifting environmental baselines in Eastern Brazil. Fisheries Management and Ecology 20: 58–67.
25. Bender M, Floeter S, Mayer F, Vila-Nova D, Longo G, et al. (2013) Biological attributes and major threats as predictors of the vulnerability of species: a case study with Brazilian reef fishes. Oryx 47: 259–265.
26. Knowlton N, Jackson JB (2008) Shifting baselines, local impacts, and global change on coral reefs. PLoS biology 6: e54.
27. Pauly D (1995) Anecdotes and the shifting baseline syndrome of fisheries. Trends in ecology & evolution 10: 430.
28. Pinnegar JK, Engelhard GH (2008) The 'shifting baseline'phenomenon: a global perspective. Reviews in Fish Biology and Fisheries 18: 1–16.
29. Sáenz-Arroyo A, Roberts C, Torre J, Cariño-Olvera M, Enríquez-Andrade R (2005) Rapidly shifting environmental baselines among fishers of the Gulf of California. Proceedings of the Royal Society B: Biological Sciences 272: 1957–1962.
30. Sáenz-Arroyo A, Roberts CM, Torre J, Cariño-Olvera M (2005) Using fishers' anecdotes, naturalists' observations and grey literature to reassess marine species at risk: the case of the Gulf grouper in the Gulf of California, Mexico. Fish and Fisheries 6: 121–133.
31. Fortibuoni T, Libralato S, Raicevich S, Giovanardi O, Solidoro C (2010) Coding early naturalists' accounts into long-term fish community changes in the Adriatic Sea (1800–2000). PloS one 5: e15502.
32. Sáenz-Arroyo A, Roberts CM, Torre J, Cariño-Olvera M, Hawkins JP (2006) The value of evidence about past abundance: marine fauna of the Gulf of California through the eyes of 16th to 19th century travellers. Fish and Fisheries 7: 128–146.
33. Roberts CM (2003) Our shifting perspectives on the oceans. Oryx 37: 166–177.
34. Drew JA (2005) Use of traditional ecological knowledge in marine conservation. Conservation Biology 19: 1286–1293.
35. Huntington HP (2000) Using traditional ecological knowledge in science: methods and applications. Ecological applications 10: 1270–1274.
36. Berkes F, Colding J, Folke C (2000) Rediscovery of traditional ecological knowledge as adaptive management. Ecological applications 10: 1251–1262.
37. Silvano RA, MacCord PF, Lima RV, Begossi A (2006) When does this fish spawn? Fishermen's local knowledge of migration and reproduction of Brazilian coastal fishes. Environmental Biology of Fishes 76: 371–386.
38. Johannes RE (1981) Working with fishermen to improve coastal tropical fisheries and resource management. Bulletin of Marine Science 31: 673–680.
39. Johannes RE, Freeman MM, Hamilton RJ (2000) Ignore fishers' knowledge and miss the boat. Fish and Fisheries 1: 257–271.
40. Shackeroff JM, Campbell LM (2007) Traditional ecological knowledge in conservation research: problems and prospects for their constructive engagement. Conservation and Society 5: 343.
41. Turvey ST, Barrett LA, Yujiang H, Lei Z, Xinqiao Z, et al. (2010) Rapidly shifting baselines in Yangtze fishing communities and local memory of extinct species. Conservation Biology 24: 778–787.
42. Jones JP, Andriamarovololona MM, Hockley N, Gibbons JM, Milner-Gulland E (2008) Testing the use of interviews as a tool for monitoring trends in the harvesting of wild species. Journal of Applied Ecology 45: 1205–1212.
43. van der Hoeven CA, de Boer WF, Prins HH (2004) Pooling local expert opinions for estimating mammal densities in tropical rainforests. Journal for nature conservation 12: 193–204.
44. Papworth SK, Rist J, Coad L, Milner-Gulland EJ (2009) Evidence for shifting baseline syndrome in conservation. Conservation Letters.
45. Silvano RAM, Begossi A (2010) What can be learned from fishers? An integrated survey of fishers' local ecological knowledge and bluefish (*Pomatomus saltatrix*) biology on the Brazilian coast. Hydrobiologia 637: 3–18.
46. Freire KMF, Carvalho-Filho A (2009) Richness of common names of Brazilian reef fishes. Pan-American Journal of Aquatic Sciences 4: 96–145.
47. Brito R (1999) Modernidade e tradição. Construção da identidade social dos pescadores de Arraial do Cabo, RJ Niterói: EDUFF.
48. Craig MT, Hastings PA (2007) A molecular phylogeny of the groupers of the subfamily Epinephelinae (Serranidae) with a revised classification of the Epinephelini. Ichthyological Research 54: 1–17.
49. Bohnsack JA, Harper DE, Center SF (1988) Length-weight relationships of selected marine reef fishes from the southeastern United States and the Caribbean: National Oceanic and Atmospheric Administration, National Marine Fisheries Service, Southeast Fisheries Center.
50. Floeter SR, Krohling W, Gasparini JL, Ferreira CE, Zalmon IR (2007) Reef fish community structure on coastal islands of the southeastern Brazil: the influence of exposure and benthic cover. Environmental Biology of Fishes 78: 147–160.
51. Underwood AJ (1997) Experiments in ecology: their logical design and interpretation using analysis of variance: Cambridge University Press.
52. Ferreira C, Gonçalves J (2006) Community structure and diet of roving herbivorous reef fishes in the Abrolhos Archipelago, south-western Atlantic. Journal of Fish Biology 69: 1533–1551.
53. Bunce M, Rodwell LD, Gibb R, Mee L (2008) Shifting baselines in fishers' perceptions of island reef fishery degradation. Ocean & Coastal Management 51: 285–302.
54. Humphries P, Winemiller KO (2009) Historical impacts on river fauna, shifting baselines, and challenges for restoration. BioScience 59: 673–684.
55. Rittenhouse CD, Pidgeon AM, Albright TP, Culbert PD, Clayton MK, et al. (2010) Conservation of forest birds: evidence of a shifting baseline in community structure. PloS one 5: e11938.
56. Ainsworth CH, Pitcher TJ, Rotinsulu C (2008) Evidence of fishery depletions and shifting cognitive baselines in Eastern Indonesia. Biological Conservation 141: 848–859.
57. McClenachan L (2009) Documenting loss of large trophy fish from the Florida Keys with historical photographs. Conserv Biol 23: 636–643.
58. O'Donnell K, Pajaro M, Vincent A (2010) How does the accuracy of fisher knowledge affect seahorse conservation status? Animal Conservation 13: 526–533.
59. Paterson B (2010) Integrating fisher knowledge and scientific assessments. Animal Conservation 13: 536–537.

60. Brook RK, McLachlan SM (2008) Trends and prospects for local knowledge in ecological and conservation research and monitoring. Biodiversity and Conservation 17: 3501–3512.
61. Boudreau SA, Worm B (2010) Top-down control of lobster in the Gulf of Maine: insights from local ecological knowledge and research surveys. Marine Ecology Progress Series 403: 181–191.
62. Palumbi SR (2003) Population genetics, demographic connectivity, and the design of marine reserves. Ecological applications 13: 146–158.
63. Hughes TP, Rodrigues MJ, Bellwood DR, Ceccarelli D, Hoegh-Guldberg O, et al. (2007) Phase shifts, herbivory, and the resilience of coral reefs to climate change. Curr Biol 17: 360–365.
64. Mumby PJ, Dahlgren CP, Harborne AR, Kappel CV, Micheli F, et al. (2006) Fishing, trophic cascades, and the process of grazing on coral reefs. Science 311: 98–101.
65. IUCN (2013) IUCN Red List of Threatened Species. Version 2013.2.
66. Worm B, Barbier EB, Beaumont N, Duffy JE, Folke C, et al. (2006) Impacts of biodiversity loss on ocean ecosystem services. Science 314: 787–790.

13

The Future of the Oceans Past: Towards a Global Marine Historical Research Initiative

Kathleen Schwerdtner Máñez[1,2]*, Poul Holm[3], Louise Blight[4], Marta Coll[5,6], Alison MacDiarmid[7], Henn Ojaveer[8], Bo Poulsen[9], Malcolm Tull[10]

1 Department of Social Sciences, Leibniz Center for Tropical Marine Ecology, Bremen, Germany, **2** Asia Research Center, Murdoch University, Murdoch, Western Australia, Australia, **3** Trinity Long Room Hub Arts and Humanities Research Institute, Trinity College Dublin, Dublin, Ireland, **4** World Wildlife Fund (WWF) -Canada, Vancouver, British Columbia, Canada, **5** Renewable Marine Resources Department, Institute of Marine Science, Barcelona, Spain, **6** Laboratoire Écosystèmes Marins Exploités, Sète Cedex, France, **7** Marine Ecology, National Institute of Water and Atmospheric Research, Kilbirnie, Wellington, New Zealand, **8** Estonian Marine Institute, University Tartu, Tartu, Estonia, **9** Faculty of Social Sciences, Aalborg University, Aalborg Ø, Denmark, **10** Murdoch Business School, Murdoch University, Murdoch, Western Australia, Australia

Abstract

Historical research is playing an increasingly important role in marine sciences. Historical data are also used in policy making and marine resource management, and have helped to address the issue of shifting baselines for numerous species and ecosystems. Although many important research questions still remain unanswered, tremendous developments in conceptual and methodological approaches are expected to contribute to a comprehensive understanding of the global history of human interactions with life in the seas. Based on our experiences and knowledge from the "History of Marine Animal Populations" project, this paper identifies the emerging research topics for future historical marine research. It elaborates on concepts and tools which are expected to play a major role in answering these questions, and identifies geographical regions which deserve future attention from marine environmental historians and historical ecologists.

Editor: Richard K. F. Unsworth, Seagrass Ecosystem Research Group, Swansea University, United Kingdom

Funding: Fellowships to the BIOWEB project and by the Spanish National Program 'Ramon y Cajal'. HO was funded by the Estonian Ministry of Education and Research (grant SF0180005s10). We thank HMAP for paying the publication costs of this paper. The funders had no role in study design, data collection and analysis, decision to publish, or preparation of the manuscript.

Competing Interests: The authors have declared that no competing interests exist.

* Email: kathleen.schwerdtner@zmt-bremen.de

Introduction

In the last fifteen years marine science has taken a historical turn. Looking much further back in time than most previous studies, global initiatives such as the History of Marine Animal Populations (HMAP) and the Sea Around Us programmes, as well as individual studies have investigated diverse socio-ecological systems, from coastal European marshes to Pacific islands. In a parallel move, environmental historians have undertaken the study of human engagement with the underwater realm in a sea change for history. Historians, archaeologists, economists, sociologists and geographers, have engaged with marine scientists in an interdisciplinary effort to bring together the study of human and underwater worlds. Thanks to this collaborative effort of marine and human sciences, researchers have not only identified but, for many regions and species, resolved the problem of shifting baselines (a term popularised by Daniel Pauly in his seminal 1995 paper [1]) by pushing back the chronological limits of our knowledge [2].

While there have been tremendous advancements in marine historical research, these are yet to be assimilated into an integrated seamless view of the global history of human interactions with life in the oceans (Figure 1). Furthermore, many important questions still remain unanswered. The following are what we consider to be the five key areas of research inquiry for a global marine historical research agenda:

i. What did the sea look like before human exploitations? So far, almost all the accessible information relates to after first human contact, and by implication very little is known about pristine life in marine waters, with the possible exception of the present-day ecosystems of the most southerly Antarctic waters, and the abyssal deeps. A better understanding of unexploited seas would serve as an important ecological baseline against which the impacts of human activities could be assessed.

ii. What is the relative importance of key drivers of environmental change over historical time frames? There is good evidence for several abiotic, climatic and human factors affecting life in the sea, but knowledge about their relative importance through time and their interactions remains poor.

iii. What has been the significance of marine resources for human societies over time? The physiological benefits of a marine diet are well documented but the economic and more intangible implications for historical societies have not been teased out.

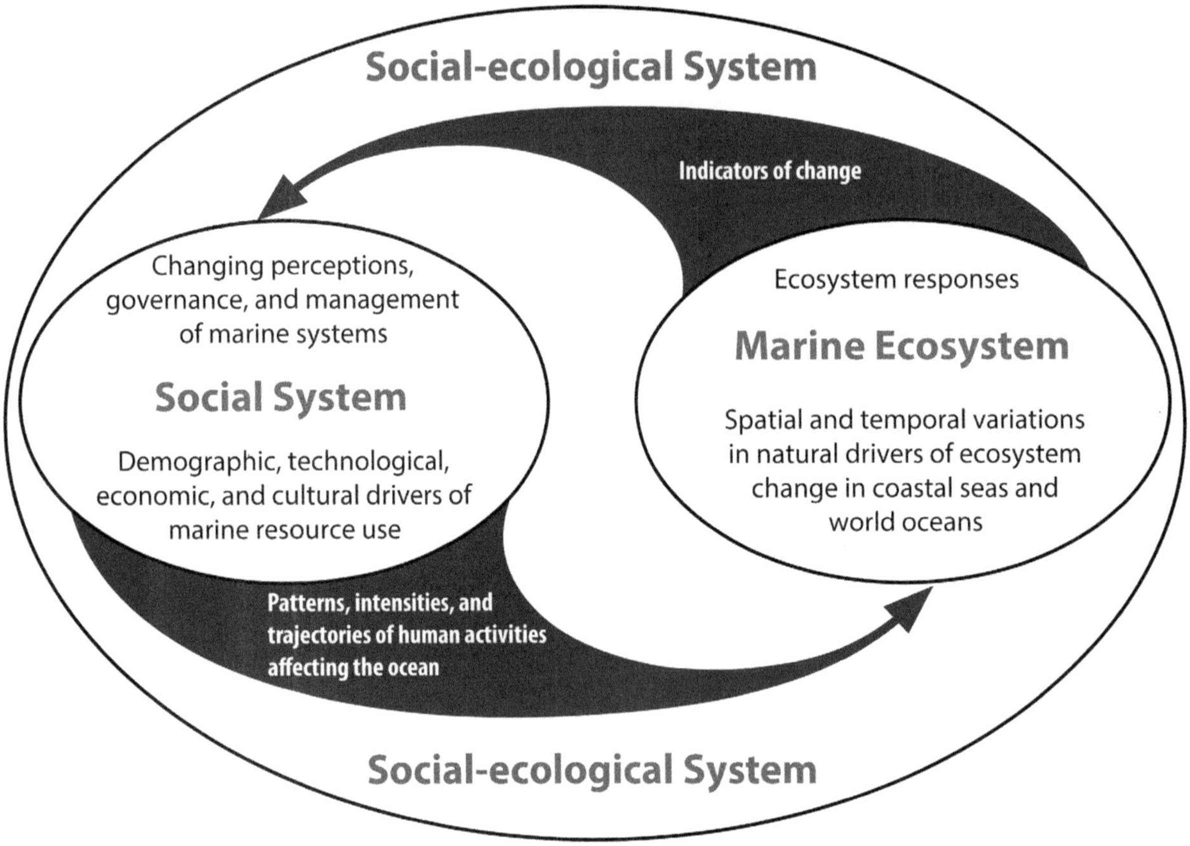

Figure 1. Integrated view of the global history of human interactions with life in the oceans.

iv. In cultural terms, societies have valued the goods and services provided by the sea very differently across various cultures and through time. What are the circumstances that have encouraged societies to exploit or give up the oceans?

v. How may historical information be used for better ocean governance and management in the future? While there are examples of the usefulness of historical baselines for fisheries management, the wider potential of historical research for ecosystem and seabed management or the development of human coastal societies remains largely unexplored.

Beginning with an elaboration of these emerging research topics in the sections below, this paper aims to develop a new research agenda for marine historical research. We also discuss potential applications and further development of existing conceptual and methodological approaches, namely the ecosystem service concept, the use of indicators and modelling approaches, new molecular methods, advanced oral history, and the need for gendered historical research. We then discuss the current coverage of historical research, and suggest further expansion in terms of regions and ecosystems to ensure a true global perspective. Finally, we summarize our thoughts on the research agenda, and suggest the foundation of a global research network: the Oceans Past Initiative (OPI) to assist in coordinating research efforts worldwide.

Impacts on Marine Systems Over Time

2.1 Untangling drivers of change

Little is known about what the oceans looked like before humans began to affect marine environments. The study of sediment cores as a library of past DNA provides one possible way of understanding what used to be swimming in the water columns above sea floors in a pre-exploitation era [3]. This could enable researchers to obtain a better understanding of pristine (pre-human) ocean life. It may also shed some light on when human activities started to impact marine systems, and how these effects can be differentiated from natural drivers of change. Similarly, a small handful of modern systems may yet be unaffected (or relatively so) by human impacts. These areas include the abyssal deeps, and the most southerly Antarctic waters [4]; the degree to which these provide clues about the functioning of other intact ecosystems prior to exploitation is currently unknown.

However, the vast majority of extant marine systems have been shaped by both natural and anthropogenic influences. Understanding the changes that these marine systems have undergone requires untangling the underlying drivers of these changes. Drivers are natural or human-induced factors that directly or indirectly force changes in ecosystems [5]. While direct drivers influence ecosystem components and processes themselves, indirect drivers alter one or more direct drivers. Physical and biological drivers such as changes in temperature regime or nutrient input, resource extraction, or the introduction of alien invasive species act directly upon marine systems and their component organisms [6–10]. Indirect drivers include human demographics, economic, sociocultural, and political processes, and scientific and technological innovation [11]. The majority of changes in ecosystems are caused by multiple interacting anthropogenic and/- or natural drivers. These interactions can be cumulative, antagonistic, synergistic, or otherwise. Decoupling the effects of single drivers to distinguish human influences from natural variability is a key interest of both scientists and ecosystem managers, and should be an integrative part of a research agenda that addresses historical change.

Determining impacts of natural drivers on pristine marine systems would improve our understanding of ecosystem functions and the role played by anthropogenic influences. Analysing the hard parts of organisms is one way to obtain information to infer the interaction of environmental drivers and ecological responses. For example, the shells of molluscs and otoliths (fish ear bones) are primarily constructed from calcium carbonate ($CaCO_3$). Stable isotopes of oxygen, carbon and nitrogen have a long history of use as geological and biological tracers [12–14]. If fish otoliths, shellfish remains or other biological samples can be obtained from a known time in the past (e.g., from dated midden horizons, museum collections), then the stable isotope and band increment evidence for environmental temperature and/or organismal growth and maturation for that period can compared to the same measures from modern individuals to examine long-term change in parameters of interest, and potential effects of human activities. Stable isotope analysis can also help researchers to examine different aspects of climate, food web structures, and other characteristics of past marine ecosystems [15]. For example, sequences of temperature estimates based on δ18O in fish otoliths can be applied to infer season at capture, which contributes to a better understanding of past fishing patterns [16]. The technique has also been used for the study of marine species and human diet. Recently, Orton and colleagues have shown the growing role of cod imports for medieval London. Using cod remains from excavations, they proof the use of this technique for the indication of changes in human diet [17]. Naturally occurring climate variations have an impact on the productivity of marine ecosystems [18]. A number of investigations have documented substantial fluctuations in fish populations related to climate forcing [19,20]. There is also well-established evidence for major climatic influences on phyto- and zooplankton communities [21,22]. Recent studies have identified ecosystem-level shifts in various ecosystems around the globe [21,23–25]. These shifts are a consequence of the interplay between climate forcing and various human impacts, primarily resource exploitation and pollution/eutrophication. Detailed investigations of reef growth and species composition in an Australian coral reef have shown that environmental degradation caused by human impacts has added additional pressure on the natural historical instability of these reefs [26]. Results from another Australian reef indicate that sediment and nutrient input following European settlement prohibited the recovery of *Acropora* assemblages after a series of acute disturbance events (SST anomalies, cyclones and flood events) [27]. One approach to detangling the impacts of such combined natural and anthropogenic drivers is the examination of paleontological, climatological, archaeological and historical evidence from localities affected by similar climate events but with differing levels of human impacts [28]. Such an approach could also provide a fresh perspective on natural variability.

Required research. Untangling natural and anthropogenic causes of environmental change to estimate the extent of human impacts on the oceans should be one of the major tasks for historical research in the near future. Coordinated multi-disciplinary research activities around the globe that focus on the same time period in areas of contrasting human exploitation of marine resources will assist in distinguishing the effects of natural and human drivers of marine environmental change.

2.2. The role of humans

Examination of the earliest human interactions with the marine environment in most parts of the world is beset by a major problem. The rise in sea level after the most recent ice-age has obscured or destroyed much of the earliest evidence of human exploitation along the margins of ancient shores [29]. Although in some areas such as the North Sea some evidence of early exploitation has been recovered from submarine deposits it has often been reworked by waves and currents making interpretation difficult [30]. However, there is evidence for an intensive use of marine resources by early modern humans in Pinnacle Point (South Africa) some 164,000 years ago. Gathering shellfish seems to have been part of a response to sea level fluctuations, which forced these people to expand their home ranges and to follow the shifting position of the coast [31]. Almost contemporarily, Neanderthals gathered shellfish in the coastal waters of Torremolinos in Southern Spain [32]. Archeological studies have revealed a general shift in marine resource exploitation from inshore towards offshore areas. Marine fish became increasingly important - as early as 1,500 years ago - on the Californian coast [33]. Similar patterns have been discovered in the Wadden Sea [34].

But for most of human history, the extent of anthropogenic impacts on natural environments was largely restricted to local and regional levels. Within the last 200 years, human actions reached the global level by affecting processes such as global nutrient and water cycles. The traces of human activities are not restricted to populated areas anymore, but can be found everywhere in the ocean [35]. For example, plastic debris is accumulating at the shorelines of even the most remote islands and in the deep sea [36]. Johan Rockström and colleagues [37] claim that human actions have accelerated to such an extent that they might push the Earth beyond its "planetary boundaries". The term planetary boundary refers to thresholds between alternate states of the global system. Once crossed, the system enters a new state, and irreversible environmental change might be unavoidable. By now, the existence of thresholds has been posited for numerous ecological and social-ecological systems, including the marine realm (see the thresholds database of the Resilience Alliance and the Santa Fee Institute, http://www.resalliance.org/index.php/thresholds_database). The world has entered a geological new era: the Anthropocene.

In this era, an important line of research will be the study of the interaction of different human activities and their impacts on specific marine areas in the past. Cross-regional comparisons of magnitude and direction of these changes to marine life in response to multiple human pressures could also be extremely helpful in teasing apart multiple long-term drivers. For example, it was only by the late 1990s that more or less the full range of harvestable fish and invertebrates in all ecological strata of Southeast Asian waters was caught [38]. Near-pristine marine systems could still be found in some parts of the Indonesian Archipelago until the 1960s, a time at which Atlantic waters had already been affected by industrial fishing activities for more than two centuries. In particular, data from large, isolated, late settled islands or regions such as Madagascar, New Zealand, and Antarctica, where significant human impacts became relevant much later than in the seas bordering the major continental landmasses, can play an important role in distinguishing among natural climatic variability and human drivers (including anthropogenic climate change). Recent work in Australian coral reefs has shown that it is indeed possible to differentiate natural variability from human-induced environmental change [27].

Despite the importance of similar drivers acting at the global level, most ecosystem changes are caused by a set of interactions that are more or less unique to a particular place [11]. To understand how resource and socio-economic endowments, in combination with other factors, have created specific exploitation patterns, bioregional histories of marine environments need to be developed. This will also provide a more comprehensive view of "what once was", e.g., historical baselines of marine animal populations, enhancing our ability to articulate desirable ecological states and management recommendations [39].

In this line of research, specific attention must be paid to the role of globalisation and changing consumption patterns. Processes of globalisation have had a significant influence on marine resource exploitation for hundreds of years. The quest for marine resources was an important aspect of European expansion into Asia, Africa and the Americas. Arctic marine mammals, sea turtles in the Caribbean, and cod from Iceland to Newfoundland are just some examples of valuable sought-after species [40]. Similarly, Asian, particularly Chinese food and medicine markets, have driven a search for marine species on a regional to global scale over the past few centuries. For example, the development of sea cucumber fishing and trading by people from Makassar, Indonesia, can be traced from its beginning more than 300 years ago to the industrialization of the fishery in the 20th century and the depletion of sea cucumbers in the 1980s [41]. Economic gain was only one side of the story. Changing preferences for certain products, the exploitation of new areas, and technical innovations all played a role in a widening search for marine resources.

Required research. How has the pattern of marine resource exploitation proceeded in each area settled by humans, and which species were affected? Was the unfolding of events in Africa fundamentally different from histories in Asia, Australia, the Americas, and Europe and recently settled remote islands? To address this will require integration of the earliest archaeological records with historical and modern information sources to determine patterns of marine resource use over the entire period of human settlement of coastal margins. Studying the role of globalisation and changing consumption patterns may improve our understanding of resource use by linking serial depletion on a global scale to local demands. We also need research into changing consumers preferences, especially towards more sustainably harvested marine products. Do such changes occur on a short or long-term scale, are they caused by public advice, by economic forces, or by other cultural factors that are to date poorly understood?

Altered Marine Ecosystems

3.1. Extinctions, species declines, and habitat changes

Extinctions and resource depletions in the marine realm have been evaluated by a number of studies (eg., [41,42,43]). We now have a relatively precise overview of globally extinct marine mammals, birds, fish, and molluscs [42]. Many species have become extirpated, or nearly so, across part of their range, such as is the case with several shark and ray species in the Mediterranean Sea [43]. Others have been depleted to such an extent that they cannot fulfil their former role in the ecosystem, as in the case of reduced abundances of some filter feeders [44]. To truly understand the extent of these changes, long-term biodiversity loss should be studied in different regions of the oceans over various periods of time. This includes the reconstruction of decadal, centennial, and millennial dynamics of marine animal populations [45,46]. A number of studies have provided important knowledge in this respect, for example on changes to shark populations in the West Atlantic [47], or on numbers of large predatory fish in different seas [48]. Most of the historical studies to date have concentrated on exploited higher trophic levels of marine animals, while intermediate and lower trophic levels have largely been ignored [49]. In large part this is because the evidence about large exploited species is more readily available from archaeological and historical sources. But there is also evidence for a sequential exploitation of invertebrate species in the past [41]. How this has shaped the status of ecosystems should be an important research topic in future.

Changes in coastal and marine habitats are a related research topic that deserves more attention. Although this line of work has been less prominent that changes in species, important contributions have already been made. For example, Lotze and colleagues reconstructed time lines, causes, and consequences of change in several estuaries and coastal seas worldwide and found similar patterns of habitat destruction, water quality degradation, and the influence of invasive species [44]. Other work has also focused on sub-tidal algae forests in the Adriatic Sea [50], or provided a two-century perspective on changes in a Scottish coastal area [51].

Required research. In the future, a major research avenue may be to link geospatial data and historical maps showing coastal and marine habitats to information about past distribution and abundance of species derived from archaeological, historical, and genetic sources. Such research could help us to understand indirect causes and effects of habitat change and be of value to marine and coastal management. Focussed research on species across levels, including intermediate and lower levels, is required to describe the patterns and magnitudes of their exploitation, preferably from the period of first human settlements to the present day.

3.2. Recoveries of populations, habitats and ecosystems

There are many examples around the world where previously affected species, habitats and ecosystems have or are responding to new management interventions. For example, over the last 40 years of protection populations of New Zealand fur seal (*Arctocephalus fosteri*) have started to rebound after 600 years of exploitation first by Polynesian settlers and later by European sealers [52]. Interestingly, the same protection regime has made little impact on the recovery of New Zealand sea lions (*Phocarctos hookeri*). In this species the return of maturing females to their beach of birth has limited the expansion of breeding populations into their historic range [53].

Research on ecological recoveries has a very clear link to providing input to decision-makers and managers, because it brings a historical perspective into the present management of marine resources and tells them what has worked so far and what not. Given that the approach to managing marine systems is evolving towards ecosystem-based adaptive management, a long-term, historical perspective to ecosystem function and change is very much needed [54,55]. An interesting example of such work is a recent reconstruction of social-ecological interactions over the last 700 years in Hawaiian coral reefs. It reports previously undetected recovery periods related to a complex set of changes in underlying social systems resulting in the release of reefs from direct anthropogenic stressors [56].

Required research. Future investigations should continue to assess the adaptive capacity and resilience of degraded ecosystems and depleted stocks to recover from human impacts. There is already work done on how marine resources have recovered from previous exploitation-related declines [45,57], but we need a stronger focus on which species, ecosystem traits, and management activities might facilitate recoveries.

Perceptions, Governance, and Management of Marine Systems

How people perceive and value marine environments and the resources they provide determines individual and collective preferences, actions and strategies in the marine realm. Historical descriptions of coastal and marine environments are prime examples of documented perceptions of the past. In the absence of other data, they have been successfully used to establish abundance changes of marine species in different time periods [58,59].

Different perceptions and valuation systems also underlie the institutional structures that govern and manage marine systems. Research on governance structures must be linked with research on perceptions and values to understand what drives and has driven approaches to marine resource exploitation and its management in different periods of time. For analytical clarity, it is important to distinguish between governance and management. Governance describes a social function centered on efforts to steer the actions of humans toward achieving desirable outcomes and avoiding undesirable ones. It covers the fundamental goals, institutional processes and structures which are the basis for planning and decision-making, and sets the stage within which management occurs. Management refers to the process by which human and material resources are used to achieve a defined goal within a known institutional structure [60].

The notion of the sea as a seemingly endless source of resources has long dominated marine governance, or rather the relative absence of institutions governing the sea. However, there are also examples of long-enduring traditional management regimes which have regulated the exploitation of valuable or scarce resources such as pearl oysters, sea cucumbers, or *Trochus* spp. shells in many coastal areas in the Pacific and parts of Southeast Asia, long before modern fisheries management was invented. For example, the Indonesian regime of "sasi laut" regulated access to marine resources by placing temporal and spatial harvesting restrictions on them. Similarly regulations in European inshore waters such as the Limford sustained fisheries for hundreds of years [61]. While such community-based management is certainly not a panacea, studies around the world have found that traditional community-based governance regimes contain elements that may support sustainable resource use [62,63].

Required research. How have different local perceptions of marine systems affected their governance and management, and how have these evolved over time? Can we identify elements of these historical, community-based approaches to ocean governance and management that are also relevant to modern marine management?

Emerging methodological approaches

A number of emerging methodological approaches are expected to have a considerable influence on future marine historical research. They include conceptual developments such as the application of the ecosystem service concept in historical analyses or the consideration of social issues like gender and equity issues, but also new biophysical tools, for example different kinds of molecular and stable isotope analyses. These approaches are not necessarily new to science, but it is their application to historical research that is expected to improve our ability to analyse and evaluate changes in marine systems and their organisms over time.

5.1 Applying the ecosystem service concept in historical analyses

Ecosystem services are an array of potential benefits derived from specific ecological components and processes, ranging from the provision of fish to the capacity of the ocean to buffer climate change. Environmental change has clearly altered the level of ecosystem services provided by marine systems, but to what extent is largely unknown. There is a need to improve our understanding of how past changes in marine systems have affected marine ecosystem productivity and marine ecosystem functioning, and how this has influenced the ability of marine systems to provide ecosystem services.

The ecosystem service concept values nature in relation to human uses. By acknowledging the role of ecosystems as providers of essential goods and services, it links ecosystem functions with the economy and social spheres, including livelihoods and well-being [64]. The ecosystem service concept thus provides an approach to assess the value of marine systems, although this does not have to be a monetary valuation. More importantly, it offers a standardized method to compare these values over time, by showing how changes in ecological functions have caused changes in benefits to society [65].

Although much historical research is clearly related to ecosystem services and benefits, the concept of ecosystem services itself has hardly been used in the discipline. One exception is the decline in marine mammals in New Zealand that followed the well-documented onset of Māori sealing soon after initial settlement, and European whaling in the early 19th century [66]. Both cultures viewed many species of marine mammals as valuable commodities to be harvested, so the numbers of these mammals declined precipitously as a result [67,68]. Now, these species are protected and their value to New Zealand as a provisioning service has declined to zero. Instead, marine mammals are currently prized for their spiritual and existence value; people enjoy directly viewing them from land, sea, and air; and they are appreciated as subjects for research and educational activities. Moreover, whales are now recognised as having important roles in ecosystem regulation [69]. Studies like these illustrate the usefulness of the ecosystem service concept as a common framework for analysing the impacts of environmental change on societies.

5.2 Developing indicators

Indicators are characteristics of ecological and social systems or their components, such as species, populations, networks, and social groups, which ideally indicate a certain state or dynamic of a system that is otherwise difficult to measure and evaluate. Because ecological and social systems are inherently complex, the use of indicators helps to describe them and their changes in simpler terms. If chosen well, indicators can track changes over time and across species or regions, and can inform managers and the general public.

One of the most fundamental ecological indicators of historical change is a change in population abundance. This can be measured as a decrease or increase in the number of individuals, their biomass, average size or age, as well as an expansion or contraction of their distribution over time [70]. These indicators have been applied to a variety of records from the past. For example, archaeological records from shell middens revealed declines in the relative abundance, size and age of white sturgeon from 2600 to 700 years ago based on their bone frequency and dentary width [71,72]. Historical whaling maps and log books have been used to reconstruct the rapid depletion of right whales (*Eubalaena japonica*) in the North Pacific by 19th-century whalers

[73]. Fisheries catch and effort data throughout the Mediterranean have been used to analyse declines in the catch-per-unit-effort (CPUE) of sharks since the 19th century [74]. In rarer cases, it has been possible to extend such analysis into the early 17th century [75]. Other fundamental ecological indicators are the basic presence or absence of a species in an ecosystem, the trends in functional groups, or changes in trophic habits and positions [70]. Future development of ecological indicators will move towards standardization and evaluation of indicators, and the development of reference levels (targets, thresholds, limits) to inform the management of marine resources and ecosystems.

Social indicators are expected to play a very important role in the future for the assessment and management of coastal and marine systems and their changes over time. Such indicators can be qualitative or quantitative and are used to describe the status of social systems, and their dynamics and processes. While status indicators measure, for example, the current perception of target species, their availability or the social networks of resource exploitation, process indicators assess specific actions, changes or functions over a defined time period, such as participation, conflict resolution or institutional change [76]. Social process indicators can be used to evaluate participation and decision-making in the management of coastal and marine resources by assessing metrics such as the number of meetings, the levels of participation, and the character of social networks involved.

Indicators are expected to play a far greater role in the analysis of social–ecological systems in the future. They allow for regular measurements of key ecological, socio-economic, and social–ecological processes to better understand system changes and their underlying causes [76].

5.3 Utilizing modelling approaches

A major research issue in this field is how to integrate historical knowledge and ecosystem modelling, including past and future applications and simulations. The future will bring developments in modelling techniques which increase the capability to hindcast ecosystem dynamics and ensure forecasting possibilities [70]. More comparative analysis of qualitative and quantitative models must be developed since the ability to model past marine ecosystems is crucial to understanding the present, and to predict the future. These analyses will be challenged by the urgent need to consider ecological, economic and social dimensions of change [70].

Qualitative and quantitative modelling techniques [77–81] have already been adapted or modified to model past marine food webs. For example, qualitative modelling was used in the Gulf of Maine to identify distinct and sequential phases in the trophic structure of kelp forests [82]. The authors used archaeological, historical, ecological, and fisheries data to document a serial targeting and depletion of abundant top consumers that led to functional loss of trophic levels, creating trophic cascades that changed the structure and function of the ecosystem [83]. In the Adriatic Sea, qualitative modelling has similarly been used to describe the historical change of marine food webs using ten historical periods that started in the pre-human period before ~100 000 BC to the global expansion of humans in AD 1950–2000 [84].

An interesting approach to modelling past ecosystems has been developing under the "Back to the Future" approach [85]. This approach aims at evaluating historic ecosystems as tools to define possible restoration goals and to design rebuilding strategies for ecosystems. Two remarkable applications were developed for Canadian marine ecosystems of Newfoundland [86] covering the years of 1450, 1900, 1985 and 1995, and in British Columbia [87], covering the periods of 1750, 1900, 1950 and 2000. Both studies highlighted the general depletion of marine resources since the first European contact and the important changes in the structure and functioning of food webs through time.

5.4 Application of molecular tools

Molecular genetics has fundamentally changed our understanding of marine ecology. Investigations have demonstrated extensive adaptive change in marine populations in response to natural and anthropogenic drivers. Research has also shown that the estimated population sizes of several species are much smaller than census sizes, which is highly relevant for management and conservation [88]. A notable example of the use of molecular techniques in historical ecology studies is Roman and Palumbi's study on North Atlantic humpback whales (*Megaptera novaeangliae*), fin whales (*Balaenoptera physalus*) and minke whales (*Balaenoptera acutorostrata*), in which they used molecular markers to estimate the pre-exploitation abundance of these species. Their results indicated that current population numbers are far below their original size [89], (although, in a recent overview, Palumbi was careful to outline the indefinite time parameter of their study and a number of uncertainties [90]). A genetic approach was also used to estimate pre-whaling abundance of Eastern Pacific gray whales. The results substantially differed from population reconstructions based on catch records, which had considered the population to be fully recovered [91]. Both examples clearly demonstrate the relevance of molecular tools in defining baselines for management of marine species, and the recovery of depleted populations.

Molecular tools also have an enormous potential to contribute to the discussion on detangling natural and anthropogenic drivers of change. A recent study using genetic data from the harbor porpoise (*Phocoena phocoena*) in the Black Sea was able to reconstruct the demographic expansion of this species some 5,000 years ago as a result of natural environmental changes, as well as its population collapse due to recent anthropogenic activity [92].

Molecular analyses are increasingly used in research on biological invasions, one of the most serious human-induced threats to marine ecosystems [49]. This has improved our ability to make inferences regarding invasion histories, enabled resolution of some of the long-standing questions regarding the cryptogenic status of marine species and provided means to recover the patterns of community structure of the ocean's biota.

5.5 Advancing methods for written documents and oral history

The provenance of documents traditionally frames modes of inquiry in the history discipline. Port records, log books, tax ledgers etc. are typical sources of quantitative information, while interviews, written memories or newspaper coverage can provide both quantitative and qualitative information.

Witness testimonies, especially, are increasingly used to acquire information on past and contemporary marine environments and fisheries. Through testimonies, individuals or groups share their perceptions and opinions about past events or experiences. Oral history helps to gather these accounts of the past, and to make use of traditional and local environmental knowledge that resource users have accumulated [93]. If oral histories have been recorded or written down, the time span for which this kind of data can be used extends far beyond living memory. The scientific literature on the use of oral history has grown rapidly over the last three decades [94]. But after Saenz-Arroyo's pioneering study on shifting baselines in fishermen's memories [95], the work by Palomares and colleagues on the establishment of abundance data from historical narratives [58], and the application of a fuzzy logic approach to data gathered from interviews [96], the methodological progress seems to have slowed, although the value of oral history has been clearly demonstrated [59].

An important aspect which certainly deserves further attention is the creation of quantitative data. Qualitative information is typically rich and insightful but difficult to control for selectivity and representativity, a problem at the very heart of the historical discipline which emphasises the critical work of uncovering problems of bias. A basic social science approach – the coding of qualitative data – can provide an important contribution in this respect. Coding comprises of the search for emerging topics and key words in documents or interview transcriptions, and their description with a unique code. This allows, for example, quantitative frequency analysis of topics.

In recent years, discourse analysis, meta-analysis and digital representation have provided historians with a wide range of new methodological and technical tools to grapple with vastly increased masses of data. "Big Data" tools which organise masses of data for geographical, temporal and discursive analysis has the potential to bring historical interpretation to a new level of breadth and precision. Marine environmental history has not yet fully embraced the potential of such new methodologies. In particular there is great potential in the application of database and geographical information tools for the study of disparate oral and documentary data.

5.6 Using a gender lens

Marine resource exploitation in general, and fisheries in particular, are often perceived to be a male domain [97]. This is largely the result of the socially constructed roles of men and women within societies: while men are typically regarded as providers, women take care of the home and family. It has also to do with the traditional perspective that fishing refers to the catching of fish with specific gears, such as lines and nets. Gleaning from shorelines and reefs has rarely been acknowledged as fishing [98]. Additionally, it has often been assumed that fisheries largely operate in the public domain, a usually male dominated sphere. The rather female dominated private domain is mostly not in the focus of attention. But especially in subsistence and other small-scale fisheries, much of the administration and logistics including financial issues, as well as the processing happens in the household and through family networks. An analytical focus on public and formal practices misses women's roles as well as a considerable part of what it takes to organize and put into practice a fishing operation [99].

This has two major implications: ignoring the role of woman in fisheries can lead to a substantial underestimation of fishing pressure, especially in coastal areas. It also leads to an underestimation of the social and economic contributions that women provide in fisheries, especially in processing and other value-adding activities [100]. A gender lens can increase understanding of the history of marine resource exploitation. Differentiating roles, responsibilities, access and opportunities of men and woman will provide a more complete picture about access to and control of marine resources [101].

Since the first consolidated publication on women in fisheries by Nadal-Klein and Davis [102], the literature has increased rapidly. While some of these publications have focussed on the role of women in marine resource collection [103,104], pre- and post-harvesting activities [105], or their role in governing and managing marine resources [106,107], others are rather conceptual contributions such as Yodanis's work on the social construction of gender in fishing communities [108]. Recent work has also high lightened women's in depth knowledge on species and environmental changes. For example, oral histories in New Zealand indicate the important role that women, often accompanied by children, once played in gathering seafood from shallow, easily accessible shores sixty to seventy years ago [109]. Women's knowledge confirmed that fish and shellfish were very abundant and could be reliably caught or gathered with little effort.

Another line of work has looked at masculinity and its linkages to fishing. For example, Fabinyi [110] has described how illegal fishing activities in the Philippines provide young men with a higher social status, while Allison has argued that physical settings and the distinct culture of fishing societies shape similar "marine masculinities" in fishing communities in different regions of the world [111]. In a recent publication, Schwerdtner Máñez & Pawelussen [99] call for a gendered perspective on the different roles of both men and women in order to better understand the history of marine resource exploitation over time.

Coverage of Historical Analyses

Under the auspices of HMAP, twelve regional and three species-based projects were developed between 2000 and 2012. These uncovered a wealth of historical information that has been published in more than 200 peer-reviewed scientific publications. However, even such a concerted effort could by no means attain a global coverage in the investigation of humankind's interactions with the sea. To understand the history of human interactions with life in the oceans on global and smaller scales historical research needs to be extended to new geographical areas.

Costello and colleagues have identified a number of areas that received little or no attention during the Census of Marine Life, such as the southern and eastern Mediterranean Sea; estuaries, coastal areas and coral reefs of the Indian Ocean; large regions of South America and the Indian Ocean; and many other habitats including ice-bound waters, ocean trenches, and some deep sea areas [49]. In terms of historical investigations, this list of gaps is even longer.

South America is one of the regions which have so far received very little attention. Although there are an increasing number of studies on the environmental history of the continent, these have had an almost exclusively terrestrial focus. One of the noteworthy exceptions is the work by Perri, who examined the Spanish overexploitation of the oyster beds around the island of Cubagua in what is today Venezulea [112]. Other work covers the ecological history of the Peruvian guano industry from 1800–1973, and its relation to the development of world's largest industrial fishery, that of anchoveta [113]. More recently, Eddy et al. have studied the change in lobster biomass over 400 years of human fishing activity on the Chilean Juan Fernández archipelago [114]. But the vast majority of the continent's coasts and marine areas remain untouched by historical studies.

For the African continent, the situation is similar. An exception is South Africa, where a number historical studies have been undertaken, including work on long-term trends on commercially exploited species [115,116], and work on the history of harbors and fisheries [117]. In an overview of the history of marine and estuarine studies in South Africa, the author argued that South Africa has some of the best data in the world on the biodiversity of rocky shores, coastal waters and estuaries exists, as well as large collections of many taxa [118]. Clearly, further historical work could take advantage of these rich collections. There are also some historical studies on the island of Rodriguez, including research on shifting baselines [119,120]. But for the rest of the continent, we are not aware of any studies in marine environmental history or historical ecology.

The polar regions certainly represent another region with a rich history in marine resource exploitation: whaling, seal hunting, and search for fish have brought people for centuries into these areas. The history of whaling has received some attention, for example in a paper by Hacquebord who analyzed the impacts of eliminating thousands of whales and walruses on other marine mammals [121]. Other studies have assessed ecological changes or population trends resulting from the mass removals of pinnipeds and whales [122]. Whaling records have also been used to establish declines in Antarctic sea-ice extent [123], and the influence of environmental factors on whaling operations has also been studied [124]. However, much research could still be done in these regions to expand our understanding of how they have been changed by resource extraction over time.

An area for which we have very little information is the third dimension of the oceans: the deep sea [125]. Whereas the coastal areas of most developed countries have been exploited for several centuries, the deep sea waters were little known by fishermen and marine scientists alike just 100 years ago. In the meantime deep sea investigations such as the Dutch Snelllius Expedition of the 1920s, and multiple Danish and Norwegian led deep sea investigations traversed the Atlantic, Indian and Pacific Oceans from the turn of the 20th century until the 1960s. *The Carlsberg Foundation's Oceanographic Expedition Round the World* from 1928–30 is an example of how global biogeography was charted well in advance of real impacts from resource exploitation in the deep seas of the World's oceans [126]. Today many of these areas are exploited for multiple marine resources. Such early endeavours deserve careful scrutiny to provide reference points for present day ocean management.

Another focus of attention should be on historically data-rich areas. For example, much documented evidence is available for large parts of maritime Southeast Asia, including the coral triangle. This region is the centre of global marine biodiversity, and the exploitation and trade of its marine resources has at least partly been documented by its former colonial regimes. Especially for the former Dutch East Indies, but also for what today is Papua New Guinea, a wealth of historical material has been collected in Dutch and German libraries and archives, and its processing has just begun [127]. Japan and China are also countries with a long tradition of record keeping, and strong connections to the sea. Working with these data might also improve our understanding of early drivers of resource exploitation, which created networks and trade routes that are still important today [41].

Outlook

Archaeological and historical research has still much to contribute to an improved understanding of the development of marine and coastal ecosystems, and their interactions with humans over time. It helps not only to identify patterns and trends, but also to quantify amounts of living marine resource exploitation. While the aim of this paper has been to establish a new research agenda, we would also like to point out that new institutional arrangements are required in order to successfully implement it. To complement individual projects, which end when their funding runs out, we suggest the establishment of a global research network for marine historical research: the Oceans Past Initiative (OPI). This could become a venue where researchers worldwide interested in historical studies could meet and discuss relevant issues virtually. OPI could provide an umbrella, under which already completed, currently running, and also planned projects and initiatives could be linked, and make their results available for decision-makers and the interested public. Such a network could also commit itself to coordinate resources and provide useful information to the research community, such as by circulating news about grant calls and funding opportunities, information on recent papers in the field, announce successful projects, and distribute information for students on courses related to marine historical research. A first major step into this direction has been undertaken through the establishment of the Oceans Past Platform (OPP), a COST action supported by the European Union which aims to measure and understand the significance and value to societies of living marine resource extraction and production to help shape the future of coasts and oceans.

Acknowledgments

We would like to thank Joseph Christensen, Meryl Williams, James Barrett and an anonymous reviewer, who provided helpful comments on earlier versions of this paper.

Author Contributions

Analyzed the data: KSM PH LB MC AMD H BP MT. Contributed reagents/materials/analysis tools: KSM PH LB MC AMD HO BP MT. Contributed to the writing of the manuscript: KSM PH LB MC AMD HO BP MT.

References

1. Pauly D (1995) Anecdotes and the shifting baseline syndrome of fisheries. Trends Ecol Evol 10: 430.
2. Holm P, Marboe AH, Poulsen B, MacKenzie BR (2010) Marine Animal Populations: A New Look Back in Time. Life in the World's Oceans. Wiley-Blackwell. 1–24.
3. Finney BP, Alheit J, Emeis K-C, Field DB, Gutiérrez D, et al. (2010) Paleoecological studies on variability in marine fish populations: a long-term perspective on the impacts of climatic change on marine ecosystems. J Mar Syst 79: 316–326.
4. Halpern BS, Walbridge S, Selkoe KA, Kappel CV, Micheli F, et al. (2008) A global map of human impact on marine ecosystems. Science 319: 948–952.
5. Carpenter SR, Bennett EM, Peterson GD (2006) Editorial: Special Feature on Scenarios for Ecosystem Services. Ecol Soc 11: 32.
6. Menge BA (2000) Top-down and bottom-up community regulation in marine rocky intertidal habitats. J Exp Mar Bio Ecol 250: 257–289.
7. Seitz RD, Lipcius RN (2001) Variation in top-down and bottom-up control of marine bivalves at differing spatial scales. ICES J. Mar. Sci 58: 689–699.
8. Philippart CM, Beukema J, Cadée G, Dekker R, Goedhart P, et al. (2007) Impacts of Nutrient Reduction on Coastal Communities. Ecosystems 10: 96–119.
9. Zeldis J, Howard-Williams C, Carter CM, Schiel DR (2008) ENSO and riverine control of nutrient loading, phytoplankton biomass and mussel aquaculture yield in Pelorus Sound, New Zealand. Mar Ecol Prog Ser 371: 131–142.
10. Powell Jr AM, Xu J (2011) Abrupt climate regime shifts, their potential forcing and fisheries impacts. Atmospheric and Climatic Sciences 1: 33–47.
11. Nelson GC, Bennett E, Berhe AA, Cassman K, DeFries R, et al. (2006) Anthropogenic Drivers of Ecosystem Change: an Overview. Ecol Soc 11: 29.
12. Campana SE, Gagne JA, MacLaren JW (1995) Elemental fingerprinting of fish otoliths using ID-ICPMS. Mar Ecol Prog Ser 122: 115–120.
13. Roelke LA, Cifuentes LA (1997) Use of stable isotopes to assess groups of king mackerel, Scomberomorus cavalla, in the Gulf of Mexico and southeastern Florida. Fisheries Bulletin 95: 540–551.
14. Campana SE, Thorrold SR (2001) Otoliths, increments and elements: keys to a comprehensive understanding of fish populations? Can J Fish Aquat Sci 58: 30–38.
15. Orton DC (2014) Archaeology as a tool for understanding past marine resource use and its impact. In: K. Schwerdtner Máñez and B Poulsen, editors. Perspectives on Oceans Past: A Handbook of Marine Environmental History. Springer Publishers. (In press).
16. Hufthammer AK, Hoie H, Folkvord A, Geffen AJ, Andersson C, et al. (2010) Seasonality of human site occupation based on stable oxygen isotope ratios of cod otoliths. J Archaeol Sci 37: 78–83.

17. Orton DC, Morris J, Locker A, Barrett JH (2014) Fish for the city: meta-analysis of archaeological cod remains and the growth of London's northern trade. Antiquity 88: 516–530.
18. Behrenfeld MJ, O'Malley RT, Siegel DA, McClain CR, Sarmiento JL, et al. (2006) Climate-driven trends in contemporary ocean productivity. Nature 444: 752–755.
19. Southward AJ, Boalch GT B, Maddock L (1998) Fluctuations in the herring and pilchard fisheries of Devon and Cornwall linked to change in climate since the 16th century. J Mar Biol Assoc U.K. 68: 423–445.
20. Alheit J, Hagen E (1997) Long-term climate forcing of European herring and sardine populations. Fish Oceanogr 6: 130–139.
21. Conversi A, Fonda Umani S, Peluso T, Molinero JC, Santojanni A, et al. (2010) The Mediterranean Sea Regime Shift at the End of the 1980s, and Intriguing Parallelisms with Other European Basins. PLoS ONE 5: e10633.
22. Klais R, Tamminen T, Kremp A, Spilling K, Olli K (2010) Decadal-Scale Changes of Dinoflagellates and Diatoms in the Anomalous Baltic Sea Spring Bloom. PloS ONE 6: e21567.
23. Drinkwater KF (2006) The regime shift of the 1920s and 1930s in the North Atlantic. Prog Oceanogr 68: 134–151.
24. Österblom H, Hansson S, Larsson U, Hjerne O, Wulff F, et al. (2007) Human-induced Trophic Cascades and Ecological Regime Shifts in the Baltic Sea. Ecosystems 10: 877–889.
25. Blamey LK, Howard JAE, Agenbag J, Jarre A (2012) Regime-shifts in the southern Benguela shelf and inshore region. Prog Oceanogr 106: 80–95.
26. Lybolt M, Neil D, Zhao J, Feng Y, Yu K-F, et al. (2010) Instability in a marginal coral reef: the shift from natural variability to a human-dominated seascape. Front Ecol Environ 9: 154–160.
27. Roff G, Clark TR, Reymond CE, Zhao J-X, Feng Y, et al. (2013) Palaeoecological evidence of a historical collapse of corals at Pelorus Island, inshore Great Barrier Reef, following European settlement. Proc Biol Sci 280: 10.1098/rspb.2012.2100.
28. Ojaveer H, Mackenzie BR (2007) Historical development of fisheries in northern Europe–Reconstructing chronology of interactions between nature and man. Fisheries Research 87: 102–105.
29. Erlandson JM, Rick TC (2008) Archeology, marine ecology and human impacts on marine environments. In: T. C Rick and J. M Erlandson, editors. Human impacts on ancient marine ecosystems. Berkely and Los Angeles, California: University of California Press. 1–19.
30. Bailey G, Barrett JH, Craig O, Milner N (2008) Historical ecology of the North Sea basin. In: T. C Rick and J. M Erlandson, editors. Human impacts on ancient marine ecosystems. Berkely and Los Angeles, California: University of California Press. 215–242.
31. Marean CW, Bar-Matthews M, Bernatchez J, Fisher E, Goldberg P, et al. (2007) Early human use of marine resources and pigment in South Africa during the Middle Pleistocene. Nature 449: 905–908.
32. Cortés-Sánchez M, Morales-Muñiz A, Simón-Vallejo MD, Lozano-Francisco MC, Vera-Peláez JL, et al. (2011) Earliest Known Use of Marine Resources by Neanderthals. PLoS ONE 6: e24026.
33. Rick TC, Erlandson JM, Braje TJ, Estes JA, Graham MH, et al. (2008) Historical ecology and human impacts on coastal ecosystems of the Santa Barbara Channel region, California. In: T. C Rick and J. M Erlandson, editors. Human impacts on ancient marine ecosystems: a global perspective. University of California Press. 77–101.
34. Lotze HK (2007) Rise and Fall of fishing and marine resource use in the Wadden Sea, southern North Sea. Fisheries Research 87: 208–218.
35. Halpern BS, Walbridge S, Selkoe KA, Kappel CV, Micheli F, et al. (2008) A global map of human impacts on marine ecosystems. Science 319: 948–952.
36. Barnes DK, Galgani F, Thompson RC, Barlaz M (2009) Accumulation and fragmentation of plastic debris in global environments. Philos Trans R Soc Lond B Biol Sci 364: 1985–1998.
37. Rockström J, Steffen W, Noone K, Persson A, Chapin FS, et al. (2009) A safe operating space for humanity. Nature 461: 472–475.
38. Butcher JG (2004) The Closing of the Frontier: A History of the Marine Fisheries of Southeast Asia c. 1850–2000. Leiden: KITLV Press.
39. Lotze HK, Worm B (2009) Historical baselines for large marine animals. Trends Ecol Evol 24: 254–262.
40. Richards JF (2006) The Unending Frontier. An Environmental History of the Early Modern World. London: University of California Press.
41. Schwerdtner Máñez K, Ferse SCA (2010) The history of Makassan trepang fishing and trade. PLoS ONE 5.
42. Monte-Luna PD, Lluch-Belda D, Serviere-Zaragoza E, Carmona R, Reyes-Bonilla H, et al. (2007) Marine extinctions revisited. Fish Fish 8: 107–122.
43. Dulvy NK, Sadovy Y, Reynolds JD (2003) Extinction vulnerability in marine populations. Fish Fish 4: 25–64.
44. Lotze HK, Lenihan HS, Bourque BJ, Bradbury RH, Cooke RG, et al. (2006) Depletion, degradation, and recovery potential of estuaries and coastal seas. Science 312: 1806–1809.
45. Lotze HK, Coll M, Magera MA, Ward-Paige C, Airoldi L (2011) Recovery of marine animal populations and ecosystems. Trends Ecol Evol 26: 595–605.
46. Jackson JBC, Kirby MX, Berger WH, Bjorndal KA, Botsford LW, et al. (2001) Historical Overfishing and the Recent Collapse of Coastal Ecosystems. Science 293: 629–638.
47. Baum JK, Myers RA, Kehler DG, Worm B, Harley SJ, et al. (2003) Collapse and conservation of shark populations in the Northwest Atlantic. Science 299: 389–392.
48. Myers RA, Worm B (2003) Rapid worldwide depletion on predatory fish communities. Nature 423: 280–283.
49. Costello MJ, Coll M, Danovaro R, Halpin P, Ojaveer H, et al. (2010) A census of marine biodiversity knowledge, resources and future challenges PLoS ONE 5: e12110.
50. Perkol-Finkel S, Airoldi L(2010) Loss and Recovery Potential of Marine Habitats: An Experimental Study of Factors Maintaining Resilience in Subtidal Algal Forests at the Adriatic Sea. PLoS ONE 5: e10791.
51. Thurstan RH, Roberts CM (2010) Ecological Meltdown in the Firth of Clyde, Scotland: Two Centuries of Change in a Coastal Marine Ecosystem. PLoS ONE 5: e11767.
52. Lalas C (2008) Recolonialisation of Otago, southern New Zealand, by fur seals and sea lions: unexpected patterns and consequences. In: B Clarkson, P Kurian, T Nachowitz and H Rennie, editors. Conservation-Vision Conference. Hamilton: University of Waikato.
53. Chilvers BL, Wilkinson IS (2008) Philopatry and site fidelity of New Zealand sea lions (*Phocarctos hookeri*). Wildlife Research 35: 463–470.
54. Pikitch EK, Santora C, Babcock EA, Bakun A, Bonfil R, et al. (2004) Ecosystem-Based Fishery Management. Science 305: 346–347.
55. FAO Fisheries Department(2003) The ecosystem approach to fisheries. FAO Technical Guidelines for Responsible Fisheries. Rome.
56. Kittinger JN, Pandolfi JM, Blodgett JH, Hunt TL, Jiang H, et al. (2011) Historical reconstruction reveals recovery in Hawaiian coral reefs. PLoS ONE 6: e25460.
57. Neubauer P, Jensen OP, Hutchings JA, Baum JK (2013) Resilience and Recovery of Overexploited Marine Populations. Science 340: 347–349.
58. Palomares MLD, Mohammed E, Pauly D (2006) Maria Lourdes D. Palomares, Elizabeth Mohammed, and Daniel Pauly on European expeditions as a source of historic abundance data on marine organisms: A case study of the Falkland Islands. Environ Hist 11: 835–847.
59. Coll M, Carreras M, Ciércoles C, Cornax M-J, Gorelli G, et al. (2014) Assessing Fishing and Marine Biodiversity Changes Using Fishers' Perceptions: The Spanish Mediterranean and Gulf of Cadiz Case Study. PLoS ONE 9: e85670.
60. Olsen SB, Page GG, Ochoa E (2009) The Analysis of Governance Responses to Ecosystem Change: A Handbook for Assembling a Baseline. LOICZ Reports & Studies. Geesthacht: GKSS Research Center.
61. Poulsen B, Holm P, MacKenzie BR (2007) A long-term (1667–1860) perspective on impacts of fishing and environmental variability on fisheries for herring, eel, and whitefish in the Limfjord, Denmark. Fish Res 87: 181–195.
62. Zerner C (1994) Through a green lense: The construction of customary environmental law in Indonesia's Maluku Islands. Law Soc Rev 28: 1079–1121.
63. Cinner JE, Basurto X, Fidelman P, Kuange J, Lahari R, et al. (2012) Institutional designs of customary fisheries management arrangements in Indonesia, Papua New Guinea, and Mexico. Marine Policy 36: 278–285.
64. Gahzoul J (2007) Recognising the complexities of ecosystem management and the ecosystem service concept. Gaia 16: 215–221.
65. Worm B, Barbier EB, Beaumont N, Duffy JE, Folke C, et al. (2006) Impacts of Biodiversity Loss on Ocean Ecosystem Services. Science 314: 787–790.
66. MacDiarmid AB, Law CS, Pinkerton M and Zeldis J (2013) New Zealand marine ecosystem services. In: J. R Dymond, editor editors. Ecosystem services in New Zealand – conditions and trends. Lincoln, New Zealand: Manaaki Whenua Press.
67. Smith I (2005) Retreat and resilience: fur seals and human settlement in New Zealand. In: G Monks, editor editors. The exploitation and cultural importance of sea mammals. Cambridge: Oxbow Books. 6–18.
68. Caroll EL (2006) The demographic and genetic bottleneck of the New Zealand southern right whale. School of Biological Sciences. The University of Auckland, New Zealand.
69. Nicol S, Bowie A, Jarman S, Lannuzel D, Meiners KM, et al. (2010) Southern ocean iron fertilisation by baleen whales and Antarctic krill. Fish Fish 11: 203–209.
70. Coll M, Lotze HK (forthcoming) Ecological indicators and food-web models as tools to study historical changes in marine ecosystems In: K. Schwerdtner Máñez and B. Poulsen, editors. Perspectives on Oceans Past: A Handbook of Marine Environmental History. Amsterdam: Springer Publishers.
71. Broughton JM (1997) Widening diet breadth declining foraging efficiency and prehistoric harvest pressure: ichthyofaunal evidence from the Emeryville Shellmound, California. Antiguity 71: 845–862.
72. Broughton JM (2002) Prey spatial structure and behavior affect archaeological tests of optimal foraging models: examples from the Emeryville Shellmound vertebrate fauna. World Archeol 34: 60–83.
73. Josephson E, Smith TD, Reeves RR (2008) Depletion within a decade: The American 19th-century North Pacific right whale fishery. In: D. J Starkey, P Holm and M Barnard, editors. Oceans Past: Management insights from the history of marine animal populations. London: Earthscan Research Edition. 133–147.
74. Ferretti F, Myers RA, Serena F, Lotze HK (2008) Loss of large predatory sharks from the Mediterranean Sea. Conserv Biol 22: 952–964.

75. Poulsen B (2008) Dutch Herring: An Environmental History, c. 1600–1860. Amsterdam: Aksant.
76. Glaser M, Christie P, Diele K, Dsikowitzky L, Ferse S, et al. (2012) Measuring and understanding sustainability-enhancing processes in tropical coastal and marine social–ecological systems. Curr Opin Environ Sustain 4: 300–308.
77. Christensen V, Walters C (2004) Ecopath with Ecosim: methods, capabilities and limitations. Ecol Modell 72: 109–139.
78. Dambacher JM, Li HW, Rossignol PA (2003) Qualitative predictions in model ecosystems. Ecol Modell 161: 79–93.
79. Fulton EA (2010) Approaches to end-to-end ecosystem models. J Mar Syst 81: 171–183.
80. Plagányi ÉE, Punt AE, Hillary R, Morello EB, Thébaud O, et al. (2012) Multispecies fisheries management and conservation: tactical applications using models of intermediate complexity. Fish Fish.
81. Williams RJ, Martinez ND (2000) Simple Rules Yield Complex Food Webs. Nature 404: 180–183.
82. Steneck RS, Vavrinec J and Leland AV (2004) Accelerating Trophic-Level Dysfunction in Kelp Forest Ecosystems of the Western North Atlantic. Ecosystems 7: 323–332.
83. Pauly D, Christensen V, Daalsgard J, Froese R, Torres FC (1998) Fishing down marine food webs. Science 279: 860–863.
84. Lotze HK, Coll M, Dunne J (2011) Historical changes in marine resources, food-web structure and ecosystem functioning in the Adriatic Sea. Ecosystems 14: 198–222.
85. Pitcher TJ (2005) Back-to-the-future: a fresh policy initiative for fisheries and a restoration ecology for ocean ecosystems. Philos Trans R Soc Lond B Biol Sci 360: 107–121.
86. Pitcher T, Heymans JJ, Vasconcellos M (2002) Ecosystem models of Newfoundland for the time periods 1995, 1985, 1900, 1450. Fisheries Centre Research Report.
87. Ainsworth C, Pitcher T, Heymans J, Vasconcellos M (2008) Reconstructing historical marine ecosystems using food web models: Northern British Columbia from Pre-European contact to present. Ecol Modell 216: 354–368.
88. Hauser L, Carvalho GR (2008) Paradigm shifts in marine fisheries genetics: ugly hypotheses slain by beautiful facts. Fish Fish 9: 333–362.
89. Roman J, Palumbi SR (2003) Whales Before Whaling in the North Atlantic. Science 301: 508–510.
90. Palumbi SR (2011) Whales, logbooks, and DNA. In: J. B. C Jackson, K Alexander and E Sala, editors. Shifting baselines: the past and the future of ocean fisheries. Washington DC: Island Press. 163–174.
91. Alter SE, Rynes E, Palumbi SR (2007) DNA evidence for historic population size and past ecosystem impacts of gray whales. Proc Natl Acad Sci U S A 104: 15162–15167.
92. Fontaine MC, Snirc A, Frantzis A, Koutrakis E, Öztürk B, et al. (2012) History of expansion and anthropogenic collapse in a top marine predator of the Black Sea estimated from genetic data. Proc Natl Acad Sci 109: E2569–E2576.
93. Thurstan RH, Buckley SM (2014) Oral histories: informing natural resource management using perceptions of the past. In: K. Schwerdtner Máñez and B Poulsen, editors. Perspectives on Oceans Past: A Handbook of Marine Environmental History. Amsterdam: Springer Publishers. (In Press).
94. Johannes RE, Neis B (2007) The value of anecdote. In: N Haggan, B Neis and I. G Baird, editors. Fishers Knowledge in Fisheries Science and Management. Paris: UNESCO. 41–58.
95. Saenz-Arroyo A, Roberts CM, Torre J, Carino-Olvera M, Enriquez-Andrade RR (2005) Rapidly shifting environmental baselines among fishers of the Gulf of California. Proc R Soc Lond Ser B-Biol Sci 272: 1957–1962.
96. Ainsworth CH, Pitcher TJ, Rotinsulu C (2008) Evidence of fishery depletions and shifting cognitive baselines in Eastern Indonesia. Biol Conserv 141: 848–859.
97. Choos PS, Nowak BS, Kusakabe K, Williams MJ (2008) Guest Editorial: Gender and Fisheries. Development 51: 176–179.
98. Kleiber D, Harris LM, Vincent ACJ (2014) Improving fisheries estimates by including women's catch in the Central Philippines. Can J Fish Aquat Sci: 1–9.
99. Schwerdtner Máñez K, Pawelussen A (2014) Fish is women's business too – looking at marine resource use through a gender lens. In: K. Schwerdtner Mánez and B Poulsen, editors. Perspectives on Oceans Past: A Handbook of Marine Environmental History. Springer Publishers.
100. Harper S, Zeller D, Hauzer M, Pauly D, Sumaila UR (2013) Women and fisheries: Contribution to food security and local economies. Marine Policy 39: 56–63.
101. Williams MJ (2008) Why look at fisheries through a gender lens? Development 51: 180–185.
102. Nadal-Klein J, Davis DL (1988) To work and to weep. In: I. o. S. a. E Research and M. U. o Newfoundland, editors. Social and Economic Papers. St. John's, Newfoundland: Institute of Social and Economic Research.
103. Hauzer M, Dearden P, Murray G (2013) The fisherwomen of Ngazidja island, Comoros: Fisheries livelihoods, impacts, and implications for management. Fish Res 140: 28–35.
104. Vunisea A (2005) Womens changing roles in the subsistence fishing sector in Fidji. In: I Novaczek, J Mitchell and J Vietayaki, editors. Pacific Voices: Equity and Sustainability in Pacific Island Fisheries. Suva, Fiji: Institute of Pacific Studies, University of South Pacific. 89–105.
105. Bennett E (2005) Gender, fisheries and development. Marine Policy 29: 451–459.
106. McCay B (1993) Fisherwomen, fisheries policy, and maritime anthropology. Reviews in Antropology 22: 77–89.
107. Meltzoff SK (1995) Marisquadoras of the Shellfish Revolution: The Rise of Women in Co-management on Illa de Arousa, Galicia. Journal of Political Ecology 2: 20–38.
108. Yodanis CL (2000) Constructing gender and occupational segregation: a study of women and work in fishing communities. Qual Sociol 23: 267–290.
109. Maxwell K, MacDiarmid AB (2014) Oral histories of marine fish and shellfish state and use in the Hauraki Gulf and along the Otago-Catlins coast 1940–2008. New Zealand Aquatic Environment and Biodiversity Report. (In press).
110. Fabinyi M (2007) Illegal fishing and masculinity in the Philippines. A look at the Calamianes Islands in Palawan. Philipp Stud 55: 509–529.
111. Allison EH (2013) Maritime Masculinities – and why they matter for management. People and the Sea VII. Amsterdam.
112. Perri M (2009) Ruined and Lost': Spanish Destruction of the Pearl Coast in the Early Sixteenth Century. Environ Hist Camb 15: 129–161.
113. Cushmann GT (2003) The lords of guano: Science and the management of Peru's marine environment, 1800–1973. PhD thesis Austin: University of Texas.
114. Eddy TD, Gardner JPA, Pérez-Matus A (2010) Applying Fishers' Ecological Knowledge to Construct Past and Future Lobster Stocks in the Juan Fernández Archipelago, Chile. PLoS ONE 5: e13670.
115. Griffiths MH (2000) Long-term trends in catch and effort of commercial linefish off South Africa's Cape Province: snapshots of the 20th century. Afr. J. Mar. Sci. 22: 81–110.
116. Attwood CG, Farquhar M (1999) Collapse of linefish stocks between Cape Hangklip and Walker Bay, South Africa. Afr. J. Mar. Sci. 31: 415–432.
117. McKenzie MJ (1946) The historical development of harbours and fisheries in South Africa.
118. Day JA (2000) Marine and estuarine studies in South Africa: a historical perspective. Transactions of the Royal Society of South Africa 55: 101–105.
119. Bunce M, Rodwell LD, Gibb R, Mee L (2008) Shifting baselines in fishers' perceptions of island reef fishery degradation. Ocean Coastal Manage 51: 285–302.
120. Oliver PG, Lynch T (2004) A historical perspective of the marine biota of Rodrigues. Journal of Natural History 38: 2927–2935.
121. Hacquebord L (2001) Three centuries of whaling and walrus hunting in Svalbard and its impact on the Arctic ecosystem. Environ Hist Camb: 169–185.
122. Emslie SD, Patterson WP (2007) Abrupt recent shift in δ13C and δ15N values in Adélie penguin eggshell in Antarctica. Proc Natl Acad Sci U S A 104: 11666–11669.
123. William K (1997) Abrupt mid-twentieth-century decline in Antarctic sea-ice extent from whaling records. Nature 389: 57–60.
124. Sanger CW (1991) Environmental factors affecting 17th–19th century whaling in the Greenland Sea. Polar Rec 27: 77–86.
125. Ramirez-Llodra E, Brandt A, Danovaro R, De Mol B, Escobar E, et al. (2010) Deep, diverse and definitely different: unique attributes of the world's largest ecosystem. Biogeosciences 7: 2851–2899.
126. Tåning ÅV (1929) The Danish Circumnavigation Expedition. Science 69: 191–194.
127. Boomgaard P (2005) Resources and people of the sea in and around the Indonesian Archipelago, 900–1900 In: P. Boomgaard and D. Henley, editors. Muddied Waters Historical and contemporary perspectives on management of forest and fisheries in island Southeast Asia Leiden: KITLV Press Koninklijk Instituut voor Taal-, Land- en Volkenkunde. 98–119.

14

An Ecosystem-Based Approach to Assess the Status of a Mediterranean Ecosystem, the *Posidonia oceanica* Seagrass Meadow

Sébastien Personnic[1]*, Charles F. Boudouresque[1], Patrick Astruch[2], Enric Ballesteros[3], Sylvain Blouet[4], Denise Bellan-Santini[5], Patrick Bonhomme[2], Delphine Thibault-Botha[1], Eric Feunteun[6], Mireille Harmelin-Vivien[1], Gérard Pergent[7], Christine Pergent-Martini[7], Jérémy Pastor[8], Jean-Christophe Poggiale[1], Florent Renaud[5], Thierry Thibaut[1], Sandrine Ruitton[1]

1 Aix-Marseille University, Mediterranean Institute of Oceanography (MIO), Université de Toulon, CNRS/INSU, IRD, UM 110, Marseille, France, **2** GIS Posidonie, Pytheas Institute, Aix-Marseille University, Marseille, France, **3** Centre d'Estudis Avançats de Blanes - CSIC, Blanes, Spain, **4** Aire marine protégée de la côte Agathoise, site natura 2000, Agde, France, **5** Aix-Marseille University, Institut Méditerranéen de Biodiversité et d'Ecologie (IMBE), UMR 7263, Station Marine d'Endoume, Marseille, France, **6** Museum National d'Histoire Naturelle, UMR 7208, Station Marine de Dinard, France, **7** Equipe Ecosystèmes Littoraux, FRES 3041, University of Corsica, Corte, France, **8** Université de Perpignan, Via Domitia, Centre de Formation et de Recherche sur les Environnements Méditerranéens, UMR 5110, Perpignan, France

Abstract

Biotic indices, which reflect the quality of the environment, are widely used in the marine realm. Sometimes, key species or ecosystem engineers are selected for this purpose. This is the case of the Mediterranean seagrass *Posidonia oceanica*, widely used as a biological quality element in the context of the European Union Water Framework Directive (WFD). The good quality of a water body and the apparent health of a species, whether or not an ecosystem engineer such as *P. oceanica*, is not always indicative of the good structure and functioning of the whole ecosystem. A key point of the recent Marine Strategy Framework Directive (MSFD) is the ecosystem-based approach. Here, on the basis of a simplified conceptual model of the *P. oceanica* ecosystem, we have proposed an ecosystem-based index of the quality of its functioning, compliant with the MSFD requirements. This index (EBQI) is based upon a set of representative functional compartments, the weighting of these compartments and the assessment of the quality of each compartment by comparison of a supposed baseline. The index well discriminated 17 sites in the north-western Mediterranean (French Riviera, Provence, Corsica, Catalonia and Balearic Islands) covering a wide range of human pressure levels. The strong points of the EBQI are that it is easy to implement, non-destructive, relatively robust, according to the selection of the compartments and to their weighting, and associated with confidence indices that indicate possible weakness and biases and therefore the need for further field data acquisition.

Editor: Arga Chandrashekar Anil, CSIR- National institute of oceanography, India

Funding: This study has been funded by the French Centre National de la Recherche Scientifique (CNRS), the French Réseau des Stations et Observatoires Marins (RESOMAR), the Muséum National d'Histoire Naturelle (MNHN, Paris), the French Agence de l'eau Rhône Méditerranée Corse through the French part of the European MSFD project (in French: Directive Cadre Stratégie pour le Milieu Marin, DCSMM). The funders had no role in study design, data collection and analysis, decision to publish, or preparation of the manuscript.

Competing Interests: The authors have declared that no competing interests exist.

* E-mail: sebastien.personnic@univ-amu.fr

Introduction

Human activities can deeply alter the environment, species composition and functioning of ecosystems. These alterations can be tracked by the use of biotic indices, i.e. species or groups of species whose function, population, or status reflect the environmental quality. Thus, biotic indices are monitored for changes in presence and abundance. The occurrence of an organism in a specific environment indicates that overall, its biological requirements are satisfied, whereas its disappearance suggests a change in the environment. Species are also monitored for changes in biochemistry, physiology or behaviour induced by environmental conditions. Biotic indices are used in terrestrial, freshwater and marine habitats, because they enable the quality of an environment to be characterized in an integrated way [1–7].

Biotic indices are widely used in the marine realm to **(i)** assess the quality of a water body, **(ii)** assess processes such as currents, sedimentation and climate under natural and anthropogenic forcing, and **(iii)** monitor the status of species or communities of interest, either emblematic species, indicators of some ecosystemic processes or indicators of pollution. Sometimes, key species and ecosystem engineers [8–10] are selected for this purpose. This is the case of the seagrass *Posidonia oceanica* [11–17].

In the European Union (EU), the so called Habitats Directive of 1992 (92/43/ECC) listed habitats and species that are used to designate areas ('Natura 2000 sites'), where they are strictly protected. While the Habitats Directive also considered the marine realm, more recently, the EU Marine Strategy Framework Directive (MSFD: 2008/56/EC) established a framework for conservation in the field of marine environmental policy. The

MSFD is considered to be the environmental pillar of the Integrated Maritime Policy adopted in 2010 by the European Commission (2010/477/EU). This directive established eleven criteria, based on the descriptors set out in the MSFD, to determine 'good environmental status': **(i)** Biological diversity is maintained, **(ii)** Introduced species are at levels that do not adversely alter the ecosystems, **(iii)** Populations of all exploited fish and shellfish are within safely biological limits, **(iv)** All elements of the food webs occur at levels capable of ensuring the long-term abundance of the species, **(v)** Human-induced eutrophication is at a minimum, **(vi)** Sea-floor integrity is at a level that ensures that the structure and functions of the ecosystems are safeguarded, **(vii)** Permanent alteration of hydrographical conditions does not adversely affect marine ecosystems, **(viii)** Concentration of contaminants are at levels not giving rise to pollution effects, **(ix)** Contaminants in fish and other seafood do not exceed levels established by Community legislation, **(x)** Properties and quantities of marine litter do not cause harm to the coastal and marine environment, and **(xi)** Introduction of energy is at levels that do not adversely affect the marine environment. The EU MSFD established a framework within which Member States agreed to take the appropriate measures to achieve or maintain good environmental status in the marine realm by the year 2020 at the latest.

A key point of the MSFD is the ecosystem-based approach. Most previous strategies only dealing with Biological Quality Elements (BQE: species and/or communities) used in the European Union Water Framework Directive (WFD) were not indicative of ecosystem status, but merely indicative of the environmental status of water bodies [18–22,17,23] for a review of the approaches for classifying and assessing quality of benthic habitats. As an example of the possible shortcomings that may arise from the WFD descriptors, a seagrass meadow characterized by normal leaf growth, shoot density and the absence of mechanical injuries would be ranked as 'good', even if deprived of some basic compartments of the ecosystem.

According to the MSFD, 'good environmental status' means that the marine environment is at a level that allows uses and activities by current and future generations, i.e. the structure, functions and processes of the constituent marine ecosystems, together with the associated physiographic, geographic, geological and climatic factors, allow those ecosystems to function fully and to maintain their resilience to human induced environmental change. Overall, marine species and marine habitats are protected, human-induced decline of biodiversity is prevented and diverse biological components function in balance.

While the ecosystem-based approach constitutes the background and the guidelines of the MSFD, the pristine state of an ecosystem is not clearly defined. Obviously, such a notion is very complex and prone to anthropocentrism; reference conditions, as observed in areas distant from known human impact, may constitute a more realistic notion. Here we try to define: **(i)** how the status of an ecosystem can be measured and how much its current status differs from 'reference conditions', and **(ii)** which parameters we need to monitor to ascertain these targets. The *Posidonia oceanica* seagrass meadow was chosen for this attempt because: **(i)** it is widely present in almost the whole of the European Mediterranean; **(ii)** it is the only marine ecosystem considered as 'priority habitat' by the EU Habitats Directive; **(iii)** its functioning is relatively well known [11,24–30]; and **(iv)** like many seagrass ecosystems in the world ocean, *P. oceanica* meadows have been impacted or lost under the influence of direct and indirect effects of human activities and are therefore regarded as threatened [31–33]. Similar ecosystem-based approaches have been attempted for fisheries (e.g. [34–38,40,41]) and for the management and conservation of ecosystem services (e.g. [34,41,42]).

Materials and Methods

The conceptual model

Posidonia oceanica (L.) Delile is a seagrass species (Magnoliophyta, kingdom Archaeplastida) endemic to the Mediterranean Sea [43], which dwells in the sublittoral zone, from mean sea level down to 30 to 40 m depth, depending upon water transparency [27,28,32]. Due to the length, up to 120 cm, and density of the leaves, the seagrass canopy decreases water movement and traps sediments [44–47]. Rhizomes resist burial by vertical growth, so that the sea bottom slowly rises. Within the sediment, the deeper parts of the rhizomes, attached leaf sheaths and roots die, but their decay is extremely slow, so that they can persist for millennia [24,27]. The terrace constituted by live and dead intertwined rhizomes, together with the sediment, which fills the interstices, is named 'matte' [27,48,49]. When *P. oceanica* dies, the matte persists (hereafter 'dead matte'), since the decay of the rhizomes proceeds extremely slowly [44,50]. *P. oceanica* is the engineer of an ecosystem of major importance in the Mediterranean Sea [27,28]. A conceptual model of the functioning of the *P. oceanica* ecosystem has been proposed [11,27,28]. Here, we use an updated version of this conceptual model (Fig. 1; C.F. Boudouresque, unpublished).

The simplified conceptual model of the functioning of the *P. oceanica* ecosystem in the north-western Mediterranean Sea used in the present study (Fig. 1) encompasses the following compartments (boxes); these compartments are listed hereafter and detailed later on:

- *Posidonia oceanica* roots and rhizomes (box 1).
- Multicellular Photosynthetic Organisms (MPOs) epibiotic on *P. oceanica* rhizomes.
- *P. oceanica* leaves (box 2).
- MPO leaf epibiota (box 3).
- Dissolved Organic Carbon (DOC).
- Pelagic microbial loop.
- Filter- and suspension-feeder leaf epibiota (box 4).
- Filter- and suspension-feeder benthic epibiota on *P. oceanica* rhizomes: the bivalve *Pinna nobilis* (box 5) and other species, e.g. bryozoans and ascidians (box 6).
- The litter detritus (essentially dead *P. oceanica* leaves and some broken rhizomes) (box 7).
- Detritus feeders 1, e.g. Amphipoda, Isopoda (crustaceans) and *Psammechinus microtuberculatus* (sea urchin).
- Detritus feeders 2 and 3, e.g. Amphipoda, Isopoda (crustaceans) and *Holothuria* spp. (sea cucumber) (box 8).
- Decomposers, namely *Bacteria*, *Archaea*, Fungi and heterotrophic stramenopiles (BAFHS) such as Labyrinthulomycota and Oomycota.
- The benthic microbial loop.
- The matte endofauna, e.g. annelids and mollusks.
- Herbivores 1, e.g. *Sarpa salpa* (teleost), *Paracentrotus lividus* (sea urchin), *Idotea* spp. and *Pisa* spp. (crustaceans) (box 9).
- Herbivores 2, e.g. Amphipoda (crustaceans), *Jujubinus* spp. and *Rissoa* spp. (gastropods).
- Predatory teleosts (e.g. *Diplodus* spp., *Sparus aurata*, *Labrus* spp. and *Symphodus* spp.), cephalopods and seastars (e.g. *Marthasterias glacialis*) (box 10).

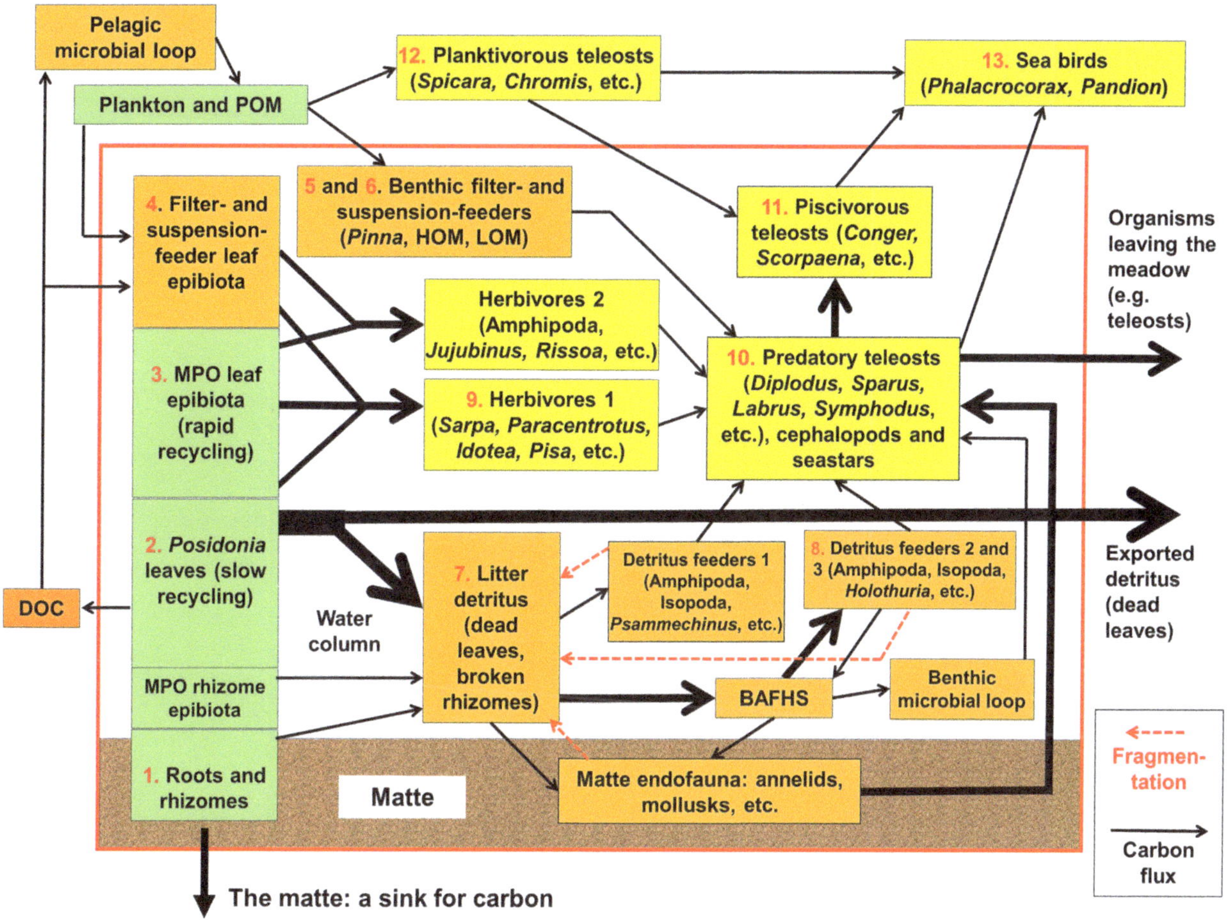

Figure 1. Conceptual model of the functioning of *Posidonia oceanica* seagrass ecosystem. For functional compartments and box numbers, see text. Primary producers are in green; filter-feeders, suspension-feeders, litter, detritus feeders, Dissolved Organic Carbon (DOC) and microbial loops are in orange; predators (including herbivores) are in yellow. The width of the arrows roughly represents the importance of the carbon flow. The proper *P. oceanica* ecosystem is included within the red rectangle. MPO: Multicellular Photosynthetic Organisms. POM: Particulate Organic Matter. From C.F. Boudouresque, unpublished.

- Piscivorous teleosts, e.g. *Conger conger*, *Scorpaena* spp. and *Serranus* spp. (box 11).
- Planktivorous teleosts of the water column, e.g. *Spicara* spp. and *Chromis chromis* (box 12).
- Sea birds, e.g. *Phalacrocorax aristotelis* ssp. *desmarestii* and *Pandion haliaetus* (box 13).
- Plankton (photosynthetic plankton and zooplankton) and non-living Particulate Organic Matter (POM).

Considered functional compartments (boxes)

Whenever possible, non-destructive methods were chosen to measure the parameters of the status of the functional compartments, as suggested by [51]. In the absence of further indications, measures are performed at a depth of ca. 15 meters (10 to 20 m), which is then considered as representative of the whole depth range [14,52]. The season of the sampling is indicated only for compartments which present seasonal variability. In the ranking of the status of the selected compartments (boxes), two cases were encountered; **(i)** a steady trend of the parameter, from very good to a low quality state; **(ii)** an upward/downward slope parameter, when the very good state corresponds to intermediate values.

***Posidonia* roots and rhizomes.** (box 1, Fig. 1). This compartment was estimated by the growth rate of vertical (orthotropic) rhizomes. The matte compartment results in carbon sequestration within the matte, which acts as a carbon sink [27,44,53,54] and was measured by means of lepidochronology [55,56]. Lepidochronology describes the annual cycle of leaf formation. A cycle includes a suite of dead leaf bases (called dead sheaths or scales) of increasing then decreasing thickness. An average of 7.5 leaves (generally between 6 and 9) are produced every year [57–60]. We considered that high and low rhizome growth rates are indicative of over sediment input or of deficit in sediment, respectively (Table 1). The highest growth rate of an orthotropic rhizome is 7 cm year^{-1}; higher sedimentation rates result in the death of the buried leaf bundle [44,61]. In contrast, deficit in sediment results in bare, non-sediment protected rhizomes, which are hence very vulnerable to storms, anchoring and trawling [27]. The nature of the substratum, namely meadows settled on rock, sand or matte, only slightly influences the growth rate of rhizomes [62] and no major differences were found at the community level [63] so that it was not necessary to adapt the scale. Thirty random *in situ* measures (growth of the rhizome corresponding to the last 8 dead leaf bases) are recommended. The obtained value was multiplied by 1.5 in order to take into account

Table 1. Relative weighting of each functional compartment (boxes; see Fig. 1 for box number) and ranges of each parameter defined for each grade of ecosystem status.

Box number	Functional compartment	Weighting (W)	Parameter	4	3	2	1	0
1	Roots and rhizomes ('rhizomes')	3	Growth of orthotropic rhizomes (mm a^{-1})	9 to 19	20 to 40	3 to 8	>40	<3
2	*Posidonia oceanica* leaves ('leaves')	5	- Density (shoots m^{-2})	≥490	489 to 370	369 to 250	249 to 130	<130
			- Cover (%)	>80	80 to 61	60 to 41	40 to 20	<20
3–4	MPOs, filter- and suspension-feeders leaf epibiota ('leaf epibiota')	4	Biomass (g DM $shoot^{-1}$) (only the 2 oldest - external - leaves)	0.3 to 0.7	0.1 to 0.2	0.8 to 1.5	<0.1	>1.5
5	Benthic filter-feeder: *Pinna nobilis* (bivalve) ('Pinna')	2	Density (individuals 100 m^{-2})	>3.0	3.0 to 1.1	1.0 to 0.6	0.5 to 0.1	<0.1
6	Other benthic filter- and suspension-feeders ('HOM LOM')	2						
	- HOM		- Density (m^{-2})	<0.1	0.1 to 0.9	1.0 to 1.9	2.0 to 5.0	>5.0
	- LOM		- Density (m^{-2})	>10.0	10.0 to 5.1	5.0 to 1.1	1.0 to 0.1	<0.1
7	Litter detritus: dead leaves and broken rhizomes ('litter')	2	g DM m^{-2}	>350	350-251	250-151	150-51	≤50
8	Detritus-feeders 2 and 3 (*Holothuria* spp.) ('Holothuria')	2	Individuals 10 m^{-2}	1.0 to 4.9	0.2 to 0.9	5.0 to 24.9	<0.2	≥25.0
9	Herbivores 1 ('herbivores')	5	- Density of *Paracentrotus lividus* (individuals m^{-2})	1.0 to 4.9	0.1 to 0.9	5.0 to 9.9	<0.1	≥10
			- Grazing index (% leaves)	30 to 59%	5 to 29%	60 to 95%	<5%	>95%
10	Predatory teleosts, cephalopods and seastars ('predators')	5	kg teleosts WM 100 m^{-2}	>1.5	1.5 to 1.1	1.0 to 0.6	0.5 to 0.3	<0.3
11	Piscivorous teleosts ('piscivores')	5	kg teleosts WM 100 m^{-2}	>1.0	1.0 to 0.6	0.5 to 0.3	0.2 to 0.1	<0.1
12	Planktivorous teleosts ('planktivores')	3						
	- Zooplankton feeders		- kg teleosts WM 100 m^{-2}	>3.0	3.0 to 1.6	1.5 to 0.8	0.7 to 0.3	<0.3
	- Omnivores		- kg teleosts WM 100 m^{-2}	>3.0	3.0 to 1.6	1.5 to 0.8	0.7 to 0.3	<0.3
9–12	All teleosts ('SRDI')	3	Specific Relative Diversity Index (SRDI)	>10	10 to 8	7 to 5	4 to 3	<3
13	Sea birds	1						
	- *Phalacrocorax* spp.		- Distance to the nearest nesting site (km)	<4	4 to 7	8 to 12	13 to 17	>17
	- *Pandion haliaetus*		- Distance to the nearest nesting site (km)	<4	4 to 7	8 to 12	13 to 17	>17

Between inverted commas: abbreviated name of the functional compartment in tables 4 and 7. DM: dry mass, including calcium carbonate. HOM: indicators of high level of organic matter. LOM: indicators of low level of organic matter. MPO: Multicellular Photosynthetic Organisms. POM: Particulate Organic Matter. WM: wet mass. Status: 4 (very good) through 0 (very low).

the fact that the rhizome continues to grow slowly during the following two years (Gérard Pergent, unpublished data).

***Posidonia* leaves.** (box 2, Fig. 1). A shoot is a rhizome tip with a bundle of living leaves. Shoot density is correlated with annual leaf primary production at local scale (patch) [64,65]. Primary production is a basic parameter for the functioning of the *P. oceanica* ecosystem. It was estimated by the number of shoots per square meter, measured within a small square frame (0.16 m^2 [66,67]) with at least 20 random replicates [5]. At a less local scale, the cover rate of the meadow is rarely 100%: it is broken by more or less extensive patches of either sand or dead matte, which reduce the overall cover. Cover rate was estimated by visual observation *via* a see-through plastic slide [68,69] or *via* vertical photographs [67,70]. Thirty random measures are recommended. Low cover rate is believed to characterize a poor condition of the meadow [5]. Some types of *P. oceanica* meadow, e.g. the hill meadow and the striped meadow, exhibit a low cover rate while in pristine state; these types of meadows do not, however, occur in the study area [27]. The final *P. oceanica* leaf index, for a given site, was the arithmetical mean between the density index, and the cover index.

MPOs, filter- and suspension-feeder leaf epibiota. (boxes 3 and 4, Fig. 1). *P. oceanica* leaf epibiota, both primary producers (MPOs and diatoms) and animals, share the same habitat (the leaf surface) and are usually co-consumed by the same species. In addition, some consumers eat simultaneously epibiota and the supporting leaf [68,71]. For these reasons, epibiota will be considered here as a single compartment. The colonisation of leaves by epibiota is a function of leaf age, the youngest leaves, in the center of the shoot, being less colonized, while the oldest, external leaves are the most colonized; in addition, leaf tips are more colonized than the lower parts of the leaves [72]. Leaf epibiota cover is believed to provide information on water quality, especially nutrient concentration in seawater [5,73,74]; however, it also reflects the herbivore pressure, epibiota biomass decreasing when macrograzer abundance increases [26]. By convention, the epibiota biomass was estimated on the two oldest (external) leaves, in July, on 30 randomly localized shoots (Table 1).

Filter- and suspension-feeder epibiota on *P. oceanica* rhizomes. (boxes 5 and 6, Fig. 1). A number of benthic filter- and suspension-feeders dwell on *P. oceanica* rhizomes, sometimes within the matte. They belong to bryozoans, hydroids, sponges, annelids (e.g. *Sabella spallanzani*), ascidians (e.g. *Halocynthia papillosa*, *Phallusia mammillata*, *P. fumigata*), gastropods and bivalves (such as the fan mussel *Pinna nobilis*) [75–79]. *Pinna nobilis* (box 5) density was estimated along 20 transects 10-m long and 1-m wide. Filter- and suspension-feeders other than *P. nobilis* (box 6) are indicators either of: **(i)** high level of organic matter in the water (hereafter HOM; e.g. *Sabella spallanzani*, *Phallusia mammillata*, *P. fumigata*, Didemnidae [80]); or **(ii)** low level of organic matter (hereafter LOM; e.g. bryozoans, sponges, *Halocynthia papillosa*, *Antedon mediterranea* [80]). HOM and LOM indicators are assessed within the same quadrats as the sea urchin *P. lividus* (see below, box 9). For colonial species, the number of colonies was taken into account, as suggested by [81]. For non-colonial species, the number of individuals was counted. Only individuals and colonies over 5 cm in diameter and/or height were considered. The final filter- and suspension-feeder epibiota (other than *P. nobilis*; box 6) on rhizomes index, for a given site, was estimated as the arithmetical mean between HOM and LOM indices.

Litter detritus. (box 7, Fig. 1). The litter detritus mass corresponds essentially to shed dead *P. oceanica* leaf blades and the epibiota they harbour, and some broken rhizomes. It therefore represents a kind of necromass [82]. Drift MPOs, exported from sublittoral reef habitats, can also occur within the litter. The litter detritus mass was estimated in July, within 5 randomly localized 0.1 m^2 quadrats; the litter was sucked up by an underwater vacuum cleaner. Litter detritus were dried at 50°C in an oven to constant weight.

Detritus-feeders 2 and 3. (box 8, Fig. 1). Detritus-feeders constitute a complex set of compartments. Here, the macro-detritus feeders *Holothuria* spp. were used as a proxy of detritus-feeders 2 and 3, as they are easy to count. Several species can be present, e.g. *H. polii* and *H. tubulosa* [83]. The abundance of *Holothuria* spp. was measured within the same quadrats as the sea urchin *P. lividus* (see below, box 9).

Herbivores 1. (box 9, Fig. 1). Macro-herbivores considered in this compartment were the sea urchin *Paracentrotus lividus* and the teleost *Sarpa salpa*, at 5 meters depth or, if the meadow is not present at this depth, at the upper limit of the meadow. They consume *P. oceanica* leaves, together with their epibiota, if present [68,71,84]. Other herbivores graze *P. oceanica* leaves, such as the spider crabs *Pisa* spp. and the isopod *Idotea hectica* [27,84], but their reduced size and habit make them more difficult to quantify and they were not considered. The abundance of *P. lividus* was assessed within 1-m^2 quadrats, with 30 replicates randomly localized. The census only considered individuals >3 cm (test diameter without spines), because small individuals can be hidden within the matte, generating bias in the census. A second proxy of the macro-herbivore pressure was the grazing index, i.e. the percentage of intermediate and adult leaves (*sensu* [85,86]) exhibiting bite scars due to *S. salpa* (all the intermediate and adult leaves of 30 shoots randomly localized). Bite scars were carefully distinguished from broken leaves, the latter being related to hydrodynamism. Most bite scars are due to the *S. salpa* browsing, and they are easy to distinguish from those due to other macro-herbivores [26,49]. The final macro-herbivore index was, for a given site, the arithmetical mean between the *Paracentrotus* index and the grazing index.

Predatory teleosts and cephalopods, piscivorous teleosts, planktivorous teleosts. (boxes 9 in part, 10, 11 and 12, Fig. 1). Teleost fishes associated with seagrass beds occupy different positions in and above the canopy during the day and at night, spend more or less time in this habitat depending on their life cycle, and naturally fluctuate in abundance and biomass with depth, season and years [87–90]. They participate actively in the functioning of the *P. oceanica* ecosystem (Fig. 1), but the perception of the composition and trophic structure of their assemblages largely depends on the methodology used [91]. Moreover, species richness, abundance and biomass of teleosts are favoured by the ecotones induced by the presence of rocky or sandy substrates in the middle of *Posidonia* beds. So these variations and methodological biases have to be taken into account to when measuring these compartments in order to assess the 'environmental status' of a site. These compartments were estimated *via* visual censuses at a standardized day time (10:00 to 16:00 UT) during the warm season (summer-autumn), preferentially in uniform beds (at least a long way from rocky substrates). All teleosts were counted within ten linear and 5-m wide transects, each census lasting 5 minutes. Around 50 m were covered so that each transect represents a surface area of nearly 250 m^2. Total length (at the nearest 2 cm) of individuals and the number of individual per species were noted. The Specific Relative Diversity Index (SRDI) is the mean number of species met with per transect. These data enable calculation of **(i)** predator biomass (predatory teleosts other than piscivorous and planktivorous; box 10), **(ii)** top predator biomass (piscivorous teleosts; box 11), **(iii)** planktivorous teleost biomass (box 12) and **(iv)** the Specific Relative Diversity Index (SRDI). The planktivorous teleost biomass is divided into 2 categories: the exclusive zooplankton feeders (*Chromis chromis*, *Spicara smaris*, *S. maena*, *Atherina* spp.) and the omnivorous feeders, which consume both zooplankton and POM (*Boops boops*, *Oblada melanura*). Some top predators (e.g. *Conger conger*, *Scorpaena* spp.) are active only during night time while *Serranus* spp. are active by day and thus more present in our visual counts. Some predators (e.g. *Symphodus rostratus*), though diurnal, are often hidden within the *P. oceanica* leaf canopy. As a result, their counts were underestimated to a greater or lesser extent. The parameter ranges within the status scale (Table 1) took into account these biases.

Sea birds. (box 13, Fig. 1). Most sea birds do not directly interact with the *P. oceanica* ecosystem, as they feed on offshore pelagic species, such as *Larus* spp. and *Puffinus* spp. The only exceptions are shags *Phalacrocorax* spp. and the osprey *Pandion haliaetus*. Shags can dive down to the benthic seagrass meadow; they mainly feed on pelagic planktivorous teleosts (e.g. *Spicara smaris*, *Chromis chromis*), but benthic teleosts (e.g. *Diplodus* spp., *Lithognathus mormyrus*, *Scorpaena notata*, *Serranus scriba* and *Symphodus mediterraneus*) account for 35% of the diet [92–94]. Osprey is an opportunistic fish-eating bird of prey. In Corsica, which harbours the only population within the study area, it mainly consumes mugilids (73% of the captures), together with *Diplodus* spp. (13%) and *Sarpa salpa* (11%) [95–97]. The sea bird compartment was estimated *via* the distance of the nearest *Phalacrocorax* spp. and *Pandion haliaetus*, respectively, nesting sites from the study site. As far as shags are concerned, the mean maximum foraging range is 16 km [95–97] (Table 1).

Some of the above-mentioned compartments correspond to inputs into the *P. oceanica* ecosystem from the pelagic habitat: plankton, POM and planktivorous teleosts. Outputs are also to be considered: **(i)** ca. 15–30% of the net primary production (NPP) corresponds to roots, rhizomes and dead sheaths buried and sequestrated within the matte [24,28,98,99]; **(ii)** ca. 6 to 50% of the NPP represents dead leaves that are exported as detritus towards beaches and adjacent habitats [28,31,98,100,101]; and **(iii)** a number of organisms (e.g. teleosts, crustaceans) leave the meadow, either temporarily, to feed in adjacent habitats, or permanently, to reach their adult quarters [87,102–104].

The Ecosystem-Based Quality Index (EBQI)

The rationale governing our ecosystem-based approach is trying to quantify and assess each compartment (box) of the conceptual model by means of a set of parameters, to balance their relative weighting and by using a simple algorithm to calculate a rank for the ecosystem status within a given area, matching the five Ecological Statuses of the Water Framework Directive (WFD), from bad to high. A great variety of parameters are available for the assessment of each compartment (box). Many of them are redundant. Others have been poorly used so that data are not available for most areas. For this reason a limited set of relevant parameters was selected for a restricted set of compartments.

Each parameter was assessed by means of a semi-quantitative scale (4 through 0), from very good (4) to very low (0). Calibration of the scale was based upon the available literature (e.g. [105]), including grey literature and expert judgment (the personal knowledge of the authors) based upon a Delphi process [106]. The highest grade (4) corresponds to the ecosystem status in the best-protected areas of well implemented MPAs, e.g. the Medes Islands reserve (Catalonia, Spain), the Port-Cros National Park (continental France), the Scandola reserve and the Bouches de Bonifacio reserve (Corsica, France).

Boxes were balanced, according to their relative weighting (W) in the ecosystem functioning, from 5 (highest weighting) to 1 (lowest weighting). The general principle in the ranking of the weighting of a box was that the boxes localized at the very base (bottom up control by primary producers), the herbivores (box 9) and the boxes localized at the very top (top down control via cascade effect) of the model were regarded as of major importance (with the exception of box 13), while intermediate ones were less weighted. Wasp-waist control has not been evidenced in that ecosystem [11,107]. The grade of each considered box was given by multiplying its status S (0 through 4) and weighting W (1 through 5), and therefore they are graded from 0 through 20. The grades of all considered boxes were added up, which gave the final grade of the ecosystem status (Ecosystem Based Quality Index, hereafter EBQI) at a given site. For practical purposes, the EBQI was converted to a scale from 0 to 10:

$$EBQI = \left[\sum_{i=1}^{13} (W_i \times S_i) / \sum_{i=1}^{13} (W_i \times S_{\max}) \right] \times 10$$

Where: W_i is the weighting of the box i, S_i the status of the box i, S_{max} the highest possible grade (= 4) for a box and i is the number of the box (1 through 13).

Five ecological status classes, from high to bad, according to the practice of the WFD, were delineated: bad (EBQI<3.5), poor (3.5≥EBQI<4.5), moderate (4.5≥EBQI<6), good (6.0≥EBQI<7.5) and high (EBQI≥7.5).

Since the box weightings were supported by partly subjective arguments (see above), we aimed to analyse the effect of the weighting choice on the EBQI and the ranking of the sites. In order to achieve our aim, we perturbed each weighting value and determined the new ranking obtained with the perturbed weightings. The perturbation on each weighting was obtained as follows. We first defined the maximum amplitude of the perturbation for all weightings and we then defined, for each weighting, a random perturbation, according to a uniform law between 0 and the maximum amplitude of perturbations. We then added or subtracted this perturbation term to or from the corresponding weighting. If the new weighting was less than 1, it was then set as equal to 1. If the new weighting was greater than 5, it was then set as equal to 5. We ended up with weightings between 1 and 5, close to the original ones if the maximum perturbation amplitude was low, and randomly selected otherwise. This has been calculated as follows: the perturbation method described previously was repeated 1,000 times and, for each site, we determined whether the rank of the site was the same as the initial one or if it had changed. An index of similarity was produced for each site, which was equal to 100% when the rank of the site was always unchanged after perturbation and 0% if it always changed.

For each compartment (box) status at each site, a Confidence Index (CI) was proposed (Table 2). The reason for the CI is **(i)** that data for one or several compartments may be missing or of poor quality in some sites, **(ii)** the reliability of available data may be different between compartments and sites, and **(iii)** it is worth drawing the attention of managers and scientists to those compartments that are poorly known and which merit further field studies. The grade of each considered compartment was given by its CI (0 through 4) and by its weighting (1 through 5), and therefore they are graded from 0 through 20. The grades of all considered compartments were added up, which gave the final mark of the CI at a given site. For practical purposes, the CI was converted to a scale from 0 to 4:

Table 2. Criteria to assess the Confidence Index (CI) of the status of a compartment (box).

CI	Criteria
4	Field data available, recent and suitable with the recommended methods
3	Field data recent, partially completed with expert judgment
2	No quantitative field data but recent expert judgment
1	No quantitative field data, but ancient expert judgment
0	No quantitative field data and no suitable expert judgment

$$CI_{EBQI} = \left[\sum_{i=1}^{13} (W_i \times CI_i) / \sum_{i=1}^{13} (W_i \times CI_{\max}) \right] \times 4$$

Where W_i is the weighting of the box i, CI_i the Confidence Index of the box i, CI_{max} the highest possible Confidence Index (=4) for a compartment (box) and i is the number of the box (1 through 13).

In order to test the efficiency of the proposed method, it was applied to seventeen sites (Table 3) with a variety of available data (published, unpublished, expert judgment). A site is defined as a *P. oceanica* meadow, from its upper limit down to the lower limit, covering in the order of a dozen to several hundred hectares. The sites are localized in the north-western Mediterranean Sea, including continental France (French Riviera, Provence, French Catalonia), Corsica, Balearic Islands and Spanish Catalonia. This area is considered as homogenous and is consistent with the marine subregions, as defined in the MSFD. These sites also cover of a wide range of human pressures, from lesser impact, within well implemented Marine Protected Areas (MPAs), such as the Port-Cros National Park, to highly impacted areas due to different disturbance and/or stress sources (pollution, overfishing, fish farms, port facilities, anchoring and mooring; see table 3).

Results

EBQI assessment

The results of the EBQI application for assessing the status of the *P. oceanica* ecosystem functioning in the sites across the NW Mediterranean area (Table 3) are presented in Table 4. An example of calculation of the EBQI is given (Table 5). The content of Table 4 is based upon unpublished observations from the authors of the present article, together with some published data [108–111] and whenever necessary (CI<4) on expert judgment; when several co-authors of the present work were involved in the assessment of a site, the Delphi method [106] was used for assessing the status of each compartment (box). According to the EBQI, the 17 study sites were placed within five Ecological Status classes, from Bad to High:

- Bad (EBQI<3.5): Sitges.
- Poor (3.5≥EBQI<4.5): Niolon, Gulf of Giens and Porquerolles Island northern coast.
- Moderate (4.5≥EBQI<6): Villefranche Bay, Saint-Cyr Bay, Plateau des Chèvres (Marseilles), Prado Bay (Marseilles), Valincu Gulf, Tossa de Mar, Scandola (Elbu Bay) and Peyrefite Bay.

Table 3. Sites used to test the proposed ecosystem-based approach to assess the status of the *P. oceanica* ecosystem. MPA: Marine Protected Area. NTZ: No Take Zone.

Site	Region	Protection status	Pressure
Espardell	Balearic Islands (Spain)	MPA, Natura 2000	Artisanal fishery
Sitges	Spanish Catalonia	Natura 2000	Pollution, artisanal and recreational fishery, sedimentation
Tossa de Mar	Spanish Catalonia	Natura 2000	Artisanal and recreational fishery
Medes Islands	Spanish Catalonia	MPA, NTZ, Natura 2000	River mouth
Peyrefite Bay	French Catalonia	MPA[a]	Artisanal fishery, anchoring[b]
Niolon (Côte Bleue)	Provence (France)	MPA, Natura 2000	River mouth, artisanal fishery, trawling
Prado Bay, Marseilles	Provence (France)		Coastal development, artisanal fishery, nutrients
Plateau des chèvres, Marseilles	Provence (France)	MPA[c], Natura 2000	Artisanal fishery, sewage outfall,
Saint Cyr Bay	Provence (France)		Coastal development, artisanal fishery, sewage outfall
Gulf of Giens	Provence (France)		Sewage outfall, artisanal fishery, trawling
Porquerolles Island, northern coast	Provence (France)	MPA, Natura 2000	Trawling, artisanal fishery, anchoring
Porquerolles Island, southern coast	Provence (France)	MPA, Natura 2000	Trawling, artisanal fishery
Bagaud Pass, Port-Cros Island	Provence (France)	National Park, MPA, Natura 2000	Artisanal fishery, anchoring
Port-Cros Island, southern coast	Provence (France)	National Park, MPA, Natura 2000	Artisanal fishery
Villefranche-sur-Mer Bay	French Riviera		Coastal development, sewage outfall artisanal fishery, anchoring
Scandola, Elbu Bay	West Corsica (France)	MPA, Natura 2000	Artisanal fishery
Valincu Gulf	West Corsica (France)		Artisanal fishery

[a]This area is close to the Natural Marine Reserve of Cerbère-Banyuls. Since October 2011, this area has been included within a natural marine park ('Parc naturel marin du golfe du Lion').
[b]Since 2010, anchoring is banned.
[c]Since May, 2012, this area has been included within a National Park ('Parc national des Calanques'). However, it is unlikely that this new status would have already resulted in a perceptible effect.

Table 4. Ecosystem status (EBQI) and Confidence Index (CI) for each sampling site, and scale value and CI of each parameter used to obtain them.

Box	1 Rhizomes	2 Leaves	3–4 Leaf epibiota	5 Pinna	6 HOM LOM	7 Litter	8 Holothuria	9 Herbivores	10 Predators	11 Piscivores	12 Planktivores	9–12 (SRDI)	13 Sea birds	EBQI and (CI)
Espardell	4(0)	4(4)	3(3)	3(4)	3(0)	3(0)	3(0)	3(3)	1(4)	1(4)	1(4)	3(4)	2(3)	6.4(2.9)
Sitges	2(0)	0(4)	0(3)	0(3)	2(0)	2(0)	2(3)	3.5 (3.5)	0(3)	0(3)	1(3)	0(3)	0(3)	2.3(2.7)
Tossa de Mar	2(3)	3(4)	4(4)	0(3)	2(0)	2(0)	4(2)	3(2.5)	2(4)	0(4)	2(4)	3(4)	1(2)	5.6(3.2)
Medes Islands	2(3)	3.5(4)	4(4)	3(3)	2(0)	2(0)	2(0)	2.5(3)	4(2)	4(2)	3(2)	4(2)	2(2)	7.9(2.4)
Peyrefite Bay	2(0)	3.5(4)	2(0)	4(4)	2(0)	2(0)	2(0)	2(4)	3(3)	1(3)	1.5(3)	4(3)	0(2)	5.8(2.3)
Niolon	2(2)	2.5(4)	2(2)	0(2)	1.5(2)	1(0)	3(0)	2(1.5)	1(2)	0(0)	2(0)	2(2)	1(4)	3.9(1.7)
Prado Bay	2(4)	2.5(3)	2(2)	0(2)	2.5(2)	2(2)	3(2)	2.5 (1.5)	3(2)	1(2)	1.5(2)	3(2)	2(4)	5.3(2.3)
Plateau des Chèvres	2(4)	2.5(4)	4(3)	0(2)	1.5(2)	2(2)	3(2)	2.5 (1.5)	2(4)	1(4)	0.5(2)	2(4)	2(2)	5.0(3.1)
Saint-Cyr Bay	1(2)	3(3)	2(0)	1(0)	2(0)	2(2)	2(0)	2(0)	2(0)	2(0)	2(0)	2(0)	0.5(4)	4.9(0.7)
Gulf of Giens	3(4)	4(4)	2(1)	2(0)	2(0)	1(0)	3(2)	1.5(2)	1(4)	0(2)	1(2)	1(3)	0.5(4)	4.3(2.4)
Porquerolles Island northern coast	3(3)	2(4)	3(3)	2(3)	2(3)	0(2)	1(2)	1.5(1)	1(2)	1(2)	2(2)	2(2)	1(4)	4.3(2.4)
Porquerolles Island southern coast	3(3)	4(5)	4(3)	3(3)	3(3)	3(2)	2(2)	2(1)	2(2)	2(2)	3(2)	3(2)	1(4)	6.9(2.6)
Bagaud Pass, Port-Cros Island	4(2)	3(4)	2(4)	4(4)	3(2)	4(2)	4(2)	3(2.5)	3(2)	2(2)	3(2)	4(2)	1(4)	7.6(2.6)
Port-Cros Island southern coast	4(0)	4(4)	4(3)	4(2)	3(2)	4(0)	3(0)	3.5(0)	4(2)	4(0)	3(2)	4(2)	1.5(4)	9.3(1.6)
Villefranche-sur-Mer Bay	2 (1)	1.5 (4)	2(1)	1(2)	3(2)	0(0)	0(2)	2(1)	3(0)	2(1)	1.5(1)	4(0)	0(4)	4.8(1.3)
Scandola, Elbu Bay	4(4)	3(4)	1(3)	4(4)	3(2)	2(2)	2(2)	2(1.5)	2(2)	1(2)	1.5(2)	3(3)	4(4)	5.7(2.6)
Valincu Gulf	(4.4)	3(4)	2(3)	2(2)	2(0)	2(2)	3(2)	2(0)	2(0)	2(0)	2(0)	2(0)	1(0)	5.4(1.4)

Boxes 1 through 13 and SRDI. The status and the Confidence Index (in brackets; CI) is indicated for each box. EBQI ranges from 0 to 10, while the CI is graded from 0 to 4).

Table 5. Example of calculation of the EBQI at the site Espardell (Balearic Islands).

Box number	Functional compartment	Weighting (W)	Parameter	Status and mean status when 2 parameters (S)	CI	Status grade: W×S	CI grade: W×CI
1	Roots and rhizomes ('rhizomes')	3	Growth of orthotropic rhizomes (mm a^{-1})	4	0	12	0
2	*Posidonia oceanica* leaves ('leaves')	5	- Density (shoots m^{-2})	(4)	(4)	(20)	(20)
			- Cover (%)	(4)	(4)	(20)	(20)
				4	4	20	20
3–4	MPOs, filter- and suspension-feeders leaf epibiota ('leaf epibiota')	4	Biomass (g DM $shoot^{-1}$)	3	3	12	12
5	Benthic filter-feeder: *Pinna nobilis* (bivalve) ('Pinna')	2	Density (individuals 100 m^{-2})	3	4	6	8
6	Other benthic filter- and suspension-feeders ('HOM LOM')	2					
	- HOM		- Density (m^{-2})	(3)	(0)	(6)	(0)
	- LOM		- Density (m^{-2})	(3)	(0)	(6)	(0)
				3	0	6	0
7	Litter detritus: dead leaves and broken rhizomes ('litter')	2	g DM m^{-2}	3	0	6	0
8	Detritus-feeders 2 and 3 (*Holothuria* spp.) ('Holothuria')	2	Individuals 10 m^{-2}	3	0	6	0
9	Herbivores 1 ('herbivores')	5	- Density of *Paracentrotus lividus* (individuals m^{-2})	(3)	(3)	(15)	(15)
			- Grazing index (% leaves)	(3)	(3)	(15)	(15)
				3	3	15	15
10	Predatory teleosts, cephalopods and seastars ('predators')	5	kg teleosts WM 100 m^{-2}	1	4	5	20
11	Piscivorous teleosts ('piscivores')	5	kg teleosts WM 100 m^{-2}	1	4	5	20
12	Planktivorous teleosts ('planktivores')	3					
	- Zooplankton feeders		- kg teleosts WM 100 m^{-2}	(1)	(4)	(3)	(12)
	- Omnivores		- kg teleosts WM 100 m^{-2}	(1)	(4)	(3)	(12)
				1	4	3	12
9–12	All teleosts ('SRDI')	3	Specific Relative Diversity Index (SRDI)	3	4	9	12
13	Sea birds	1					
	- *Phalacrocorax* spp.		- Distance to the nearest nesting site (km)	(4)	(3)	(4)	(3)
	- *Pandion haliaetus*		- Distance to the nearest nesting site (km)	0	(3)	0	(3)
				2	3	2	3
					Sum of the weighted status grades (left) and of the weighted CI grades (right) of the 13 boxes[a]	107	122
					EBQI[b] (left) and CI[c] (right) of the site	6.4	2.9

[a]The maximum value of the sum is 168 (see text).
[b]Maximum value: 10 (see text).
[c]Maximum value: 4 (see text).
CI: Confidence Index. DM: dry mass, including calcium carbonate. HOM: indicators of high level of organic matter. LOM: indicators of low level of organic matter. MPO: Multicellular Photosynthetic Organisms. POM: Particulate Organic Matter. WM: wet mass.

- Good (6.0≥EBQI<7.5): Espardell and Porquerolles Island southern coast.
- High (EBQI≥7.5): Bagaud Pass (Port-Cros Island), Medes Islands and Port-Cros Island (southern coast).

Redundancy of EBQI with already existing indices

Is the EBQI superfluous, i.e. redundant with already existing indices? A number of indices (EQR, Ecological Quality Ratio) based upon *P. oceanica* (the species itself, sometimes leaf epibiota,

Table 6. Comparison of EBQI with Ecological Quality Ratios (EQRs) based mainly upon *P. oceanica* (the organism itself) and aimed at establishing the ecological status of the seawater body.

Site	EBQI/10 (CI)	EQR/1	Type of index	Reference
Port-Cros Island, southern coast	9.3 (1.6)	0.802	PREI[a]	[14]
Medes Islands	7.9 (2.4)	0.752	POMI[b]	[12]
Scandola, Elbu Bay	5.7 (2.6)	0.802	BiPo[c]	[15]
Tossa de Mar	5.6 (3.2)	0.682	POMI	[12]
Valincu Gulf	5.4 (1.4)	0.386	PREI	[14]
		0.729	BiPo	[15]
Prado Bay, Marseilles	5.3 (2.3)	0.636	PREI	[14]
Plateau des chèvres, Marseilles	5.0 (3.1)	0.477	PREI	[14]
Saint Cyr Bay	4.9 (0.7)	0.682	PREI	[14]
Villefranche-sur-Mer Bay	4.8 (1.3)	0.280	PREI	[14]
Gulf of Giens	4.3 (2.4)	0.708	PREI	[14]
Porquerolles Island, northern coast	4.3 (2.4)	0.819	PREI	[14]
Niolon (Côte Bleue)	3.9 (1.7)	0.465	PREI	[14]
Sitges	2.3 (2.7)	0.238	POMI	[12]

[a]The metrics of PREI are: shoot density, shoot leaf surface area, ratio between epibiota biomass and leaf biomass, depth of the lower limit and type of this limit [14].
[b]The metrics of POMI are: shoot density, meadow cover, percentage of plagiotropic rhizomes, shoot leaf surface area, percentage of foliar necrosis, P, N and sucrose content in rhizomes, $\delta^{15}N$ and $\delta^{34}S$ isotopic ratio in rhizomes, N content in epiphytes, Cu, Pb and Zn content in rhizomes [12].
[c]The metrics of BiPo are: shoot density, shoot leaf surface area, lower depth limit and lower limit type [15].
PREI, POMI and BiPo indices are based upon distinct but similar metrics and range from 0 (lowest ecological status) to 1 (highest ecological status).

not the ecosystem) have been proposed and are currently routinely used, generally for the purpose of monitoring the ecological status of a water body, e.g. POMI [12], PREI [14] and BiPo [15]. They are based on different combinations of a number of parameters, e.g. shoot density, leaf surface area per shoot, depth of the lower limit of the meadow, percent cover of the dead matte and epibiota biomass. One or several of these indices are available from the literature for 13 out of the 17 study sites (Table 6). As these indices are more or less congruent [16], we have investigated the possible correlation between EBQI and the available EQR (POMI, BiPo or PREI) in these 13 sites, and between EBQI and PREI. There is no significant correlation (Spearman coefficient), either when all available EQRs are considered ($n = 13$, $r_s = 0.469$, p value $= 0.11$) or when only PREI data are concerned ($n = 9$, $r_s = -0.017$, p value $= 0.97$). This result is not unexpected: EQRs mainly assess the health of the seagrass, linked to e.g. the water quality, while EBQI measures the status of the ecosystem, linked not only to water quality but also to e.g. the overfishing. The contrasting ranks of Porquerolles Island (northern coast) through EBQI (poor) and EQR (high, first ranking, see Table 6), together with those of Valincu Gulf and Gulf of Giens, can be due to impacts other than the water quality, e.g. artisanal and recreational overfishing and anchoring of pleasure boats.

Redundancy of parameters to calculate the EBQI

Are some of the parameters used (functional compartments, boxes) redundant with other boxes or with the EBQI? Should such a redundancy exist, this could mean either that the box alone is sufficient to estimate the whole ecosystem's status, or that some boxes are useless. Obviously, the higher weighting is given to a box, the stronger is the probability of such a correlation. Some boxes (e.g. box 5 'Pinna' and box 10 'predators') are significantly correlated with others and/or with the EBQI (Table 7). However, the removal of the boxes correlated with others (boxes 5, 9, 11 and SRDI) results in changes (loss of accuracy?) in the EBQI: the five-class ranking (see below) of 53% of the study sites moves from one class to the next (result not presented). In addition, the removal of these possibly superfluous boxes would not save field time (data acquisition), since data for these boxes are acquired simultaneously with those of non-removed compartments.

Robustness of the EBQI

Is the EBQI robust, i.e. only slightly influenced by arbitrary choices in its conception and biases? Possible arbitrary choices are e.g. choice of the considered compartments (boxes) and the weighting (1 through 5) of the boxes. As far as the choice of the compartments taken into account is considered, it should be first emphasized that the choice is anything but arbitrary; boxes encompass the whole spectrum of an ecosystem functioning, from primary producers, herbivores, filter-feeders, suspension-feeders and detritus-feeders to top predators (Fig. 1). It is worth noting that the removal of up to 4 boxes (see above) only slightly changes the five-class ranking of the study sites: at the most, half of the sites move from one class to the next (e.g. good to moderate, poor to bad; results not presented). Finally, the changing of the weighting of the boxes (hereafter 'perturbation') alters the ranking of the sites; the changes increase with the importance of the perturbation (±1, ±2, fig. 2; ±3 and ±4, not presented). However, the changes due to weighting perturbation are relatively slight. In addition, an index of similarity was produced for each site, which is equal to 100% when the rank of the site is always unchanged after perturbation and 0% if it is always changed (Fig. 3). Some sites are very robust (e.g. Porquerolles Island northern coast and Bagaud Pass), other are more sensitive to the choice of weightings (e.g. Medes Islands and Tossa de Mar). The similarity ranges between 100 and 95% (perturbation ±1), 100 and 71% (±2), 100 and 57% (±3) and 100 and 12% (±4). The low values of the similarity index for high levels of perturbation of the weighting emphasize the fact that weighing the boxes is useful, despite the robustness of this parameter.

Table 7. Spearman's rank correlations between the status of each compartment (boxes 1 through 13 and SRDI), the other compartments (boxes) and the EBQI (see Table 5).

	1.Rhizomes	2. Leaves	3–4. Leaf epibiota	5. Pinna	6. HOM LOM	7. Litter	8. Holothuria	9. Herbivores	10. Predators	11. Piscivores	9–12 (SRDI)	12. Planktivores	13. Sea birds
1. Rhizomes	-												
2. Leaves	0.44	-											
3–4. Leaf epibiota	0.03	0.41	-										
5. Pinna	0.71	0.64	0.02	-									
6. HOM-LOM	0.57	0.33	−0.02	0.63	-								
7. Litter	0.44	0.55	0.29	0.48	0.51	-							
8. Holothuria	0.27	0.27	0.20	−0.09	−0.08	0.44	-						
9. Herbivores	0.07	0.03	0.20	−0.08	0.23	0.67	0.46	-					
10. Predators	−0.02	0.23	0.29	0.37	0.36	0.34	0.02	0.24	-				
11. Piscivores	0.22	0.30	0.33	0.51	0.46	0.42	−0.22	0.10	0.73	-			
12. Planktivores	0.18	0.28	0.38	0.34	0.24	0.34	0.05	0.06	0.48	0.62	-		
9–12 (SRDI)	0.16	0.32	0.30	0.56	0.59	0.41	−0.02	0.29	0.85	0.60	0.46	-	
13. Sea birds	0.37	0.22	0.37	0.12	0.19	0.36	0.30	0.26	0.20	0.18	0.08	0.19	-
EBQI	0.45	0.69	0.50	0.72	0.55	0.78	0.24	0.40	0.69	0.65	0.54	0.77	0.47

Bold italic characters: significant ($p<0.01$).

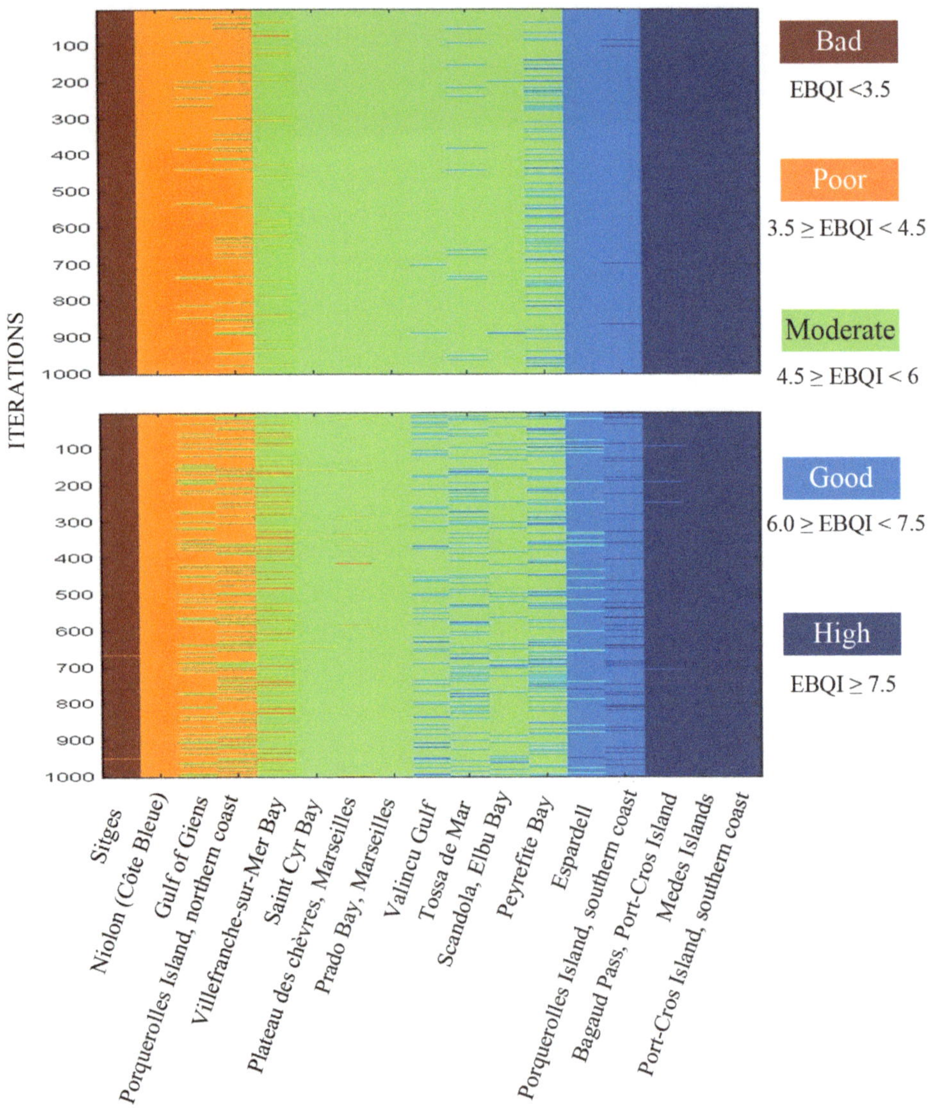

Figure 2. Robustness of the quality index (EBQI) with regard to the weighting of compartments (boxes). The 17 sites are arranged from left to right according to their growing EBQI (see Table 4) and ecological status (bad through high). Deep red = bad, orange = poor, green = moderate, light blue = good, deep blue = high. In order to test the effect of the weighting of the compartments (boxes) on the EBQI (robustness), weighting values have been randomly perturbed (above, ±1; below, ±2). 1000 iterations were performed. The change of the ecological status (bad through high) of a site, for a given iteration, is shown by the color of the new class within which it falls.

EBQI implementation

Is the EBQI excessively time-consuming or easy and rapid to implement? The acquisition of the parameters requires SCUBA diving field work. Considering that it is possible to work for around 1 hour at 15 m depth, we estimate that 6 dives involving 2 scientific divers are necessary to acquire all of the data for the assessment of one site (Table 8). For safety reasons, a diver can perform a maximum of two dives per day, so this will require 3 days of field work for a two man diving team. This field work requires expertise in seagrass and fish visual censuses. In addition, for a rapid and provisional assessment of the EBQI, already available data and expert judgment can be used (provided that the CI is specified).

Within the northern Mediterranean, the study sites are spread within three eco-regions, namely Corsica, Provence and French Riviera and Languedoc-Catalonia. Is the EBQI biased by regional environmental conditions, e.g. a higher mean temperature in Corsica and more water turbidity in Languedoc-Catalonia? The present study cannot answer this question as the sites were not randomly selected but chosen according to data availability. However, the EBQI range within each one of the three eco-regions does not exhibit obvious differences (Table 4), as far as the EBQI mean is concerned (~5.6 within each one of the three eco-regions).

Discussion

In the context of European Directives (Habitats Directive and WFD) and national regulations, a number of species and indices, based upon one or several species, have been designated to assess the quality of water bodies, the health status of emblematic species and habitats and to establish Marine Protected Areas (e.g. PMN [112], CI [113], EEI [2,3], SI [114], POMI [12], PREI [14], BiPo [15], PoSte [15], ZoNI [22] for European seagrasses

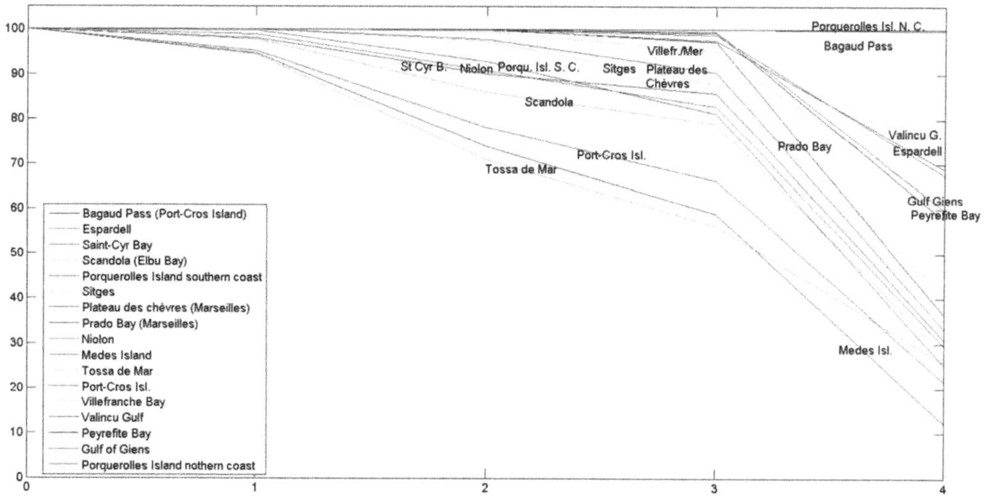

Figure 3. Robustness (index of similarity) of the quality index (EBQI) to the weighting of compartments (boxes). Percentage of times the ecological status of each site was unaltered by random perturbation of the weighting of the boxes (1,000 iterations) by ±1 through ±4. The similarity is equal to 100% when the class of the site is never changed after perturbation and 0% if it is always changed.

([17,51,115,116] for comprehensive reviews). These indices provide a valuable body of tools and information to coastal waters managers and make it possible to assess the status of a water body. However, the good quality of a water body and the apparent health of a species, whether emblematic or not, such as the seagrass *P. oceanica*, is not always indicative of the good structure and functioning of the whole ecosystem, a network of compartments, fluxes, functions, inputs and exportations. The most original contribution of the new MSFD European Directive is to provide an ecosystem-based approach to assess the ecological

Table 8. Estimation of time and diving effort needed for data acquisition within each box used in the EBQI.

Box	Proxy	Time	Dive organization
1 – *P. oceanica* roots and rhizomes	30 random *in situ* measures (growth of the rhizome corresponding to the last 8 leaf bases)	30 min	1 dive for 2 scientific divers
2 - *P. oceanica* leaves	20 random measures of shoot density in a square frame (0.16 m^2)	60 min	
	30 random measures of cover	10 min	
3 - MPO leaf epibiota	Sampling of the two oldest external leaves on 30 shoot randomly localized to estimate the epibiota biomass	10 min	
4 - Filter- and suspension-feeder leaf epibiota	Sampling of the two oldest external leaves on 30 shoot randomly localized to estimate the epibiota biomass	Same as box 3	
5 – Filter feeder benthic epibiota	Density of *Pinna nobilis* along 20 transects 10-m long and 1-m wide	120 min	1 dive for 2 scientists divers
7 - Litter detritus	Litter detritus mass collected in July, within 5 randomly localized 0.1-m^2 quadrats	90 min	1 dive for 2 scientist divers
6 – Other filter- and suspension-feeder benthic epibiota	Abundance of filter- and suspension-feeder benthic epibiota other than *Pinna nobilis* within 1-m^2 quadrats, with 30 replicates randomly localized	120 min	1 dive for 2 scientist divers
8 - Detritus feeders	Abundance of *Holothuria* spp. within 1-m^2 quadrats, with 30 replicates randomly localized	Same as box 6	
9 - Herbivores 1	Abundance of *Paracentrotus lividus* within 1-m^2 quadrats, with 30 replicates randomly localized at 5 m depth	60 min	1 dive for 2 scientist divers
	Percentage of intermediate and adult leaves exhibiting bite scars due to *S. salpa* (all the intermediate and adult leaves of 30 shoots randomly localized) at 5 m depth	30 min	
10 - Predatory teleosts	All teleosts counted within ten linear 50-m long and 5-m wide transects, each census lasting 5 minutes.	60 min	1 dive for 2 scientist divers
11 - Piscivorous teleosts	All teleosts counted within ten linear 50-m long and 5-m wide transects, each census lasting 5 minutes	Same as boxes 10 and 12	
12 - Planktivorous teleosts of the water column	All teleosts counted within ten linear 50-m long and 5-m wide transects, each census lasting 5 minutes	Same as boxes 10 and 11	
9–12 - SRDI	All teleosts counted within ten linear 50-m long and 5-m wide transects, each census lasting 5 minutes	Same as boxes 10, 11 and 12	

status of marine regions. The present approach constitutes a contribution towards this goal, and focuses on the *P. oceanica* seagrass meadow.

Here, we have developed, applied and tested an ecosystem-based index of the structure and functioning of the most emblematic and best-known Mediterranean coastal ecosystem, the *P. oceanica* meadow. This index (EBQI) is based upon a set of representative functional compartments, the weighting of these compartments and the assessment of each compartment quality by comparison of a supposed baseline based upon the available literature. The strong points of the EBQI are: **(i)** It is easy to implement, not too time-consuming and therefore relatively cheap; **(ii)** It is non-destructive, which is particularly important, dealing with protected species (e.g. *P. oceanica* and *Pinna nobilis*); sampling only concerns old leaves (not the shoot) and the litter, i.e. shed dead leaves; **(iii)** It is relatively robust, according to the selection of the compartments and to their weighting; **(iv)** It is associated with confidence indices CI (each compartment, and the overall mark) which indicate possible weakness and biases and therefore the need for further field data acquisition; **(v)** It can draw the attention of managers to sites whose CI is high and where routine monitoring can therefore be implemented, and conversely those with a low CI indicating a lack of knowledge.

The Integrated Coastal Zone Management (ICZM) contributes towards maximizing the benefits provided by the coastal zone and minimizing conflicts and the harmful effects of activities upon each other (e.g. [117]). ICZM, together with Ecosystem-Based Management (EBM; e.g. [118]) and, more generally, the implementation of European Directives, are in need of indices to assess ecological quality, either based upon a single or a few species, or ecosystem-based. The sites corresponding to well implemented MPAs, such as the Port-Cros National Park (Port-Cros Island southern coast, Bagaud Pass) and the Medes Islands marine reserve, not unexpectedly get the best marks, which is congruent with the overall excellent status of their habitats [108]. In contrast, sites localized in areas highly impacted by most human activities and waste, such as Sitges (South of Barcelona) and Niolon (close to the port of Marseilles and outfall of wastewater), obtain a very low mark. Though sites with obviously good or bad ecosystem status naturally obtain a good and bad, EBQI mark, respectively, some sites (e.g. Saint-Cyr Bay and Villefranche-sur-Mer Bay) do not obtain the mark they would seemingly have deserved through the apparent health of the seagrass itself. This confirms the usefulness of an index, EBQI, based upon the whole ecosystem rather than only upon the seagrass (sometimes also epibiota and other parameters). It is worth noting that the Confidence Index (CI) is overall relatively low. This is especially the case of Saint-Cyr Bay (CI = 0.7; table 4) and Villefranche-sur-Mer Bay (CI = 1.3), which could also account for the surprisingly low and not so low, EBQI, respectively, of these sites, according to literature data [119,120].

The weak points of the EBQI are: **(i)** The baselines used to assess the compartments, which may prove to be biased either by the poor knowledge or availability of totally non-human impacted areas, or by climate change which renders the supposed baseline obsolete; **(ii)** The probable need for testing and/or adapting the EBQI in areas distant from the study area -the north-western Mediterranean- and more generally to other seagrass ecosystems; (iii) The conceptual model which constitutes the basis for the present ecosystem-based approach is obviously oversimplified. For example, small-sized predatory 'invertebrates' were not taken into consideration. Some of them belong to the matte endofauna box, which encompasses detritus feeders, suspension-feeders and predators. Other are creeping or clinging organisms on the *P. oceanica* leaves. However, considering them would have made the model more complex and weighed down the assessment of ecosystem status. Moreover, other compartments are likely to provide redundant information, so that no improvement in accuracy is to be expected.

A number of attempts to review and compare the biotic indices using benthic (e.g. macrophytes) and pelagic marine, lagoon and estuarine organisms, have been performed [16,39,51,116,121]. For example, [116] analysed the strengths and weaknesses of 90 published indices. They identified several weaknesses: **(i)** problems of applicability due to practical and conceptual difficulties, affecting most of the indices related to ecosystem function, **(ii)** the failure of many indices using e.g. the taxonomic composition of the community to connect its condition with the stressors, and **(iii)**, as far as indices based upon the sub-individual level are concerned (e.g. multi-biomarkers), their poor strength to assess the ecological integrity of the habitat. They concluded that the most promising approach would be to aggregate indices with complementary strengths. In a sense, the EBQI could be considered as such an aggregative index. Although the goal of the EBQI is not to compete with the already available biotic indices utilizing the *P. oceanica* seagrass meadow for assessing the quality of a water body, but to assess the status of the proper seagrass ecosystem, the comparison of the EBQI with some other biotic indices will be of relevance. An attempt has been made here, using some published data; a more accurate comparison would require data acquisition at the same sites as those used for the EBQI, and is therefore beyond the scope of the present work. In addition, the correlation between the EBQI and anthropogenic gradients, including overfishing, will be useful, in order to assess which human impacts are of major relevance for the status of the ecosystem. This will constitute the next step.

Conclusions

Overall, the Ecosystem Based Quality Index (EBQI) is easy to implement, relatively robust and does not seem to be redundant with existing indices based upon the seagrass itself. In addition, the non-congruence of the EBQI, i.e. the quality of the ecosystem functioning, with the empirical idea one may have of the *P. oceanica* ecosystem at a given site, due e.g. to the health of the seagrass, the clearness of the water (and even the beauty of the landscape and the seascape) confirms the usefulness of the EBQI index, based upon the whole ecosystem rather than upon only the seagrass. Other anthropogenic impacts, e.g. overfishing, are putatively more important than the above mentioned parameters for the structure and functioning of the ecosystem. Finally, the EBQI that is proposed here for the *P. oceanica* ecosystem could constitute a model for similar indices designed for other ecosystems, such as coralligenous outcrops, underwater caves, soft bottoms and sublittoral reefs.

Acknowledgments

The authors are indebted to Fabien Morat for data concerning *Phalacrocorax aristotelis*, Philippe Lenfant, Boris Daniel (AAMP) and Pierre Boissery (AERMC) for contributions and valuable comments, Michael Paul for improvement of the English text and two anonymous reviewers for their highly valuable suggestions and improvement of the manuscript.

Author Contributions

Conceived and designed the experiments: CFB CPM DTB EB FR GP MHV SP SR. Performed the experiments: CPM EB GP JP PA PB SB SR TT. Analyzed the data: CFB JCP SP SR. Wrote the paper: DBS EB EF CFB JCP MHV SP SR.

References

1. Karr JR (1981) Assessment of biotic integrity using fish communities. Fisheries 6: 21–27.
2. Orfanidis S, Panayotidis P, Stamatis N (2001) Ecological evaluation of transitional and coastal waters: a marine benthic macrophytes-based model. Medit Mar Sci 2: 45–65.
3. Orfanidis S, Panayotidis P, Stamatis N (2003) An insight to the ecological evaluation index (EEI). Ecol Indic 3: 27–33.
4. White J, Irvine K (2003) The use of littoral mesohabitats and their macroinvertebrates in the assemblages of lakes. Aquat Conserv 13: 331–351.
5. Pergent-Martini C, Leoni V, Pasqualini V, Ardizzone GD, Balestri E, et al. (2005) Descriptors of *Posidonia oceanica* meadows: Use and application. Ecol Indic 5: 213–230.
6. Orfanidis S, Panayotidis P, Ugland KI (2011) Ecological Evaluation Index continuous formula (EEI-c) application: a step forward for functional groups, the formula and reference condition values. Medit Mar Sci 12: 199–231.
7. Hoare JM, Monks A, O'Donnel CFJ (2013) Do population indicators work? Investigating correlated responses of bird populations in relation to predator management. Ecol Indic 25: 23–34.
8. Jones CG, Lawton JH, Shachak M (1994) Organisms as ecosystem engineers. Oikos 69: 373–386.
9. Lawton JH (1994) What do species do in ecosystems? Oikos 71: 367–374.
10. Bond W (2001) Keystone species – Hunting the snark? Science 292: 63–64.
11. Boudouresque CF, Mayot N, Pergent G (2006) The outstanding traits of the functioning of the *Posidonia oceanica* seagrass ecosystem. Biol Mar Medit 13(4): 109–113.
12. Romero J, Martínez-Crego B, Alcoverro T, Pérez M (2007) A multivariate index based on the seagrass *Posidonia oceanica* (POMI) to assess ecological status of coastal waters under the water framework directive (WFD). Mar Poll Bull 55: 196–204.
13. Fernández-Torquemada Y, Díaz-Valdés M, Colilla F, Luna B, Sánchez-Lizaso JL, et al. (2008) Descriptors from *Posidonia oceanica* (L.) Delile meadows in coastal waters of Valencia, Spain, in the context of the EU Water Framework Directive. ICES J Mar Sci 65: 1492–1497.
14. Gobert S, Sartoretto S, Rico-Raimondino V, Andral B, Chery A, et al. (2009) Assessment of the ecological status of Mediterranean French coastal waters as required by the Water Framework Directive using the *Posidonia oceanica* Rapid Easy Index: PREI. Mar Poll Bull 58: 1727–1733.
15. Lopez y Royo C, Casazza G, Pergent-Martini C, Pergent G (2010) A biotic index using the seagrass *Posidonia oceanica* (BiPo), to evaluate ecological status of coastal waters. Ecol Indic 10: 380–389.
16. Lopez y Royo C, Pergent G, Alcoverro T, Buia MC, Casazza G, et al. (2011) The seagrass *Posidonia oceanica* as indicator of coastal water quality: experimental intercalibration of classification systems. Ecol Indic 11: 557–563.
17. Marbà N, Krause-Jensen D, Alcoverro T, Birk S, Pedersen A (2012) Diversity of European seagrass indicators: patterns within and accross regions. Hydrobiologia 704: 265–278.
18. Panayotidis P, Feretopoulos J, Montesanto B (1999) Benthic vegetation as an ecological quality descriptor in an Eastern Mediterranean coastal area (Kalloni Bay, Aegean Sea, Greece). Estuar Coast Shelf Sci 48: 205–214.
19. Ballesteros E, Torras X, Pinedo S, Garcia M, Mangialajo L, et al. (2007) A new methodology based on littoral community cartography by macroalgae for the implementation of the European Water Framework Directive. Mar Poll Bull 55: 172–180.
20. Guinda X, Juanes JA, Puente A, Revilla JA (2008) Comparison of two methods for quality assessment of macroalgae assemblages, under pollution types. Ecol Indic 8: 743–753.
21. Oliva S, Mascaró O, Llagostera I, Pérez M, Romero J (2012) Selection of metrics based on the seagrass Cymodocea nodosa and development of a biotic index (CYMOX) for assessing ecological status of coastal and transitional waters. Estuar Coast Shelf Sci 114: 7–17.
22. García-Marín P, Cabaço S, Hernández I, Vergara JJ, Silva J (2013) Multi-metric index based on the seagrass Zostera noltii (ZoNI) for ecological quality assessment of coastal and estuarine systems in SW Iberian Peninsula. Mar Poll Bull 68: 46–54.
23. Diaz R, Solan M, Valente R (2004) A review of approaches for classifying benthic habitats and evaluating habitat quality. J Environ Manage 73: 165–181.
24. Mateo MA, Romero J, Pérez M, Littler M, Littler D (1997) Dynamics of millenary organic deposits resulting from the growth of the Mediterranean seagrass *Posidonia oceanica*. Estuar Coast Shelf Sci 44: 103–110.
25. Alcoverro T, Manzanera M, Romero J (2001) Annual metabolic carbon balance of the seagrass *Posidonia oceanica*: the importance of carbohydrate reserves. Mar Ecol Prog Ser 211: 105–116.
26. Prado P, Tomas F, Alcoverro T, Romero J (2007) Extensive direct measurements of *Posidonia oceanica* defoliation confirm the importance of herbivory in temperate seagrass meadows. Mar Ecol Prog Ser 340: 63–71.
27. Boudouresque CF, Bernard G, Bonhomme P, Charbonnel E, Diviacco G, et al. (2012) Protection and conservation of *Posidonia oceanica* meadows. Marseilles: RAMOGE, RAC/SPA and Gis Posidonie publ. pp. 202 p.
28. Pergent G, Bazairi H, Bianchi CN, Boudouresque CF, Buia MC, et al. (2012) Mediterranean seagrass meadows: resilience and contribution to climate change mitigation. A short summary. Gland, Málaga: IUCN. 40 p.
29. Mazzuca S, Björk M, Beer S, Felisberto P, Gobert S, et al. (2013) Establishing research strategies, methodologies and technologies to link genomics and proteomics to seagrass productivity, community metabolism, and ecosystem carbon fluxes. Front Plant Sci 4(38): 1–19.
30. Sala E (2004) The past and present topology and structure of Mediterranean subtidal rocky-shore food webs. Ecosystems 7: 333–340.
31. Bellan-Santini D, Picard J (1984) Conséquences de la régression des herbiers de Posidonies pour le benthos animal des milieux méditerranéens. In: Boudouresque CF, Jeudy de Grissac A, Olivier J, editors. International Workshop on *Posidonia oceanica* beds. Marseilles: GIS Posidonie publ. pp. 423–429.
32. Boudouresque CF, Bernard G, Pergent G, Shili A, Verlaque M (2009) Regression of Mediterranean seagrasses caused by natural processes and anthropogenic disturbances and stress: a critical review. Botanica Marina 52: 395–418.
33. Coles R, Grech A, McKenzie L (2013) Seagrass under pressure. Seagrass-Watch 47: 2–6.
34. Ward T, Hegerl E (2003) Marine protected areas in ecosystem-based management of fisheries. Canberra: Commonweath of Australia publ. 66 p.
35. Pikitch EK, Santora C, Babcock EA, Bakun A, Bonfil R, et al. (2004) Ecosystem-based fishery management. Science 305: 346–347.
36. Turrell B (2004) The policy basis of the "Ecosystem Approach" to fisheries management. Norrköping, Sweden: EuroGOOS publ. 21: 1–28.
37. Rice JC (2005) Implementation of the Ecosystem Approach to Fishery Management – asynchronous co-evolution at the interface between science and policy. Mar Ecol Prog Ser 300: 265–270.
38. Tudela S, Short K (2005) Paradigm shifts, gaps, inertia, and political agendas in ecosystem-based fisheries management. Mar Ecol Prog Ser 300: 281–286.
39. Martínez-Crego B, Vergés A, Alcoverro T, Romero J (2008) Selection of multiple seagrass indicators for environmental biomonitoring. Mar Ecol Progr Ser 361: 93–109.
40. Christensen V, Maclean J (eds) (2011) Ecosystem approach to fisheries. A global perpective. Cambridge University Press. 342 p.
41. Kenchington E, Heino M, Nielsen EE (2003) Managing marine genetic diversity: time for action? ICES J Mar Sci 60: 1172–1176.
42. Rosenberg AA, McLeod KL (2005) Implementing ecosystem-based approaches to management for the conservation of ecosystem services. Mar Ecol Prog Ser 300: 270–274.
43. Ruíz JM, Boudouresque CF, Enríquez S (2009) Mediterranean seagrasses. Botanica Marina 52: 369–382.
44. Boudouresque CF, Jeudy de Grissac A (1983) L'herbier à *Posidonia oceanica* en Méditerranée: les interactions entre la plante et le sédiment. J Rech Océanogr 8: 99–122.
45. Gacia E, Granata TC, Duarte CM (1999) An approach to measurement of particule flux and sed-iment retention within seagrass (*Posidonia oceanica*) meadows. Aquat Bot 65: 255–268.
46. Sánchez-González JF, Sánchez-Rojas V, Memos CD (2011) Wave attenuation due to *Posidonia oceanica* meadows. J Hydraul Res 49: 503–514.
47. Infantes E, Orfila A, Simarro G, Terrados J, Luhar M, et al. (2012) Effect of a seagrass (*Posidonia oceanica*) meadow on wave propagation. Mar Ecol Prog Ser 456: 63–72.
48. Molinier R, Picard J (1952) Recherches sur les herbiers de Phanérogames marines du littoral méditerranéen français. Ann Inst Océanogr 27: 157–234.
49. Boudouresque CF, Meinesz A (1982) Découverte de l'herbier de posidonie. Hyères: Parc National de Port-Cros publ. iii + 79 p.
50. López-Sáez JA, López-Merino L, Mateo MÁ, Serrano Ó, Pérez-Díaz S, et al. (2009) Paleoecological potential of marine organic deposits of *Posidonia oceanica*: a case study in the NE Iberian Peninsula. Palaeogeogr Palaeoclimatol Palaeoecol 271: 215–224.
51. Montefalcone M (2009) Ecosystem health assessment using the seagrass *Posidonia oceanica*: a review. Ecological Indicators 9: 595–604.
52. Sartoretto S, Beaucour C (2011) Habitats particuliers de l'infralittoral : Herbier à *Posidonia oceanica* de la sous-région marine Méditerranée occidentale DCSMM/EI/MO. Ministère de l'Écologie, du Développement Durable, des Transports et du Logement DCSMM/EI/EE/MO/2.2.4.c. 14p.
53. Pergent-Martini C, Pergent G (1994) Lepidochronological analysis in the Mediterranean seagrass *Posidonia oceanica*: state of the art and future development. Oceanol Acta 17: 673–681.
54. Fourqurean JW, Duarte CM, Kennedy H, Marbà N, Holmer M, et al. (2012) Seagrass ecosystems as a globally significant carbon stock. Nat Geosci 5:505–509.
55. Boudouresque CF, Crouzet A, Pergent G (1983) Un nouvel outil au service de l'étude des herbiers à *Posidonia oceanica*: la lépidochronologie. Rapp PV Réun Commiss Internatl Explor Sci Médit 28(3): 111–112.
56. Pergent G, Boudouresque CF, Crouzet A, Meinesz A (1989) Cyclic changes along *Posidonia oceanica* rhizomes (lepidochronology): present state and perspectives. PSZN I Mar Ecol 10: 221–230.

57. Thélin I, Mossé A, Boudouresque CF, Lion R (1985) Le benthos littoral d'El Dabaa (Méditerranée, Égypte). II. L'herbier à *Posidonia oceanica*. Rapp PV Réun Commiss Internatl Explor Sci Médit 29(5): 247–248.
58. Pergent G, Pergent-Martini C (1991) Leaf renewal cycle and primary production of *Posidonia oceanica* in the bay of Lacco Ameno (Ischia, Italy) using lepidochronological analysis. Aquat Bot 42: 49–66.
59. Pergent G, Pergent-Martini C, Boudouresque CF (1995) Utilisation de l'herbier à *Posidonia oceanica* comme indicateur biologique de la qualité du milieu littoral en Méditerranée: état des connaissances. Mésogée 54: 3–27.
60. Martín MA, Sánchez Povedo M, Marhuenda E, Martínez JM, Sánchez Lizaso JL (1994) Estudio del estrato hipogeo de la pradera de *Posidonia oceanica* (L.) Delile de la reserva marina de Tabarca. In: Ramos Esplá A, editor. Trabajos de campo en la reserva marina de Tabarca (Alicante). Alicante: Universidad de Alicante publ. pp. 25–31.
61. Boudouresque CF, Jeudy de Grissac A, Meinesz A (1984) Relations entre la sédimentation et l'allongement des rhizomes orthotropes de *Posidonia oceanica* dans la baie d'Elbu (Corse). In: Boudouresque CF, Jeudy de Grissac A, Olivier J, editors. International Workshop on *Posidonia oceanica* Beds. Marseilles: GIS Posidonie publ. pp. 185–191.
62. Di Maida G, Tomasello A, Sciandra M, Pirrotta M, Milazzo M, et al. (2013) Effect of different substrata on rhizome growth, leaf biometry and shoot density of *Posidonia oceanica*. Mar Environ Res 87–88: 96–102.
63. Giovannetti E, Lasagna R, Montefalcone M, Bianchi CN, Albertelli G, et al. (2008) Inconsistent responses to substratum nature in *Posidonia oceanica* meadows: an integration through complexity levels? Chemistry and Ecology 24 (S1): 145–153.
64. Bay D (1984) A field study of the growth dynamics and productivity of *Posidonia oceanica* (L.) Delile in Calvi Bay, Corsica. Aquat Bot 20: 43–64.
65. Pergent-Martini C, Rico-Raimondino V, Pergent G (1994) Primary production of *Posidonia oceanica* in the Mediterranean basin. Mar Biol 120: 9–15.
66. Panayotidis P, Boudouresque CF, Marcot-Coqueugniot J (1981) Microstructure de l'herbier de *Posidonia oceanica* (Linnaeus) Delile. Botanica Marina: 115–124.
67. UNEP-MAP-RAC/SPA (2011) Draft guidelines for the standardization of mapping and monitoring methods of marine magnoliophyta in the Mediterranean. Working document, prepared by C Pergent-Martini for the tenth meeting of Focal Points for SPAs, Marseilles, France, 17–20 May 2011, UNEP(DEPI)/MED WG 359/9. Tunis: RAC/SPA Publ. 59 p.
68. Boudouresque CF, Verlaque M (2007) Ecology of *Paracentrotus lividus*. In: Lawrence J, editor. Edible sea urchins: biology and ecology, Second edition. Amsterdam: Elsevier. pp. 243–285.
69. Boudouresque CF, Bernard G, Bonhomme P, Charbonnel E, Le Diréach L, et al. (2007) Monitoring methods for *Posidonia oceanica* seagrass meadows in Provence and the French Riviera. Sci Rep Port-Cros Natl Park 22: 17–38.
70. Short FT, McKenzie LJ, Coles RG, Vidler KP (2002) SeagrassNet manual for scientific monitoring of seagrass habitat. Cairns: QDPI, QFS. 56 p.
71. Boudouresque CF, Verlaque M (2013) *Paracentrotus lividus*. In: Lawrence JM, editor. Sea urchins: Biology and ecology, Third Edition. Amsterdam: Elsevier. pp. 297–327.
72. van der Ben D (1971) Les épiphytes des feuilles de *Posidonia oceanica* Delile sur les côtes françaises de la Méditerranée. Mém Inst R Sci Nat Belg 168 : 1–101 + 6 tables.
73. Giovannetti E, Montefalcone M, Morri C, Bianchi CN, Albertelli G (2010) Early warning response of *Posidonia oceanica* epiphyte community to environmental alterations (Ligurian Sea, NW Mediterranean). Mar Poll Bull 60 (7): 1031–1039.
74. Mabrouk L, Hamza A, Ben Brahim M, Bradai MN (2013) Variability in the structure of epiphyte assemblages on leaves and rhizomes of *Posidonia oceanica* in relation to human disturbances in a seagrass meadow of Tunisia. Aquat Bot 108: 33–40.
75. Pérès JM, Picard J (1964) Nouveau manuel de bionomie benthique de la Mer Méditerranée. Rec Trav Stat Mar Endoume 31(47): 3–137.
76. Vicente N, De Gaulejac B (1993) *Pinna nobilis* L. (Mollusque Bivalve), indicateur biologique du littoral méditerranéen. In : Boudouresque CF, Avon M, Pergent-Martini C, editors. Qualité de milieu marin. Indicateurs biologiques et physico-chimiques. Marseilles: GIS Posidonie publ. pp. 141–150.
77. García-March JR, García-Carrascosa AM, Peña Cantero AL, Wang YG (2007) Population structure, mortality and growth of Pinna nobilis Linnaeus, 1758 (Mollusca, Bivalvia) at different depths in Moraira Bay (Alicante, Western Mediterranean). Mar Biol 150: 861–871.
78. García-March JR, Vicente N (2011) *Pinna nobilis* (Mollusca, Bivalvia) indicateur biologique de la qualité du littoral méditerranéen: densité des populations. In: Francour P, Bodilis P, Arceo H, editors. Actes du Congrès 2011 de la Société Zoologique de France. Nice: Université de Nice publ. pp. 80–81.
79. Urra J, Ramirez AM, Marina P, Salas C, Gofas S, et al. (2013) Highly diverse molluscan assemblages of *Posidonia oceanica* meadows in northwestern Alboran Sea (W Mediterranean): seasonal dynamics and environmental drivers. Estuar Coast Shelf Sci 117: 136–147.
80. Naranjo SA, Carball JL, Garcia-Gomez JC (1996) Effects of environmental stress on ascidian populations in Algeciras Bay (southern Spain). Possible marine bioindicators? Mar Ecol Progr Ser 144: 119–131.
81. Bianchi CN, Bedulli D, Morri C, Occhipinti Ambrogi A (1989) L'herbier de Posidonies: écosystème ou carrefour éco-éthologique? In: International Workshop on Posidonia Beds (CF . Boudouresque, A . Meinesz, E . Fresi & V . Gravez, eds). GIS Posidonie, Marseille 2: 257–272.
82. Romero J, Pergent G, Pergent-Martini C, Mateo M.A, Régnier C (1992). The detritic compartment in a *Posidonia oceanica* meadow: litter features, decomposition rates and mineral stocks. Mar Ecol PSZNI 13(1): 69–83.
83. Mezali K, Chekaba B, Zupo V, Asslah B (2003) Comportement alimentaire de cinq espèces d'holothuries aspidochirotes (*Holothuria*: Echinodermata) de la presqu'île de Sidi-Fredj (Algérie). Bull Soc Zool Fr 128: 49–62.
84. Verlaque M (1981) Preliminary data on some *Posidonia* feeders. Rapp PV Réun Commiss Internatl Explor Sci Méditerr 27(2): 201–202.
85. Giraud G (1979) Sur une méthode de mesure et de comptage des structures foliaires de *Posidonia oceanica* (Linnaeus) Delile. Bull Mus Hist Nat Marseille 39: 33–39.
86. Peirano A, Niccolai I, Mauro R, Bianchi CN (2001) Seasonal grazing and food preference of herbivores in a *Posidonia oceanica* meadow. Scientia Marina 65 (4): 367–374.
87. Francour P (1997) Fish assemblages of *Posidonia oceanica* beds at Port-Cros (France, NW Mediterranean): assessment of composition and long term fluctuations by visual census. PSZN I Mar Ecol 18: 157–173.
88. Bell JD, Harmelin-Vivien ML (1982) Fish fauna of French Mediterranean *Posidonia oceanica* seagrass meadows. 1. Community structure. Téthys 10: 337–347.
89. Harmelin-Vivien M (1982) Ichtyofaune des herbiers de posidonies du Parc national de Port-Cros: I. Composition et variations spatio-temporelles. Trav Sci Parc Natl Port-Cros 8: 69–92.
90. Francour P, Harmelin-Vivien M, Harmelin JG, Duclerc J (1995) Impact of *Caulerpa taxifolia* colonization on the littoral ichthyofauna of north-western Mediterranean Sea: preliminary results. Hydrobiologia 300/301: 345–353.
91. Harmelin-Vivien M, Francour P (1992) Trawling or visual censuses? Methodological bias in the assessment of fish populations in seagrass beds. PSZN I Mar Ecol 13: 41–51.
92. Guyot I (1990) Le cormoran huppé de Corse: biologie et interactions avec la pêche professionnelle. Trav Sci Parc Nat Rég Rés Nat Corse 28: 1–40.
93. Morat F (2007) Régime alimentaire de la population de cormoran huppé de Méditerranée (*Phalacrocorax aristotelis desmarestii*) de Riou. Marseilles: CEEP and Centre d'Océanologie de Marseille. 23 + 6 p.
94. Morat F, Mante A, Drunat E, Dabat J, Bonhomme P, et al. (2011) Diet of the Mediterranean European shag, *Phalacrocorax aristotelis desmaretii*, its ecological significance and interaction with local fisheries in the Riou Archipelago (Marseilles, France). Vie Milieu 61: 77–86.
95. Francour P, Thibault JC (1995) Le régime alimentaire du balbuzard pêcheur (*Pandion haliaetus*) dans la Réserve naturelle de Scandola. Trav Sci Parc Nat rég Rés Nat. Corse 53: 11–27.
96. Francour P, Thibault JC (1996) The diet of breeding osprey *Pandion haliaetus* on Corsica: exploitation of a coastal marine environment. Bird Study 43: 129–133.
97. Thibault JC, Bretagnolle V, Dominici JM (2001) Le balbuzard pêcheur en Corse. Ajaccio: Alain Piazzola. 184 p.
98. Pergent G, Romero J, Pergent-Martini C, Mateo MA, Boudouresque CF (1994) Primary production, stocks and fluxes in the Mediterranean seagrass *Posidonia oceanica*. Mar Ecol Prog Ser 106: 139–146.
99. Mateo MA, Romero J (1997) Detritus dynamics in the seagrass *Posidonia oceanica*: elements for an ecosystem carbon and nutrient budget. Mar Ecol Prog Ser 151: 43–53.
100. Pergent G, Rico-Raimondino V, Pergent-Martini C (1997) Fate of primary production in *Posidonia oceanica* meadows of the Mediterranean. Aquat Bot 59: 307–321.
101. Mateo MA, Sánchez-Lizaso JL, Romero J (2002) *Posidonia oceanica* 'banquettes': a preliminary assessment of the relevance for meadow carbon and nutrient budget. Estuar Coast Shelf Sci 56: 85–90.
102. Macquart-Moulin C (1968) Les amphipodes bentho-planctoniques du golfe de Marseille. Analyse des captures faites au cours de pêches planctoniques nocturnes régulières (années 1963–1964). Rec Trav Stat Mar Endoume 43(59): 311–332.
103. Harmelin-Vivien ML, Harmelin JG, Leboulleux V (1995) Microhabitat requirements for settlement of juvenile sparid fishes on Mediterranean rocky shores. Hydrobiologia 300–301: 309–320.
104. Jiménez S, Cano R, Bayle J, Ramos A, Sánchez Lizaso JL (1996) Las praderas de *Posidonia oceanica* (L.) Delile como zona de protección de juveniles de especies de interés comercial. Real Soc Esp Hist Nat, tomo extraordinario: 375–378.
105. UNEP-MAP-RAC/SPA (2011) Proposal of standard methods for inventorying and monitoring coralligenous and rhodoliths populations. Working document, prepared by G Pergent for the tenth meeting of Focal Points for SPAs, Marseilles, France, 17–20 May 2011, UNEP(DEPI)/MED WG 359/10. Tunis: RAC/SPA Publ. 23 p.
106. Dalkey N, Helmer O (1963) An experimental application of the Delphi method to the use of experts. Manage Sci 9: 458–467.
107. Ferrari B, Raventos N, Planes S (2008) Assessing effects of fishing prohibition on *Posidonia oceanica* seagrass meadows in the Marine Natural Reserve of Cerbère-Banyuls. Aquatic Botany 88: 295–302.
108. Astruch P, Boudouresque CF, Bonhomme D, Goujard A, Antonioli PA, et al. (2012) Mapping and state of conservation of benthic marine habitats and assemblages of Port-Cros national Park (Provence, France, northwestern Mediterranean Sea). Sci Rep Port-Cros Natl Park 26: 45–90.

109. Bazin N, Imbert M (2012) Le cormoran huppé de Méditerranée *Phalacrocorax aristotelis desmarestii*. Etat des connaissances et de la conservation actualisé des populations nicheuses des petites îles de Méditerranée. Marseilles: Initiative PIM. 18 p.
110. Romero J, Pérez M, Alcoverro T (2012) L'alguer de *Posidonia oceanica* de les illes Medes: més de trenta anys d'estudi. In: Quintana X, Hereu B, editors. El fons marí de les illes Medes i el Montgrí: quatre dècades de recerca per la conservatió. Toroella de Montrgrí: Càtedra d'Ecosistems Marins Mediterranis. pp. 81–100.
111. Rouanet E, Astruch P, Bonhomme D, Bonhomme P, Rogeau E, et al. (2012) Suivi de l'herbier de posidonie de la passe de Bagaud, impact de l'ancrage (Parc national de Port-Cros, Var, France). Marseilles: GIS Posidonie publ. 82 p.
112. Boudouresque CF, Charbonnel E, Meinesz A, Pergent G, Pergent-Martini C, et al. (2000) A monitoring network based on the seagrass *Posidonia oceanica* in the nortwestern Mediterranean Sea. Biol mar medit 7 (2): 328–331.
113. Moreno D, Aguilera PA, Castro H (2001) Assessment of the conservation status of seagrass (*Posidonia oceanica*) meadows: implications for monitoring strategy and the decision-making process. Biol Conservation 102: 325–332.
114. Montefalcone M, Albertelli G, Morri C, Bianchi CN (2007) Urban seagrass: status of *Posidonia oceanica* facing Genoa city waterfront (Italy) and implications for management. Mar Poll Bull 54: 206–213.
115. Van Hoey G, Borja A, Birchenough S, Buhl-Mortensen L, Degraer S, et al. (2010) The use of benthic indicators in Europe: from the Water Framework Directive to the Marine Strategy Framework Directive. Mar Poll Bull 60: 2187–2196.
116. Martínez-Crego B, Alcoverro T, Romero J (2010) Biotic indices for assessing the status of coastal waters: a review of strengths and weaknesses. J Environm Monit 12 (5): 1013–1028.
117. Pergent-Martini C, Pasqualini V, Ferrat L, Pergent G (2006) Ecological data in Integrated Coastal Zone Management: case study of *Posidonia oceanica* meadows along the Corsican coastline (Mediterranean Sea). Environ Manage 38: 889–895.
118. Parravicini V, Rovere A, Vassallo P, Micheli F, Montefalcone M, et al. (2012) Understanding relationships between conflicting human uses and coastal ecosystem status: a geospatial modelling approach. Ecol Indic 19: 253–263.
119. Meinesz A, Lefevre JR, Astier JM (1991) Impact of coastal development on the infralittoral zone along the southern Mediterranean shore of continental France. Mar Poll Bull 23: 343–347.
120. Leriche A, Pasqualini V, Boudouresque CF, Bernard G, Bonhomme P, et al. (2006) Spatial, temporal and structural variations of a *Posidonia oceanica* seagrass meadow facing human activities. Aquat Bot 84: 287–293.
121. Mascaró O, Alcoverro T, Dencheva K, Díez I, Gorostiaga JM, et al. (2013) Exploring the robustness of macrophyte-based classification methods to assess the ecological status of coastal and transitional ecosystems under the Water Framework Directive. Hydrobiologia 704: 279–291.

15

Fortuitous Encounters between Seagliders and Adult Female Northern Fur Seals (*Callorhinus ursinus*) off the Washington (USA) Coast: Upper Ocean Variability and Links to Top Predator Behavior

Noel A. Pelland[1]*, Jeremy T. Sterling[2], Mary-Anne Lea[3], Nicholas A. Bond[4], Rolf R. Ream[2], Craig M. Lee[5,1], Charles C. Eriksen[1]

1 School of Oceanography, University of Washington, Seattle, Washington, United States of America, **2** National Marine Mammal Laboratory, National Oceanographic and Atmospheric Administration, Seattle, Washington, United States of America, **3** Institute for Marine and Antarctic Studies, University of Tasmania, Hobart, Tasmania, Australia, **4** Joint Institute for the Study of the Atmosphere and Ocean, Seattle, Washington, United States of America, **5** Applied Physics Laboratory, University of Washington, Seattle, Washington, United States of America

Abstract

Behavioral responses by top marine predators to oceanographic features such as eddies, river plumes, storms, and coastal topography suggest that biophysical interactions in these zones affect predators' prey, foraging behaviors, and potentially fitness. However, examining these pathways is challenged by the obstacles inherent in obtaining simultaneous observations of surface and subsurface environmental fields and predator behavior. In this study, migratory movements and, in some cases, diving behavior of 40 adult female northern fur seals (NFS; *Callorhinus ursinus*) were quantified across their range and compared to remotely-sensed environmental data in the Gulf of Alaska and California Current ecosystems, with a particular focus off the coast of Washington State (USA) – a known foraging ground for adult female NFS and where autonomous glider sampling allowed opportunistic comparison of seal behavior to subsurface biophysical measurements. The results show that in these ecosystems, adult female habitat utilization was concentrated near prominent coastal topographic, riverine, or inlet features and within 200 km of the continental shelf break. Seal dive depths, in most ecosystems, were moderated by surface light level (solar or lunar), mirroring known behaviors of diel vertically-migrating prey. However, seal dives differed in the California Current ecosystem due to a shift to more daytime diving concentrated at or below the surface mixed layer base. Seal movement models indicate behavioral responses to season, ecosystem, and surface wind speeds; individuals also responded to mesoscale eddies, jets, and the Columbia River plume. Foraging within small scale surface features is consistent with utilization of the inner coastal transition zone and habitats near coastal capes, which are known eddy and filament generation sites. These results contribute to our knowledge of NFS migratory patterns by demonstrating surface and subsurface behavioral responses to a spatially and temporally dynamic ocean environment, thus reflecting its influence on associated NFS prey species.

Editor: Andreas Fahlman, Texas A&M University-Corpus Christi, United States of America

Funding: Northern fur seal tagging, monitoring, and analysis was funded by NOAA (www.noaa.gov), the North Pacific Research Board (www.nprb.org), the National Cooperative Research Program of the NOAA Fisheries Office of Science and Technology (www.st.nmfs.noaa.gov/cooperative-research/index), and the Joint Institute for the Study of the Atmosphere and Ocean at the University of Washington (jisao.washington.edu). Northern fur seal research was performed under National Marine Mammal Laboratory permits 782–1455 and 782–1708. Seaglider field work and analysis was supported by National Science Foundation (www.nsf.gov) grants OCE9911036, OCE0095414, and OCE0526634, with assistance in Seaglider development from the Office of Naval Research (www.onr.navy.mil) and National Oceanographic Partnership Program (www.nopp.org). The funders had no role in study design, data collection and analysis, decision to publish, or preparation of the manuscript.

Competing Interests: The authors have declared that no competing interests exist.

* Email: nap2@u.washington.edu

Introduction

Northern fur seal (NFS; *Callorhinus ursinus*) migration and overwinter foraging represents a critical portion of its annual life cycle. For the Eastern Stock (animals breeding on the Pribilof Islands and Bogoslof Island, Alaska, USA) this migration begins at the onset of subarctic fall as most animals leave the breeding grounds in the Bering Sea for pelagic overwinter habitat at more southerly latitudes, remaining away from land and foraging on the open sea for the following ~8 months. During the migration, foraging success is of fundamental importance to pregnant females, who must invest energy in their fetus and improve their own physical condition prior to the return trip to the breeding grounds in the summer [1–5]. Reproductive females concentrate their overwintering activity mainly in the productive coastal transition zones of the Gulf of Alaska (GA) and California Current (CC) ecosystems, though some make their way to the mid-ocean Transition Zone Chlorophyll Front [1,6].

Within the CC and GA ecosystems and along migratory pathways to and from the breeding grounds, environmental variability that affects the abundance, distribution, availability, and quality of prey (biotic factors) and/or NFS metabolism or swimming energetics (abiotic factors) could potentially have important consequences for overwinter survival and reproductive success of adult female NFS. Since female recruitment to breeding age and annual adult female survivorship are two of the most important determinants of age structure and long-term stability in the population [3,7–11], environmental variability affecting overwintering females could thus potentially exert significant influence on demography and population trends in the Eastern Stock as a whole. However, despite this potential importance, the pathways by which changes in ocean surface patterns influence foraging opportunities and success of individual adult female NFS outside of the Bering Sea, and how this is reflected and expressed in patterns of their horizontal movement, diving frequency, and vertical localization in the marine environment, are not fully understood. This lack of understanding hinders the effort to explain population declines since the 1980s within the Eastern Stock, which may be due to changes in Pacific Ocean climate, human-related causes, predation, or interactions between these factors [3,12–15]. These declines are isolated to the Pribilof Islands (St. Paul and St. George Islands), collectively the largest NFS breeding sites [3,9,11,16,17]. In contrast to trends at the Pribilof Islands, pup production on other breeding islands in the eastern North Pacific Ocean (NP) and Bering Sea has been stable or increasing since the 1980s. The Bogoslof Island population has experienced exponential growth, in part due to immigration, though this flux is of insufficient magnitude to solely account for the Pribilof Islands population decline [18–20]. Pup production on San Miguel Island, California (USA), the largest breeding island in the California Stock, has also increased over a similar time period, though with large interannual fluctuations that are mostly explained by El Niño-Southern Oscillation (ENSO) events [21,22].

Studies undertaken during and immediately prior to the decline of the Pribilof Islands population raised important questions surrounding the role that large and small scale environmental oceanic variability plays in influencing fitness and survivorship during the pelagic phase for juvenile and adult female NFS [3,9,14,23]. At the large scales, basin-wide patterns of hydrography and marine ecosystems in the NP could play a role in the location of overwintering habitat of the NFS population and its segregation by sex and age class. It is commonly assumed that the extreme size dimorphism observed in NFS means that larger adult males are physiologically capable of remaining in the central NP, GA, or southern Bering Sea during the overwintering period, exploiting prey fields that are in colder water and located deeper in the water column [24–26]. The smaller females, juveniles, and pups cannot dive as deeply and may contend with greater mass-specific body heat loss in cold water (<2°C; [27]). While the thermal tolerance of juvenile and adult females allows them to exploit a wide range of winter habitats in the NP, they are likely unable to remain in the southern Bering Sea; this is especially true for pups, whose thermoregulatory abilities are not fully developed [27,28]. For these components of the population it is assumed that overwintering habitat is more suitable off the west coast of North America where they benefit from the productive CC and GA boundary current ecosystems [29]. Here, energy is transferred from lower trophic levels to pelagic schooling fishes and squids that comprise the bulk of NFS prey [1,30–32]. The idea that basin-scale patterns of surface ocean biophysical conditions, and interannual perturbations to these patterns, are important to the NFS Eastern Stock is supported by the fact that they exert a strong influence on San Miguel Island NFS and other pinniped species in the CC, where many females and pups from the Eastern Stock overwinter. For example, strong ENSO events have significantly affected NFS pup, juvenile, and adult survivorship at San Miguel Island [22] and both the abundance and feeding habits of another Eastern Pacific otariid, the California sea lion (*Zalophus californianus*; [33,34]). Warm ENSO events cause elevated sea level height, sea surface temperatures, and a deepening of the mixed-layer depth (MLD), thermocline and nutricline that results in reduced ocean productivity and abundance and availability of pinniped prey. Along the South American west coast, ENSO also affects pup production of Galapagos fur seals (*Arctocephalus galapagoensis*) and sea lions (*Z. wollebaeki*; [35,36]). Furthermore, periods of growth and decline in monk seal (*Monachus schauinslandi*) abundance are associated with positive and negative phases of the Pacific Decadal Oscillation – a low-frequency pattern of basin-wide ocean-atmosphere variability in the NP [37,38].

Embedded within these broad patterns are smaller-scale or higher-frequency environmental features, such as storms, ocean eddies, or bathymetric features, to which individuals or groups of NFS have demonstrated behavioral responses. Satellite tagging of NFS during the summer breeding period and winter migration has revealed that eddy edges and surface fronts can act as a movement and/or foraging cue for adult females, males, and juvenile males [1,24,39,40]. These behavioral responses have also been detected for other pinnipeds and seabirds [41–46]. Once out of the Bering Sea, adult female NFS are often sighted in a cross-shore band 70 to 130 km from the coast, or quoting Kajimura [31] near "sea valleys, submarine canyons, seamounts, and along the continental shelf and slope where abrupt changes...in depths and upwellings of nutrient-rich water occur." This is presumably due to aggregations of prey in these areas. Atmospheric variability is also important during the end of the breeding season and the early fall migration. For example, pup dispersal from the Pribilof Islands in the fall tends to be abrupt and is often triggered by single storms [47–50]. Dispersal during transient large-scale atmospheric events or foraging near bathymetric features, eddies, or ocean fronts suggests that physical interactions in these areas are affecting behavior and potentially reproductive success of NFS [24].

The goal of this study was to investigate how variability in the ocean environment affects adult female NFS distribution, movement, and diving behavior during their overwintering phase. To do so, we examine satellite tracks for 40 Eastern Stock (St. Paul and Bogoslof Islands) seals collected in seven migratory seasons. Our approach combines remotely sensed and *in situ* environmental data to compare to seal distribution and individual behaviors at both large and small scales. At large scales, we provide a general description of the spatial distribution of females overwintering in the CC and GA and compare seal diving depths to a time series of ocean profiles taken off the coast of Washington State (WA, USA), an important overwinter foraging ground for adult females. We use statistical models to examine the effect of different environmental variables on adult female behavioral state, diving frequency, and dive depth. Furthermore, we quantify the amount of time spent in coherent mesoscale oceanographic features by ecosystem, and examine differences in habitat utilization relative to these features between behavioral modes. At small scales we combined fortuitous encounters between three individual seals and Seagliders, which are a type of long-range, profiling ocean glider [51]. These females, equipped with dive recorders, passed near profiling Seagliders, providing an opportunity to examine behavior of these individuals relative to mesoscale hydrography and subsurface structure off the WA coast. These results are supplemented by satellite measurements of sea level anomaly (SLA) and chlorophyll.

Methods

Ethics Statement

All work was conducted in accordance with and under the authority of the United States Marine Mammal Protection Act (National Marine Fisheries Service [NMFS] Permits 782–1455 and 782–1708). The Marine Mammal Protection Act was established in 1972 requiring all research conducted on marine mammals in the United States be done under the authority of federal permits issued by either NMFS or the United States Fish and Wildlife Service (USFWS). All applications for a permit to conduct research on marine mammals have gone through a four-stage review process that includes: 1) agency review (either NMFS or USFWS); 2) a public notice and review period; 3) review and recommendation from the Scientific Advisors to the United States Marine Mammal Commission; and 4) a final action by the reviewing agency. All capture and handling activities described in this manuscript have gone through and been approved by this process. At the time this work was conducted there was no additional requirement for review of these procedures by an institutional review board or ethics committee. In 2010, a NMFS Institutional Animal Care and Use Committee was established for the Alaska Fisheries and Northwest Fisheries Science Centers and the capture and handling protocols described here were reviewed and approved by this committee.

Adult Female Satellite-Telemetry Data

Adult female NFS satellite tracking and diving data were included in the analysis if the animal migrated into the GA or CC ecosystems and spent some time between 40°N–55°N latitude (Fig. 1). It is important to note that in any NFS migratory season, many seals migrate to regions outside our ecosystem selection criteria and that our goals specifically aimed to integrate seal behavior and ecosystem processes that took place in the GA and CC ecosystems [1,52]. To spatially define the GA and CC ecosystems we used boundaries from the National Oceanographic and Atmospheric Administration (NOAA) Large Marine Ecosystems (LMEs) of the World (http://www.lme.noaa.gov; [53]). Forty-one migratory routes from forty seals (one seal was monitored during two separate migrations) met the criteria and the dataset spanned seven migratory seasons between 2002/03 and 2009/10 (Table 1). Of the forty seals, eight were equipped with satellite transmitters that also recorded and transmitted diving behavior.

Seal capture and satellite transmitter deployments took place on two islands, Bogoslof (n = 10 deployments; 53.94°N, 168.04°W) and St. Paul (n = 31 deployments; 57.11°N, 170.29°W) Islands, Alaska. Satellite transmitter types used included KiwiSat 101 and 202 Platform Terminal Transmitters (PTTs; Sirtrack Limited, Havelock North, New Zealand) and ST10 and ST16 Satellite Dive Recorder, SPLASH, and SPOT 5 PTTs (Wildlife Computers, Redmond, WA). All instruments were consistently programmed to transmit during two periods every 24 h. The time of day at which these periods took place varied between years. Descriptions of how each transmitter type was programmed can be found in Loughlin et al. [54] and Ream et al. [1] for the KiwiSat 101, ST10 and ST16. Lea et al. [49,55] and Sterling et al. [25] describe instrument programming methods and dive data processing for SPLASH, SPOT 5, and KiwiSat 202 satellite transmitters. Transmitted tag information, location estimates, and in some cases, dive behavior summaries were received and processed by Service Argos (http://www.argos-system.org).

Wildlife Computers ST10, ST16 and SPLASH satellite-dive recorders were programmed to collect data in 6 h time periods and distribute dive depths among 14 pre-defined dive depth bins (2, 4, 6, 10, 20, 34, 50, 74, 100, 124, 150, 174, 200, >200 m). Dives >6 m were analyzed and the average dive depth and total number of dives for each 6 h period were used as response variables in generalized linear mixed-effects models (GLMM) of diving behavior (see Methods section *Behavior Statistical Analyses*).

Modeling Seal Movement

Raw location data for each tag, calculated by Service Argos, were obtained at irregular time intervals within a deployment and with varying degrees of spatial error. However, for analysis purposes, it is desirable to interpolate these location data to a time base with regular intervals, such that they are aligned with the time base of environmental variables to be used as explanatory variables in GLMMs. To do so, we fitted seal Argos location data with a switching state-space model (SSSM), which estimated the evolution of each animal's position and behavior through time by modeling seal movement as a finite-difference correlated random walk process [25,56–59]. Given the animal's release position, and assumptions about the distribution of turning angle and correlation between direction and move speed during transit movements [60], SSSMs use a Bayesian approach with Markov chain Monte Carlo (MCMC) estimation to fit a model to each animal track. The MCMC procedure was performed using WinBUGS and implemented with R statistical software (http://www.r-project.org; V2.14.1) and the R2WinBUGS package. The SSSMs estimate seal locations uniformly spaced in time every 6 h and allow for all observations to be compared at the same temporal scale. These models have additional advantages of accounting for the spatial error associated with Argos positions, regardless of the listed quality class of each position measurement, and providing an estimate of seal horizontal movement behavior for each 6 h period on a continuous scale from 0 ("transient," fast, somewhat linear or directed movements) to 1 ("resident," area-restricted search or foraging movements) [61]. Estimates of seal horizontal behavior defined in this way are hereafter referred to as "behavioral state" [25] and indicated by the mathematical symbol b (for a glossary of mathematical symbols and acronyms used in this manuscript, refer to Table 2).

Estimates of Horizontal Habitat Utilization

Two-Dimensional, Alongshore, and Cross-Shore Utilization. Using the estimated 6 h female locations, we analyzed two-dimensional overwintering habitat utilization with kernel-smoothed estimates of adult female range in the eastern Pacific Ocean destination zone, defined as the region between 140°W–120°W and 30°N–55°N. This domain was divided into a grid with 0.1° resolution, and a bivariate normal kernel density estimate of female utilization distribution was computed at each grid point using a fixed 15 km smoothing parameter [62]. The smoothing parameter length was chosen arbitrarily in order to resolve large-scale features of the overwintering distribution while still providing adequate detail around prominent coastal features and female distribution peaks. The kernel-smoothed density grid allows calculation of the 95% utilization contour, which is the minimum area that could be drawn to encompass 95% of female habitat utilization in the destination region. This method was also used to calculate range extent for various lower utilization percentiles, which illustrate smaller-scale peaks in adult female

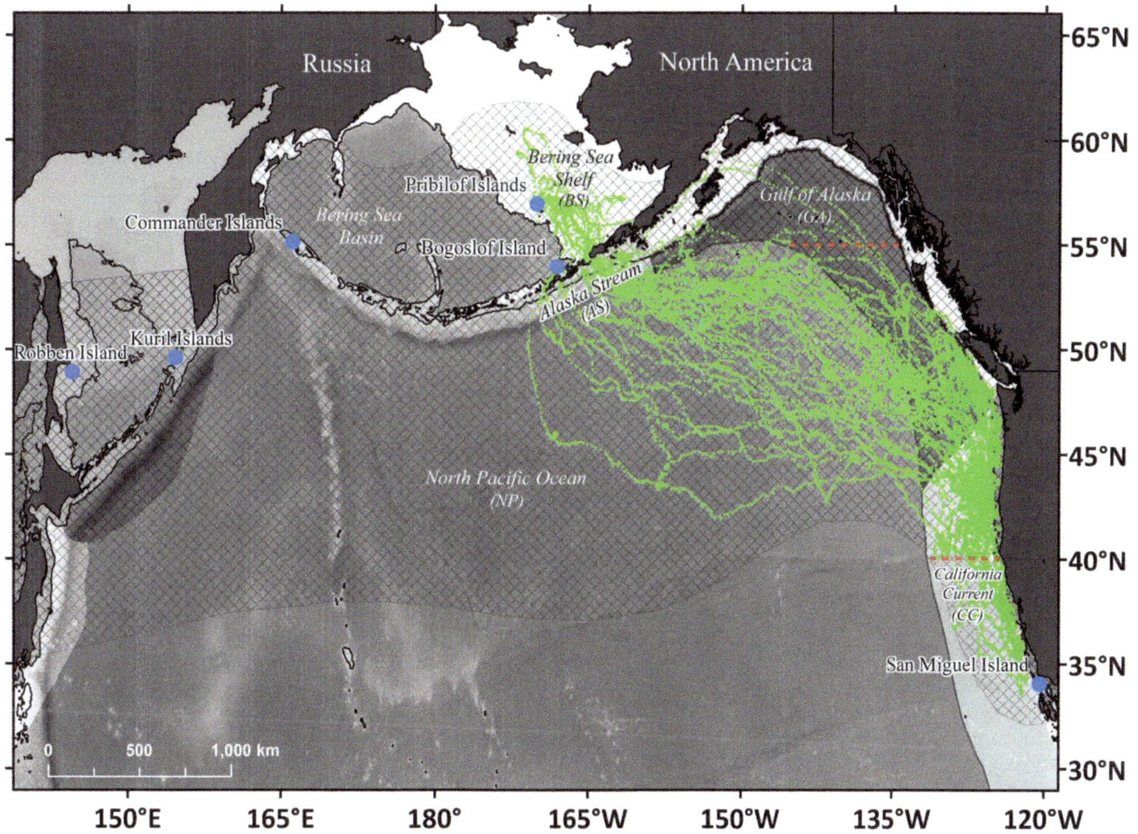

Figure 1. Northern fur seal habitat within the North Pacific Ocean. Blue circles indicate breeding site locations and cross-hatching corresponds to approximate range extent. *Italic* labels denote Large Marine Ecosystems (LMEs) used in this study. California Current (CC) and Gulf of Alaska (GA) LMEs are indicated by shaded polygons. Within these LMEs, shelf habitat is indicated by filled polygons adjacent to the coast. Other LMEs used in this study include the Bering Sea Shelf (BS, filled polygon within the Bering Sea), Alaska Stream (AS, light-shaded coastal habitat extending west from GA), and North Pacific Ocean (NP, all habitat seaward of the CC, GA, and AS). The Pribilof Islands (St. Paul and St. George Islands, Alaska) are collectively the largest northern fur seal breeding sites. Satellite tags were applied to 40 adult female northern fur seals on St. Paul Island and Bogoslof Island, yielding 41 satellite tracks. One animal was tracked in two separate migratory seasons. Tracking data used in this study are shown by green dots. To be considered for use, tracks were required to have entered the CC or GA LMEs and to have spent some time between 40°N–55°N (latitude boundaries indicated by dashed red lines).

foraging distribution. Utilization contours were computed for the full dataset as well as for a subset of 6 h locations corresponding only to area-restricted search behavior, defined as those with $b > 0.75$ following Jonsen et al. [60].

We constructed empirical alongshore distributions for the females as they traveled through the CC and GA LMEs (Fig. 1) from January to June. Total time spent by tracked females within 0.5° latitudinal bands between 33°N to 61°N was estimated monthly. The northern boundary of this domain was extended relative to the two-dimensional utilization analysis in order to display a small amount of time spent in the northern GA ecosystem early in the overwintering period. Interannual variability was not explored, since sampling effort and tag duration varied markedly both within and between years, making it difficult to isolate real differences in foraging distribution between years. The empirical cross-shore distribution (time spent versus distance from the shelf break) of adult female distribution in 20 km bins was also computed over the entire study period. The shelf break was defined as the 200 m isobath and distances from the shelf break were computed along a great circle line perpendicular to the regional orientation of the continental shelf edge. The cross-shore distribution considered only estimated 6 h locations within a subset of the CC and GA LMEs between 41°N and 52°N. These latitudes were chosen to represent the region where the majority of female foraging occurred (see Results section *Migratory Distribution*).

Utilization of Mesoscale Features by State. We investigated the habitat utilization of adult female NFS relative to mesoscale eddies as identified from altimetric measurements of SLA by Chelton et al. [63]. Specifically, we explored whether the utilization distribution relative to eddies was conditioned by behavioral state. Chelton et al. [63] provide eddy estimates at 7 d time steps. For each 6 h adult female estimated location during the overwintering period, at the closest 7 d time step, the nearest mesoscale eddy center position to the female's location was identified. The distance to this eddy center, divided by the eddy's reported length scale L_s (see below in section Supplementary Environmental Data), was defined as the normalized radius $\hat{r}$. We computed kernel-smoothed probability density functions (PDFs; smoothing parameter = 0.1) for adult female utilization as a function of $\hat{r}$ to the nearest eddy, for two categories of estimated behavioral state: area-restricted search ($b > 0.75$) and transit ($b < 0.25$). These categorical definitions follow Jonsen et al. [60]. We quantify the difference between these two distributions using measures of both distribution location and shape. As a measure of distribution location, we calculate the median $\hat{r}$ at each state, and

Table 1. Migration summary of satellite tagged adult female northern fur seals (NFS) from St. Paul and Bogoslof Islands used in this study.

Year	ID	Island	Departure date	Date of entry into NP	Date of entry into GA	Date of entry into CC	Days to arrival in the NP	Days to arrival in the GA	Days to arrival in the CC	Total tracking days
2005	283	Bogoslof	11/16/05	11/16/05	12/31/05	01/02/06	0.5	45.5	46.8	60.5
2005	285	Bogoslof	11/03/05	11/05/05	01/02/06	01/07/06	2.0	60.3	65.0	72.3
2005	291	Bogoslof	11/14/05	11/15/05	12/30/05		1.3	46.0		47.5
2005	293	Bogoslof	11/16/05	11/17/05	01/13/06	01/25/06	1.3	57.8	70.5	97.0
2005	295	Bogoslof	11/14/05	11/16/05	01/06/06	02/02/06	2.3	52.5	79.5	131.3
2005	296	Bogoslof	11/14/05	11/15/05	01/28/06	01/29/06	1.8	75.3	76.3	132.3
2005	297	Bogoslof	11/19/05	11/22/05		02/18/06	2.8		91.0	107.3
2006	450	Bogoslof	11/09/06	11/12/06	12/20/06	01/06/07	3.0	41.5	58.5	96.3
2006	456	Bogoslof	11/06/06	11/10/06	01/03/07	01/07/07	3.8	57.5	62.3	92.3
2006	460*	Bogoslof	11/17/06	11/18/06	01/04/07	01/04/07	1.3	48.0	48.3	211.5
		Mean	11/12	11/14	01/04	01/18	2.0	53.8	66.5	104.8
		(SD)	(4.9)	(4.7)	(10.7)	(16.6)	(1.0)	(10.3)	(14.5)	(46.5)
2002	189	St. Paul	11/23/02	12/04/02	02/24/03	03/12/03	11.0	93.0	108.5	240.0
2002	192	St. Paul	11/25/02	12/23/02	01/20/03	01/28/03	28.3	56.5	64.0	194.5
2002	193	St. Paul	11/23/02	12/02/02	01/22/03	01/31/03	9.3	60.5	69.0	155.0
2002	197	St. Paul	11/29/02	12/07/02	12/22/02		8.3	23.5		213.3
2002	198	St. Paul	11/26/02	11/30/02	01/16/03	01/19/03	3.8	51.0	53.8	71.0
2002	200	St. Paul	11/29/02	12/08/02	01/14/03	01/22/03	9.3	46.3	54.0	87.0
2002	201	St. Paul	11/28/02	12/05/02	02/12/03	03/26/03	6.8	76.0	117.3	195.0
2004	243	St. Paul	11/10/04	11/20/04	12/24/04		10.8	44.3		138.8
2004	246	St. Paul	11/13/04	11/19/04	01/27/05		6.0	75.3		148.3
2004	251	St. Paul	11/21/04	11/29/04	02/02/05	02/03/05	7.8	72.5	74.0	114.8
2004	254	St. Paul	11/14/04	12/05/04	01/04/05	03/23/05	21.0	51.3	129.5	198.0
2004	256	St. Paul	11/17/04	12/07/04	03/04/05		19.8	107.0		194.3
2004	257	St. Paul	11/13/04	11/20/04	12/18/04	01/02/05	6.8	35.5	50.5	111.5
2005	305	St. Paul	11/06/05	11/13/05	01/07/06		7.8	62.8		64.3
2005	317	St. Paul	11/08/05	11/14/05	12/20/05	01/24/06	6.0	42.5	77.5	82.3
2006	427	St. Paul	11/29/06	12/13/06	01/16/07	02/09/07	13.3	48.0	71.8	159.5
2006	439	St. Paul	11/12/06	11/16/06	12/20/06	12/21/06	4.8	38.0	39.5	132.8
2006	442	St. Paul	11/12/06	11/19/06	01/03/07	01/11/07	7.0	52.3	60.5	66.5
2007	626*	St. Paul	11/17/07	11/23/07	01/15/08	04/15/08	5.3	58.3	149.5	249.5
2007	627	St. Paul	11/18/07	11/23/07	01/10/08	01/16/08	5.0	53.5	59.3	81.8
2007	628*	St. Paul	11/17/07	11/23/07	12/26/07	01/07/08	6.3	39.0	51.0	217.8
2007	630	St. Paul	11/18/07	11/28/07	12/02/07	01/28/08	10.0	13.8	70.8	182.3
2008	661	St. Paul	11/21/08	11/25/08	01/09/09	01/11/09	4.0	49.3	51.3	220.5

Table 1. Cont.

Year	ID	Island	Departure date	Date of entry into NP	Date of entry into GA	Date of entry into CC	Days to arrival in the NP	Days to arrival in the GA	Days to arrival in the CC	Total tracking days
2008	662	St. Paul	11/19/08	11/29/08	12/05/08	01/02/09	9.8	15.5	43.8	178.5
2008	663	St. Paul	11/20/08	11/27/08	12/27/08	01/05/09	7.5	37.5	46.5	153.0
2008	666	St. Paul	11/21/08	11/28/08	01/04/09	01/06/09	7.3	44.5	46.3	56.0
2008	668	St. Paul	11/21/08	11/28/08	03/18/09	03/23/09	6.8	116.8	122.0	225.0
2009	662B	St. Paul	11/11/09	11/19/09	12/30/09	01/19/10	8.0	48.8	68.8	122.0
2009	670	St. Paul	11/05/09	11/09/09	11/26/09		4.8	21.3		110.0
2009	676	St. Paul	11/12/09	11/16/09	01/05/10	01/11/10	4.0	54.5	60.3	119.0
2009	677	St. Paul	10/15/09	10/29/09	01/02/10	01/08/10	14.3	79.0	85.3	139.0
		Mean	11/17	11/26	01/09	01/29	9.1	53.8	73.0	149.1
		(SD)	(9.1)	(10.9)	(25.4)	(31.0)	(5.4)	(23.8)	(29.6)	(56.8)

Table columns indicate deployment year, seal identification number, breeding island of origin, departure date from breeding island, dates of entry into the North Pacific (NP), Gulf of Alaska (GA), and California Current (CC) Large Marine Ecosystems, travel time to these ecosystems, and total tracking days.
*Female NFS equipped with dive recorders who foraged near Seagliders.

compute the difference between these values to establish which distribution is more localized towards the center of eddy features in our observations [$\Delta\hat{r} = \hat{r}_{1/2}(b>0.75) - \hat{r}_{1/2}(b<0.25)$, where ½ subscript indicates the median value]. For distribution shape, we calculate the difference between the probability density functions as a function of $\hat{r}$, to illustrate radial positions where probability densities are higher or lower between states [$\Delta f(\hat{r}) = f(\hat{r}|b>0.75) - f(\hat{r}|b<0.25)$, where $f(\hat{r}|\ldots)$ denotes the probability density at radius $\hat{r}$, conditioned by behavioral state].

The two-sample Kolmogorov-Smirnov (K-S) test is a standard statistical test that could be used to test for differences between the area-restricted search and transit radial distributions. This test evaluates the significance of a single statistic that is sensitive to both differences in location and shape between the distributions of two sample populations. However, the high degree of autocorrelation within tracks of the satellite dataset reduces the robustness of the two-sample K-S test and introduces difficulty when computing critical test statistics based on the number of independent samples within the dataset. Instead, we test for differences between the area-restricted search and transit $\hat{r}$ distributions using a bootstrap method. The bootstrap method is designed to measure the sensitivity of the observed differences in distribution shape and location to a random sample of individual fur seals. In other words, this test assumes that the dominant random effects in our dataset are between rather than within tracks, and explores whether between-track variance in $\Delta\hat{r}$ and $\Delta f(\hat{r})$ in our dataset is large compared to the observed values. In each bootstrap iteration, a random sample of 41 adult female satellite tracks was drawn with replacement from the observed 41 tracks, to create a synthetic dataset. The quantities $\Delta\hat{r}$ and $\Delta f(\hat{r})$ were computed for this synthetic dataset and this was repeated for 10,000 iterations. Confidence bounds for the observed values of $\Delta\hat{r}$ and $\Delta f(\hat{r})$ were computed from the bootstrap distributions using bias-correction and acceleration [64].

Behavior Statistical Analyses

Generalized Linear Mixed-Effects Models. We followed methods detailed in Sterling et al. [25] and used GLMMs to investigate the effects of several environmental variables on three seal response variables – behavioral state, and the average dive depth and total number of dives in each 6 h dive period. In a large tracking dataset like the one employed in this study, between-animal variability in the degree and character of behavioral responses can confound the ability of ordinary linear models to detect correlations between predictor and response variables [65]. Generalized linear mixed-effects models were chosen for their flexibility in allowing us to specify individual animals as a random effect in our dataset, and for their established use in modeling behavioral responses of this and other pinniped species [25,55,57]. For seal behavioral state, all 41 seal migration tracks consisting of 22,597 estimated locations and behavioral states were used. We assumed an AR(1) autocorrelation structure within each track and to assist with normality, we added 0.0001 to behavioral state values of 0 and subtracted 0.0001 from behavioral state values of 1 prior to logit transforming all the behavioral state values. With respect to seal dive behavior, only 8 of the 41 female seal tracks had corresponding dive data sets. These tracks were collected in four migratory seasons (Table 1) and all eight seals traveled to the GA ecosystem, while seven of these seals entered the CC ecosystem. From these eight seals we received 1,888 dive summaries, which we then used to calculate the average dive depth and the number of dives for each 6 h dive histogram period [25,55]. Both the average dive depth and number of dives in each 6 h period were log-transformed prior to model fitting.

Table 2. Glossary of acronyms and mathematical symbols used in this manuscript.

Acronym	Definition
AIC	Akaike's Information Criterion
AVISO	Archiving, Validation, and Interpretation of Satellite Oceanographic data (source for SLA products)
BC	British Columbia, Canada
BS	Bering Sea Shelf
CC	California Current
CF	Cape Flattery, USA
Chl*a*	Chlorophyll-*a*
DESW1	National Buoy Data Center Destruction Island station
EKE	Eddy kinetic energy – see Methods section *Supplementary Environmental Data*
ENSO	El Niño-Southern Oscillation
GA	Gulf of Alaska
GAM	Generalized additive model – *see* Methods section *Statistical Analysis of Behavior Relative to Seaglider Data*
GH	Grays Harbor, USA
GLMM	Generalized linear mixed-effects model – *see* Methods section *Behavioral Statistical Analyses*
K-S	Kolmogorov-Smirnov
LME	NOAA Large Marine Ecosystem
MCMC	Markov chain Monte Carlo
MLD	Mixed-layer depth
MODIS Aqua	Moderate Resolution Imaging Spectroradiometer – Aqua (ocean color satellite)
NCEP2	National Centers for Environmental Prediction-Department of Energy Reanalysis 2 product
NFS	Northern fur seal (*Callorhinus ursinus*)
NMFS	National Marine Fisheries Service
NOAA	National Oceanographic and Atmospheric Administration
NP	North Pacific Ocean
OR	Oregon, USA
PDF	Probability density function
PTT	Platform Terminal Transmitter
SD	Standard deviation
SeaWiFS	Sea-viewing Wide Field-of-view Sensor (ocean color satellite)
SG	Seaglider
SLA	Sea level anomaly
SSSM	Switching state-space model – *see* Methods section *Modeling Seal Movement*
USFWS	United States Fish and Wildlife Service
UTC	Coordinated Universal Time
WA	Washington State, USA
WET	Western Environmental Technologies
Symbol	**Definition**
a_0	GAM intercept term
b	Adult female NFS behavioral state [0 = transit, 1 = area-restricted search], estimated using SSSM
$f(\hat{r})$	Probability density of adult female NFS habitat utilization as a function of $\hat{r}$
$F(\hat{r})$	Cumulative distribution of adult female NFS habitat utilization as a function of $\hat{r}$
$g_1(d)$	GAM term quantifying MLD vs. day of year
$F_i \times g_2(d)$	GAM term quantifying depth offset of adult female NFS day diving from MLD vs. day of year
L_s	Length scale of an altimetry-identified coherent mesoscale eddy reported by Chelton et al. [63]
$\hat{r}$	Normalized radial position – distance of NFS position from nearest mesoscale eddy center, divided by that eddy's L_s
(u,v)	East-west, north-south components of wind at 10 m height
(u_g,v_g)	East-west, north-south components of geostrophic surface ocean current anomalies
y_i	GAM response variable – MLD or adult female 6 h average day dive depth

Table 2. Cont.

Symbol	Definition
$\Delta f(\hat{r})$	Difference between $f(\hat{r})$ calculated for locations identified as area-restricted search ($b>0.75$) and $f(\hat{r})$ calculated for locations identified as transit ($b<0.25$)
$\Delta\hat{r}$	Difference between median value of $\hat{r}$ calculated for $b>0.75$ and median value of $\hat{r}$ calculated for $b<0.25$
ε_i	GAM residual term
σ_t	Density anomaly [$\sigma_t=\rho(S,T,0)-1000$ kg m^{-3}, where $\rho(S,T,P)$ is seawater density at measured salinity S, temperature T, and pressure P]

Movement and dive behavior were assessed with respect to several environmental and habitat fixed effects variables. For behavioral state, these included ecosystem, season, surface wind speed, and surface ocean kinetic energy. For the average dive depth and number of dives in each 6 h period, we added the effects of light from both the sun and moon. As a proxy for sunlight, we calculated the fraction of daylight time in each 6 h interval (hereafter, "proportion daylight") using the NOAA Sunrise/Sunset and Solar Position calculators (http://www.srrb.noaa.gov/highlights/sunrise/sunrise.html). For moonlight level, we used calculations of the lunar fractions (illuminated area divided by total area) extracted from the United States Naval Observatory website (http://aa.usno.navy.mil/data/docs/MoonFraction.php). To consider ecosystem-specific effects, we used the same modification of NOAA's LMEs described in Sterling et al. [25] as explanatory variables. These included the Alaska Stream (AS) and Bering Sea Shelf (BS), in addition to the NP, CC and GA (Fig. 1). There were very few estimated locations within the Bering Sea Basin (n = 19 dive summaries, n = 72 behavioral state estimates) and these were excluded from the behavioral statistical analyses. Season was defined as days since 1 October.

Surface wind speeds were obtained for each 6 h seal location from the National Centers for Environmental Prediction-Department of Energy Reanalysis 2 product (hereafter, "NCEP2"). These data are distributed by the NOAA Office of Oceanic and Atmospheric Research, Earth Sciences Research Laboratory Physical Sciences Division, Boulder, Colorado (USA), and made available from their web site at http://www.esrl.noaa.gov/psd/. The NCEP2 product gives surface (10 m height) east-west (u) and north-south (v) wind velocity components at four daily time steps (0000, 0600, 1200, and 1800 Coordinated Universal Time [UTC]) on a 2.5° resolution global grid. Seal location data were intentionally aligned with these time intervals using SSSMs. At each time point, u and v estimates from NCEP2 were interpolated from the global grid to seal estimated locations and converted to wind speed (m s^{-1}) and direction. In addition to wind, previous studies have identified surface ocean mesoscale circulation as an influence on horizontal behavior [1]. We calculated eddy kinetic energy (EKE) from satellite estimates of surface geostrophic velocity anomaly (see below in section *Supplementary Environmental Data*) and evaluated this variable as a predictor of estimated behavioral state [66]. For the reasons outlined above in section *Estimates of Horizontal Habitat Utilization*, interannual variability was not explored in analyses of movement and diving behavior, which likely contributes to unexplained variance in each model.

Models were built and calculated using the nlme V3.1-103 package within the R statistical software. For all models the NP ecosystem response was used as the base or reference model. Several model configurations for each response variable were constructed using the main effects and interaction terms, fit by maximum likelihood methods, and then contrasted and ranked using Akaike's Information Criterion (AIC; [67]), where the top models were judged by lowest AIC score. The configuration of main effects and interaction terms for the top three models of each response variable are presented in Table 3. For the top-ranked model, the coefficients and significance level of each term are also presented (see Results section *Diving and Movement Behavior*).

Washington Coast Seaglider Sampling

We compared adult NFS female behavioral patterns and satellite-tracked movements to a time series of remote *in situ* hydrography collected by Seagliders operating off the WA coast in the same years and same area that adult female NFS overwintered. The Seaglider time series included data collected from two test deployments in late 2002 and early 2003 and nearly continuous data from late 2003 to early 2009 (Fig. 2; [68–70]). Seaglider is a long-range ocean glider, which operates by decreasing (increasing) its buoyancy in order to descend (ascend) from the surface to 1000 m [51]. It is equipped with small wings that generate lift (downforce) during the descent (ascent) phase, allowing the vehicle to move horizontally in a sawtooth pattern with an average horizontal speed through the water of approximately 0.2 m s^{-1}. A dive-climb cycle to full depth lasts 8 hours and covers 6 km horizontally. Seagliders were programmed to surface and transmit profile data after every dive-climb cycle. Washington coast deployments typically lasted several months over which time a Seaglider performed several hundred of these cycles. Each Seaglider was equipped with a Paine 211-75-710-05 pressure transducer and custom-fitted SeaBird Electronics SBE-3 thermistor and SBE-4 conductivity cell mounted in a dorsal sting in order to sample temperature, salinity, and pressure. Conductivity cells and thermistors on WA coast Seagliders were calibrated by SeaBird before and after each deployment. Each vehicle also carried a Western Environmental Technologies (WET) Labs ECO-BB2F optical "puck"-style sensor, which sampled chlorophyll-*a* (Chl*a*) fluorescence and optical backscatter (proxies for phytoplankton pigment and particle concentration, respectively) in the top 150 m of each vertical profile [68].

During the 5.5-year time series, Seagliders collected hydrographic data while navigating along two cross-shore transect lines. These 210–225 km long transects extended from two points along the shelf break and were joined at their offshore end at 47°N, 128°W (Fig. 2A). A single crossing usually required two weeks of transit time and recorded 50–75 dive-climb cycles, thus yielding 100–150 vertical profiles. Over each profile, all sensors sampled every 10 seconds in the top 150 m of the water column, which corresponds to a vertical resolution of 0.6–1.0 m. Adult female NFS dives rarely approach this depth in the CC [1,25]. Samples were collected every 20 seconds to 300 m and every 30 seconds between 300 m and 1000 m. Seagliders made 63 crossings along

Table 3. Top-ranked generalized linear mixed-effects models (GLMMs) result for number of dives per 6 h period, average maximum dive depth per 6 h period, and estimated behavioral state.

Response Variable	Model Terms	AIC	ΔAIC
Number of Dives			
(n = 1888)			
	(a) DAYL+LME+SEAS +(DAYL×LME)	5909.3	0.0
	(b) DAYL+LME+LUN +(DAYL×LME)+(DAYL×LUN)	5970.5	61.2
	(c) DAYL+LME+LUN+SEAS +(SEAS×DAYL)	5970.7	61.5
Dive Depth			
(n = 1888)			
	(a) DAYL+LME+LUN+SEAS +(DAYL×LME)+(DAYL×LUN)+(LUN×LME)	2857.1	0.0
	(b) DAYL+LME+LUN+SEAS +(DAYL×LUN)+(LUN×LME)	2910.5	53.4
	(c) DAYL+LME+LUN +(DAYL×LME)+(DAYL×LUN)+(LUN×LME)	2920.1	63.0
Behavioral State			
(n = 22597)			
	(a) WIND+SEAS+LME +(WIND×SEAS)	61974.3	0.0
	(b) WIND+SEAS+LME +(WIND×LME)	61979.6	5.3
	(c) WIND+SEAS +(WIND×SEAS)	61984.8	10.5

Columns indicate response variable, model terms, Akaike's Information Criterion (AIC) score, and difference in AIC score from the best model (ΔAIC). Model terms are abbreviated as follows: DAYL (proportion daylight in each 6 h period), LME (NOAA Large Marine Ecosystem), SEAS (days since 1 October), LUN (lunar fraction in each 6 h period), WIND (NCEP2 interpolated 10 m wind speed).

the northern transect (Cape Flattery; CF) and 62 along the southern transect (Grays Harbor; GH) and achieved near-continuous coverage from August 2003 to January 2009 (Fig. 2). Hardware faults resulted in some data gaps, the longest of which was from mid-January to early April 2008 (Fig. 2B). Seaglider data coverage overlaps with the period during which the most satellite-tagged females were present in the CC and GA LMEs, providing good coverage of the overwinter periods between 2003–04 and late 2008-early 2009 (Fig. 2B).

Seaglider Data Processing

Seaglider observations of seawater conductivity, temperature, and pressure were used to derive profiles of salinity and density as described in [70]. In contrast to shipboard measurements, flow of seawater past the Seaglider conductivity sensor is not driven by a pump, and is instead flushed by the motion of the vehicle. This can produce large spikes or biases in salinity if not properly accounted for. The post-processing procedure outlined in [70] makes corrections for this where possible or discards portions of profiles that cannot fully be corrected. A detailed description of the correction applied to Seaglider salinity measurements is forthcoming [Eriksen, CC; unpublished data]. Seaglider measurements of salinity are accurate to 0.03 (parts per thousand, hereafter presented without units) in regions of strong vertical temperature gradient or 0.01 in other regions of the water column, and temperature sample accuracy is 0.003°C.

Mixed-layer depth for each profile was calculated based on the density step algorithm of de Boyer Montegut et al. [71], using a density step equivalent to a 0.2°C temperature decrease from a 10 m reference value. Although profile data are used where possible, observations for each 2-week crossing were also interpolated to a grid with regular horizontal and vertical spacing as described in [70].

The fluorescence and optical backscatter sensors (collectively referred to as the optical sensors) fielded on WA Seagliders provide qualitative information about the distribution of phytoplankton pigment and biomass. We report results from the fluorescence sensor only. Perry et al. [68] and Sackmann [69] analyzed the Seaglider optical data through the year 2007, including comparison to satellite-inferred surface Chl*a*, and we followed their procedures for processing and interpretation. Starting from reported digital sensor counts at each sample, we subtracted a background offset unique to each sensor, determined from *in situ* measurements in clear, dark water, and then applied the manufacturer's calibration formula to convert fluorescence counts above background to Chl*a* concentration (mg m^{-3}). We used night values of Chl*a* only in order to avoid the effects of fluorescence quenching, which produces a low bias in daytime measurements collected near the surface [68,69,72]. It should be emphasized that

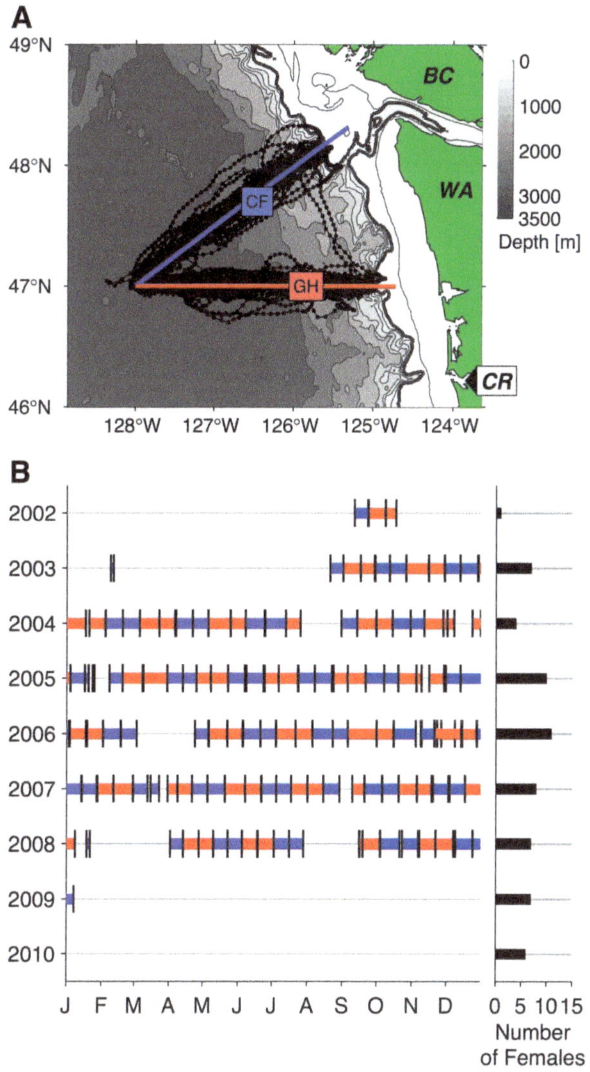

Figure 2. Seaglider time series over the Washington continental slope region. (A) Intended navigational track pattern for Cape Flattery (blue, labeled CF) and Grays Harbor (red, labeled GH) transects, with Seaglider tracks (black lines) and dive mid-point locations (black circles). Isobaths are contoured at 100 and 200 m, at 200 m intervals to 1000 m, and at 500 m intervals at depths greater than 1000 m (depth scale at top right). The 200 m isobath is highlighted in dark gray. Black labels denote Washington (WA), Vancouver Island, British Columbia (BC), and the mouth of the Columbia River (CR). (B) Yearly Seaglider pattern occupation and overlap with satellite-tagged northern fur seal adult female presence in the California Current and Gulf of Alaska ecosystems. The left portion of panel B shows Seaglider transit along the Cape Flattery line (blue), Grays Harbor line (red), and section boundaries (turnaround points, indicated by black vertical lines). Black bars at right show the number of individual satellite-tagged adult females present in the California Current and Gulf of Alaska ecosystems during each year.

absolute concentrations derived using this method, and reported herein, are unreliable due to the unavailability of contemporary shipboard *in situ* data for comparison. However, relative temporal and spatial structure measured by the sensors is robust [68,69,73]. For an extended description of the fluorescence processing, see the supporting information (Supporting Methodology S1).

Seaglider Analyses

For physical and optical variables, we generated an average annual cycle for the surface ocean zone (top 150 m) within a cross-shore band 60 to 80 km from the shelf break. Observations of Chl*a*, salinity, and density anomaly (σ_t; kg m^{-3}) in this zone were sorted into bins by depth and days since 1 January of each year. Bins had 6 m vertical and three-week temporal width and 66% overlap between adjacent bins, i.e., adjacent bins have their center one week apart in time and 2 m in depth. The overlap between bins acts to smooth the resulting averages. The amount of overlap was chosen in order to reduce noise in the average annual cycle that is generated by interannual variability in the timing of events such as the spring mixed layer shoaling and phytoplankton bloom. The mean value in each bin was taken first for all observations within each year and then the median value of these means was taken across years.

Statistical Analysis of Behavior Relative to Seaglider Data

The upper-ocean properties observed from Seagliders were used to investigate seasonal trends in adult female daytime diving in the CC ecosystem. The choice of daytime diving characteristics was motivated by a parallel study of NFS migration during a single year, in which Sterling et al. [25] observed a single adult female enter the CC ecosystem and increase the proportion of dives which occurred during daytime relative to all other ecosystems. Specifically, we examined the average depth of female dives in 6 h daytime periods, defined as those with >80% proportion daylight, along with Seaglider observations of MLD in the region 60–80 km from the shelf break, versus days since 1 January using a generalized additive model (GAM). A GAM is a model in which the assumption of a linear response to predictor variables, even in transformed space, is relaxed and the predictor terms may take arbitrary form [65]. The GAM is an appropriate choice for modeling these response variables since we expect that some portions of their winter-summer evolution may be nonlinear or discontinuous, particularly in the case of MLD which shoals abruptly after the spring transition [68]. In our case, we used a GAM to model log-transformed depth – either depth of the surface mixed layer or depth of day dives averaged in 6 h periods – using an intercept and two predictor terms that are functions of yearday. The model takes the form

$$\log(y_i) = a_0 + g_1(d_i) + F_i \times g_2(d_i) + \varepsilon_i \qquad (1)$$

where y_i is the depth (either MLD or seal dive depth) and d_i the day of observation i, a_0 the intercept, ε_i the residual, and g_1 and g_2 are arbitrary functions to be estimated. F_i is the interaction coefficient, which is set to 0 for observations of MLD and 1 for observations of average day dive depth. Thus, the function $g_1(d)$ is a fit to the annual cycle of MLD over the months January–June while $g_2(d)$ quantifies a possible time-dependent offset of adult female NFS day dive depths relative to the MLD.

Each predictor term is in practice a locally weighted regression of the observations and thus acts similar to a smoothing filter or running average. The model fit is accomplished by minimizing the negative penalized log likelihood function, which considers not only the model disagreement from data but also the "roughness" of the empirical functions $g_1(d)$ and $g_2(d)$. Thus the GAM is intermediate between a function that produces an exact fit to all observed MLD/average day dive depth observations (e.g., spline interpolation) and one that applies a linear fit to all observations in each category versus yearday (analysis of covariance). The degree of compromise between these two – the relative weight applied to

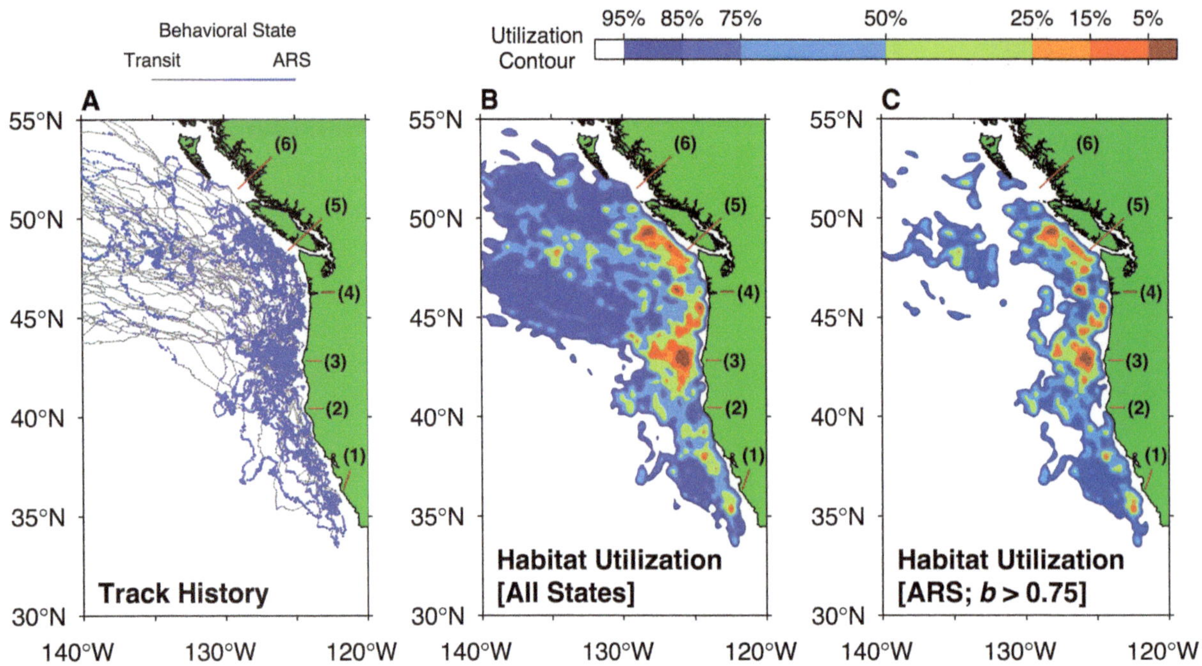

Figure 3. Track density and habitat utilization of adult female northern fur seals in the California Current and southern Gulf of Alaska. (A) Satellite-tracked migratory routes of adult female northern fur seals. Tracks are colored and weighted by estimated behavioral state: thin gray lines correspond to transit behavior, while thick blue lines indicate area-restricted search (ARS, scale at top left). (B) Density-kernel estimate of horizontal habitat utilization in the California Current and southern Gulf of Alaska. Contours indicate the range size enclosed by each rank percentile of habitat utilization; i.e., the 50% contour is the minimum possible area that could be drawn to enclose 50% of female time spent within the domain. These contours are generated from a fixed-kernel habitat density estimate using a 15 km spatial smoothing scale (scale at top right). (C) Area-restricted search habitat utilization estimate – as for panel B, but obtained using a subset of tracking points with seal behavioral state (indicated by variable *b*) greater than 0.75. In each panel, prominent coastal features are labeled as follows: (1) Point Sur; (2) Cape Mendocino; (3) Cape Blanco; (4) Columbia River; (5) Strait of Juan de Fuca mouth; (6) Queen Charlotte Sound.

model misfit versus model roughness – is determined through cross-validation. The GAM was implemented in R using the *gam* package V1.09. Residuals from the GAM fit were approximately normally distributed and did not show significant evidence of nonstationarity (Fig. S1).

Supplementary Environmental Data

To compare seal tracks to mesoscale oceanographic circulation, we utilized gridded SLA and surface geostrophic velocity anomaly products obtained from the Archiving, Validation, and Interpretation of Satellite Oceanographic data (AVISO) Reference Series merged delayed-time product (http://www.aviso.oceanobs.com). From these we calculated EKE $[(u_g{}^2+v_g{}^2)/2$ (m^2 s^{-2}), where u_g and v_g are zonal (positive eastward) and meridional (positive northward) geostrophic velocity anomalies, respectively]. To identify coherent mesoscale features and compare female movements to eddy locations in regions outside of the Seaglider transect pattern, we used eddy trajectories of Chelton et al. [63], available online from http://cioss.coas.oregonstate.edu/eddies/. This dataset also utilizes the AVISO Reference Series gridded SLA product, giving eddy center latitudes and longitudes, length scales (degrees of arc length), polarities (cyclonic or anticyclonic), and strengths (in cm of SLA) at 7 d time steps for the period October 1992 to January 2011 at the time of download. Chelton et al. [63] report eddy length scales (L_s) as the radius of a circle with area equal to that enclosed by the contour of maximum velocity around each eddy. It should be noted that an individual eddy is in fact rarely circular, though consistent radial structure does appear when averaging over a large number of eddies [63]. The gridded SLA data used for qualitative interpretation in our analysis, and also used for eddy detection by Chelton et al., [63] are produced by interpolating and smoothing the output of two altimeters. Chelton et al. [63] analyzed the approximate spatial and temporal filtering characteristics and found that eddies with spatial scales of 30–40 km were the smallest that the gridded product could resolve at the latitudes considered in this study. Altimetry-resolved eddies in the CC have horizontal radii typically ranging from 60–90 km, and for long-lived coherent features, lifetimes on the order of weeks [74]. Mean propagation speeds are ~0.05 m s^{-1} or less at the latitudes considered in this study, which indicates that eddies move relatively slowly in comparison to the 7 d gridded altimetry time scale.

Several environmental variables were utilized for qualitative comparison and discussion relative to seal behavior. For spatial representation of surface Chl*a* and temperature patterns relative to Seaglider and seal foraging data we utilized Sea-viewing Wide Field-of-view Sensor (SeaWiFS) and Moderate Resolution Imaging Spectroradiometer – Aqua (MODIS Aqua) Level-2 surface ocean color and temperature swaths, processed and made available online by the National Aeronautics and Space Administration Ocean Color project (http://oceancolor.gsfc.nasa.gov/; [75]). All bathymetry data shown were sourced from the NOAA Earth Topography One Arc-Minute Global Relief Model (ETOPO1) gridded dataset (http://www.ngdc.noaa.gov/mgg/bathymetry/relief.html). We obtained profile data from one Argo float from the United States Global Ocean Data Assimilation Experiment Argo Page (http://www.usgodae.org/argo/argo.html). Wind data for the WA coast were obtained from National

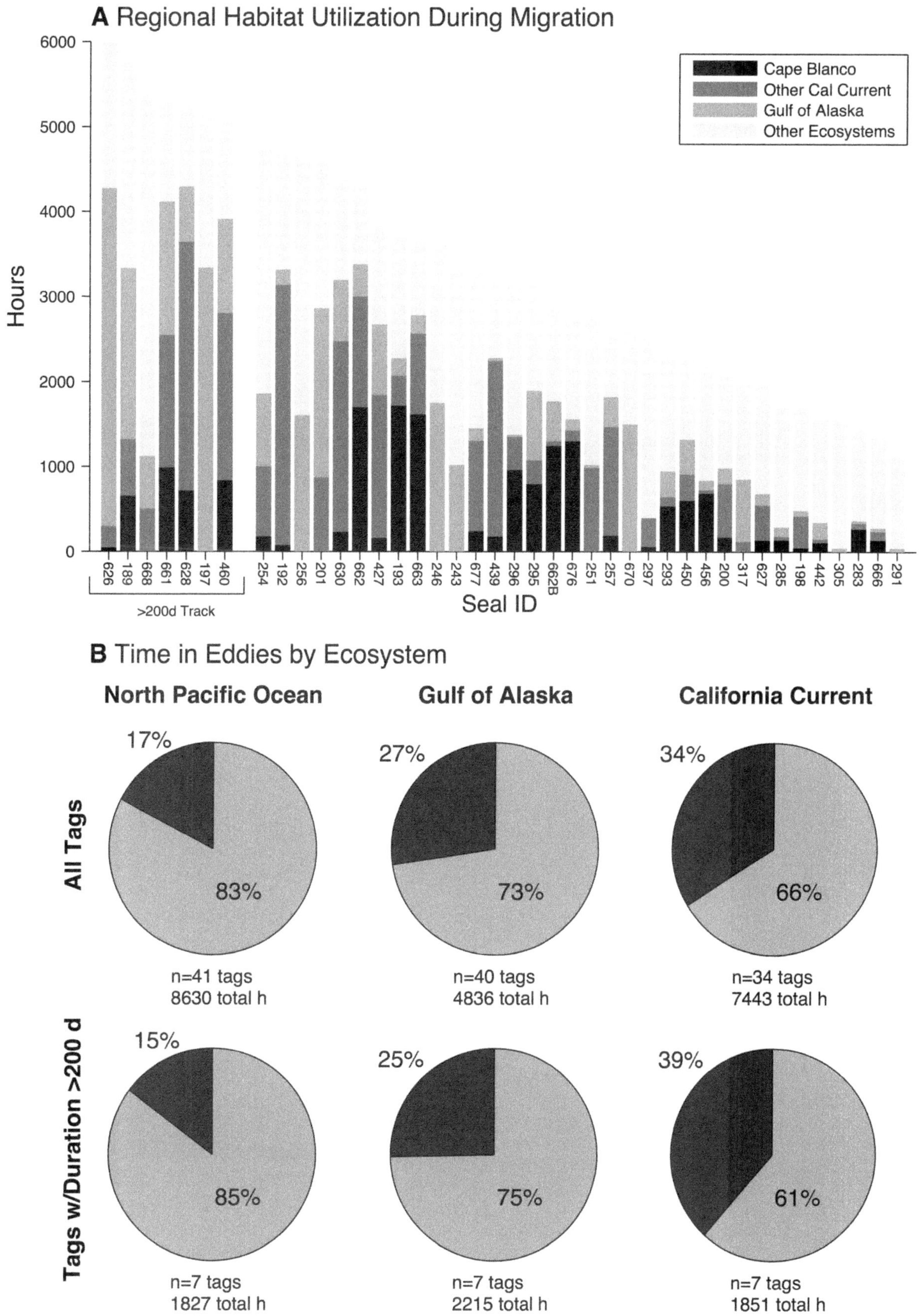

Figure 4. Regional habitat utilization and time spent in eddies by adult female northern fur seals. For each satellite-tracked migratory seal, panel (A) shows number of hours spent in the Cape Blanco region (black, defined as 41.5°N to 44.5°N and east of 128.5°W), other portions of the California Current (dark gray), the Gulf of Alaska ecosystem (medium gray), and all other ecosystems (light gray). (B) The fraction of total female time

spent within L_s+20 km of the center of mesoscale eddies as identified by Chelton et al. [63] in each ecosystem, where L_s is each eddy's reported radial length scale. The first row displays results for the full dataset, while the second row shows results for only seals with a tracking duration > 200 d.

Data Buoy Center station Destruction Island (DESW1), available at http://www.ndbc.noaa.gov. Wind vector measurements at DESW1, made at a height of 52.7 m above mean sea level, were daily averaged and rotated into an alongshore/cross-shore coordinate system, with positive alongshore direction (poleward winds) defined as being along 338°T, the local orientation of the coastline near DESW1.

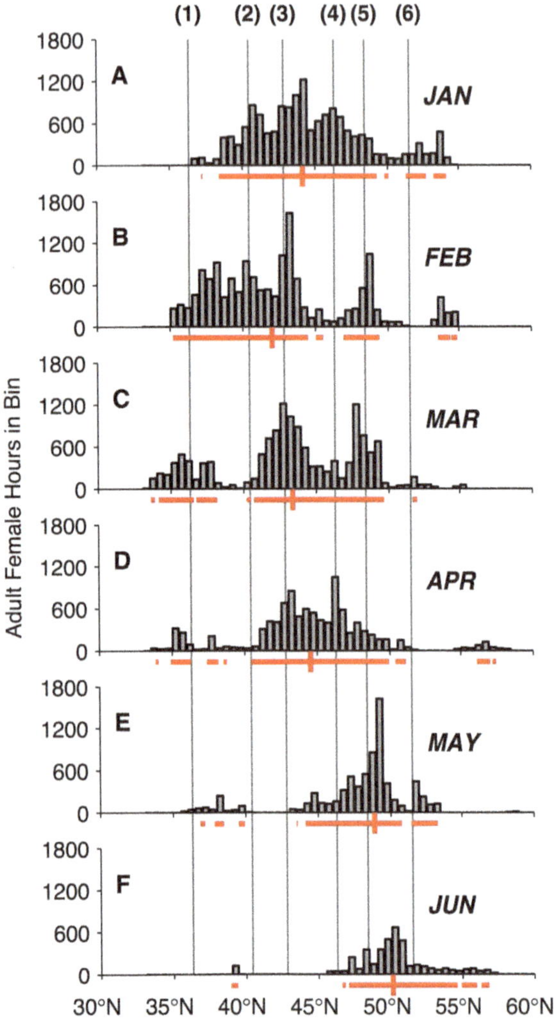

Figure 5. Monthly evolution of adult female northern fur seal alongshore distribution. January (A) through June (F). Gray bars in each plot are a histogram of time spent in that month in 0.5° latitude bins within the California Current and Gulf of Alaska Large Marine Ecosystems east of 145°W, during the winters 2002–03 to 2009–10. Red horizontal lines below A–F indicate the latitudes in which 95% of time spent occurred in that month. Vertical red lines indicate median latitude of the alongshore distribution. Vertical gray lines extending through panels A–F correspond to the latitudes of prominent coastal features, illustrated in Figure 3, which are also numbered above panel A in this figure as follows: (1) Point Sur; (2) Cape Mendocino; (3) Cape Blanco; (4) Columbia River; (5) Strait of Juan de Fuca mouth; (6) Queen Charlotte Sound.

Results

Migratory Distribution

Migratory tracks from all 40 females (41 tracks total) are shown in Figure 1. Seals departing St. Paul Island averaged 9.1 d (n = 31, standard deviation [SD] = 5.4 d) to reach the NP (Table 1), while seals from Bogoslof Island took only 2.0 d (n = 10, SD = 1.0 d) due to their closer proximity to the NP (Fig. 1). The earliest arrival in the GA ecosystem was 26 November (mean: 8 January) and the earliest arrival in the CC was 21 December (mean: 26 January). Mean time between departure and first entry to the GA and CC ecosystems was 53.8 d (n = 40, SD = 21.4 d) and 71.3 d (n = 34, SD = 26.4 d) respectively and mean track duration was 138.3 d (n = 41, SD = 57.2 d).

Individuals in this study typically entered the CC and GA ecosystems at latitudes between 45°N–55°N and, once in the destination region, foraged within a broad zone between 140°W–120°W, 30°N–55°N (Fig. 3A–C). The 95% habitat utilization contour calculated using only locations within this region extends from ~54°N to as far south as 34°N (Fig. 3A–B). South of 45°N, fur seal habitat utilization was primarily confined between the coast and 130°W (Fig. 3B). North of 45°N, significant utilization extended west of 130°W, but this was primarily driven by seals arriving into the GA and CC ecosystems during their early migration transit phase (Fig. 3A). Confining the two-dimensional distribution analysis to only locations exhibiting area-restricted search behavior ($b>0.75$) showed that these locations were more closely confined to the coast, with some limited area-restricted search utilization near 135°W, near the eastern terminus of the North Pacific Current, the broad eastward-flowing current that forms the boundary between the northeast Pacific subtropical and subarctic gyres (Fig. 3C; [76]).

The area-restricted search-only distribution shows several distinct regions of concentrated adult female habitat utilization, the largest of which is near Cape Blanco, Oregon (OR, USA; 42.84°N). For each track, Figure 4 displays the amount of time

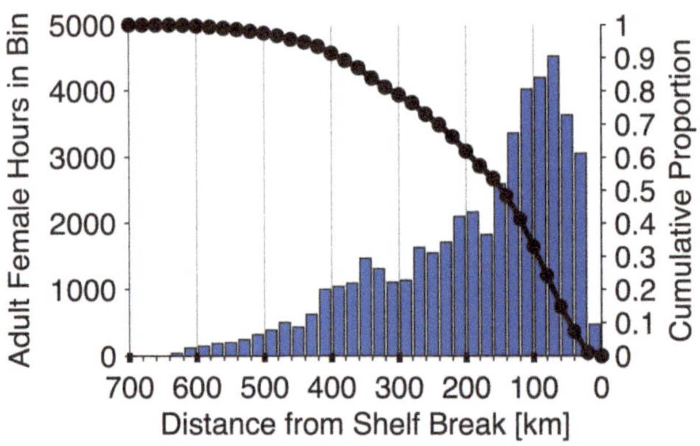

Figure 6. Cross-shore distribution of adult female northern fur seals. Figure depicts time spent versus distance from the shelf break, binned in 20 km segments (blue bars), by seals from 41°N to 52°N over the study duration. These latitudes were chosen to represent the region where the majority of adult female habitat utilization occurred. Cumulative proportion of female time spent from the shelf break (200 m isobath) shown in black.

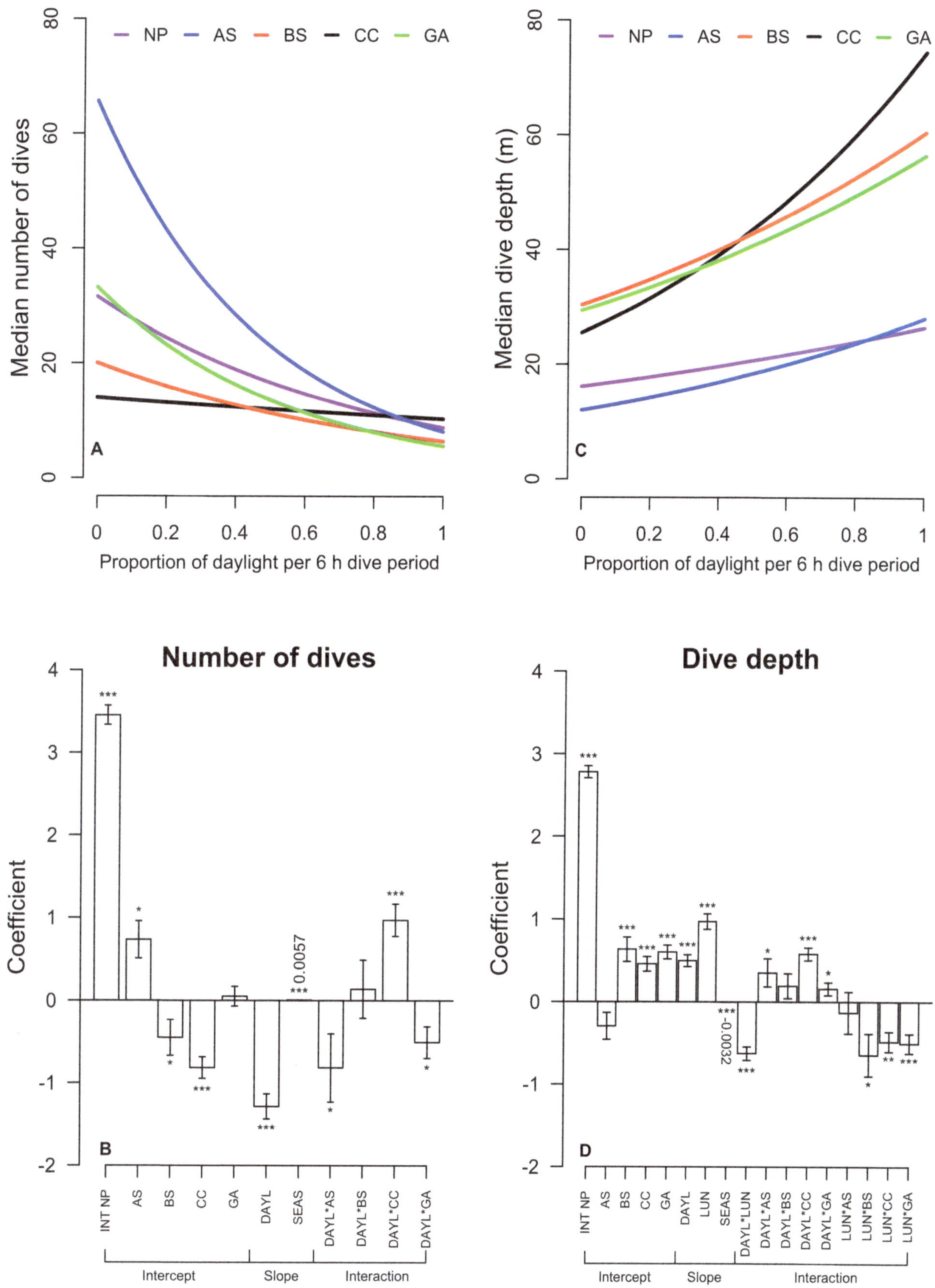
NP
AS
BS
CC
GA
Median number of dives
Median dive depth (m)
A
C
Proportion of daylight per 6 h dive period
Number of dives
Dive depth
Coefficient
B
D
0.0057
-0.0032
INT NP
AS
BS
CC
GA
DAYL
LUN
SEAS
DAYL*LUN
DAYL*AS
DAYL*BS
DAYL*CC
DAYL*GA
LUN*AS
LUN*BS
LUN*CC
LUN*GA
Intercept
Slope
Interaction

Figure 7. Generalized linear mixed-effects model (GLMM) results for adult female northern fur seal dive behavior. (A) Predicted model response for median number of dives per 6 h period as a function of proportion daylight and ecosystem [North Pacific (NP) = purple, Alaska Stream (AS) = blue, Bering Sea Shelf (BS) = red, California Current (CC) = black, and Gulf of Alaska (GA) = green]. The CC was the only ecosystem to show a nearly equal amount of dives throughout the diurnal cycle, contrasting all other ecosystems where more dives occurred at night and fewer with increasing proportion daylight. (B) Plot of the coefficients of predictor terms in the best GLMM for number of dives per 6 h period. Each vertical bar corresponds to one term. Size of the bar indicates its magnitude (positive = increasing number of dives per 6 h period, scale on y-axis) and whiskers indicate standard error. Terms are labeled below the x-axis and grouped by intercept, main effects or slope, and interaction terms. Asterisks above or below each bar indicate the significance level of each term: $* = p < 0.05$; $** = p < 0.001$; $*** = p < 0.0001$. Coefficients are labeled where bars are too small to see. Abbreviations, in addition to ecosystems labeled above, are as follows: INT NP (intercept, North Pacific base model), DAYL (proportion daylight per 6 h period), SEAS (season, days since 1 October). Note that while the season term is small, its effect may be large since season has values that range from 0 to 297 d, while proportion daylight, as an example, varies from 0 to 1. (C) Modeled response for average dive depth per 6 h period as a function of proportion daylight and ecosystem, symbols as in panel A. The CC also showed the greatest response of increasing average dive depth with increasing proportion daylight. (D) Schematic plot of coefficients for terms in the best GLMM for average dive depth per 6 h period, symbols as in panel B. Lunar fraction is abbreviated by LUN.

spent in four migratory regions: the coastal zone proximate to Cape Blanco (defined as between 41.5°N–44.5°N, and east of 128.5°W), portions of the CC ecosystem excluding Cape Blanco, the GA ecosystem, and all other ecosystems. Twenty migratory females spent at least one week in the Cape Blanco zone, while 9 spent more than one month, confirming that the elevated utilization of this zone was not driven exclusively by a relatively small number of animals (Fig. 4A). Figure 4A also demonstrates the consistency across the animals in this study of migratory transit duration between the breeding grounds and destination ecosystems – this is shown by the length of time spent in ecosystems outside the CC or GA in each track (labeled "Other Ecosystems" and indicated by the lightest gray shading in Fig. 4A). For tracks with shorter tag lifetimes, time spent in the CC or GA was reduced, but time spent in other ecosystems, presumably during the transit phase, remained relatively constant. As females moved closer to the continental margin, and eddy generation regions within the CC and GA eastern boundary current systems, their fraction of time spent within $L_s + 20$ km of the center of identified coherent eddy features increased by a factor of 2 (Fig. 4B). This signal was most pronounced in the 7 of 41 tracks with duration > 200 d (Fig. 4B bottom).

Adult female distribution as a function of latitude along the continental margin was not static throughout the overwintering period, reflecting large-scale ecosystem seasonal patterns and migratory pressure to return to the breeding grounds late in the overwintering period (Fig. 5). Adult female time spent off California peaked in February and declined from March onwards. As females began traveling northward for the return leg of their migration, the median latitude of their distribution shifted northward to the OR and WA coasts in March and April and British Columbia (BC, Canada) in May and June (Fig. 5D–F). Females occupied latitudes off the OR, WA, and southern BC coasts consistently from January–April and some remained off the WA and BC coasts into May and June. In addition to Cape Blanco (January–April), the latitudinal distribution exhibited local peaks of female time spent near other prominent coastal topographic, riverine, or inlet features such as the Columbia River mouth (April), the Strait of Juan de Fuca, (February–March, May), Queen Charlotte Sound, BC (May–June), and Point Sur/Monterey Bay, California (February–April). Females also spent elevated time near latitudes corresponding to Cape Mendocino, California (January–February), though examination of the two-dimensional habitat utilization shows that this time was farther offshore than near other coastal capes.

Seals in the CC and GA ecosystems spent the majority of their time between 41°N and 52°N (72.0% of 68,016 total off-shelf hours in the months January–June; Fig. 5). From March onwards, females spent 80.6% of their time in these latitudes. The cross-shore distribution of time spent within this zone peaked between 60 and 80 km from the shelf break (Fig. 6). Females spent 33% of their off-shelf time within 100 km of the shelf break and 62% within 200 km, the latter of which is the approximate cross-shore zone sampled by Seagliders off WA (Fig. 2). The median (50% of time spent) of the cross-shore distribution occurred at 146 km offshore from the shelf break. When evaluating the cross-shore distribution by month (Fig. S2), the median value had a maximum of 260 km (January) and minimum of 71 km (May).

Diving and Movement Behavior

The top-ranking GLMM for seal number of dives per 6 h period identified proportion daylight, ecosystem, season, and the interaction between ecosystem and proportion daylight as significant predictors (Table 3). In the NP (the base model) increasing proportion daylight was a predictor of fewer dives per 6 h period (Fig. 7A–B). However, the intercepts for the AS, BS, and CC ecosystems differed, resulting in more dives in complete darkness in the AS ecosystem and fewer dives in complete darkness in the BS and CC ecosystems (Fig. 7A–B). The slope of the ecosystem and proportion daylight interaction term in the CC differed from the NP ecosystem due to more female dives occurring during the daytime as they entered coastal transition zone foraging habitat off the coast of North America (Fig. 7A–B). This result is contrary to other ecosystems considered – in the CC, adult female dives were evenly spread throughout the day and showed virtually no correlation with proportion daylight per 6 h period (Fig. 7A).

For seal average dive depth per 6 h period, the top-ranking model identified proportion daylight, ecosystem, lunar fraction, season, and the interactions between proportion daylight and ecosystem, proportion daylight and lunar fraction, lunar fraction and ecosystem as significant predictors (Table 3). Proportion daylight and lunar fraction were both associated with increased average dive depth per 6 h period (Fig. 7C–D). Both of these terms affect the depth distributions of diel migrating prey fields, which respond to increasing light levels regardless of whether they are due to the sun or moon. Ecosystem interactions were important in altering the modeled response of average dive depth to increasing proportion daylight, with the CC ecosystem having the strongest response and deepest average depth during daytime (Fig. 7C). Furthermore, season was a negative predictor of average dive depth per 6 h period, as dives tended to shoal later in the overwintering period (Fig. 7C–D). When compared to the NP, seal average dive depths in all other ecosystems except the AS were deeper at night and were less affected by lunar fraction (Fig. 7C–D).

Surface wind speed, season, and ecosystem were factors consistently identified in the top-ranked models for seal behavioral state (Table 3). Eddy kinetic energy was not found to be an

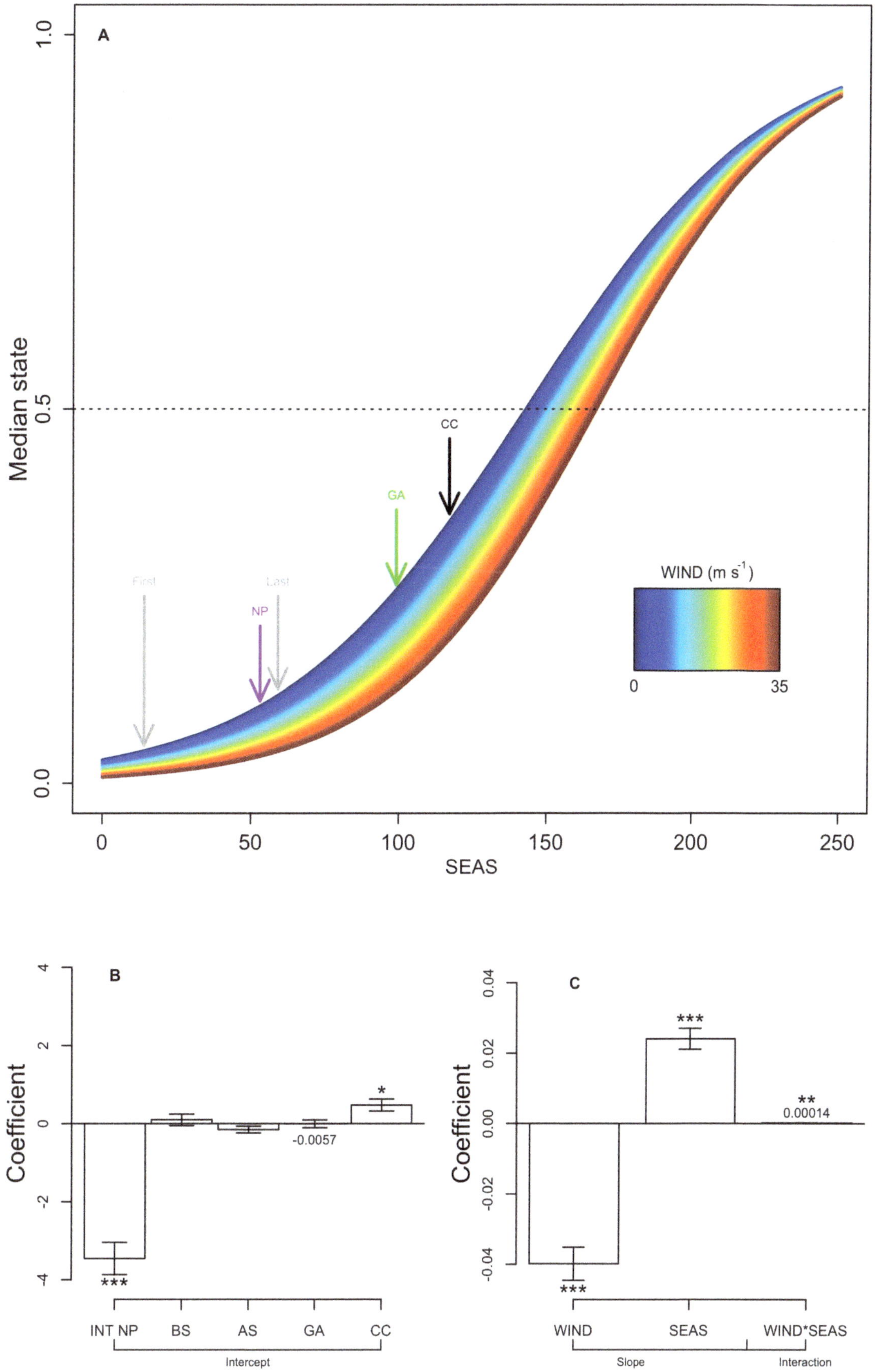
A
1.0
0.5
0.0
Median state
CC
GA
First
Last
NP
WIND (m s-1)
0
35
0
50
100
150
200
250
SEAS
B
4
2
0
-2
-4
Coefficient
***
*
-0.0057
INT NP
BS
AS
GA
CC
Intercept
C
0.04
0.02
0.00
-0.02
-0.04
Coefficient
***
**
0.00014
WIND
SEAS
WIND*SEAS
Slope
Interaction

Figure 8. Generalized linear mixed-effects model (GLMM) results for adult female northern fur seal estimated behavioral state. (A) Predicted model response for behavioral state (0 = transient, 1 = area-restricted search) as a function of season (SEAS, days since 1 October, x-axis) and surface wind speed (WIND, m s^{-1}, indicated by width of curve and color scale). Note that the surface wind speed effect diminishes with increasing days since 1 October, as indicated by curve width decreasing after reaching its maximum at ~125 d. Vertical arrows indicate first and last dates of departure from the breeding grounds (gray), and mean date of first entry into North Pacific (NP, purple), Gulf of Alaska (GA, green), and California Current (CC, black) ecosystems. (B) Plot of intercept terms in the best GLMM for behavioral state. Abbreviations as follows: INT NP (intercept, North Pacific base model), BS (Bering Sea Shelf), AS (Alaska Stream), other abbreviations as in panel A. Bar size gives coefficient magnitude and whiskers indicate standard error. Asterisks indicate the significance level of each term: $* = p < 0.05$; $** = p < 0.001$; $*** = p < 0.0001$. Coefficients are labeled where bars are too small to see. (C) Plot of the coefficients for main effects or slope and interaction terms in the best GLMM for behavioral state – symbols as in panel B, though note difference in y-axis scale. Positive values indicate increasing state (towards area-restricted search).

important predictor in the top three models. Increasing surface wind speed was a predictor of reduced behavioral state, which indicates a tendency away from area-restricted search and towards transient behavior with higher wind speeds (Fig. 8A). Of the ecosystem intercept terms in the best model for behavioral state, only the CC was significantly different from the NP ecosystem,

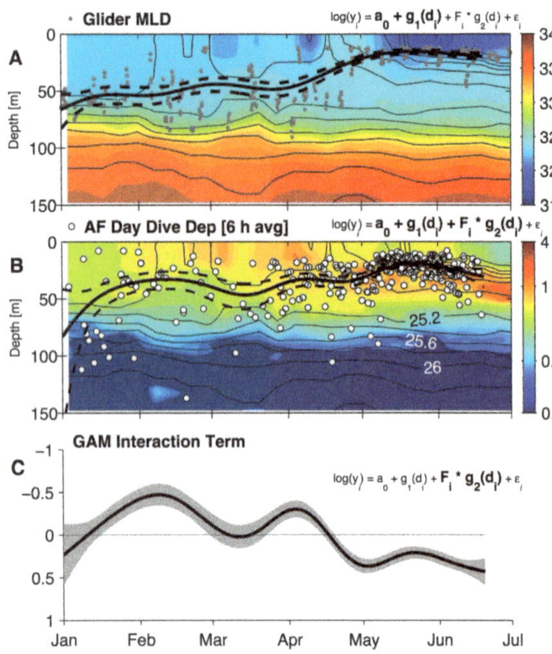

Figure 9. Composite annual cycle of upper-ocean characteristics, mixed-layer depth (MLD) and adult female northern fur seal daytime diving. (A)–(B) Colors show average annual cycles of salinity (panel A) and chlorophyll-*a* (Chl*a*, panel B) collected 60–80 km off the Washington shelf break 2003–09 from Seagliders. This region was chosen to correspond with the cross-shore peak of adult female northern fur seal foraging as shown in Figure 6. Only the overwintering portion of the annual cycle is shown. Contours in each panel are average annual cycle of density anomaly (σ_t, kg m^{-3}) in intervals of 0.2 kg m^{-3}. Gray filled circles in panel A indicate MLD calculated from individual Seaglider profiles, while white circles in panel B indicate average dive depths in day (>80% proportion daylight) 6 h periods in the California Current ecosystem. Solid black lines in panels A and B display generalized additive model (GAM) results for MLD and female average day dive depth, respectively (dashed black lines are 95% confidence intervals). The GAM model form, equation (1) in the main text, is shown above each panel and terms being displayed in each panel are bold. (C) GAM interaction term (solid line) and 95% confidence interval (gray shading) in log space. This term models the time-dependent depth offset of adult female seal day diving relative to the MLD; a negative interaction term indicates mean diving depths shallower than the MLD. Note y-scale direction in panel C is set positive downwards, in order to match the direction of depth axes in panels A and B.

reflecting increased behavioral state within the CC (Fig. 8A–B). Season was a positive predictor of behavioral state, reflecting the strong tendency towards transit during the initial movement across the NP and more area-restricted search behavior after mid-February when most females had arrived to their destination areas (Fig. 8A, C). The season/wind speed interaction term shows that this effect is strongest early in the migratory period, as high surface winds have a diminished effect on behavioral state with time (Fig. 8A, C).

Vertical Localization

Generalized linear mixed-effects models indicated that the rate of female NFS daytime diving was greater in the CC than in other ecosystems. The average annual cycle of upper-ocean physical structure, as described by Seaglider data off the WA coast, suggests that the depth of the surface mixed layer may influence seasonal patterns in vertical localization of these daytime dives. The average annual cycle of surface ocean physical properties and Chl*a* obtained from the Seagliders in a region 60–80 km seaward of the shelf break is shown in Figure 9A–B. This zone corresponded with the cross-shore distribution peak of female fur seals. The MLD in this zone off WA reached maximum depths of 45–75 m in January–February and was most variable in March and April. This is partially driven by the occasional presence of Columbia River plume water 60–80 km seaward of the shelf break in these months, which is reflected by low values of average near-surface salinity prior to the average mixed layer shoaling in May (Fig. 9A). The MLD off WA shoals in late April or early May, and average Chl*a* patterns in these months show the onset of the spring bloom in the mixed layer followed by development of a subsurface Chl*a* maximum in June.

Female daytime diving depths shoal and become more vertically localized late in the overwintering period; this pattern suggests an association with the MLD, which exhibits a similar winter-summer progression (Fig. 9A–B). Recall that the GAM quantifies this relationship by seeking a time-varying fit to MLD and female dive depth using two predictor terms in (1). The first term $g_1(d)$ fits only the MLD (Fig. 9A), while the second (interaction) term $F_i \times g_2(d)$ estimates the additive effect of female daytime diving response above or below the MLD and the change of this response with yearday. The total adult female response, MLD+interaction, is shown in Figure 9B; the isolated interaction term is shown in Figure 9C. The GAM results suggest a rough partition of the overwintering period into two regimes, separated by the shoaling of the MLD around 1 May. Prior to this time, the MLD is deep, shoals slowly, and is more variable (Fig. 9A). Female day dive depths are on average shallower than the MLD though the interaction term is not significant for the entirety of this period (Fig. 9C). After shoaling, the MLD is uniformly 15–20 m depth, with much less variability (smaller confidence bounds in Fig. 9A). Female dive depths are also much shallower during this time and are less variable (Fig. 9B). The interaction term changes sign prior to the shoaling and indicates that late spring day diving is on

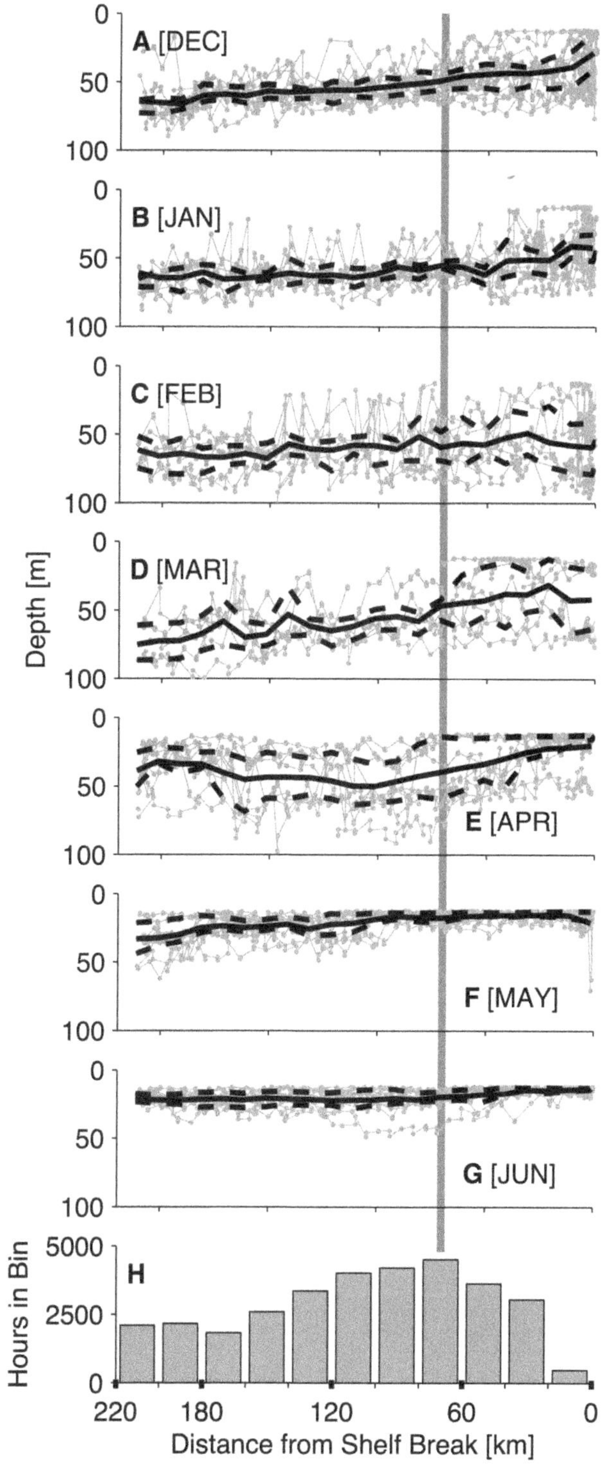

Figure 10. Cross-shore mixed-layer depth profile by month. (A) December to (G) June. Mixed-layer depth calculated from individual Seaglider crossings in gray, mean profile in 10 km bins in black. Dashed black lines in A–G indicate the 1st and 3rd quartile of the observations within each bin. Panel H shows a portion of the cross-shore distribution of adult female northern fur seal time spent versus distance from the shelf break, truncated at 220 km offshore (the full distribution is shown in Fig. 6). The vertical gray line through all panels corresponds to the station of the bin with the greatest amount of female time spent in panel H.

average 5–10 m below the MLD (Fig. 9B–C). The GAM smoothing scale was long enough to avoid effects of the occasional shoaling of MLD due to freshwater plumes, which is not an important part of the annual cycle in other locations in the CC at which female dives were collected.

Seaglider surveys observed MLDs that were on average shallower near the coast throughout much of the overwintering period (Fig. 10A–G). The offshore-inshore difference in MLD varied throughout the year and was greatest in both March (Fig. 10D) and September–October (not shown). The peak in the adult female cross-shore distribution (Fig. 10H) corresponded to the region of steepest onshore shoaling in March and April. It should be noted that the late winter cross-shore difference in average MLD off WA is probably driven in part by freshwater plumes from the Columbia River or other smaller rivers along the WA coast [77], and may not be representative of all locations in the CC. Washington coast Seagliders recorded negligible cross-shore gradients in summer (Fig. 10F–G), when strong surface heating and weaker wind stress prevail over the region.

Links to Mesoscale Circulation and Coastal Topography

The distribution results show that female habitat utilization during the overwintering period is concentrated off the shelf but within the inner coastal transition zone, and that the fraction of time spent near eddies increases twofold in the CC ecosystem (Fig. 4B). Examination of individual animal tracks suggests that females respond behaviorally to some (though not all) mesoscale eddies and jets by altering movement and behavioral state. Examples of this are shown below for tracks of four animals, one of which additionally foraged near a Seaglider, which will be described separately in the section *Individual Case Studies*.

Female seal 460, tracked during the 2006–07 overwintering migration, was one of seven individuals with track duration of > 200 d (Fig. 4A); she entered the CC ecosystem from a more southerly route than was typical of other tracks (Figs. 3A, 11A). Female 460 transitioned to area-restricted search behavior coincident with her encounter with the edge of a mesoscale cyclonic eddy (counterclockwise-rotating, locally low SLA) on 9 January 2007. This followed a prolonged period of transit, during which time 460 bypassed [25] an elongated anticyclonic (clockwise, locally high SLA) feature immediately to the northwest of where she began foraging on 9 January (Fig. 11A–B). Panels B–G of Figure 11 illustrate weekly portions of the following two-month period in which 460 foraged directly west of Cape Mendocino. On 9 February, 460 began transit movement to the SE, possibly initiated by increased surface wind speeds (Fig. 11H), and began following currents in an adjacent meander, before foraging near the NNE edge of a cyclone/anticyclone dipole (Fig. 11E–G). Female seal 460 later moved north, spent some time to the west of Cape Blanco, and foraged for two months off the WA coast. Figure 11H shows other instances in 460's track in which alterations in behavioral state over time periods of ~2–3 d were associated with increases in surface wind speed. Additionally, this panel illustrates that female 460 experienced overall lower wind speeds while foraging in the CC and GA in comparison to the early migratory transit phase (prior to 9 January). Similar instances of altering movement behavior in response to some mesoscale features were found in other satellite tracks in this dataset and two are included in the supporting information (Figs. S3, S4).

Observations of movement and behavioral state changes toward area-restricted search in association with mesoscale eddies suggests one possible reason for increased female density near coastal capes such as Cape Blanco and Cape Mendocino: these irregularities in the continental shelf influence alongslope currents, creating

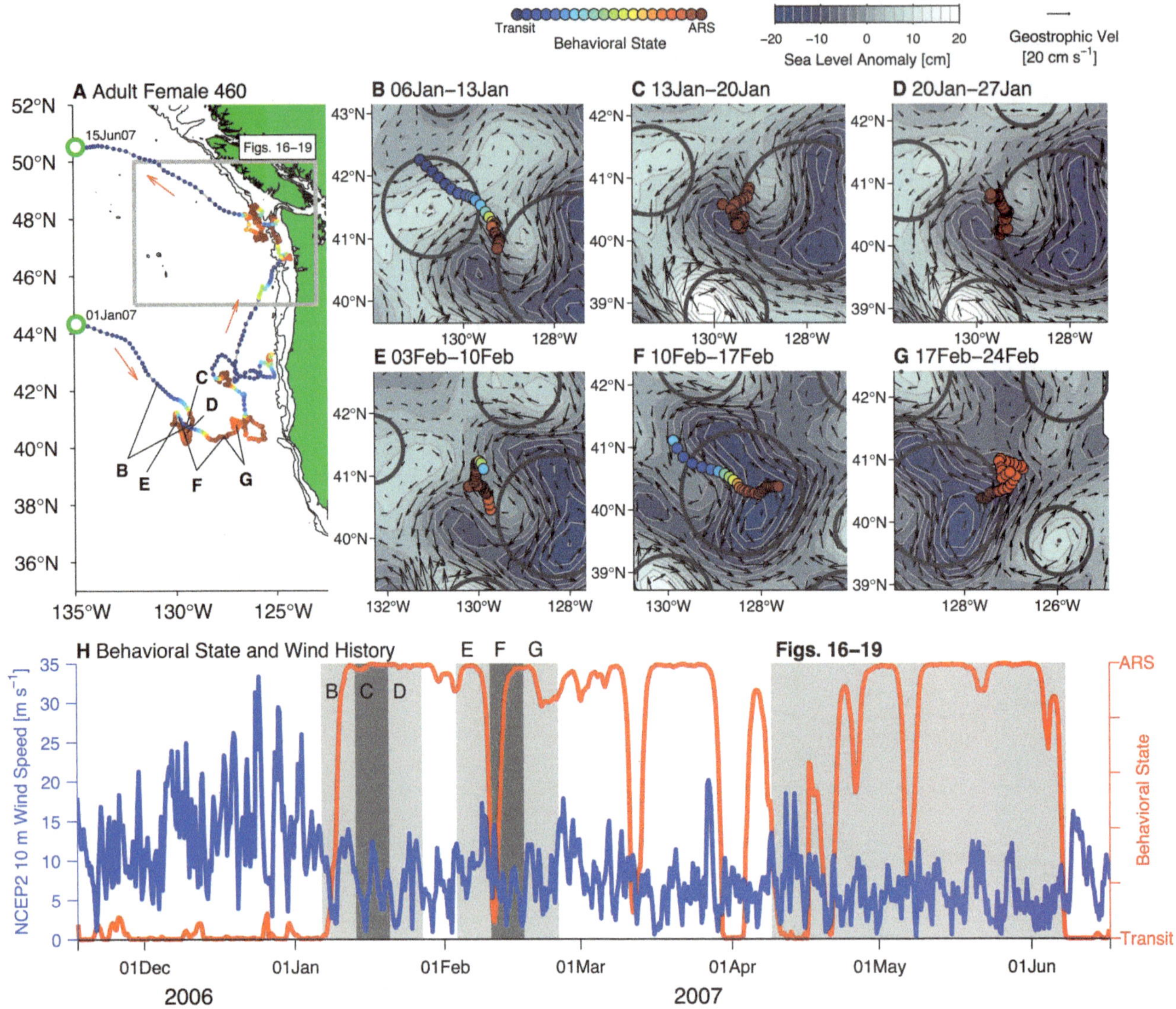

Figure 11. Behavioral responses of adult female northern fur seal 460 to mesoscale circulation and surface wind speed. (A) Overview of satellite-tracked locations of female 460 in the California Current and Gulf of Alaska ecosystems from 01 January 2007 to 15 June 2007. Six h locations are colored according to estimated behavioral state (scale at top, ARS = area-restricted search). Gray box in panel A and gray shading in panel H indicates the spatial and temporal extent covered by Figures 16A–19A, when female seal 460 foraged in close proximity to the Seaglider region. (B)–(G) show weekly intervals of 460's estimated locations and behavioral state in 6 h periods (filled color circles, scale at top) plotted over sea level anomaly (color contours) and surface geostrophic velocity anomaly (Vel, black arrows, scale at top right). Each plot is centered on female seal 460's locations over the weekly period. Note that one weekly interval is omitted between panels D and E, during which time female seal 460 continued to forage in the same location as in panel D. Thick gray circles indicate the locations and approximate spatial extent of altimetry-identified mesoscale eddies from Chelton et al. [63]. Eddies are plotted as circular features though this is intended for illustration purposes only. (H) Plot of estimated behavioral state (red line, scale on right y-axis) and 10 m height wind speed (m s^{-1}) at seal 460's location (blue line, scale on left y-axis) versus time for the overwintering period 2007–08. Wind speed estimates are obtained by interpolating National Centers for Environmental Prediction Reanalysis 2 (NCEP2) product to 460's estimated locations at each 6 h time point. Gray bars in alternating shading display the extent of time covered by panels B–G.

instabilities, meanders, and eddies. Figure 12 displays an example of observed eddy generation at Cape Blanco in the 2009–10 overwintering period, and utilization of this region by three separate adult female NFS. One of these individuals (track 662B; triangles) carried a dive recorder and entered the Cape Blanco region on 21 January, stopping in a small, weak anticyclonic feature on the northern edge of a larger cyclonic eddy centered at 43°N, 126.5°W (Fig. 12B–D). Another individual (677; squares) had passed through this region a week prior but remained in transit mode (Fig. 12A). Female 662B foraged for three weeks on the NW side of the cyclone before moving across the eddy to the east and foraging within a strong poleward jet that formed the boundary between the offshore cyclonic feature and a newly developing anticyclone on the inshore side of the jet (Fig. 12E–I). Another satellite-tagged female (track 676; circles) also foraged within a similar spatial extent from 6 February onwards, appearing to utilize the same mesoscale habitat feature (Fig. 12E–I).

Dive patterns recorded by female 662B during this period are characteristic of patterns apparent in the statistical analyses of number of dives and average dive depth per 6 h period.

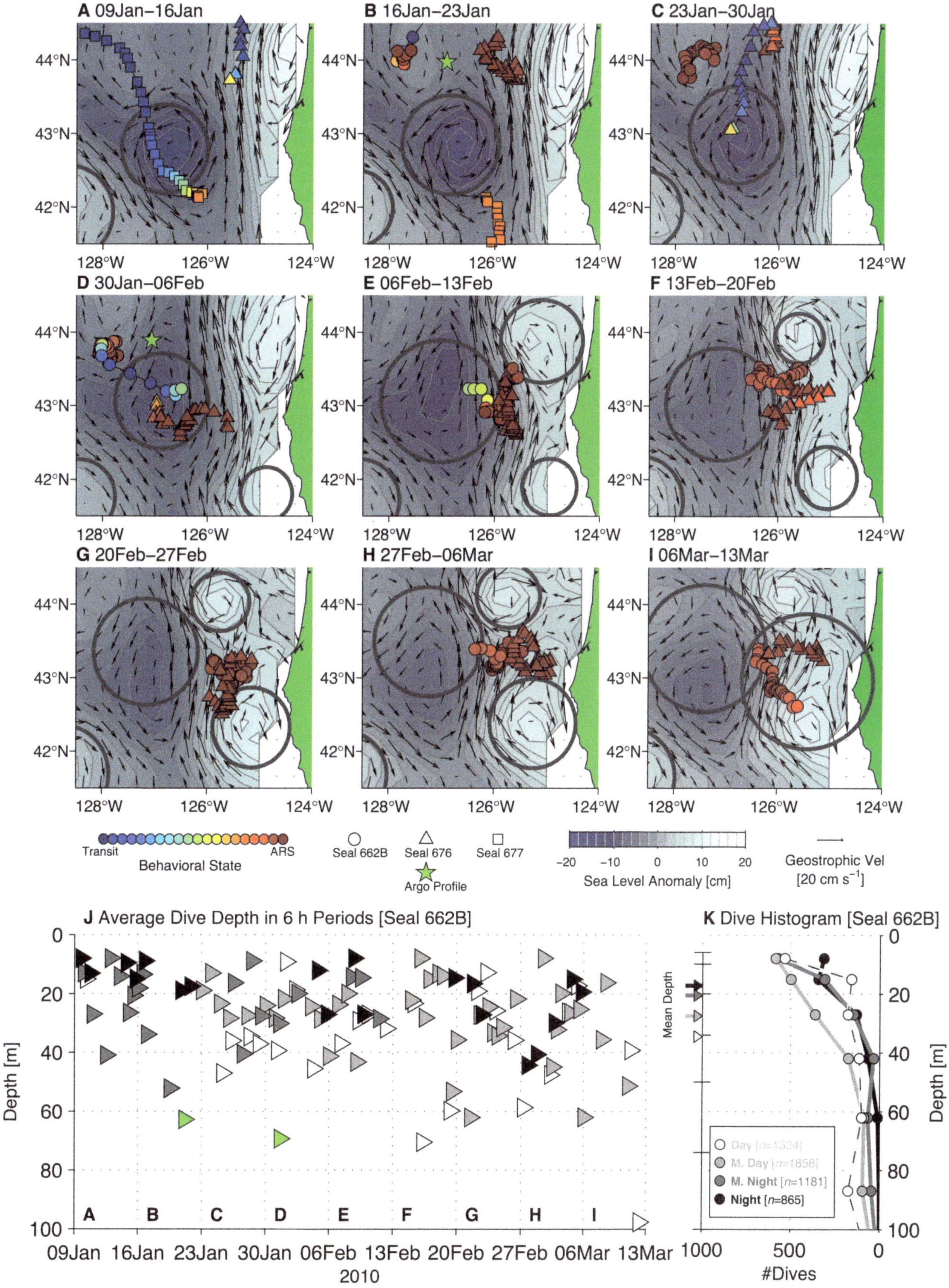
A 09Jan–16Jan
B 16Jan–23Jan
C 23Jan–30Jan
D 30Jan–06Feb
E 06Feb–13Feb
F 13Feb–20Feb
G 20Feb–27Feb
H 27Feb–06Mar
I 06Mar–13Mar
44°N
43°N
42°N
128°W
126°W
124°W
Transit
ARS
Behavioral State
Seal 662B
Seal 676
Seal 677
Argo Profile
-20 -10 0 10 20
Sea Level Anomaly [cm]
Geostrophic Vel
[20 cm s⁻¹]
J Average Dive Depth in 6 h Periods [Seal 662B]
Depth [m]
09Jan 16Jan 23Jan 30Jan 06Feb 13Feb 20Feb 27Feb 06Mar 13Mar
2010
K Dive Histogram [Seal 662B]
Mean Depth
Day [n=1324]
M. Day [n=1858]
M. Night [n=1181]
Night [n=865]
#Dives
1000 500 0

Figure 12. Winter surface circulation at Cape Blanco in 2010 and habitat utilization by three adult female northern fur seals. (A)–(I) Weekly intervals of sea level anomaly (color contours, scale below panel H), surface geostrophic velocity anomaly (Vel, black arrows, scale below panel I), and estimated locations of three satellite-tagged adult females (colored by estimated behavioral state, scale below panel G, ARS = area-restricted search) in the Cape Blanco region in 2010. Symbol types for each individual are also shown below panels G–H. As in Figure 11, thick gray circles indicate the locations and approximate spatial extent of altimetry-identified mesoscale eddies from Chelton et al. [63]. In panels B and D, green stars show the location of two Argo float profiles collected during these time periods. (J) Mixed-layer depth calculated from Argo profiles (green triangles), along with average dive depth in 6 h periods from female seal 662B track. Each 6 h period is sorted into one of four categories based on its proportion daylight (black triangles = night, dark gray = mostly night, light gray = mostly day, white = day, as defined in Results section *Links to Mesoscale Circulation and Coastal Topography*). (K) Histograms, one for each proportion daylight category, showing number of dives in that category that have maximum depth within each dive tag pre-programmed depth bin. Depth bin bounds are indicated on the left y-axis by whiskers. Arrows at left of panel K indicate average dive depth by proportion daylight category. Symbol shading follows that of panel J.

Figure 12J–K display average dive depth in 6 h periods versus time and depth histograms of female 662B diving, with each 6 h dive period classified into one of four categories of proportion daylight: day (proportion daylight $\geq$80%), mostly day (50% $\leq$ proportion daylight $<$80%), mostly night (20% $\leq$ proportion daylight $<$50%), and night (proportion daylight $<$20%). The majority of diving during 9 January–13 March was collected during 6 h periods classified as day or mostly day (Fig. 12J–K). Average dive depth increased with increasing proportion daylight (Fig. 12K). Two temperature/salinity profiles were collected near female seal 662B's foraging on 21 January and 1 February from Argo float ID 4900574. Mixed-layer depth, as calculated from these two profiles, was 20–30 m deeper than the average depths of 662B's day dives during this period (Fig. 12J). This is consistent with the results of the GAM for average depth in daytime 6 h period during this portion of the season – i.e., within the mixed layer prior to the transition to spring conditions in late April/early May.

Though efforts to model the effect of eddies on estimated behavioral state using an EKE term were unsuccessful, adult female habitat utilization relative to coherent eddy features identified by Chelton et al. [63] showed some differences with estimated behavioral state. For both area-restricted search ($b>0.75$) and transit ($b<0.25$), the PDF of female habitat utilization as a function of $\hat{r}$ (normalized radial distance from the nearest eddy center) was unimodal with a peak near $\hat{r}=1.5$ and a long positive tail (Fig. 13A). However, the PDF of area-restricted search points peaked closer to the center and had a weaker positive tail; i.e., area-restricted search locations were more closely aligned with identified eddy features than transit.

For area-restricted search locations, 55.8% were within 2 radii of the nearest eddy center, while this was true of 47.8% of transit locations (Fig. 13A). Differences in PDF shape between states were significant near $\hat{r}=1$ (higher search density; $p=0.0211$, Fig. 13B) and $\hat{r}=4$ (higher transit density; $p=0.0029$). A similar pattern, though with greater differences in PDF shape, was found when performing the same analysis but exclusively for tracks with length >200 d (Fig. S5). These differences indicate that, when in area-restricted search, individuals were more likely to utilize areas near $\hat{r}=1$ and less likely to utilize areas near $\hat{r}=4$ than when in transit. The median value of $\hat{r}$ was 1.85 for area-restricted search and 2.07 for transit, though the difference between these two values was not significant based on the bootstrap method ($p=0.0614$, Fig. 13C). The observed difference in median values was enhanced in tracks >200 d and was significant ($p<0.0001$, Fig. S5).

Individual Case Studies

Three females equipped with dive recorders (individuals numbered 460, 626, and 628, Table 1) foraged near Seagliders off the WA coast. The following sections compare the movement and diving records of each individual to subsurface physical and bio-optical oceanographic structure as revealed by the Seaglider vertical profiles. Two animals (female seals 626 and 628; this section) foraged for periods ~1 week, exhibiting diel diving patterns consistent with those revealed in the statistical analysis. A third animal, female seal 460, foraged for a month within the Columbia River freshwater plume.

Diel Diving Patterns. Two individuals (female seals 626 and 628) recorded brief bouts of area-restricted search movement near Seaglider transects as the females made their way northward along the OR and WA coasts in late spring and early summer 2008. Female 626 was first to arrive in the WA area in late April, making her way northward roughly 200 km from the shelf break and parallel to the coast while slowing or stopping in irregular fashion, presumably to forage (Fig. 14A). On 28 April, she slowed movement near 46.8°N, 127.5°W, within 50 km of the offshore end of the Seaglider 101 (SG101) transect (Fig. 14A). She departed to the north a week later, crossing 47°N at 127.42°W on 6 May. At that time SG101 was heading offshore from the shelf break along 47°N and crossed 127.42°W on 7 May at 1900 UTC. The mapped SG101 data for the transect (Fig. 14B) show moderate concentrations of Chla and MLDs that had begun to shoal from the winter maximum to 30–40 m. Female 626's diving during 6 h periods classified as day (mean = 29.6 m, n = 391; Fig. 14C) showed correspondence with the MLD measured by SG101 (mean = 32.7 m of n = 19 profiles collected >175 km from shelf break). Night dives (black) were more numerous and shallower (mean = 16.6 m, n = 1603; Fig. 14C). The partition of diving effort (night/day dives ratio = 4.1) is more typical of diving in the in the offshore migratory transit period in the NP, while the correspondence between day dives and the depth of the mixed layer is typical of statistical results in the CC.

Female seal 628 arrived in May, a month after 626, and beginning 6 June spent a week near 47.5°N, 125.75°W, 50 km north of the inshore portion of the SG101 transect, before departing to the northwest on 11 June (Fig. 15A). Female seal 628 spent nearly 200 hours in an area 60–75 km from the shelf break and the movement model indicated area-restricted search behavior within this zone (Fig. 15A). Seaglider 101 was near the offshore waypoint of the its navigation pattern when 628 first arrived, and subsequently headed inshore along 47°N, crossing 125.75°W on 14 June, three days after 628's departure. Chlorophyll-*a* concentrations were moderate and mesoscale eddies were absent. Seaglider 101 sampled the edge of a freshwater plume as the vehicle reached the shelf break (Fig. 15B) where the MLD shoaled to 20 m depth. Female 628 foraged about 20 km west from the edge of this plume. Day dives (n = 419) were nearly as numerous as night dives (n = 462). The vertical distribution of day diving (mean = 38.7 m) of female 626 was centered on the MLD (mean = 38.3 m of n = 16 profiles between 50 and 100 km from the shelf break), though with a broader vertical range (Fig. 15C). Night dives were surface-intensified and mostly shallower than the MLD (mean = 19.59 m).

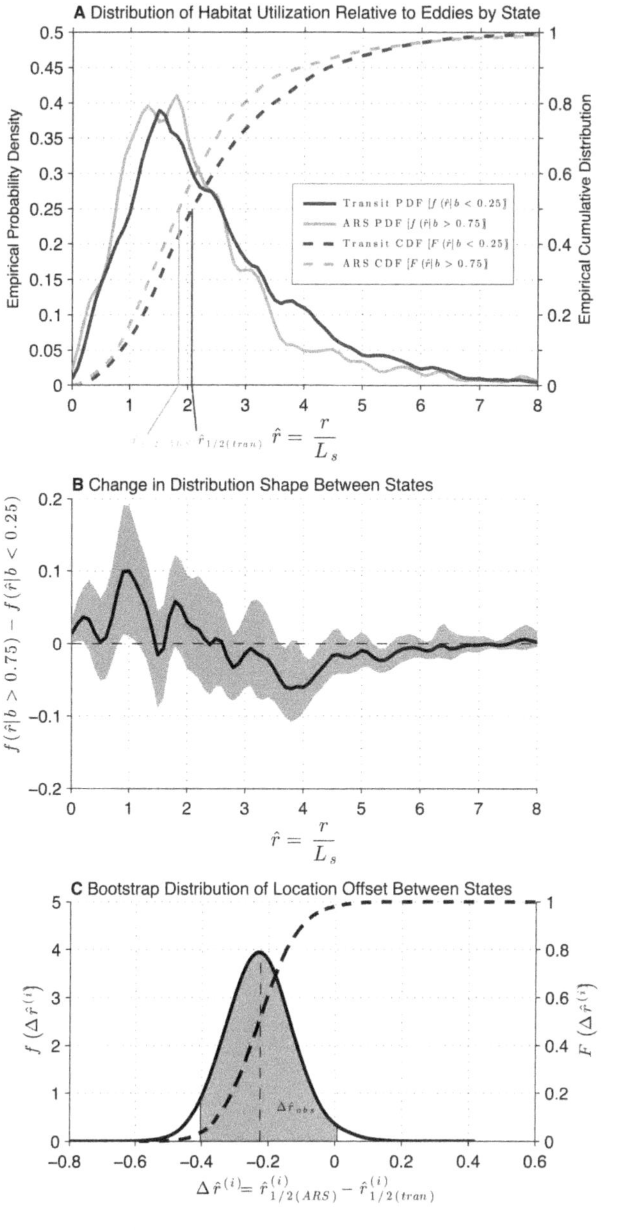

Figure 13. Adult female northern fur seal habitat utilization relative to eddy features, conditioned by estimated behavioral state. (A) One-dimensional radial probability density functions [PDFs, $f(\hat{r})$] and cumulative distribution functions [CDFs, $F(\hat{r})$] for habitat utilization relative to eddy features at two categories of behavioral state b (transit = dark gray, area-restricted search [ARS] = light gray). "Transit" periods are those with $b<0.25$; "ARS" periods are those with $b>0.75$. Solid lines show PDFs, dashed lines are CDFs (scale on right y-axis). Distributions are computed as a function of normalized distance to the nearest eddy center $\hat{r}$, defined as absolute distance to the nearest eddy divided by that eddy's radial length scale L_s [63]. (B) Observed difference between area-restricted search PDF and transit PDF (solid black line) with 95% confidence bounds computed using a bootstrap method (gray shading). (C) Bootstrap distribution of $\Delta\hat{r}^{(i)}$, the difference of median values of $\hat{r}$ between area-restricted search and transit states on bootstrap iteration i. Solid black line shows the PDF of these values, thick dashed line shows CDF (scale at right). Thin vertical dashed line is the observed value of $\Delta\hat{r}$. The gray shaded area denotes the bias-corrected/accelerated 95% confidence interval on the observed value of $\Delta\hat{r}$. For the observed value of $\Delta\hat{r}$ to be significantly different from zero using a two-tailed test, this confidence interval must not include zero.

Columbia River Plume. Freshwater outflow from the Columbia River is another important source of mesoscale variability in the WA coastal transition zone that influences NFS foraging. This is illustrated by Seaglider data taken in spring 2007, during which time female 460 foraged from the outer shelf to the outer slope between GH and CF (Figs. 16–19; for track overview see Fig. 11A). Female 460 spent the most time in close proximity to the Seaglider transects of any satellite-tagged individual equipped with a dive recorder. She arrived at 47°N on 21 April 2007 and departed from the CF line on 5 June, moving seaward and generally northward while in the Seaglider area. During this time Seaglider 014 (SG014) made four cross-shore transects.

As female 460 transited from pelagic waters off the OR coast toward the WA shelf in mid-April, SG014 performed a transit along the GH line (Fig. 16A). During this time, late-winter conditions prevailed. Seaglider 014 results along the GH transect line reflect moderate Chl*a* concentrations and a 50 m MLD offshore (Fig. 16B). During this time, seal 460 was mostly in a transitory state and performed few dives to depths at or below the MLD (Fig. 16B–C). Seaglider 014 also crossed a portion of a low-salinity freshwater plume that was straddling the shelf edge. This is evident in the narrow 15 km low-salinity band that was found at the surface just off the shelf break, with a shallow MLD and elevated concentrations of Chl*a* (Fig. 16B). The plume's cross-shore position was consistent with the alongshore component of surface winds observed at DESW1 (47.675°N, 124.485°W), an island on the WA inner shelf between GH and CF. From late March to mid-April winds were strong and predominantly poleward (Fig. 16D), typical of late-winter winds off WA prior to a shift to upwelling-favorable winds known as the "spring transition" [78]. Under late-winter conditions, frictional surface currents, background shelf flow, and buoyancy-driven geostrophic currents keep the Columbia River plume over the shelf and in a meridional orientation extending northward from the mouth of the Columbia estuary [78–82]. Female 460 moved onto the WA shelf around 22 April, and, roughly coincident with her encounter of the plume, began foraging outside the mouth of Willapa Bay before moving northward parallel to the shelf break (Fig. 17A).

The next SG014 transect, taken from 22 April to 6 May, showed the plume off CF with low salinity and elevated Chl*a* concentrations at depths less than 25 m and within 60 km of the shelf break (Fig. 17A–B). During this time, 460 spent most of her hours in an area over the shelf break and within the plume. Behavioral state estimates during this time indicate area-restricted search. The number of total dives increased substantially ($n_{day} = 349$; $n_{night} = 1011$). The salinity-stratified MLD was shallow enough that both day (mean = 36.0 m) and night dives (mean = 20.6 m) were at or slightly below the MLD (mean = 14.3 m of 39 profiles within 60 km of the shelf break; Fig. 17B–C). The alongshore component of wind from 22 April to 6 May remained poleward (Fig. 17D) but began to relax on 7 May, after which followed an abrupt, sustained reversal to equatorward winds (Fig. 18D). Shifts to equatorward (upwelling) winds, following poleward (downwelling) winds, advect low-salinity plume water westward and southward offshore from the WA shelf. As a result, SG014's return trip offshore along the CF line from 6 May to 21 May observed the plume offshore extent increase to roughly 100 km from the shelf break at its maximum (Fig. 18A–B). It is also possible that freshwater outflow from the Strait of Juan de Fuca contributed to freshwater in the plume off CF. Chlorophyll-*a* concentrations within the plume were still elevated relative to surface concentrations offshore and during this time female seal 460's foraging effort was the most spatially concentrated, contained entirely within 50 km of the shelf break and

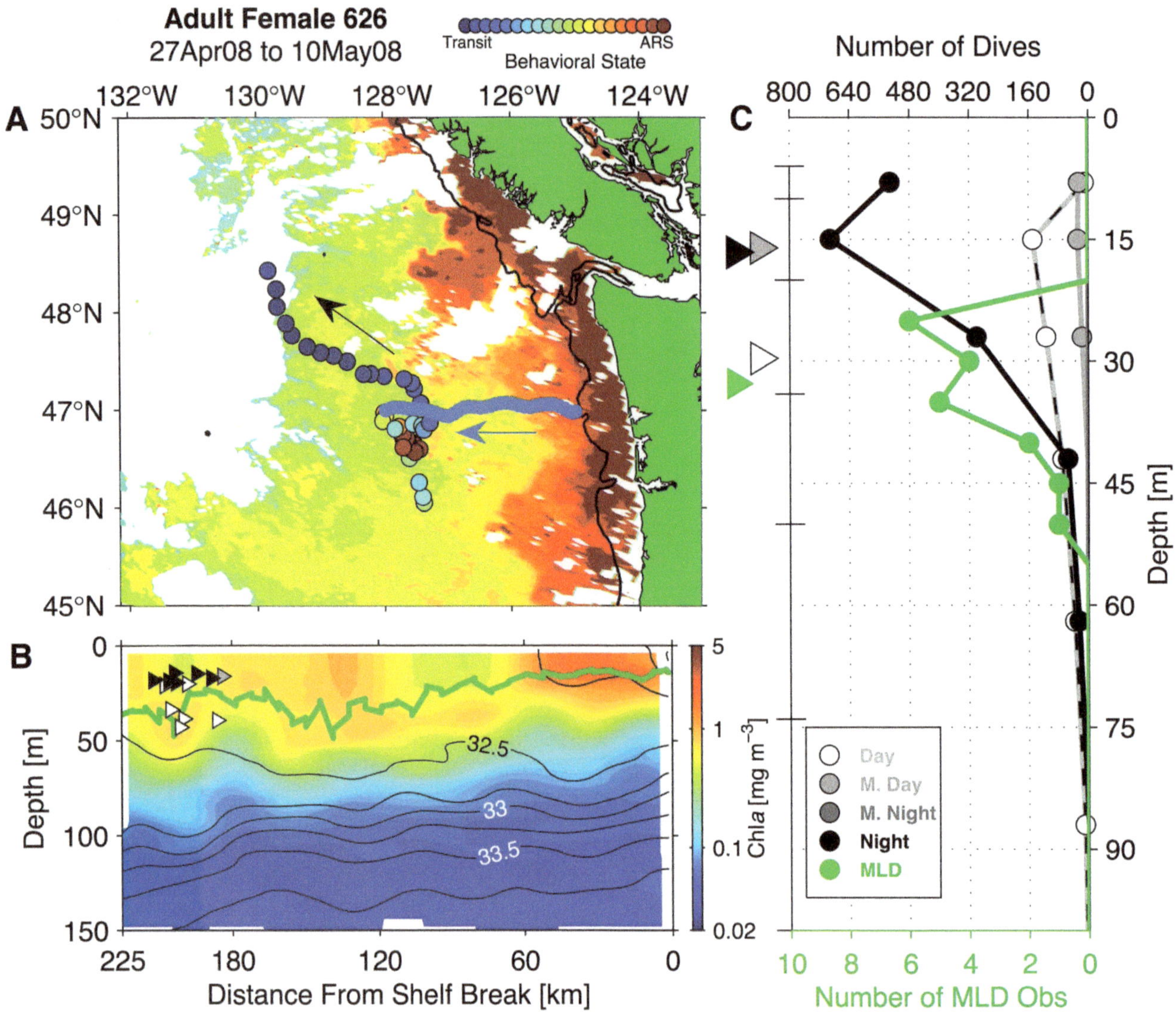

Figure 14. Adult female northern fur seal 626 near Seaglider 101 (SG101) during the time period 27 April to 10 May 2008. (A) Estimated locations of female 626 (filled color circles) and reported GPS locations of SG101 (thick blue line) off the Washington and British Columbia coast (green). The 200 m isobath is indicated by thin black line. Arrows indicate direction of travel of seal 626 and SG101. Colors of filled circles indicate behavioral state (scale at top, ARS = area-restricted search). Background colors in panel A indicate surface chlorophyll-*a* (Chl*a*) in cloud-free regions on 16 May 2008. Chlorophyll color scaling is the same as that in panel B. Blank regions are portions of the image removed due to cloud cover (16 May was the clearest available image from the surrounding time period). (B) Upper-ocean conditions by depth and distance from the shelf break recorded by SG101 during its crossing: Chl*a* (background colors), salinity (black contours), and mixed-layer depth (MLD, dark green line). Triangles indicate average depths of female northern fur seal 626's dives in each 6 h period. Triangles are plotted by cross-shore distance along the Seaglider line, and triangle fill colors indicate proportion daylight category of each 6 h period (black triangles = night, dark gray = mostly night, light gray = mostly day, white = day, as defined in Results section *Links to Mesoscale Circulation and Coastal Topography*). (C) Histograms of seal 626's diving, one for each proportion daylight category, showing number of dives within each dive tag pre-programmed depth bin. Depth bin bounds are indicated on the left y-axis by whiskers. Green line in panel C is a histogram of MLD observations collected >175 km from the shelf break using 5 m depth bins. Filled triangles to the left of Panel C indicate the average dive depth in each category of proportion daylight and the average MLD.

50 km from the Seaglider survey (Fig. 18A–B). Foraging was offshore in comparison to the previous time period, roughly tracking the plume's movement. Day diving (mean = 24.6 m, n = 493) was mostly to depths near the MLD but with some diving averaging 30–75 m below (Fig. 18C). Night diving (mean = 11.3 m, n = 936) was more vertically concentrated and some night diving extended to the MLD, which deepened in the plume between early and mid-May (mean MLD = 16.8 m of 35 profiles within 60 km of the shelf break; Fig. 18B). During this time, MLDs began to shoal offshore and a subsurface Chl*a* maximum outside of the stratified freshwater layer developed (150–200 km from the shelf break; Fig. 18B).

From 21 May onwards, female NFS 460 continued foraging within the plume, tracking southward and away from the shelf break while in between the two Seaglider transect locations, before departing around 5 June (Fig. 19A). Seaglider 014 returned to the GH line from 21 May to 8 June and recorded very low concentrations of Chl*a* within the freshwater layer, which indicated that the bloom there was depleted (Fig. 19B). In contrast, Chl*a* concentrations were high offshore, where the

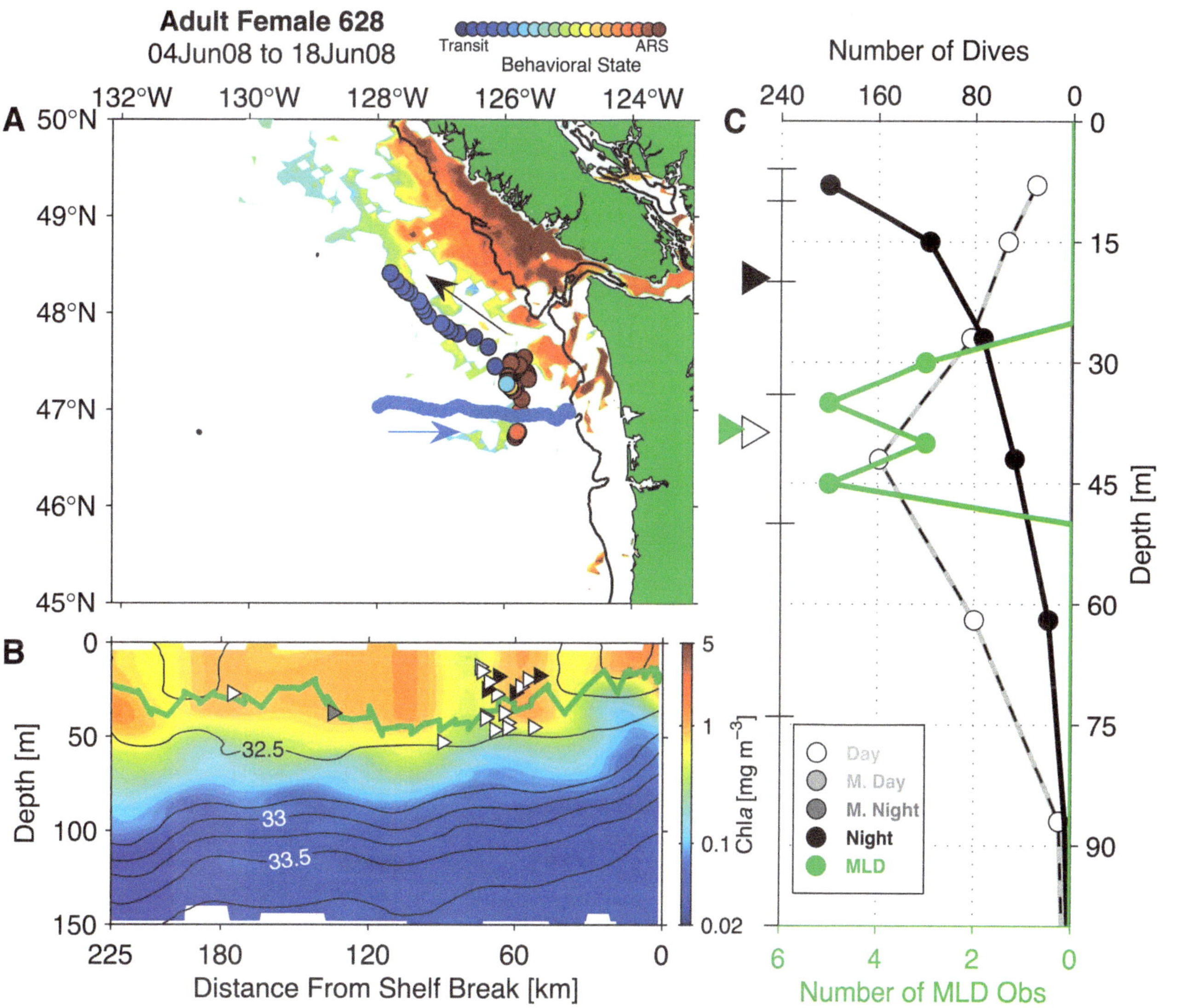

Figure 15. Adult female northern fur seal 628 near Seaglider 101 (SG101), 4 June to 18 June 2008. (A) Estimated locations of female 628 (filled color circles) and reported GPS locations of SG101 (thick blue line) off the Washington and British Columbia coast (green). Colors of filled circles indicate behavioral state (scale at top, ARS = area-restricted search). The 200 m isobath is indicated by thin black line. Arrows indicate direction of travel of 628 and SG101. Background colors in panel A indicate surface chlorophyll-*a* (Chl*a*) in cloud-free regions on 13 June 2008, plotted according to the color scale in panel B. (B) Upper-ocean conditions by depth and distance from the shelf break recorded by SG101 during its crossing: Chl*a* (background colors), salinity (black contours), and mixed-layer depth (MLD, thick green line). Triangles indicate average depths of female northern fur seal 628's dives in each 6 h period. Triangles are plotted by cross-shore distance along the Seaglider line, and triangle fill colors indicate proportion daylight category of each 6 h period (black triangles = night, dark gray = mostly night, light gray = mostly day, white = day, as defined in Results section *Links to Mesoscale Circulation and Coastal Topography*). (C) Histograms of seal 628's diving, one for each proportion daylight category, showing number of dives within each dive tag pre-programmed depth bin. Depth bin bounds are indicated on the left y-axis by whiskers. Green line in panel C is a histogram of MLD observations collected between 50 and 100 km from the shelf break using 5 m depth bins. Filled triangles to the left of Panel C indicate the average dive depth in each category of proportion daylight and the average MLD.

spring bloom had just begun. Persistent southward winds, driving near-surface frictional currents, had by this time advected the Columbia River plume offshore with its edge 125 km from the shelf break (Fig. 19B, D). Female 460's final foraging effort fell within this region before she departed to the northwest (Fig. 19A). Consistent with late-season GAM results for depth of diving, 460's day dives (mean = 23.1 m, n = 507) were concentrated ~10 m deeper than the MLD (mean MLD = 14.5 m, n = 58 profiles collected between 25 and 150 km from the shelf break; Fig. 19B–C). Night dives (mean = 15.0 m, n = 936) were concentrated near the MLD and more vertically localized than day dives (Fig. 19B–C).

Discussion

The timing of migration and the alongshore and cross-shore distribution of female NFS observed in this study are largely consistent with previous studies derived from pelagic scientific takes or anecdotal knowledge. Historical data indicate that after transiting the Aleutian passes that connect the Bering Sea and NP, female NFS cross the NP and arrive on the North American west coast beginning in late November [1,4,23,52]. Pregnant individ-

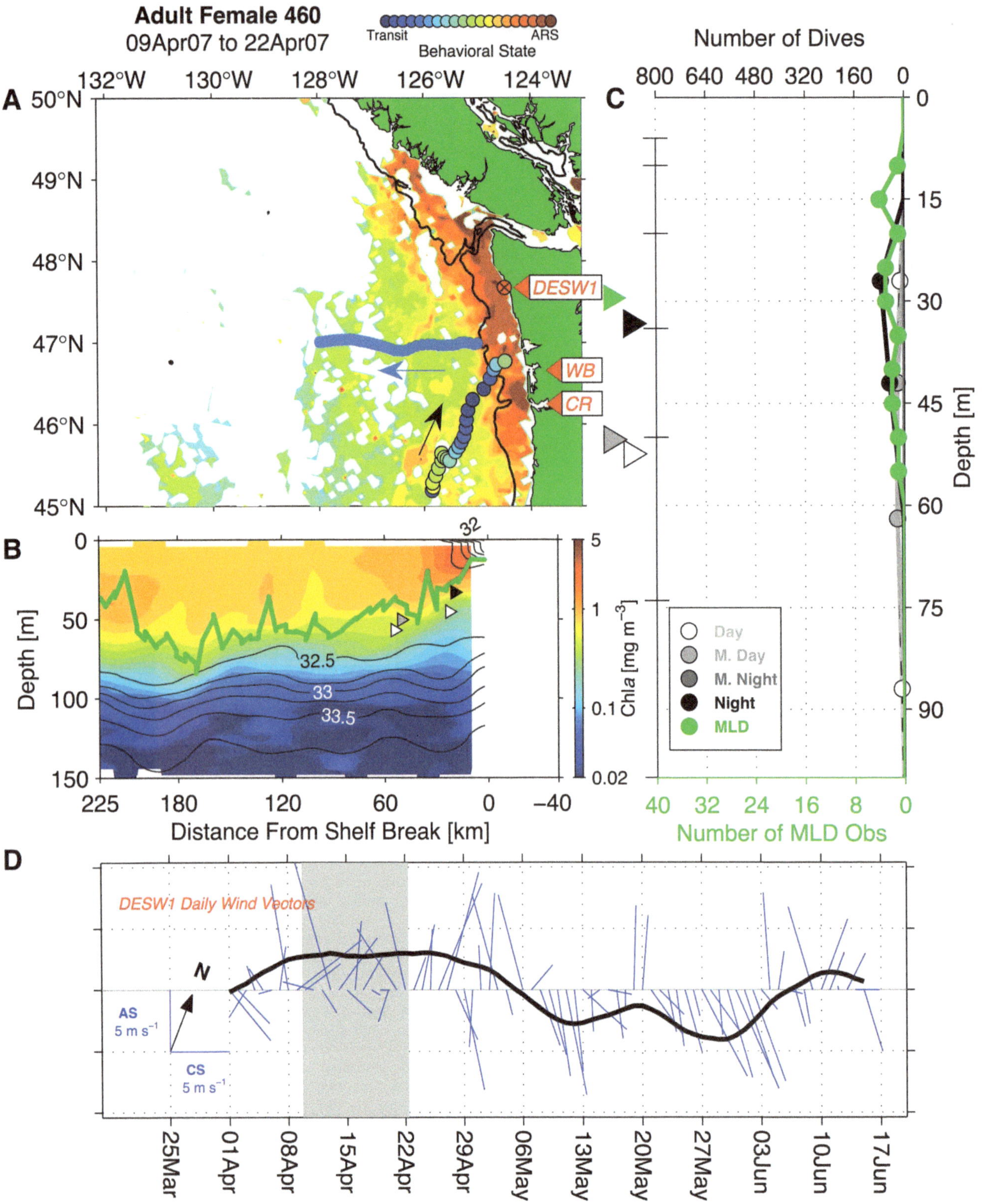

Figure 16. The initial approach of adult female northern fur seal 460 to Seaglider 014 (SG014), 9 April to 22 April 2007 (1 of 4). (A) Estimated locations of female 460 (filled color circles) and reported GPS locations of SG014 (thick blue line) off the Washington and British Columbia coast (green). Red labels indicate the Columbia River mouth (CR), Willapa Bay coastal inlet feature (WB), and the location of National Data Buoy Center station Destruction Island (DESW1, red cross-circle at station). Colors of filled circles indicate behavioral state (scale at top, ARS = area-restricted search). The 200 m isobath is indicated by thin black line. Arrows indicate direction of travel of 460 and SG014. Background colors in panel A indicate surface chlorophyll-*a* (Chl*a*) in cloud-free regions on 22 April 2007, plotted according to the color scale in panel B. (B) Upper-ocean conditions by depth and distance from the shelf break recorded by SG014 during its crossing: Chl*a* (background colors), salinity (black contours), and mixed-layer depth (MLD, dark green line). Triangles indicate average depths of female northern fur seal 460's dives in each 6 h period. Triangles are plotted by

cross-shore distance along the Seaglider line, and triangle fill colors indicate proportion daylight category of each 6 h period (black triangles = night, dark gray = mostly night, light gray = mostly day, white = day, as defined in Results section *Links to Mesoscale Circulation and Coastal Topography*). (C) Histograms of seal 460's diving, one for each proportion daylight category, showing number of dives within each dive tag pre-programmed depth bin. Depth bin bounds are indicated on the left y-axis by whiskers. Green line in panel C is a histogram of MLD observations collected between 0 and 60 km from the shelf break using 5 m depth bins. Filled triangles to the left of Panel C indicate the average dive depth in each category of proportion daylight and the average MLD. (D) Daily-averaged wind vectors at 52.7 m height obtained from DESW1. Wind measurements at DESW1 have been rotated into an alongshore (AS)/cross-shore (CS) coordinate system. In panel D, the solid black line is a 10 d running average of the poleward component of wind. Gray shading indicates the time period covered by panels A–C.

uals consistently return to the Pribilof Islands in late June/early July [83], and in the intervening months, analysis of pelagic sampling suggested that concentrations of seals were highest off California in early winter with peak numbers shifting gradually northward as the overwintering period progressed [52,83,84]. The monthly patterns of alongshore distribution in this study exhibit both characteristics of this migratory pattern. The cross-shore distribution also supports the observations of Kajimura et al. [85], who noted that scientific pelagic collections found females preferentially concentrated within a 60 km band off the shelf break and over the continental slope, likely due to abundant food resources and/or increased prey availability in this area.

Daytime diving patterns of females in the CC ecosystem suggested that the MLD influences the depth of these dives. Generalized additive model results indicated that female NFS average daytime diving depths in 6 h periods tracked the winter-summer shoaling of the MLD, but with differing offsets and character between early season (day diving shallower than MLD) and late season (day diving highly concentrated and 5–10 m below MLD). Though the female average dive depths used as a response variable in the GAM were not necessarily spatially co-located with Seaglider measurements, with the exception of the presence of freshwater plumes, the Seaglider average winter-summer evolution of MLD is similar to that in other locations in the coastal CC and southern GA (e.g., Fig. 17 of [86]). Three adult females who encountered Seagliders in the WA coast region in April, May, and June dived to or slightly deeper than the MLD during the daytime, while night dive depths remained shallow and were not affected by the MLD, with the possible exception of diving within the Columbia River plume, where the MLD is very shallow (10–20 m).

These dive patterns are likely associated with the diel migration of typical fur seal prey in coastal transition zone habitats. Diet analysis of the scientific pelagic catch concluded that off WA, OR, and BC, NFS are opportunistic foragers that feed on a wide range of species, including various squids (market squid *Loligo opalescens*, also *Gonatus* spp., *Onychoteuthis* spp., especially important off the shelf), northern anchovy (*Engraulis mordax*), Pacific hake (*Merluccius productus*), rockfish (*Sebastes* spp.) and Pacific herring (*Clupea pallasii*) [1,31]. Juveniles of the latter species feed in the Strait of Juan de Fuca and Hecate Strait where they are preyed upon by salmon (*Oncorhynchus* spp.), which in turn are also fur seal prey. All of these species have been observed undertaking diel vertical migrations, feeding near the surface at night (corresponding to shallow nighttime female dive depths in the CC) and moving deeper during the day. Female diving behavior observed in this study indicates that during daylight hours, some prey species must congregate in the high-stratification transition layer immediately below the well-mixed surface layer rather than in the deep scattering layer at depths of 100 m or below. This places them within the vertical range of adult female physiological diving capabilities.

Ecosystem effects in the GLMM results for number of dives per 6 h period reflect the differing environmental conditions that seals encounter as well as the time of year at which each ecosystem is occupied. Fewer dives in the BS ecosystem at night is a similar result to [25] and likely reflects adult females rapidly exiting the Bering Sea and making their way to the CC and GA ecosystems. The interaction between proportion daylight and the CC ecosystem showed that as females moved closer to the coast, daytime dives represented a greater percentage of total diving effort. This resulted in a nearly even distribution of diving versus proportion daylight in the CC ecosystem. Sterling et al. [25] tracked a single female NFS who entered the CC in one overwintering period, and this individual showed an increase in day diving after entering the CC, though it was unclear in their results if this was an ecosystem response or simply individual variability. The results presented here for number of dives versus proportion daylight per 6 h period confirm that this is a consistent behavioral response to ecosystem traits, present in multiple individuals across multiple years. This is contrary to the pelagic behavior displayed in the NP ecosystem and in previous studies where adult females concentrate most of their effort in shallow, night or crepuscular diving [1,25,29,55,87–92]. Diving behavior of adult females in the CC in this study is, instead, similar to adult male diving behavior expressed during overwintering in the northern NP, where seals dived mostly during the daytime and to the MLD and below [25].

The MLD in the northern NP is deeper (100–125 m) in winter when compared to the MLD in the Seaglider survey region in winter (20–75 m; [77,93], this study). The majority of female overwinter habitat utilization was concentrated within 150 km of the shelf break, and Seaglider results demonstrate that MLD in this region was 60–70 m on average in mid-winter and rarely exceeded 80 m (Figs. 9–10). If the MLD represents the minimum depth of daytime seal prey fields, the deeper winter MLD in the NP implies that daytime prey fields are located deeper in the water column there than in the CC. The dive duration constraint of female fur seals, owing to their smaller mass, may mean that in the central and northern NP, the depth of winter daytime prey fields could place them outside the aerobic diving capabilities of the smaller females or require energy expenditure and recovery times that are disadvantageous compared to foraging closer to the shelf break. Thus, the cross-shore variability of the Pacific Ocean winter MLD could lead to a lateral gradient in accessibility of daytime prey fields. In the coastal ecosystem, wind-driven upwelling and subsequent productivity creates an abundance and diversity of prey [30]. The diving behavior observed in this study indicates that some of these prey species are localized at shallower depths where female physiology allows access at all times of day. This is consistent with the hypothesis that the location of female overwinter foraging grounds is dictated by interactions between prey availability and fur seal physiology [25,94].

The greater proportion of dives during daytime in the CC habitat resulted in fewer total dives in the CC LME. The rest periods required between dives increase with dive depth for adult female NFS [88]. As a result, the rate of dives per hour decreases nonlinearly with increasing dive depth. Costa and Gentry [95] compared the energy budgets of deep- and shallow-diving adult females in the Bering Sea and found that although deep divers

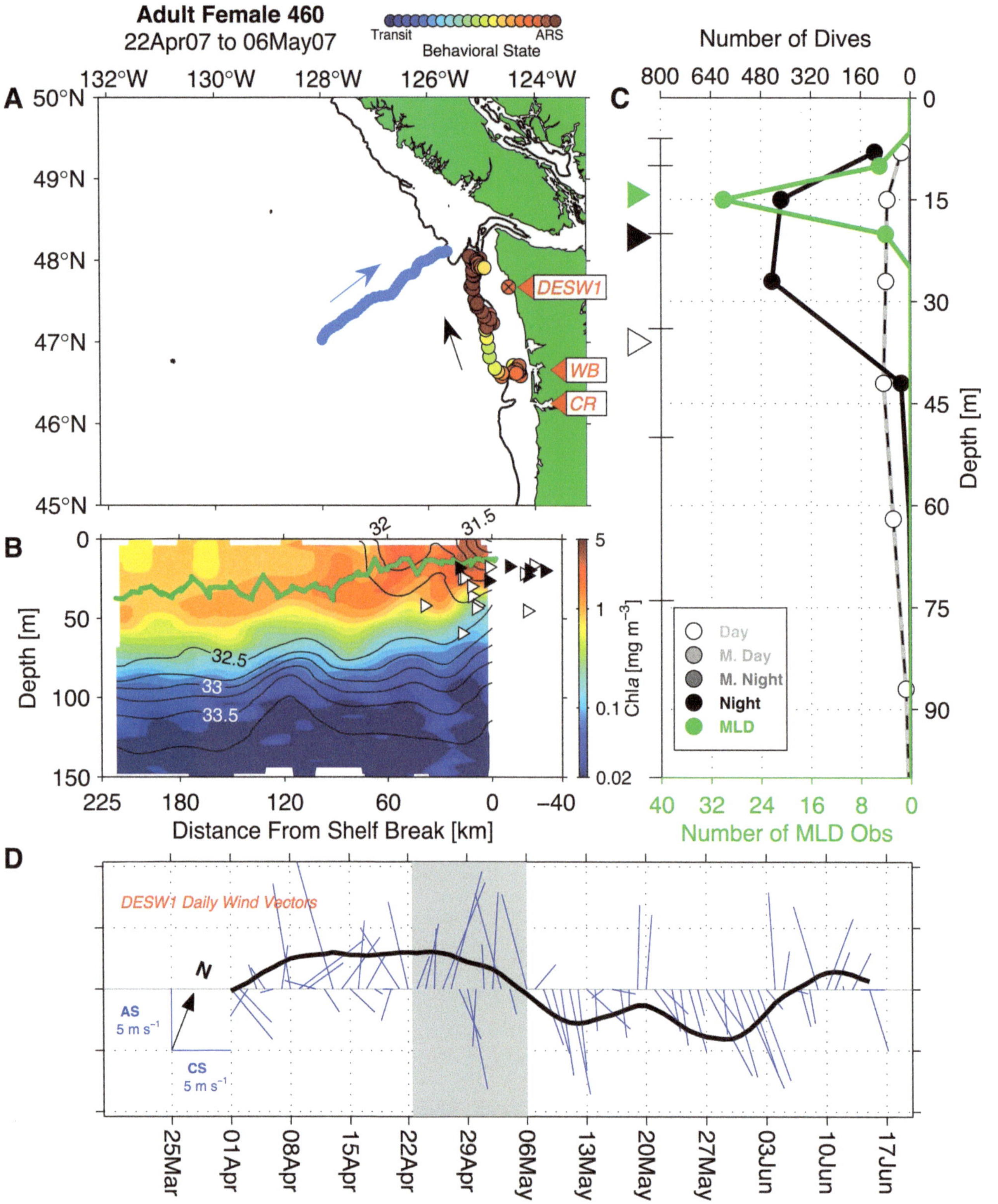

Figure 17. Adult female northern fur seal 460 in area-restricted search off the Washington coast, continued: 22 April to 06 May (2 of 4). (A) Estimated locations of female 460 (filled color circles) and reported GPS locations of Seaglider 014 (SG014, thick blue line) off the Washington and British Columbia coast (green). Red labels indicate the Columbia River mouth (CR), Willapa Bay coastal inlet feature (WB), and the location of National Data Buoy Center station Destruction Island (DESW1, red cross-circle at station). Colors of filled circles indicate behavioral state (scale at top, ARS = area-restricted search). The 200 m isobath is indicated by thin black line. Arrows indicate direction of travel of 460 and SG014. A sufficiently cloud-free surface chlorophyll-*a* (Chl*a*) image was not found for this time period. (B) Upper-ocean conditions by depth and distance from the shelf break recorded by SG014 during its crossing: Chl*a* (background colors), salinity (black contours), and mixed-layer depth (MLD, dark green line). Triangles indicate average depths of female northern fur seal 460's dives in each 6 h period. Triangles are plotted by cross-shore distance along the

Seaglider line, and triangle fill colors indicate proportion daylight category of each 6 h period (black triangles = night, dark gray = mostly night, light gray = mostly day, white = day, as defined in Results section *Links to Mesoscale Circulation and Coastal Topography*). (C) Histograms of seal 460's diving, one for each proportion daylight category, showing number of dives within each dive tag pre-programmed depth bin. Depth bin bounds are indicated on the left y-axis by whiskers. Green line in panel C is a histogram of MLD observations collected between 0 and 60 km from the shelf break using 5 m depth bins. Filled triangles to the left of Panel C indicate the average dive depth in each category of proportion daylight and the average MLD. (D) Daily-averaged wind vectors at 52.7 m height obtained from DESW1. Wind measurements at DESW1 have been rotated into an alongshore (AS)/cross-shore (CS) coordinate system. In panel D, the solid black line is a 10 d running average of the poleward component of wind. Gray shading indicates the time period covered by A–C.

captured less total prey biomass, their net energy storage and mass gain was comparable to shallow divers over the duration of a foraging trip, due to less total energy expended in deep diving and, presumably, a higher energy content of their prey. Thus an increase in foraging efficiency (net energy stored vs. foraging metabolism) in deeper dives offsets the lower rate at which they occur. Some of these same factors may also be important for overwinter foraging of adult females in the CC, although the depth difference between deep day dives versus shallow night dives in this study (e.g., 30–60 m deep vs. 10–20 m shallow) was less than day/night differences in the Bering Sea (where day dive depths are 60–100 m; [89,95]). Whether prey energy content in the CC LME differs between near-surface and at the base of the mixed layer, and whether day and night prey fields are comprised of different species or age class composition within a species, is unknown. This may be unimportant in late migration season or in the Columbia River plume, where MLDs are shallow enough that if prey aggregate just below the mixed layer during daytime, they are accessible by dives of only 20–25 m depth as shown in Fig. 18B.

Goebel et al. [89] concluded that aside from prey size and energy density, ease of capture is also another important factor that influences foraging energetics, and this could also partially explain the shift to day diving with continental slope habitat and season. Daytime foraging, especially below the shallow MLD off WA in late spring and early summer, may offer some advantage in visual identification of prey and increase the foraging success of adult females [96]. Increased spatial density of schooling fish in the daylight hours (e.g., [97–99]) may also be a contributing factor that adds to ease of capture and thus a potentially greater assimilated energy per dive in deeper daytime bouts. An increase in feeding event size during daytime, suggestive of greater assimilated energy per dive, has been observed in free-ranging adult female gray seals (*Halichoerus grypus*) in the Atlantic Ocean although the reasons for this are unclear [100].

The GLMM results for behavioral state are consistent with Sterling et al. [25] and show that surface wind speed is an important driver of behavioral state. Animals experience stronger winds while in transit during the early migratory period in the central NP and GA, and additionally respond to strong winds in the destination ecosystems by shifting away from area-restricted search towards transitory behavior. The relationship between day dive depths and the MLD observed in this study suggests a mechanism by which strong surface winds could reduce prey accessibility, and thus influence behavioral state, since these winds can force convection and deepen the MLD. The seasonal progression of behavioral state from low indices (transit) to high (area-restricted search) reflects females transiting quickly through the Aleutian passes and across the open NP, which is consistent with historical understanding of NFS migratory patterns. The interaction between surface wind speed and season as a predictor of behavioral state is a new result, reflecting a reduced effect of surface wind speed on behavior in the late spring and early summer.

A key qualitative result of this study, consistent with other analyses of both the NFS overwintering period and summer foraging, is evidence of adult female behavioral cues in response to eddies and energetic surface circulation as revealed by satellite altimetry. Females demonstrated movement alignment with geostrophic surface currents [1] and area-restricted search within and around eddies, or within strong jets forming the boundary between adjacent features [24,25,39,40]. On aggregate, females spent twice as much time (as a fraction of total time within the ecosystem) within and near (20 km from approximate edge) altimetry-identified coherent mesoscale eddies in the CC than during transit in the NP. This effect was magnified for long-duration tracks, which may relate to increased eddy generation late in the overwintering season as upwelling winds induce a strong coastal jet, which becomes unstable and increases EKE values near the coast [101]. Adult female foraging in the eddy-rich margins of energetic boundary currents is similar to that observed for males exiting the Bering Sea in fall and winter [24]. The distribution of female habitat utilization relative to eddies showed increased probability density of habitat utilization near $\hat{r} = 1$ for area-restricted search locations relative to transit. For an axially symmetric eddy identified in the Chelton et al. [63] time series, $\hat{r} = 1$ corresponds to the radial station of maximum velocity regardless of azimuthal station (compass point) around the eddy center. It is important to note when considering the eddy results that many individual eddies demonstrate substantial asymmetry or ellipticity (e.g., Fig. 12); thus, the description of the aggregate physical environment encountered by females at $\hat{r} = 1$ is only an approximate one.

The reason(s) underlying female foraging responses within and near some eddies are unknown, but may relate to physical processes near these features that alter prey characteristics in such a way as to make them energetically advantageous foraging hotspots. Benoit-Bird et al. [102] showed that in the Bering Sea in summer, the vertical distribution of aggregate prey patches and density within those patches were strong predictors of NFS habitat utilization. As noted above, vertical distribution affects the required energy expenditure per individual dive, while within-patch density affects the efficiency of prey capture during a dive. There are a variety of ways in which eddies or fronts could lead to shallower prey vertical localization or increased density within prey patches. For example, Godø et al. [103] hypothesized that grazing zooplankton align themselves with the strong currents at the eddy edge in order to optimize their foraging environment, and that mesopelagic forage fish cue on these resources through a variety of techniques, including passive displacement by background currents. Godø et al. [103] report observations of biomass accumulation near an eddy edge during periods of negligible phytoplankton community growth, suggesting that the mechanisms that make eddies a favorable environment for forage fishes are independent of any boost in primary productivity. This is consistent with female foraging in this study within eddies in January–March, when overall productivity in the CC is low and more spatially confined to the nearshore region [104]. It is also possible that NFS prey that prefer a narrow thermal range may simply avoid eddy core regions due to the contrast in temperature between an eddy cores and surrounding water, leading to apparent

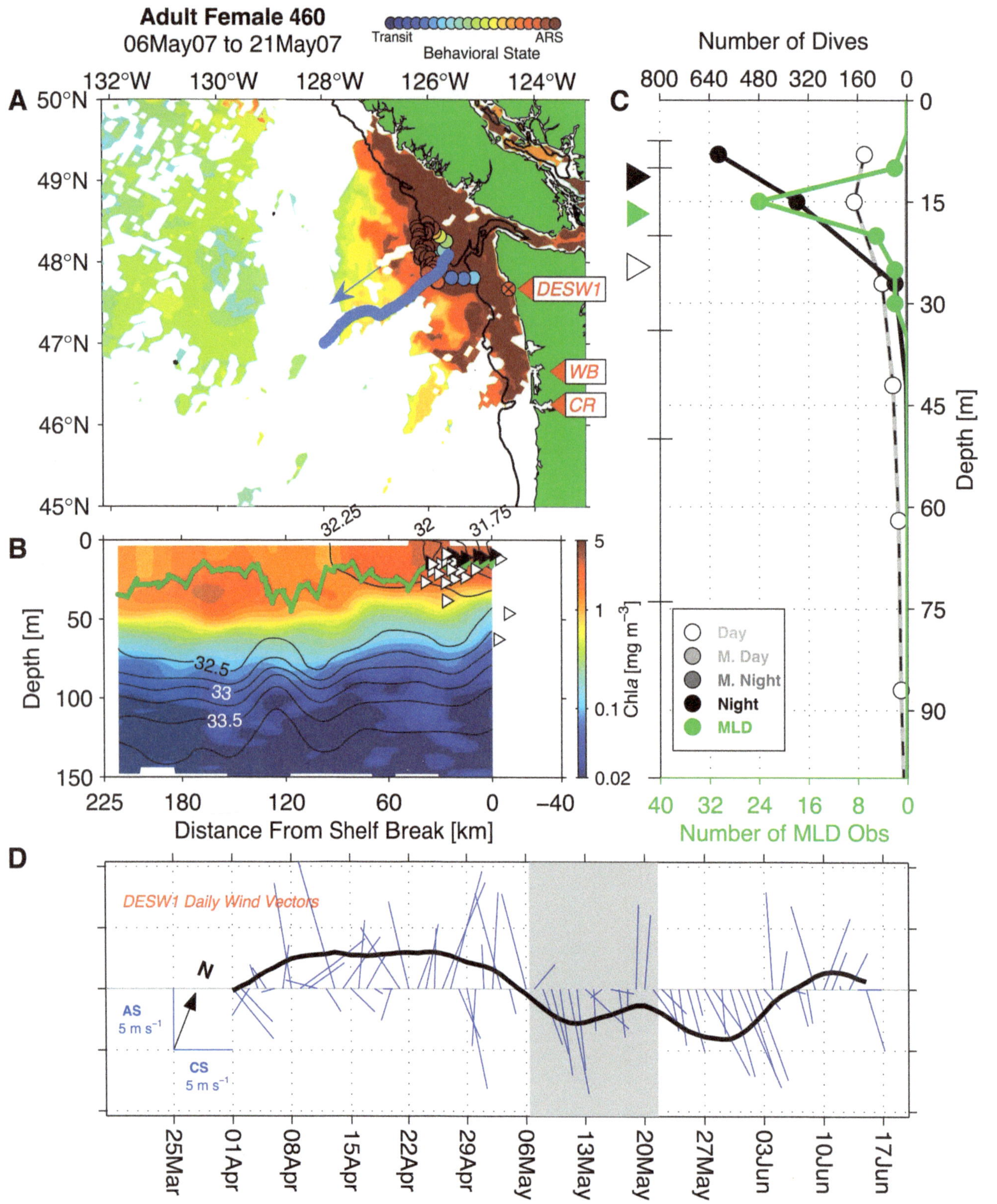

Figure 18. Adult female northern fur seal 460 in area-restricted search off the Washington coast, continued: 06 May to 21 May 2007 (3 of 4). (A) Estimated locations of female 460 (filled color circles) and reported GPS locations of Seaglider 014 (SG014, thick blue line) off the Washington and British Columbia coast (green). Red labels indicate the Columbia River mouth (CR), Willapa Bay coastal inlet feature (WB), and the location of National Data Buoy Center station Destruction Island (DESW1, red cross-circle at station). Colors of filled circles indicate behavioral state (scale at top, ARS = area-restricted search). The 200 m isobath is indicated by thin black line. Arrow indicates direction of travel of SG014. Background colors in panel A indicate surface chlorophyll-*a* (Chl*a*) in cloud-free regions on 11 May, plotted on the same scale as in panel B. (B) Upper-ocean conditions by depth and distance from the shelf break recorded by SG014 during its crossing: Chl*a* (background colors), salinity (black contours), and mixed-layer depth (MLD, dark green line). Triangles indicate average depths of female northern fur seal 460's dives in each 6 h period. Triangles are

plotted by cross-shore distance along the Seaglider line, and triangle fill colors indicate proportion daylight category of each 6 h period (black triangles = night, dark gray = mostly night, light gray = mostly day, white = day, as defined in Results section *Links to Mesoscale Circulation and Coastal Topography*). (C) Histograms of seal 460's diving, one for each proportion daylight category, showing number of dives within each dive tag preprogrammed depth bin. Depth bin bounds are indicated on the left y-axis by whiskers. Green line in panel C is a histogram of MLD observations collected between 0 and 60 km from the shelf break using 5 m depth bins. Filled triangles to the left of Panel C indicate the average dive depth in each category of proportion daylight and the average MLD. (D) Daily-averaged wind vectors at 52.7 m height obtained from DESW1. Wind measurements at DESW1 have been rotated into an alongshore (AS)/cross-shore (CS) coordinate system. In panel D, the solid black line is a 10 d running average of the poleward component of wind. Gray shading indicates the time period covered by A–C.

aggregation at the edge as an eddy moves through a uniform density of NFS prey. At ocean fronts, mixed-layer instabilities act to restratify the upper water column [105], thus reducing the MLD and potentially influencing prey vertical localization.

Elevated phytoplankton biomass at eddy edges or within frontal features could act as a secondary driver by creating a bottom-up forcing of forage fish concentrations over longer time scales, especially late in the upwelling season. Strong wind-driven vertical velocities near the region of maximum velocity at an eddy edge [106], or along the axis of a jet [107], increase the vertical transport of growth-limiting nutrients to the euphotic zone, and several studies have found elevated primary productivity, zooplankton, and fish populations associated with eddies and fronts in the CC [108–112]. This mechanism is consistent with the observed increase in habitat utilization near $\hat{r}=1$ for area-restricted search behavior. In the northern GA, eddies also play an important role in cross-shelf exchange and horizontal stirring of shelf and offshore waters in which phytoplankton growth is limited by different nutrients [113–118], thus also potentially leading to an increase in primary productivity relative to background conditions. The stirring of nutrients at the eddy edge, or redistribution of productivity around the rim of any eddy impinging on the shelf (independent of any enhancement), could in part account for NFS cueing on the outer rim of eddies in the GA [119]. A secondary benefit of eddies may be energetic savings due to swimming with eddy currents during either phase of the migration [1].

Given the variety of possible mechanisms involved, there are likely spatial and temporal differences between the migratory ecosystems regarding which factors are most important in affecting eddy-related foraging, and indeed whether an individual eddy would be a profitable foraging area. It should be noted that females avoided many features, and that the observed difference in median $\hat{r}$ value with behavioral state was not statistically significant for the entire dataset (though it was significant when considering only tracks >200 d, see Fig. S5). Furthermore, despite the qualitative evidence of alignment, efforts to quantify eddy statistical effect on behavioral state by an EKE term were unsuccessful, though EKE did not significantly degrade model performance/variance explained (not shown). This may be due to several factors. Migratory fur seals bypassed some high-EKE features while concentrating their effort in others, indicating that there may be other covariates that interact with EKE, such as subsurface water column structure, that we may not be able to measure in this analysis. Furthermore, mean eddy kinetic energy varies strongly with latitude in the eastern Pacific [120], even though EKE as a fraction of mean energy remains large [121]. Thus, an animal foraging near $\hat{r}=1$ off the WA or BC coast (e.g., Fig. S3) may be in a region of EKE that is locally elevated but weak relative to the overall dataset. Finally, migratory momentum and strong winds likely also play a role in females bypassing high EKE features during their transit across the NP ecosystem [25,122].

Despite the lack of consistent response, evidence of some female foraging within mesoscale variability in this and other studies suggests an additional driver of increased concentration of female effort within the inner coastal transition zone and near topographic features. In the CC eastern boundary current system, interaction between alongshore flow and topography occurs near canyons, sharp turns in the continental slope, and headlands including Cape Mendocino, Cape Blanco, and Point Sur. The resulting meanders in the coastal jet are unstable and produce eddies, jets, and fronts [30,86,101,123–127]. Eddy kinetic energy in these regions is highest during the upwelling season, but altimetry studies indicate that there is still significant EKE in the winter months, located at a greater distance from shore than in spring and summer [101,127]. Figure 12 illustrates that eddy generation processes are still active during winter at Cape Blanco. By the mechanisms outlined above, regions of enhanced eddy activity may also be regions with a greater likelihood of favorable foraging conditions for adult female NFS. Spatial density of eddy generation and mean EKE are greatest within 100–200 km of the shelf break in the CC and GA ecosystems [63,120,128,129] and this may be an additional driver of the observed cross-shore distribution of adult female habitat utilization beyond regional differences in prey accessibility. Aside from headlands in the central CC and the Columbia River, females congregated near or offshore of Queen Charlotte Sound, formation site for Haida eddies [130], and the entrance to the Strait of Juan de Fuca – habitat for juvenile herring (*C. pallasi*) and salmon (*Oncorhynchus* spp.) on which adult female NFS feed [31,131]. The persistent and productive Juan de Fuca eddy also appears annually at the mouth of the Strait, though after the spring transition [132]. This is near the time of year when females are probably beginning their return to Bogoslof and St. Paul Islands.

It should be noted that both the alongshore distributions of NFS derived from historical pelagic data [4] and the distribution of female NFS in this study suffer from potential biases. For example, at-sea takes of females from early pelagic sealing were largely weather-dependent and generally restricted to more coastal habitats. Pelagic scientific collections from 1958–74 were conducted with a requirement that each participant nation obtain a minimum quota; as a result, observational effort was concentrated in specific portions of their range that females were known to inhabit based on historical evidence. In this study, satellite tag longevity and retention effects likely reduced the overall number of adult female seal-hours in the late-season months, probably contributing to the apparent departure of females from the region in May–June (Fig. 5E–F). Nonetheless, the distribution of females shows good qualitative agreement with previous understanding of migratory patterns in the coastal region.

The multi-week record of female NFS 460's WA coast foraging in the spring and early summer of 2007 is the first direct evidence of the Columbia River plume as a feature on which female NFS cue to find prey. Female 460's movements tracked within the plume edge and core, moving away from the shelf break along with the plume after a switch to equatorward winds, and also departing at roughly the same time as bloom exhaustion within the plume (although the latter could be purely coincidental). In spring, off the WA coast, the plume serves as an effective lateral transporter of biomass offshore as plume orientation shifts with

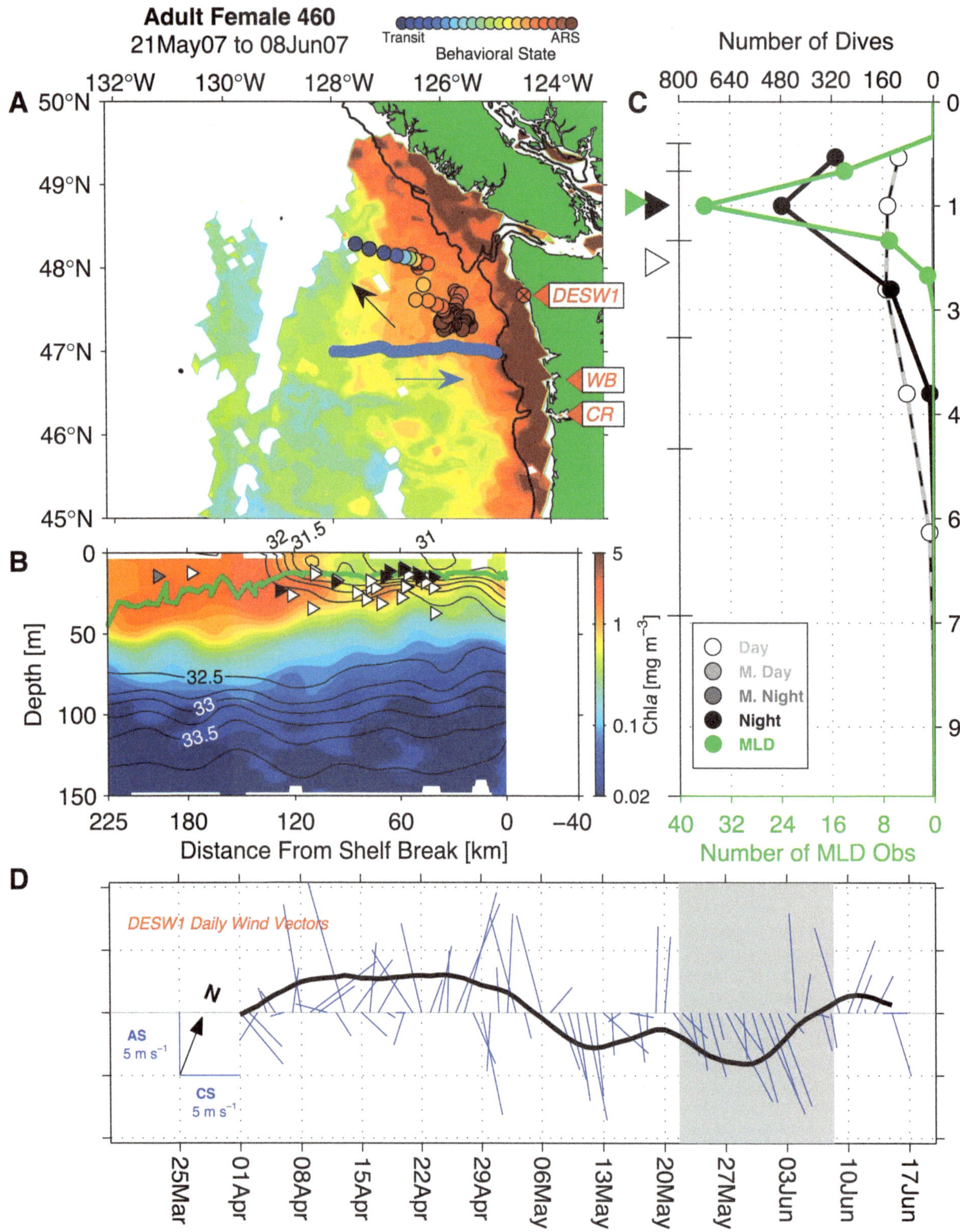

Figure 19. Adult female northern fur seal 460 in area-restricted search and departure from the Washington coast: 21 May to 08 June (4 of 4). (A) Estimated locations of female 460 (filled color circles) and reported GPS locations of Seaglider 014 (SG014, thick blue line) off the Washington and British Columbia coast (green). Red labels indicate the Columbia River mouth (CR), Willapa Bay coastal inlet feature (WB), and the location of National Data Buoy Center station Destruction Island (DESW1, red cross-circle at station). Colors of filled circles indicate behavioral state (scale at top, ARS = area-restricted search). The 200 m isobath is indicated by thin black line. Arrows indicate direction of travel of 460 and SG014. Background colors in panel A indicate surface chlorophyll-*a* (Chl*a*) in cloud-free regions on 29 May 2007, plotted according to the color scale in panel B. (B) Upper-ocean conditions by depth and distance from the shelf break recorded by SG014 during its crossing: Chl*a* (background colors), salinity (black contours), and mixed-layer depth (MLD, dark green line). Triangles indicate average depths of female northern fur seal 460's dives in each 6 h

period. Triangles are plotted by cross-shore distance along the Seaglider line, and triangle fill colors indicate proportion daylight category of each 6 h period (black triangles = night, dark gray = mostly night, light gray = mostly day, white = day, as defined in Results section *Links to Mesoscale Circulation and Coastal Topography*). (C) Histograms of seal 460's diving, one for each proportion daylight category, showing number of dives within each dive tag pre-programmed depth bin. Depth bin bounds are indicated on the left y-axis by whiskers. Green line in panel C is a histogram of MLD observations collected between 0 and 60 km from the shelf break using 5 m depth bins. Filled triangles to the left of Panel C indicate the average dive depth in each category of proportion daylight and the average MLD. (D) Daily-averaged wind vectors at 52.7 m height obtained from DESW1. Wind measurements at DESW1 have been rotated into an alongshore (AS)/cross-shore (CS) coordinate system. In panel D, the solid black line is a 10 d running average of the poleward component of wind. Gray shading indicates the time period covered by A–C.

the alongshore winds [81,133]. Additionally, a key element of the plume is the low-salinity "cap" it places on the water column [80,82]. Its strong salinity stratification reduces the size of turbulent vertical overturns and leads to a shallow MLD, potentially increasing the accessibility of daytime prey fields for adult females encountering the plume off the WA coast.

These results provide further support to the idea that as top predators, NFS respond to integrated biological changes cascading upward from the ocean physical environment through the marine ecosystem. In this study, eddies and the Columbia River plume represent sources of mesoscale variability that elicited strong behavioral responses in mature, experienced individuals, underscoring that adult females are highly flexible in altering their foraging intensity in response to ephemeral prey resources encountered during the overwintering phase [90]. At longer scales, seasonal alterations in the character and intensity of adult female NFS diving likely reflect seasonal changes in surface winds and heating, ecosystem phenology, and prey movements in the CC and GA. Females exhibited a shift in character in day dives, towards vertically concentrated foraging below the MLD, that occurred roughly in tandem with the spring shoaling of the MLD in late April and early May. This timing is coincident with the onset of increases in body size (aggregate body weight and length) in adult females collected in historical pelagic sampling and emphasizes that the processes discussed in this manuscript can potentially exert a fundamental control over adult female NFS condition during the overwintering period [134]. The association of day diving depths with the MLD suggests that short-term physical processes such as storms and atmospheric variability can influence vertical localization of NFS prey. The MLD is highly dynamic during fall in the northeast Pacific, where seasonal deepening is episodic, driven by individual storms that force convection and can increase the MLD by 10 m over a 1–2 d span [135,136]. Experienced adults contend with these biophysical alterations but it is unknown if and how event or seasonal scale biophysical variation could impact the condition of less physiologically capable juveniles or pups, and ultimately NFS demography by affecting overwinter survival. Several studies have reported numerous tagged NFS pup carcasses recovered along the coasts of OR, WA, and BC following heavy weather in the northeast Pacific [137–140]. Processes described in this study support the hypothesis that a deeper dispersed prey field altered by winter processes could disproportionately affect smaller, foraging-naive pups or juveniles lacking the physiological diving capabilities of larger adults [55], and warrant further investigation.

Conclusions

Comparison of NFS behavior to remotely-sensed oceanographic fields, both surface and subsurface, informs our understanding of how oceanographic parameters affect individuals or groups of individuals. This is an important step in exploring both the long-term role of a variable environment on a widely-distributed top marine predator as well as the mechanisms potentially underpinning its population decline. The results of this study demonstrate that in the pelagic environment off WA and in the CC ecosystem, adult female NFS exhibit a shift in their foraging effort towards daylight hours and that foraging preferentially occurs in a cross-shore band with its center 60–80 km from the shelf break. The depth of daytime foraging suggests association of prey fields with the depth of the surface mixed layer, especially late in the migratory season where daytime diving was vertically concentrated 5–10 m below the MLD. Consistent with other studies, behavioral state was found to increase towards area-restricted search as the season progressed and as surface wind speed decreased. Individual females showed behavioral responses (movement alignment and changes to behavioral state) to some mesoscale features during the overwintering period, primarily in the CC ecosystem.

Observations from Seagliders support the idea that preferential habitat utilization in the near-shelf region reflects the abundant prey resources in the CC and basin-scale lateral gradients in MLD that may play a role in increasing accessibility of prey to adult females during the daytime. Three females equipped with dive recorders who traveled through the WA region demonstrated daytime diving to the base of the mixed layer and below, and in one case, movement behavior associated with the Columbia River plume, which is associated with a shallow MLD. Foraging within eddies may contribute to enhanced adult female NFS habitat utilization near eddy generation sites and contribute to the location of the peak in cross-shore distribution at the energetic outer edge of continental slope currents. Enhancements to prey spatial concentration and/or accessibility are possible reasons why females demonstrate behavioral responses to mesoscale eddies and other circulation features.

These results suggest that surface light levels and wind speed, oceanic MLD, coastal primary productivity, freshwater plumes, and eddy activity are important environmental factors that influence the distribution of adult female NFS and their behavioral responses within the destination ecosystems for overwinter foraging. Female prey composition and depth distribution during the overwintering period are areas that require further study, which may reveal more about the functional nature of the interaction between NFS diving and the MLD. The quantitative relationship between adult female behavior and mesoscale physical features during the overwintering period remains unclear and requires further observations in which subsurface structure can be studied along with behavioral responses or lack thereof. A more focused sampling effort using autonomous vehicles and satellite tagging of NFS could address this and other outstanding questions regarding the overwintering period in the future. Increased knowledge of how ocean conditions affect adult female behavior during overwinter migration is important to addressing questions surrounding demography during the pelagic phase of the NFS life cycle, and may shed light on how long-term climate patterns in the eastern Pacific affect the population as a whole.

Supporting Information

Figure S1 Generalized additive model residuals and response diagnostic plots. (Top Left) Plot of residual quantile versus theoretical (normal) quantile. (Bottom Left) Histogram of residuals. Residuals are distributed in approximately normal fashion. (Top Right) Distribution of residuals versus a linear predictor. There is no evidence of significant change in the distribution of residuals as a function of the linear predictor value. (Bottom Right) Response variable versus modeled values.

Figure S2 Monthly evolution of adult female northern fur seal cross-shore distribution. January (A) through June (F). Blue bars in each plot are a histogram of estimated time spent in that month versus distance offshore from the shelf break in 20 km bins within the California Current and Gulf of Alaska Large Marine Ecosystems, north of 41°N and south of Haida Gwaii, during the winters 2002–03 to 2009–10. Above each plot, vertical red lines indicate the median cross-shore position in that month.

Figure S3 Behavioral responses of adult female northern fur seal 626 to mesoscale circulation and surface wind speed. (A) Overview of satellite-tracked locations of female 626 in the California Current and Gulf of Alaska ecosystems from 1 April 2008 onwards (track prior to this time was localized near the same positions as 1 April–1 May and is not shown in order to avoid visual clutter). Locations every 6 h are colored according to estimated behavioral state (scale at top, ARS = area-restricted search). Gray box in panel A indicates the spatial extent covered by Figure 14A, when seal 626 foraged in close proximity to the Seaglider region. Temporal extent covered by Figure 14A approximately overlaps with that of panel B in this figure. (B)–(G) Weekly intervals of seal 626's estimated locations and behavioral state in 6 h periods (filled color circles, scale at top) plotted over sea level anomaly (color contours, scale at top) and surface geostrophic velocity anomaly (Vel, black arrows, scale at top right). Each plot is centered on female seal 626's locations over the weekly period. Thick gray circles indicate the locations and approximate spatial extent of altimetry-identified mesoscale eddies from Chelton et al. [63]. Eddies are plotted as circular features though this is intended for illustration purposes only. After crossing 47°N on 6 May, 626 transited to the northwest, then transitioned to an area-restricted search near 49°N, 130°W. This location appeared to be at the boundary between a weak cyclonic (counterclockwise-rotating, locally low SLA values) eddy feature centered at 49.83°N, 129.25°W and a possibly-developing anticyclonic (clockwise, locally high SLA values) eddy at 48.75°N, 130.33°W. Seal 626 remained in this location until June, at which point she circuited the eddy in the same sense as its rotation, following the geostrophic surface current. (H) Plot of estimated behavioral state (red line, scale on right axis) and 10 m height wind speed at 626's location (blue, scale on left y-axis) versus time for the overwintering period 2007–08. Wind speed estimates are obtained by interpolating National Centers for Environmental Prediction Reanalysis 2 (NCEP2) product to 626's estimated locations at each 6 h time point. Gray bars in alternating shading display the extent of time covered by panels B–G. Behavioral state for 626 indicated area-restricted search almost uniformly from 10 May to 14 June. This was the longest period of prolonged area-restricted search state for 626's recorded migration. Like other prolonged search periods shown, 10 May–14 June was characterized by lower overall surface wind speeds than the remainder of the satellite record for this individual, though there were still instances of elevated (>10 m s^{-1}) winds (e.g., temporal extent corresponding to panels D, F, and G). During these times seal 626's behavioral state was temporarily reduced from the maximum area-restricted search index, though did not fully transition to transient behavior, possibly owing to the short duration of these wind events.

Figure S4 Behavioral responses of adult female northern fur seal 628 to mesoscale circulation and surface wind speed. (A) Overview of satellite-tracked locations of female 628 in the California Current and Gulf of Alaska ecosystems from 24 December 2007 onwards. Six h locations are colored according to estimated behavioral state (scale at top, ARS = area-restricted search). Gray box in panel A and gray shading in panel H indicates the spatial and temporal extent covered by Figure 15A, when 628 foraged in close proximity to the Seaglider region. (B)–(G) Weekly intervals of seal 628's estimated locations and behavioral state in 6 h periods (filled color circles) plotted over sea level anomaly (color contours, scale at top) and surface geostrophic velocity anomaly (Vel, black arrows, scale at top right). Each plot is centered on female seal 628's locations over the weekly period. Thick gray circles indicate the locations and approximate spatial extent of altimetry-identified mesoscale eddies from Chelton et al. [63]. Eddies are plotted as circular features though this is intended for illustration purposes only. These panels illustrate transit of 628 through a cyclonic eddy at 37°N, 128.5°W from 26 January to 8 March 2008. Seal 628 moved through a highly elliptical anticyclonic feature, encountered the cyclone, and like female seal 626 moved through this feature with the same sense as the cyclonic currents in panels C–E. Seal 628 then encountered a second cyclone and its movement was briefly aligned with this feature as well in panel F. (H) Plot of estimated behavioral state (red line, scale on right y-axis) and 10 m height wind speed at 628's location (blue, scale on left y-axis) versus time for the overwintering period 2007–08. Wind estimates are obtained by interpolating National Centers for Environmental Prediction Reanalysis 2 (NCEP2) product to 628's estimated locations at each 6 h time point. Gray bars in alternating shading display the extent of time covered by panels B–G. Seal 628's estimated behavioral state in panels B–G was uniformly indicative of area-restricted search, despite occasional increases in surface wind speed. The wind speed and behavioral state history of 628 illustrates higher wind speeds encountered during the transit phase in comparison to those encountered in the California Current (February–June).

Figure S5 Adult female northern fur seal habitat utilization relative to eddy features, conditioned by estimated behavioral state (evaluated for tracks >200 d only). (A) One-dimensional radial probability density functions [PDFs, $f(\hat{r})$] and cumulative distribution functions [CDFs, $F(\hat{r})$] for habitat utilization relative to eddy features at two categories of behavioral state b (transit = dark gray, area-restricted search [ARS] = light gray). "Transit" periods are those with $b < 0.25$; "ARS" (area-restricted search) periods are those with $b > 0.75$. Solid lines show PDFs, dashed lines are CDFs (scale on right y-axis). Distributions are computed as a function of normalized distance to the nearest eddy center $\hat{r}$, defined as absolute distance to the nearest eddy divided by that eddy's radial length scale L_s [63]. (B) Observed difference between search PDF and transit PDF (solid black line) with 95% confidence bounds computed using a bootstrap method (gray shading). Probability density functions show significant differences near $\hat{r}=1$ ($p<0.0001$) and $\hat{r}=4$ ($p=0.0003$). (C) Bootstrap distribution of $\Delta\hat{r}^{(i)}$, the difference of median values of $\hat{r}$ between ARS and transit states on bootstrap iteration i. Solid black line shows the PDF of these values, thick dashed line shows CDF (scale at right). Thin vertical dashed line is the observed value of $\Delta\hat{r}$. The gray shaded area denotes the bias-corrected/accelerated 95% confidence interval on the observed value of $\Delta\hat{r}$. For the observed value of $\Delta\hat{r}$ to be significantly different from zero using a two-tailed test, this confidence interval must not include zero.

Acknowledgments

The authors would like to thank the Alaska Ecosystems Programs scientists for their field efforts in capturing and instrumenting northern fur seals on Bogoslof and St. Paul Islands. Significant contributions came from Rod Towell, Jim Thomason, Brian Fadely, Carey Kuhn, Tom Gelatt, and Kate Towell. We also thank Kirk O'Donnell, James Bennett, Bill Fredericks, Troy Swanson, Neil Bogue, and Tom Lehman for tireless and dedicated work during the Seaglider field campaign. Many thanks also to Mike Twardowski and Brandon Sackmann for their helpful input and advice regarding processing of the optical data, and to Kelly Benoit-Bird for useful insight and discussion regarding eddies and fur seal prey. We are grateful to Sharon Melin, Ward Testa, and three anonymous reviewers for their constructive and thorough input that helped to improve the manuscript, and to Devin Johnson for thoughtful suggestions relevant to the statistical modeling. The findings and conclusions in the paper are those of the authors and do not necessarily represent the view of the National Marine Fisheries Service. Reference to trade names does not imply endorsement by the National Marine Fisheries Service or NOAA.

Author Contributions

Conceived and designed the experiments: JTS RRR MAL CCE CML. Performed the experiments: JTS MAL RRR CCE CML. Analyzed the data: NAP JTS MAL NAB. Wrote the paper: NAP JTS MAL NAB RRR CCE CML.

References

1. Ream RR, Sterling JT, Loughlin TR (2005) Oceanographic influences on northern fur seal migratory movements. Deep Sea Res Part II 52: 823–843.
2. Trites AW (1991) Fetal growth of northern fur seals: life-history strategy and sources of variation. Can J Zool 69: 2608–2617.
3. Trites AW (1992) Northern fur seals: why have they declined. Aquat Mamm 18: 3–18.
4. Bigg MA (1982) Migration of northern fur seals in the eastern North Pacific and eastern Bering Sea: an analysis using effort and population composition data. Submitted to the standing scientific committee, 25th annual meeting of the North Pacific Fur Seal Commission. 77 p. Available: NMML Library, Alaska Fisheries Science Center, 7600 Sand Point Way NE, Seattle WA 98115.
5. Gentry RL, Costa DP, Croxall JP, David JHM, Davis RW, et al. (1986) Synthesis and conclusions. In: Gentry RL, Kooyman GL, editors. Fur seals: maternal strategies on land and at sea. Princeton NJ: Princeton University Press. pp. 220–264.
6. Polovina JJ, Howell EA, Kobayashi DR, Seki MP (2001) The transition zone chlorophyll front, a dynamic global feature defining migration and forage habitat for marine resources. Prog Oceanogr 49: 469–483.
7. Chapman DG (1961) A critical study of Pribilof fur seal population estimates. Fish Bull 63: 657–669.
8. Eberhardt L (1981) Population dynamics of the Pribilof fur seals. In: Fowler CW, Smith TD, editors. Dynamics of large mammal populations. New York NY: John Wiley & Sons, Inc. pp. 197–220.
9. Lander RH (1979) Role of land and ocean mortality in yield of male Alaskan fur seal, *Callorhinus ursinus*. Fish Bull 77: 311–314.
10. Lander RH (1981) A life table and biomass estimate for Alaskan fur seals. Fish Res 1: 55–70.
11. York AE, Hartley JR (1981) Pup production following harvest of female northern fur seals. Can J Fish Aquat Sci 38: 84–90.
12. French DP, Reed M (1990) Potential impact of entanglement in marine debris on the population dynamics of the northern fur seal, *Callorhinus ursinus*. In: Shomura RS, Godfrey ML, editors. Proceedings of the second international conference on marine debris, 2–7 April 1989. Honolulu HI: US Department of Commerce. NOAA Technical Memorandum NMFS-SWFSC-154. pp. 431–452. Available: http://http://swfsc.noaa.gov/publications/TM/SWFSC/NOAA-TM-NMFS-SWFSC-154_TOC.PDF.
13. Springer A, Estes J, van Vliet G, Williams T, Doak D, et al. (2003) Sequential megafaunal collapse in the North Pacific Ocean: an ongoing legacy of industrial whaling? Proc Nat Acad Sci USA 100: 12223–12228.
14. York AE (1995) The relationship of several environmental indices to the survival of juvenile male northern fur seals (*Callorhinus ursinus*) from the Pribilof Islands. Can Spec Publ Fish Aquat Sci 121: 317–327.
15. NMFS (2007) Conservation plan for the Eastern Pacific Stock of northern fur seal (*Callorhinus ursinus*). Washington DC: Department of Commerce/NOAA/Protected Resources Division, Alaska Region. 136 p. Available: http://www.fakr.noaa.gov/protectedresources/seals/fur/cplan/final1207.pdf.
16. Towell RG, Ream RR, Bengtson JL, Sterling JT (2011) 2010 northern fur seal pup production and adult male counts on the Pribilof Islands, Alaska. Seattle WA: Alaska Fisheries Science Center. 6 p. Available: http://www.afsc.noaa.gov/nmml/pdf/2010-nfs-pup-estimates.pdf.
17. Melin SR, Sterling JT, Ream RR, Towell RG, Zeppelin TK, et al. (2012) A tale of two stocks: studies of northern fur seals breeding at the northern and southern extent of the range. Seattle WA: NOAA Fisheries Service. Alaska Fisheries Science Center quarterly report feature (April-May-June 2012). 11 p. Available: http://www.afsc.noaa.gov/Quarterly/amj2012/amj12featurelead.htm.
18. Ream RR, Baker JD, Towell RG (1999) Bogoslof Island studies. In: Sinclair EH, Robson BW, editors. Fur seal investigations, 1997. Seattle WA: National Marine Fisheries Service. NOAA Technical Memorandum NMFS-AFSC-106. pp. 81–92. Available: http://www.afsc.noaa.gov/Publications/AFSC-TM/NOAA-TM-AFSC-106.pdf.
19. Towell RG, Ream RR, York AE (2006) Decline in northern fur seal (*Callorhinus ursinus*) pup production on the Pribilof Islands. Mar Mamm Sci 22: 486–491.
20. Kuhn CE, Baker JD, Towell RG, Ream RR (2014) Evidence of localized resource depletion following a natural colonization event by a large marine predator. J Anim Ecol doi: 10.1111/1365-2656.12202.
21. Philander SGH (1983) El Niño southern oscillation phenomena. Nature 302: 295–301.
22. DeLong R, Antonelis G (1991) Impact of the 1982–1983 El Niño on the northern fur seal population at San Miguel Island, California. In: Trillmich F, Ono KA, editors. Pinnipeds and El Niño: responses to environmental stress. Berlin, Germany: Springer-Verlag. pp. 75–83.
23. Kenyon KW, Wilke F (1953) Migration of the northern fur seal, *Callorhinus ursinus*. J Mammal 34: 86–98.
24. Loughlin TR, Baba N, Robson BW (1999) Use of surface-current model and satellite telemetry to assess marine mammal movements in the Bering Sea. In: Loughlin TR, Ohtani K, editors. Dynamics of the Bering Sea. Fairbanks AK: University of Alaska Sea Grant. pp. 615–630.
25. Sterling JT, Springer AM, Iverson SJ, Johnson SP, Pelland NA, et al. (2014) The sun, moon, wind, and biological imperative—shaping contrasting wintertime migration and foraging strategies of adult male and female northern fur seals (*Callorhinus ursinus*). PLoS ONE 9: e93068.
26. Boyd IL, McCafferty DJ, Reid K, Taylor R, Walker TR (1998) Dispersal of male and female Antarctic fur seals (*Arctocephalus gazella*). Can J Fish Aquat Sci 55: 854–852.
27. Dalton AJM, Rosen DAS, Trites AW (2014) Broad thermal capacity facilitates the primarily pelagic existence of northern fur seals (*Callorhinus ursinus*). Mar Mamm Sci 30: 994–1013.
28. Rosen DAS, Trites AW (2014) Thermal limits in young northern fur seals, *Callorhinus ursinus*. Mar Mamm Sci 30: 1014–1018.
29. Gentry RL (1998) Behavior and ecology of the northern fur seal. Princeton NJ: Princeton University Press. 392 p.
30. Checkley DM Jr, Barth JA (2009) Patterns and processes in the California current system. Prog Oceanogr 83: 49–64.
31. Kajimura H (1982) Opportunistic feeding of the northern fur seal *Callorhinus ursinus*, in the eastern North Pacific Ocean and eastern Bering Sea [PhD Thesis]. Tokyo, Japan: University of Tokyo. 150 p.
32. Zeppelin TK, Orr AJ (2010) Stable isotope and scat analyses indicate diet and habitat partitioning in northern fur seals *Callorhinus ursinus* across the eastern Pacific. Mar Ecol Prog Ser 409: 241–253.
33. Trillmich F, Ono KA (1991) Pinnipeds and El Niño: responses to environmental stress. Berlin, Germany: Springer-Verlag. 293 p.
34. Weise MJ, Costa DP, Kudela RM (2006) Movement and diving behavior of male California sea lion (*Zalophus californianus*) during anomalous oceanographic conditions of 2005 compared to those of 2004. Geophys Res Lett 33: L22S10.

35. Trillmich F (1986) Attendance behavior of Galapagos fur seals. In: Gentry RL, Kooyman GL, editors. Fur seals: maternal strategies on land and at sea. Princeton NJ: Princeton University Press. pp. 168–185.
36. Trillmich F, Dellinger T (1991) The effects of El Niño on Galapagos pinnipeds. In: Trillmich F, Ono KA, editors. Pinnipeds and El Niño: responses to environmental stress. Berlin, Germany: Springer-Verlag. pp. 66–74.
37. Baker JD, Howell EA, Polovina JJ (2012) Relative influence of climate variability and direct anthropogenic impact on a sub-tropical Pacific top predator, the Hawaiian monk seal. Mar Ecol Prog Ser 469: 175–189.
38. Mantua N, Hare S, Zhang Y, Wallace J, Francis R (1997) A Pacific interdecadal climate oscillation with impacts on salmon production. Bull Am Meteorol Soc 78: 1069–1079.
39. Sterling JT (2009) Northern fur seal foraging behaviors, food webs, and interactions with oceanographic features in the eastern Bering Sea [Ph.D. Thesis]. Seattle WA: University of Washington. 219 p.
40. Nordstrom CA, Battaile BC, Cotté C, Trites AW (2013) Foraging habitats of lactating northern fur seals are structured by thermocline depths and submesoscale fronts in the eastern Bering Sea. Deep Sea Res Part II 88–89: 78–96.
41. Bailleul F, Cotté C, Guinet C (2010) Mesoscale eddies as foraging area of a deep-diving predator, the southern elephant seal. Mar Ecol Prog Ser 408: 251–264.
42. Campagna C, Piola AR, Rosa Marin M, Lewis M, Fernandez T (2006) Southern elephant seal trajectories, fronts and eddies in the Brazil/Malvinas confluence. Deep Sea Res Part I 53: 1907–1924.
43. Nel DC, Lutjeharms JRE, Pakhomov EA, Ansorge IJ, Ryan PG, et al. (2001) Exploitation of mesoscale oceanographic features by grey-headed albatross *Thalassarche chrysostoma* in the southern Indian Ocean. Mar Ecol Prog Ser 217: 15–26.
44. Yen PPW, Sydeman WJ, Bograd SJ, Hyrenbach KD (2006) Spring-time distributions of migratory marine birds in the southern California current: oceanic eddy associations and coastal habitat hotspots over 17 years. Deep Sea Res Part II 53: 399–418.
45. Yoda K, Shiomi K, Sato K (2014) Foraging spots of streaked shearwaters in relation to ocean surface currents as identified using their drift movements. Prog Oceanogr 122: 54–64.
46. Haney JC, McGillivary PA (1985) Aggregations of Cory's shearwaters (*Calonectris diomedea*) at Gulf Stream fronts. Wilson Bull 97: 191–200.
47. Baker JD (2007) Post-weaning migration of northern fur seal *Callorhinus ursinus* pups from the Pribilof Islands, Alaska. Mar Ecol Prog Ser 341: 243–255.
48. Hooper CL (1895) Report of Captain Hooper, dated November 21, 1892. In: US Senate. 53rd Congress, 2nd session. Fur-seal arbitration, proceedings of the Tribunal of Arbitration, volume VII, Senate executive document 177 (serial set 3166). Washington DC: Government Printing Office. pp. 228–233.
49. Lea MA, Johnson D, Ream R, Sterling J, Melin S, et al. (2009) Extreme weather events influence dispersal of naive northern fur seals. Biol Letters 5: 252–257.
50. Peterson RS (1965) The behavior of the northern fur seal [Ph.D. Thesis]. Baltimore MD: John Hopkins University. 214 p.
51. Eriksen CC, Osse TJ, Light RD, Wen T, Lehman TW, et al. (2001) Seaglider: a long-range autonomous underwater vehicle for oceanographic research. IEEE J Oceanic Eng 26: 424–436.
52. US Senate. 54th Congress, 1st session. (1896) Seal life on rookeries of the Pribilof Islands, 1893–95, part II, Senate document 137 (serial set 3351). Washington DC: Government Printing Office. 154 p.
53. Sherman KR (1991) The large marine ecosystem concept: research and management strategy for living marine resources. Ecol Appl: 350–360.
54. Loughlin TR, Sterling JT, Merrick RL, Sease JL, York AE (2003) Diving behavior of immature Steller sea lions (*Eumetopias jubatus*). Fish Bull 101: 566–582.
55. Lea MA, Johnson D, Melin S, Ream RR, Gelatt T (2010) Diving ontogeny and lunar responses in a highly migratory mammal, the northern fur seal *Callorhinus ursinus*. Mar Ecol Prog Ser 419: 233–247.
56. Block BA, Jonsen ID, Jorgensen SJ, Winship AJ, Shaffer SA, et al. (2011) Tracking apex marine predator movements in a dynamic ocean. Nature 475: 86–90.
57. Breed GA, Jonsen ID, Myers RA, Bowen WD, Leonard ML (2009) Sex-specific, seasonal foraging tactics of adult grey seals (*Halichoerus grypus*) revealed by state-space analysis. Ecology 90: 3209–3221.
58. Jonsen I, Myers R, Flemming J (2003) Meta-analysis of animal movement using state-space models. Ecology 84: 3055–3063.
59. Jonsen ID, Flemming JM, Myers RA (2005) Robust state-space modeling of animal movement data. Ecology 86: 2874–2880.
60. Jonsen ID, Myers RA, James MC (2007) Identifying leatherback turtle foraging behaviour from satellite telemetry using a switching state-space model. Mar Ecol Prog Ser 337: 255–264.
61. Jonsen ID, Basson M, Bestley S, Bravington MV, Patterson TA, et al. (2013) State-space models for bio-loggers: a methodological road map. Deep Sea Res Part II 88–89: 34–46.
62. Worton BJ (1995) Using Monte Carlo simulation to evaluate kernel-based home range estimators. J Wildl Manag 59: 794–800.
63. Chelton DB, Schlax MG, Samelson RM (2011) Global observations of nonlinear mesoscale eddies. Prog Oceanogr 91: 167–216.
64. DiCiccio TJ, Efron B (1996) Bootstrap confidence intervals. Stat Sci 11: 189–212.
65. Venables WN, Dichmont CM (2004) GLMs, GAMs and GLMMs: an overview of theory for applications in fisheries research. Fish Res 70: 319–337.
66. Benson SR, Eguchi T, Foley DG, Forney KA, Bailey H, et al. (2011) Large-scale movements and high-use areas of western Pacific leatherback turtles, *Dermochelys coriacea*. Ecosphere 2: art84.
67. Akaike H (1998) Information theory and an extension of the maximum likelihood principle. In: Parzen E, Tanabe K, Kitagawa G, editors. Selected papers of Hirotugu Akaike. New York NY: Springer. pp. 199–213.
68. Perry M, Sackmann B, Eriksen C, Lee C (2008) Seaglider observations of blooms and subsurface chlorophyll maxima off the Washington coast. Limnol Oceanogr 53: 2169–2179.
69. Sackmann B (2007) Remote assessment of 4-D phytoplankton distributions off the Washington coast [Ph.D. Thesis]. Orono ME: University of Maine. 204 p.
70. Pelland NA, Eriksen CC, Lee CM (2013) Subthermocline eddies over the Washington continental slope as observed by Seagliders, 2003–09. J Phys Oceanogr 43: 2025–2053.
71. de Boyer Montégut C, Madec G, Fischer AS, Lazar A, Iudicone D (2004) Mixed layer depth over the global ocean: an examination of profile data and a profile-based climatology. J Geophys Res-Oceans 109: C12003.
72. Sackmann BS, Perry MJ, Eriksen CC (2008) Seaglider observations of variability in daytime fluorescence quenching of chlorophyll-*a* in northeastern Pacific coastal waters. Biogeosciences Discuss 5: 2839–2865.
73. Carroll M, Chigounis D, Gilbert S, Gundersen K, Hayashi K, et al. (2006) Performance verification statement for the WET Labs ECO FLNTUSB fluorometer. Solomons MD: Alliance for Coastal Technologies. 36 p.
74. Chaigneau A, Eldin G, Dewitte B (2009) Eddy activity in the four major upwelling systems from satellite altimetry (1992–2007). Prog Oceanogr 83: 117–123.
75. Feldman GC, McClain CR (2013) Ocean color web, SeaWiFS/MODIS Aqua reprocessing level-2. Kuring N, Bailey SW editors. Greenbelt MD: NASA Goddard Space Flight Center. Accessed: 1 January 2013.
76. Freeland HJ, Cummins PF (2005) Argo: a new tool for environmental monitoring and assessment of the world's oceans, an example from the NE Pacific. Prog Oceanogr 64: 31–44.
77. Li M, Myers PG, Freeland H (2005) An examination of historical mixed layer depths along Line P in the Gulf of Alaska. Geophys Res Lett 32: L05613.
78. Hickey BM (1989) Patterns and processes of circulation over the Washington continental shelf and slope. In: Landry MR, Hickey BM, editors. Coastal oceanography of Washington and Oregon. Elsevier oceanography series, volume 47. Amsterdam, Netherlands: Elsevier Science. pp. 41–115.
79. Fong DA, Geyer WR (2001) Response of a river plume during an upwelling favorable wind event. J Geophys Res-Oceans 106: 1067–1084.
80. Hickey BM (1998) Coastal oceanography of western North America from the tip of Baja to Vancouver Island. In: Robinson A, Brink S, editors. The global coastal ocean: regional studies and syntheses. The Sea, volume 11. New York NY: Wiley. pp. 345–393.
81. Hickey BM, Kudela RM, Nash JD, Bruland KW, Peterson WT, et al. (2010) River influences on shelf ecosystems: introduction and synthesis. J Geophys Res-Oceans 115: C00B17.
82. García Berdeal I, Hickey BM, Kawase M (2002) Influence of wind stress and ambient flow on a high discharge river plume. J Geophys Res-Oceans 107: 13-1–13-24.
83. Bigg MA (1986) Arrival of northern fur seals, *Callorhinus ursinus*, on St. Paul Island, Alaska. Fish Bull 84: 383–394.
84. Olesiuk PF (2012) Habitat utilization by northern fur seals (*Callorhinus ursinus*) in the northeastern Pacific Ocean and Canada. Ottawa, Canada: Department of Fisheries and Oceans. Canadian Science Advisory Secretariat Research Document 2012/040. 27 p. Available: http://publications.gc.ca/site/eng/436921/publication.html.
85. Kajimura H, Lander RH, Perez MA, York AE, Bigg MA (1980) Further analysis of pelagic fur seal data collected by the United States and Canada during 1958–74, part I. Submitted to the standing scientific committee, 23rd annual meeting of the North Pacific fur seal commission. 94 p. Available: NMML Library, Alaska Fisheries Science Center, 7600 Sand Point Way NE, Seattle WA 98115.
86. Kosro PM, Huyer A, Ramp SR, Smith RL, Chavez FP, et al. (1991) The structure of the transition zone between coastal waters and the open ocean off northern California, winter and spring 1987. J Geophys Res-Oceans 96: 14707–14730.
87. Bigg MA (1990) Migration of northern fur seals (*Callorhinus ursinus*) off western North America. Can Tech Rep Fish Aquat Sci 1764.
88. Gentry RL, Kooyman GL, Goebel ME (1986) Feeding and diving behavior of northern fur seals. In: Gentry RL, Kooyman GL, editors. Fur seals: maternal strategies on land and at sea. Princeton NJ: Princeton University Press. pp. 61–78.
89. Goebel ME, Bengtson JL, Delong RL, Gentry RL, Loughlin TR (1991) Diving patterns and foraging locations of female northern fur seals. Fish Bull 89: 171–179.
90. Kajimura H (1984) Opportunistic feeding of the northern fur seal, *Callorhinus ursinus*, in the eastern North Pacific Ocean and eastern Bering Sea. Seattle WA: National Marine Fisheries Service. NOAA Technical Report NMFS SSRF-779. 49 p. Available: http://spo.nmfs.noaa.gov/SSRF/SSRF779.pdf.

91. Sinclair E, Loughlin T, Pearcy W (1994) Prey selection by northern fur seals (*Callorhinus ursinus*) in the eastern Bering Sea. Fish Bull 92: 144–156.
92. Sterling JT, Ream RR (2004) At-sea behavior of juvenile male northern fur seals. Can J Zool 82: 1621–1637.
93. Suga T, Motoki K, Aoki Y, Macdonald AM (2004) The North Pacific climatology of winter mixed layer and mode waters. J Phys Oceanogr 34: 3–22.
94. Staniland IJ, Robinson SL (2008) Segregation between the sexes: Antarctic fur seals, *Arctocephalus gazella*, foraging at South Georgia. Anim Behav 75: 1581–1590.
95. Costa DP, Gentry RL (1986) Free-ranging energetics of northern fur seals. In: Gentry RL, Kooyman GL, editors. Fur seals: maternal strategies on land and at sea. Princeton NJ: Princeton University Press. pp. 79–101.
96. Schusterman RJ, Balliet RF (1971) Aerial and underwater visual acuity in the California sea lion (*Zalophus californianus*) as a function of luminance. Ann NY Acad Sci 188: 37–46.
97. Alverson DL, Larkins H (1969) Status of knowledge of the Pacific hake resource. Calif Coop Oceanic Fish Invest Rep 13: 24–31.
98. Appenzeller A, Leggett W (1992) Bias in hydroacoustic estimates of fish abundance due to acoustic shadowing: evidence from day-night surveys of vertically migrating fish. Can J Fish Aquat Sci 49: 2179–2189.
99. Bertrand A, Gerlotto F, Bertrand S, Gutiérrez M, Alza L, et al. (2008) Schooling behaviour and environmental forcing in relation to anchoveta distribution: an analysis across multiple spatial scales. Prog Oceanogr 79: 264–277.
100. Austin D, Bowen WD, McMillan JI, Boness DJ (2006) Stomach temperature telemetry reveals temporal patterns of foraging success in a free-ranging marine mammal. J Anim Ecol 75: 408–420.
101. Kelly KA, Beardsley RC, Limeburner R, Brink KH, Paduan JD, et al. (1998) Variability of the near-surface eddy kinetic energy in the California current based on altimetric, drifter, and moored current data. J Geophys Res-Oceans 103: 13067–13083.
102. Benoit-Bird KJ, Battaile BC, Heppell SA, Hoover B, Irons D, et al. (2013) Prey patch patterns predict habitat use by top marine predators with diverse foraging strategies. PLoS ONE 8: e53348.
103. Godø OR, Samuelsen A, Macaulay GJ, Patel R, Hjøllo SS, et al. (2012) Mesoscale eddies are oases for higher trophic marine life. PLoS ONE 7: e30161.
104. Collins CA, Pennington JT, Castro CG, Rago TA, Chavez FP (2003) The California current system off Monterey, California: physical and biological coupling. Deep Sea Res Part II 50: 2389–2404.
105. Boccaletti G, Ferrari R, Fox-Kemper B (2006) Mixed layer instabilities and restratification. J Phys Oceanogr 37: 2228–2250.
106. Mahadevan A, Thomas LN, Tandon A (2008) Comment on "Eddy/wind interactions stimulate extraordinary mid-ocean plankton blooms". Science 320: 448.
107. Niiler PP (1969) On the Ekman divergence in an oceanic jet. J Geophys Res 74: 7048–7052.
108. Ainley DG, Dugger KD, Ford RG, Pierce SD, Reese DC, et al. (2009) Association of predators and prey at frontal features in the California current: competition, facilitation, and co-occurrence. Mar Ecol Prog Ser 389: 271–294.
109. Ainley DG, Spear LB, Tynan CT, Barth JA, Pierce SD, et al. (2005) Physical and biological variables affecting seabird distributions during the upwelling season of the northern California current. Deep Sea Res Part II 52: 123–143.
110. Logerwell EA, Smith PE (2001) Mesoscale eddies and survival of late stage Pacific sardine (*Sardinops sagax*) larvae. Fish Oceanogr 10: 13–25.
111. Nishimoto MM, Washburn L (2002) Patterns of coastal eddy circulation and abundance of pelagic juvenile fish in the Santa Barbara Channel, California, USA. Mar Ecol Prog Ser 241: 183–199.
112. Lara-Lopez AL, Davison P, Koslow JA (2012) Abundance and community composition of micronekton across a front off southern California. J Plankton Res 34: 828–848.
113. Henson SA, Thomas AC (2008) A census of oceanic anticyclonic eddies in the Gulf of Alaska. Deep Sea Res Part I 55: 163–176.
114. Ladd C, Crawford WR, Harpold CE, Johnson WK, Kachel NB, et al. (2009) A synoptic survey of young mesoscale eddies in the eastern Gulf of Alaska. Deep Sea Res Part II 56: 2460–2473.
115. Ladd C, Kachel NB, Mordy CW, Stabeno PJ (2005) Observations from a Yakutat eddy in the northern Gulf of Alaska. J Geophys Res-Oceans 110: C03003.
116. Ladd C, Mordy CW, Kachel NB, Stabeno PJ (2007) Northern Gulf of Alaska eddies and associated anomalies. Deep Sea Res Part I 54: 487–509.
117. Okkonen SR, Weingartner TJ, Danielson SL, Musgrave DL, Schmidt GM (2003) Satellite and hydrographic observations of eddy-induced shelf-slope exchange in the northwestern Gulf of Alaska. J Geophys Res-Oceans 108: 3033.
118. Whitney FA, Crawford WR, Harrison PJ (2005) Physical processes that enhance nutrient transport and primary productivity in the coastal and open ocean of the subarctic NE Pacific. Deep Sea Res Part II 52: 681–706.
119. Crawford WR, Brickley PJ, Peterson TD, Thomas AC (2005) Impact of Haida eddies on chlorophyll distribution in the eastern Gulf of Alaska. Deep Sea Res Part II 52: 975–989.
120. Marchesiello P, McWilliams JC, Shchepetkin A (2003) Equilibrium structure and dynamics of the California current system. J Phys Oceanogr 33: 753–783.
121. Paduan JD, Niiler PP (1993) Structure of velocity and temperature in the northeast Pacific as measured with Lagrangian drifters in fall 1987. J Phys Oceanogr 23: 585–600.
122. Mackas DL, Tsurumi M, Galbraith MD, Yelland DR (2005) Zooplankton distribution and dynamics in a North Pacific eddy of coastal origin: II. Mechanisms of eddy colonization by and retention of offshore species. Deep Sea Res Part II 52: 1011–1035.
123. Barth JA (1994) Short-wave length instabilities on coastal jets and fronts. J Geophys Res-Oceans 99: 16095–16115.
124. Barth JA, Pierce SD, Cowles TJ (2005) Mesoscale structure and its seasonal evolution in the northern California current system. Deep Sea Res Part II 52: 5–28.
125. Barth JA, Pierce SD, Smith RL (2000) A separating coastal upwelling jet at Cape Blanco, Oregon and its connection to the California current system. Deep Sea Res Part II 47: 783–810.
126. Hayward TL, Mantyla AW (1990) Physical, chemical and biological structure of a coastal eddy near Cape Mendocino. J Mar Res 48: 825–850.
127. Strub PT, James C (2000) Altimeter-derived variability of surface velocities in the California current system: 2. Seasonal circulation and eddy statistics. Deep Sea Res Part II 47: 831–870.
128. Kurian J, Colas F, Capet X, McWilliams JC, Chelton DB (2011) Eddy properties in the California current system. J Geophys Res-Oceans 116.
129. Ladd C (2007) Interannual variability of the Gulf of Alaska eddy field. Geophys Res Lett 34: L11605.
130. Crawford W (2002) Physical characteristics of Haida eddies. J Oceanogr 58: 703–713.
131. Miller BS, Simenstad CA, Cross JN, Fresh KL, Steinfort SN (1980) Nearshore fish and macroinvertebrate assemblages along the Strait of Juan de Fuca including food habits of the common nearshore fish. Seattle WA: University of Washington. Fisheries Research Institute Report 8001. 213 p. Available: http://hdl.handle.net/1773/3939.
132. MacFadyen A, Hickey BM, Cochlan WP (2008) Influences of the Juan de Fuca eddy on circulation, nutrients, and phytoplankton production in the northern California current system. J Geophys Res-Oceans 113: C08008.
133. Banas NS, MacCready P, Hickey BM (2009) The Columbia River plume as cross-shelf exporter and along-coast barrier. Cont Shelf Res 29: 292–301.
134. Trites AW, Bigg MA (1996) Physical growth of northern fur seals (*Callorhinus ursinus*): seasonal fluctuations and migratory influences. J Zool (Lond) 238: 459–482.
135. Large WG, McWilliams JC, Niiler PP (1986) Upper ocean thermal response to strong autumnal forcing of the northeast Pacific. J Phys Oceanogr 16: 1524–1550.
136. Dohan K, Davis RW (2010) Mixing in the transition layer during two storm events. J Phys Oceanogr 41: 42–66.
137. Scheffer VB (1950) Winter injury to young fur seals on the northwest coast. Calif Fish Game 36: 378–379.
138. Kenyon KW, Scheffer VB, Chapman DG (1954) A population study of the Alaska fur seal herd. Washington DC: US Fish and Wildlife Service. Special Scientific Report - Wildlife, No. 12. 77 p.
139. Ichihara T (1974) Possible effect of surface wind force on the sex-specific mortality of young fur seals in the eastern Pacific. Bull Far Seas Fish Res Lab 11: 1–8.
140. Niggol K, Fiscus CH Jr, Wilke F (1959) Pelagic fur seal investigations California-Oregon-Washington, 1959. Seattle WA: US Fish and Wildlife Service. 92 p. Available: http://www.afsc.noaa.gov/nmml/library/nfsi/Pelagic-Fur-Seal-Investigations-1959.pdf.

Assessing the Health of the U.S. West Coast with a Regional-Scale Application of the Ocean Health Index

Benjamin S. Halpern[1,2,3*], Catherine Longo[1], Courtney Scarborough[1], Darren Hardy[1,9], Benjamin D. Best[1], Scott C. Doney[4], Steven K. Katona[5], Karen L. McLeod[6], Andrew A. Rosenberg[7], Jameal F. Samhouri[8]

1 National Center for Ecological Analysis and Synthesis, Santa Barbara, California, United States of America, **2** Bren School of Environmental Science and Management, University of California, Santa Barbara, California, United States of America, **3** Imperial College London, Silwood Park Campus, Ascot, United Kingdom, **4** Marine Chemistry & Geochemistry Department, Woods Hole Oceanographic Institution, Woods Hole, Massachusetts, United States of America, **5** Conservation International, Arlington, Virginia, United States of America, **6** COMPASS, Oregon State University, Department of Zoology, Corvallis, Oregon, United States of America, **7** Union of Concerned Scientists, Cambridge, Massachusetts, United States of America, **8** Conservation Biology Division, Northwest Fisheries Science Center, National Marine Fisheries Service, National Oceanic and Atmospheric Administration, Seattle, Washington United States of America, **9** Digital Library Systems & Services, Stanford University, Stanford, California, United States of America

Abstract

Management of marine ecosystems increasingly demands comprehensive and quantitative assessments of ocean health, but lacks a tool to do so. We applied the recently developed Ocean Health Index to assess ocean health in the relatively data-rich US west coast region. The overall region scored 71 out of 100, with sub-regions scoring from 65 (Washington) to 74 (Oregon). Highest scoring goals included tourism and recreation (99) and clean waters (87), while the lowest scoring goals were sense of place (48) and artisanal fishing opportunities (57). Surprisingly, even in this well-studied area data limitations precluded robust assessments of past trends in overall ocean health. Nonetheless, retrospective calculation of current status showed that many goals have declined, by up to 20%. In contrast, near-term future scores were on average 6% greater than current status across all goals and sub-regions. Application of hypothetical but realistic management scenarios illustrate how the Index can be used to predict and understand the tradeoffs among goals and consequences for overall ocean health. We illustrate and discuss how this index can be used to vet underlying assumptions and decisions with local stakeholders and decision-makers so that scores reflect regional knowledge, priorities and values. We also highlight the importance of ongoing and future monitoring that will provide robust data relevant to ocean health assessment.

Editor: James P. Meador, Northwest Fisheries Science Center, NOAA Fisheries, United States of America

Funding: Beau and Heather Wrigley generously provided the founding grant. Additional financial and in-kind support was provided by the Pacific Life Foundation, Thomas W. Haas Fund of the New Hampshire Charitable Foundation, the Oak Foundation, Akiko Shiraki Dynner Fund for Ocean Exploration and Conservation, Darden Restaurants Inc. Foundation, Conservation International, New England Aquarium, National Geographic, and the University of California Santa Barbara's National Center for Ecological Analysis and Synthesis, which supported the Ecosystem Health Working Group as part of the Science of Ecosystem-Based Management project funded by the David and Lucile Packard Foundation. Individual authors also acknowledge support from the U.S. National Science Foundation. The funders had no role in study design, data collection and analysis, decision to publish, or preparation of the manuscript.

Competing Interests: The authors have declared that no competing interests exist.

* Email: halpern@bren.ucsb.edu

These authors contributed equally to this work.

Introduction

As decision-makers shift towards more comprehensive approaches to managing ecosystems [1–3], management goals and targets increasingly focus on overall ecosystem health rather than on single sectors or stressors. This trend is particularly apparent for marine systems where efforts to implement ecosystem-based management (EBM) often have the stated objective of improving ocean health [3–6]. Along the United States west coast this emphasis exists in the regional governing body (West Coast Governor's Alliance on Ocean Health; [2]), NOAA's National Marine Sanctuaries' regular assessments of condition [7], ecosystem based approaches to fisheries management plans [8], and state-level and local efforts such as the Marine Life Protection Act, Puget Sound Action Agenda, and the west coast EBM network [9–11]. Until recently a standard tool to measure and track changes in ocean health in a repeatable, transparent, quantitative and goal-driven manner was lacking, although it is key to informing management and policy [3,12–13]. Within the United States, both federal and state agencies must make decisions regarding changing ocean uses, new regulations, balancing needs of multiple stakeholders, and supporting coastal economies, along with many other issues. We developed the Ocean Health Index (hereafter, the Index) in part to help address these needs [14].

Public policy necessarily serves multiple interests and goals, such as biodiversity conservation, food production and many others, and thus relies on assessments of ecosystem health through the human lens of meeting societal goals and delivering desired benefits (see File S1). This perspective on ecosystem health is a departure from traditional conservationist views that focus on health as a measure of pristineness (recently debated by [15–17]). Assessments of ocean health are thus measured and bounded by

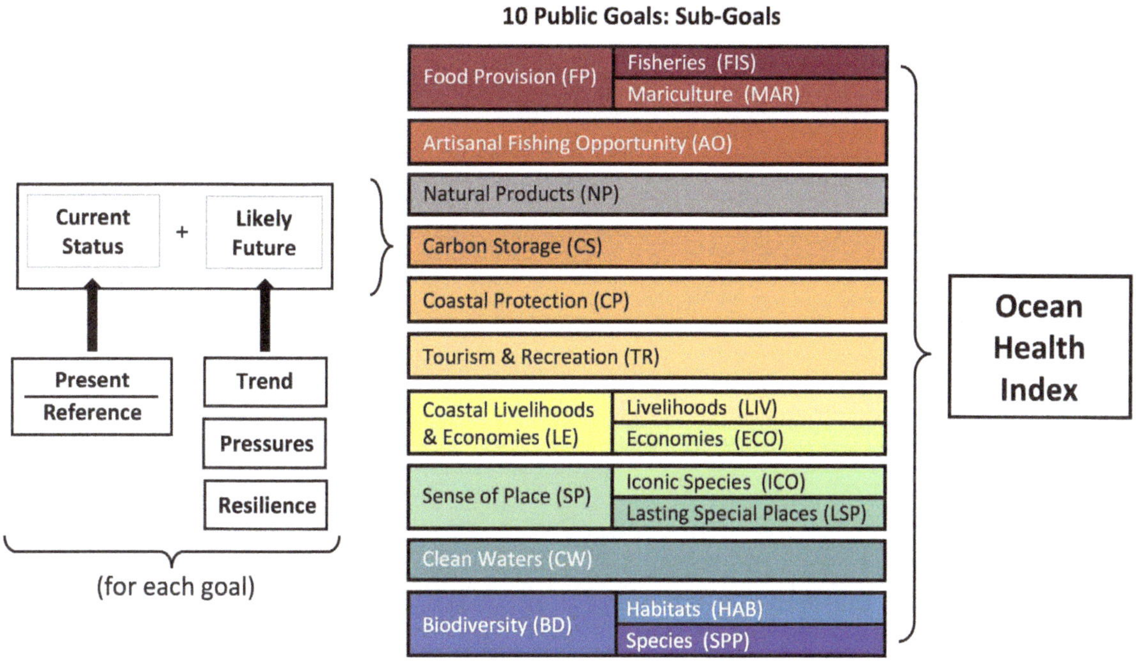

Figure 1. Schematic of the Ocean Health Index showing the 10 goals that comprise it, some with sub-goals, and broadly how each goal is calculated. Natural products is grey to indicate that it was not relevant in the U.S. west coast and thus not assessed.

human interactions with the ecosystem, not solely by the state of the natural ecosystem. A growing scientific literature also focuses on ecosystems as coupled social-ecological systems (e.g., [18–20]). We thus define a healthy ocean as one that sustainably delivers a range of benefits to people now and in the future (Fig. 1; [14]).

As a consequence of this human focus, individuals may perceive and understand ecosystems differently and have different views and sets of values that influence their assessment of ecosystem health. These differences require addressing questions such as: which aspects of ecosystem health are more or less important, how does one set reference points used to quantify health (e.g., ambitious versus practical), and which proxy measures can and cannot be used to estimate the status of each dimension of ecosystem health. Answering these kinds of questions exemplifies the types of inherently subjective decisions that must be made when developing an indicator of ecosystem health. The role of subjectivity cannot be ignored or avoided, but instead it should be made fully transparent, allowing subjective decisions to be modified to suit specific applications in a way that reflects regional values. The Ocean Health Index was designed in part to address this challenge with respect to measuring the health of marine ecosystems.

The Index has many potential applications. One, comparison of performance among regions, requires data consistency across regions. A second, comparison of ocean health within a region over time, makes using the best regionally (as opposed to globally) available data paramount. Halpern et al. [14] exemplifies the first type of assessment. Here we demonstrate an application of both types of assessment, exploring how well the Index performs at a regional scale in a relatively data-rich setting and how sub-regional comparisons might inform local and regional management decisions along the west coast of the United States.

We addressed three core questions: 1) How can the Index framework be adapted to a regional setting and make the best use of locally-available information, i.e., calculating components based on new and different data? 2) What is the health of the U.S. west coast and how is it changing? 3) How can the Index and its underlying framework tie into regional policy decisions? In addressing these questions we also hope to provide guidance and insight to other scientists and practitioners who may want to apply the Index to a new region.

Methods

Case study region

We calculated Index scores for five coastal sub-regions of the U.S. west coast –Washington, Oregon, Northern California, Central California, and Southern California – as well as an area-weighted average of these sub-regions to produce an overall regional score. These sub-regions were chosen based on a compromise between political and ecological boundaries. Our focus here is primarily on political boundaries, as most data are gathered and reported by agencies and organizations based on jurisdictional boundaries. However, because California's large geographical extent and ecological, social and economic diversity justified further subdivision, we used county boundaries, which closely align with biogeographic boundaries (Fig. 2), to define three sub-region boundaries. We did this because many data are reported at the county level, which facilitated calculating the Index scores within these sub-regions. The Index could be applied to ecological regions (such as marine ecoregions; [21]), but insufficient data are currently reported at these scales to make this feasible.

From a socio-economic perspective, the five sub-regions differ substantially (e.g., [22]). Southern California contains the heavily urbanized and densely populated coasts of San Diego and Los Angeles but also includes eight large, nearly uninhabited offshore islands. Central California is more sparsely populated but also includes the densely-populated San Francisco Bay. Northern California is even less densely populated, except in Sonoma county bordering San Francisco Bay. The coast of Oregon is uniformly

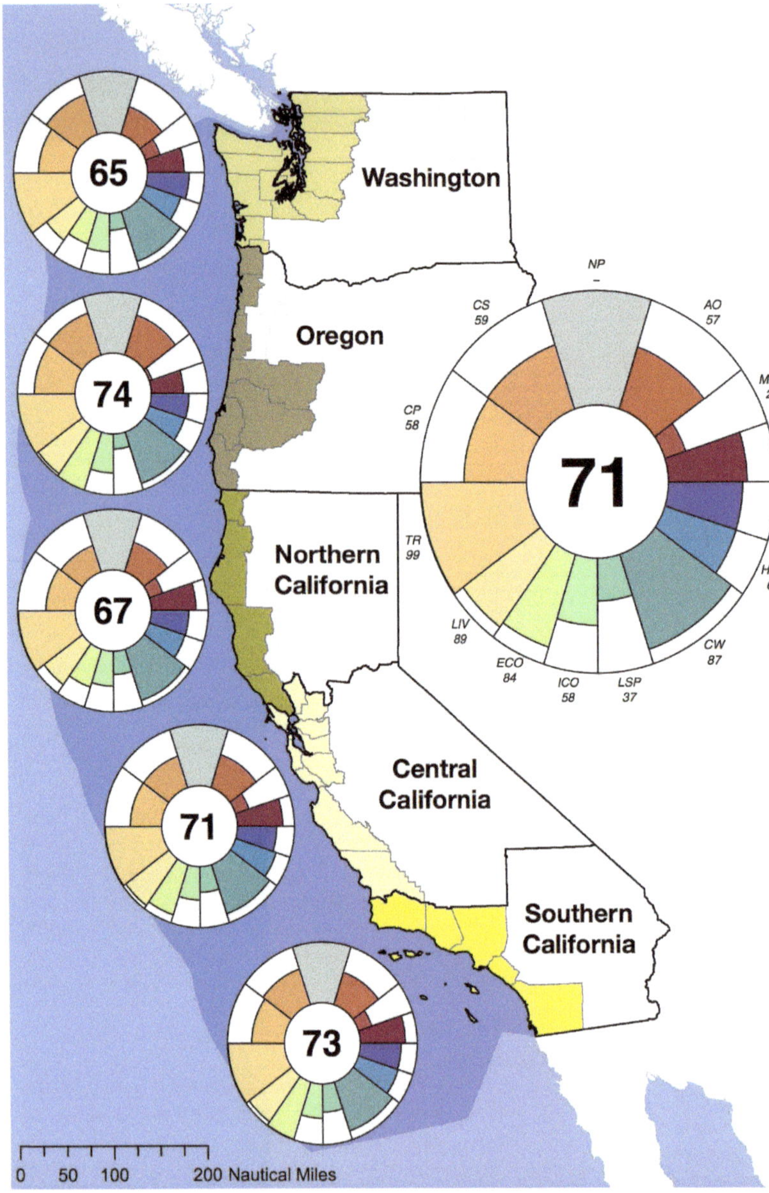

Figure 2. Map of the study region with each goal score per sub-region (left) and for the overall U.S. west coast (right). Each petal in the plots represents the score (radius) and weight (width) for the goal or sub-goal; see Fig. 1 for color legend and goal names. The number in the center is the overall Index score. Natural products is not assessed. Regions are depicted with coastal counties and the 200 nm exclusive economic zone is shaded in darker blue for reference only; regional scores are the area-weighted average of sub-region scores.

sparsely populated, with just 16,000 inhabitants in the largest town (Coos Bay). Washington state includes two distinct regions – the outer coast, which is similar to Oregon's coast, and Puget Sound, which includes many urbanized areas, most notably Seattle and Tacoma.

Index calculations

The Index is comprised of ten widely-held goals for healthy oceans that capture the different aspects of how people use, benefit from and value ocean ecosystems: food provision, artisanal fishing opportunities, natural products, carbon storage, coastal protection, sense of place, tourism & recreation, coastal livelihoods & economies, clean waters, and biodiversity (see Table 1 for goal definitions). These goals broadly map to the set of ecosystem services described by others (e.g., [23]), but have important differences (such as inclusion of coastal livelihoods and economies, which is not an ecosystem service) that motivate calling them public goals rather than ecosystem services.

An overall Index score, I, is calculated as the weighted sum of the scores for each goal assessed in the Index ([14] and Fig. 1), such that:

$$I = \sum_{i=1}^{N} \alpha_i I_i, \quad (1)$$

where α is the importance (i.e. weight) placed on each goal i. For the U.S. west coast analysis we assumed equal weights for all N goals. In a previous application of the Index at the global scale, when a goal was not relevant to a location (for example, extractive goals for uninhabited islands or assessments of coral health in

Table 1. Details for the current status calculation of goals and sub-goals that comprise the Ocean Health Index.

Goal	Sub-goal	Definition	Reference point type	Reference point	Data used
Food provision	**Fisheries**	Harvest of sustainably caught wild seafood	Functional Relationship	Single species biomass at maximum sustainable yield (BMSY) and single species fishing mortality at maximum sustainable yield (FMSY)	B/BMSY and F/FMSY estimates from stock assessments; mean annual commercial catch per species
	Mariculture	Production of sustainably cultured seafood	Established Target	350% increase in production from 2005 levels, distributed evenly among farmable areas in all sub-regions	Tons of shellfish produced; areas deemed safe for mariculture farming by NOAA
Artisanal fishing opportunity		Opportunity to engage in artisanal-scale fishing for subsistence or and/or recreation	Established Target	**Physical Access:** One coastal access points per mile of coastline	**Physical Access:** Number of coastal access points per mile
				Economic Access: No increase in the ratio of fuel price to median income over a five-year period	**Economic Access:** Change in gas price over time
				Resource Access: Perfect sustainability score for all fish stocks	**Resource Access:** Condition of fish stocks as measured by the NOAA fish Stock Sustainability Index (FSSI)
Natural products		Sustainable harvest of natural products, such as shells, algae, and fish oil used for reasons other than food provision	N/A	N/A	N/A
Carbon storage		Conservation status of natural habitats affording long-lasting carbon storage	Temporal Comparison (historical benchmark)	**Salt Marshes:** 50% of historical areal extent (roughly since 1850s)	**Salt Marshes:** Areal extent in 2006, 2002, 1996, and the 1850s
				Seagrasses: Zero pressure to coastal areas from nutrient input	**Seagrasses:** Nutrient input model applied within the 100 m depth contour
Coastal protection		Conservation status of natural habitats affording protection of the coast from inundation and erosion	Temporal Comparison (historical benchmark)	**Salt Marshes:** 50% of historical areal extent (roughly since 1850s)	**Salt Marshes:** Areal extent in 2006, 2002, 1996, and the 1850s
				Sand Dunes: 100% of the areal extent in 1960	**Sand Dunes:** Areal extent in 2006, 2002, 1996, and 1960
				Seagrasses: Zero pressure to coastal areas from nutrient input	**Seagrasses:** Nutrient input model applied within the 100 m depth contour
Tourism & recreation		Opportunity to enjoy coastal areas for recreation and tourism	Temporal Comparison (moving target)	No net loss in participation in marine-related activities over a 10 year period	Model of per capita participation rates in 19 marine-related activities based on demographic variables
Coastal livelihoods & economies	**Coastal livelihoods**	Jobs and wages from marine-related sectors	Temporal Comparison (historical benchmark)+Spatial Comparison	**Jobs:** No net loss in the number of jobs in marine-related sectors relative to all job sectors in each region over a five-year period	Jobs and wages data for 20 marine-related sectors; total jobs (marine and non-marine sectors)
				Wages: Highest per capita average annual wages across all regions and marine sectors	
	Coastal economies	Revenues from marine-related sectors	Temporal Comparison (moving target)	No net loss in revenue in marine-related sectors relative to all economy sectors over a five-year period	Revenue data for 20 marine-related sectors; total revenue (marine and non-marine sectors)

Table 1. Cont.

Goal	Sub-goal	Definition	Reference point type	Reference point	Data used
Sense of place	**Iconic species**	Cultural, spiritual, or aesthetic connection to the environment afforded by iconic species	Established Target	All assessed species coservation status classified as of least concern	Species conservation status as determined by NatureServe criteria
	Lasting special places	Cultural, spiritual, or aesthetic connection to the environment afforded by coastal and marine places of significance	Established Target	30% of all marine and terrestrial areas protected	Marine and terrestrial areas protected and managed for conservation
Clean waters		Clean waters that are free of nutrient and chemical pollution, marine debris and pathogens	Established Target	Zero marine debris, nutrient run-off, beach closures due to pathogens, and chemical contaminants in sediments and bivalve tissue	Nutrient plume models; beach closure data; beach clean-up data; concentration of chemicals in sediment and bivalve tissue samples
Biodiversity	**Habitats**	The existence value of biodiversity measured through the conservation status of habitats	Temporal Comparison (historical benchmark)	**Salt Marshes:** 50% of historical areal extent (roughly since 1850s)	**Salt Marshes:** Areal extent in 2006, 2002, 1996, and the 1850s
				Sand Dunes: 100% of the areal extent in 1960	**Sand Dunes:** Areal extent in 2006, 2002, 1996, and 1960
				Seagrasses: Zero pressure to coastal areas from nutrient input	**Seagrasses:** Nutrient input model applied within the 100 m depth contour
				Soft-bottom: Zero pressure from bottom trawl fishing	**Soft-bottom**: Amount of fish caught using bottom-trawl methods and location of current soft-bottom habitats
	Species	The existence value of biodiversity measured through the conservation status of marine-associated species	Established Target	All assessed species extinction risk status classified as of least concern	Species extinction risk status as determined by IUCN criteria

countries that do not have coral reefs), it was dropped from the assessment as it was deemed not applicable. In other words, in these cases the community assigns no value to the goal (i.e., the goal weight is zero; [14]), thus resulting in its irrelevance to the overall assessment. In this U.S. west coast application, we exclude the natural products goal because for most products there is no recorded trade within the region, even though it likely occurs at small scales, and for kelp, limited commercial harvest existed in Southern and Central California but no longer occurs (for unknown reasons). We further explain our rationale for excluding this goal in File S1.

Goal scores were calculated as the average of current (x_i) and likely future status ($\hat{X}_{i,F}$) Current status was measured as the present value (X_i) relative to a reference value ($X_{i,R}$), such that $x_i = X_i / X_{i,R}$. Likely future status was measured as current status modified by the recent trend (T), cumulative pressures (p), and resilience (r), such that:

$$\hat{x}_{i,F} = (1+\delta)^{-1}[1+\beta T_i + (1-\beta)(r_i - p_i)] \cdot x_i, \qquad (2)$$

where δ is the discount rate (δ = 0) and β is the relative importance of trend versus the difference between pressures and resilience in determining the likely future status (we assumed β = 0.67, following Halpern et al. [14]). Note that the likely future status does not predict the future, but only estimates what the status score is likely to be approximately 5 yr hence, given what is known today about recent trends and the counterbalance of pressure versus resilience metrics. Reference points for each goal are described in Table 1 (and further detailed in File S1). These reference points are similar but not equivalent to management targets; a reference point as we define it here is the maximum sustainable level of production of each goal. In some cases management may choose a target different than these reference points for practical or sociopolitical reasons [24]. Resilience measures focused on the presence of relevant institutions, but in general could not evaluate their effectiveness due to lack of such data. Details of goal models and parameters are provided in Table 2.

Changes to Index calculations for regional application

As compared with the global application of the Index [14], higher resolution data and longer time series, along with a better understanding of the regional context, allowed for improved approaches to modeling and setting reference points for many goals. Briefly, the following changes were made (see also File S1):

(1) **Higher resolution data**. We used regional-scale data wherever possible, relying on national data used in Halpern et

Table 2. Models and parameter used to calculate each goal and sub-goal.

Goal	Sub-Goal	Status Model Equations	Variables
Food Provision (x_{FP})		$x_{FP}=w_{FP}\cdot x_{FIS}+(1-w_{FP})\cdot x_{MAR}$	
		$w_{FP}=\frac{C_T}{C_T+Y_r}$	Y_r = Total sustainable mariculture harvest; C_T = Total current wild-caught fishing yield; w_{FP} = weight per seafood sector
		$x_{FIS}=\sum_i w_i\cdot\frac{F'_i+B'_i}{2}$	w_i = weight per stock *i*; *F* = current fishing mortality of stock *i*; *B* = current biomass of stock *i*
		$w_i=\frac{\bar{C_i}}{\sum\bar{C}}$	$\bar{C}_i$ = mean catch of stock *i* throughout the time-series
	Fisheries (x_{FIS})	$B'=\begin{cases}\frac{\frac{B}{Bmsy}}{0.8} & \text{when } \frac{B}{Bmsy}<0.8\\ 1 & \text{when } 0.8\le\frac{B}{Bmsy}<1.5\\ \frac{3.35-\frac{B}{Bmsy}}{1.8} & \text{when } \frac{B}{Bmsy}\ge 1.5\end{cases}$	B_{msy} = biomass of stock *i* producing maximum sustainable yield
		$F'=\begin{cases}0 & \text{when } \frac{B}{Bmsy}<0.8 \text{ and } \frac{F}{Fmsy}>\frac{B}{Bmsy}+1.5\\ \frac{\frac{F}{Fmsy}}{\frac{B}{Bmsy}-0.2} & \text{when } \frac{B}{Bmsy}<0.8 \text{ and } \frac{F}{Fmsy}<\frac{B}{Bmsy}-0.2\\ \frac{\frac{B}{Bmsy}+1.5-\frac{F}{Fmsy}}{1.5} & \text{when } \frac{B}{Bmsy}<0.8 \text{ and } \frac{B}{Bmsy}+0.2<\frac{F}{Fmsy}<\frac{B}{Bmsy}+1.5\\ 1 & \text{when } \frac{B}{Bmsy}<0.8 \text{ and } \frac{B}{Bmsy}-0.2\le\frac{F}{Fmsy}<\frac{B}{Bmsy}+0.2\\ 1 & \text{when } \frac{B}{Bmsy}\ge 0.8 \text{ and } 0.8\le\frac{F}{Fmsy}<1.2\\ \frac{\frac{F}{Fmsy}}{0.8} & \text{when } \frac{B}{Bmsy}\ge 0.8 \text{ and } \frac{F}{Fmsy}<0.8\\ \frac{2.5-\frac{F}{Fmsy}}{1.3} & \text{when } \frac{B}{Bmsy}\ge 0.8 \text{ and } \frac{F}{Fmsy}\ge 1.2\end{cases}$	F_{msy} = Fishing mortality that yields long term maximum sustainable yield of stock *i*
	Mariculture (x_{MAR})	$x_{MAR}=\frac{Y_c}{Y_r}$	Y_C = Current total sustainable mariculture harvest
		$Y_{C_k}=\sum_i Y_{k,i}\cdot S_i$	*S* = Sustainability coefficient for species *i*; *k* = sub-region
		$Y_r=\frac{FA_k}{FA_T}\cdot 3.5\cdot Y_{2005}$	*FA* = farmable area; Y_{2005} = yield in 2005
Artisanal Fishing Opportunities (x_{AO})		$x_{AO}=\frac{\left(\frac{AP_c}{AP_r}\right)+\left(\frac{FSSI}{FSSI_r}\right)+AE_i}{3}$	*AP* = average distance between coastal access points; *c* = current; *r* = reference; *FSSI* = catch-weighted average NOAA Fish Stock Sustainability Index score
		$AE_i=\left(\frac{G_r}{I_r}-\frac{G_c}{I_c}\right)$	*AE* = economic access; *G* = gas price per gallon; *I* = median income
Natural Products (x_{NP})		Not assessed for this region	
Carbon Storage (x_{CS})		$x_{CS}=\sum_j\left(\frac{C_{c,j}}{C_{r,j}}\cdot\frac{A_j}{A_T}\right)$	A_k = area covered by habitat *j*; A_T = total area covered by all habitats; C_j = condition of habitat *j*
Coastal Protection (x_{CP})		$x_{CP}=\sum_{j=1}^{J}\left(\alpha_j\cdot\frac{C_{c,j}}{C_{r,j}}\right)$	r_j = protective ability rank of habitat *j*
		$\alpha_j=\frac{w_j\cdot A_j}{\sum_j(w_j\cdot A_j)}$	
		$w_j=\frac{r_j}{\sum_j r_j}$	
Coastal Livelihoods & Economies (x_{LE})		$x_{LE}=\frac{x_{LIV}+x_{ECO}}{2}$	

Table 2. Cont.

Goal	Sub-Goal	Status Model Equations	Variables
	Livelihoods (x_{LIV})	$x_{LIV} = \frac{\left(\frac{\sum_z j_{c,z}}{\sum_z j_{r,z}} + \frac{\sum_z g_{k,z}}{\sum_z g_{r,z}}\right)}{2}$	j = adjusted number of direct and indirect jobs within sector z; g = average PcPPP-adjusted wages within sector z; c = current year; r = reference year (j, e) or reference location (for g)
	Economies (x_{ECO})	$x_{ECO} = \sum_z \frac{e_{c,z}}{e_{r,z}}$	e = total adjusted revenue generated directly and indirectly from sector z
Tourism and Recreation (x_{TR})		$x_{TR} = \frac{\sum_{i=1}^{19} P_{c,i}}{\sum_{i=1}^{19} \mathrm{P}_{r,i}}$	P_c = current predicted participation in each recreation activity i (of 19); P_r = observed participation in recreation activity i in year 2000
Sense of Place (x_{SP})		$x_{SP} = \frac{x_{ICO} + x_{LSP}}{2}$	
	Iconic Species (x_{ICO})	$x_{ICO} = \frac{\sum_{l=1}^{6} S_l \cdot w_l}{\sum_{l=1}^{6} S_l}$	l = IUCN threat category; S_m = number of assessed iconic species in category l; w_l = weight per threat category l
	Lasting Special Places (x_{LSP})	$x_{LSP} = \frac{\frac{MPA_{3\,nm}}{0.3*A_{3\,nm}} + \frac{MPA_{EEZ}}{0.3*A_{EEZ}} + \frac{TA_{PA}}{0.3*TA}}{3}$	MPA = fully protected marine area; $_{EEZ}$ = offshore waters (3–200 nm); $3\ nm$ = coastal waters (0–3 nm); TA = area on coastal land (0–1 mi); TA_{PA} = protected area on coastal land
Clean Waters (x_{CW})		$x_{CW} = \sqrt[4]{a \cdot u \cdot l \cdot d}$	a = population without access to sanitation relative to global maximum; u = 1 - nutrient inputs; l = 1 - chemical inputs; d = 1 - marine debris
Biodiversity (x_{BD})		$x_{BD} = \frac{x_{SPP} + x_{HAB}}{2}$	
	Species (x_{SPP})	$x_{SPP} = \frac{\sum_M \left(\frac{\sum_N w_{i,k}}{N}\right) \cdot A_c}{A_T}$	n = number species per grid cell c; m = number of grid cells in the assessment region; A_c = total area of grid cell c; A_T = total area of the assessment region
	Habitats (x_{HAB})	$x_{HAB} = \sum_j \left(\frac{C_{c,j}}{C_{r,j}} \cdot \frac{A_j}{A_T}\right)$	See variables above

See File S1 for details on data and rationales for each.

al. [14] in only 20% (10 of 49) of data layers. In particular, we were able to use higher resolution data from local sources for the status and trend calculation of all goals. Most of the pressure layers, and all of the resilience layers, were also calculated using local data sets.

(2) **Models adapted to better represent regional goals for ocean health**. For several goals we modified the approach to assessing current status based on higher-quality regional data. (a) For food provision derived from wild-caught fisheries, we were able to use formal stock assessments routinely employed for local fisheries management to capture the status of major commercially-caught species. These estimates are derived from complex models developed by working groups of experienced local experts. In contrast, the global analysis had to rely on models requiring many simplifying assumptions, leading to higher uncertainty. (b) For food provision derived from mariculture, we improved our estimate of potential sustainable productivity by assuming cultivation could only increase in areas already under production. In contrast, the global analysis assumed potential productivity scaled to total coastal area and highest observed production density, i.e. China. (c) For the tourism & recreation goal, we were able to use information on participation rates in a range of coastal and marine tourism and recreational activities. Participation rates more closely match the intent of this goal and are a more robust proxy than the international tourist arrivals data used in the 2012 global study [14] and are a more direct measure than the tourism employment proxy used in the 2013 global study [25]. We also changed the reference point from spatial (used in the global analysis) to temporal, because adequate time-series data were available.

(3) **Reference points based on U.S. west coast priorities.** (a) For the mariculture sub-goal, the reference point was based on regional projections of nationally-desired economic and food security targets. (b) For the habitat health scores used in coastal protection, carbon storage, and the habitats sub-goal of biodiversity, we used reconstructions of historic extents, rather than recently recorded trends, to set targets that were more ambitious than in the global analysis. (c) For the lasting

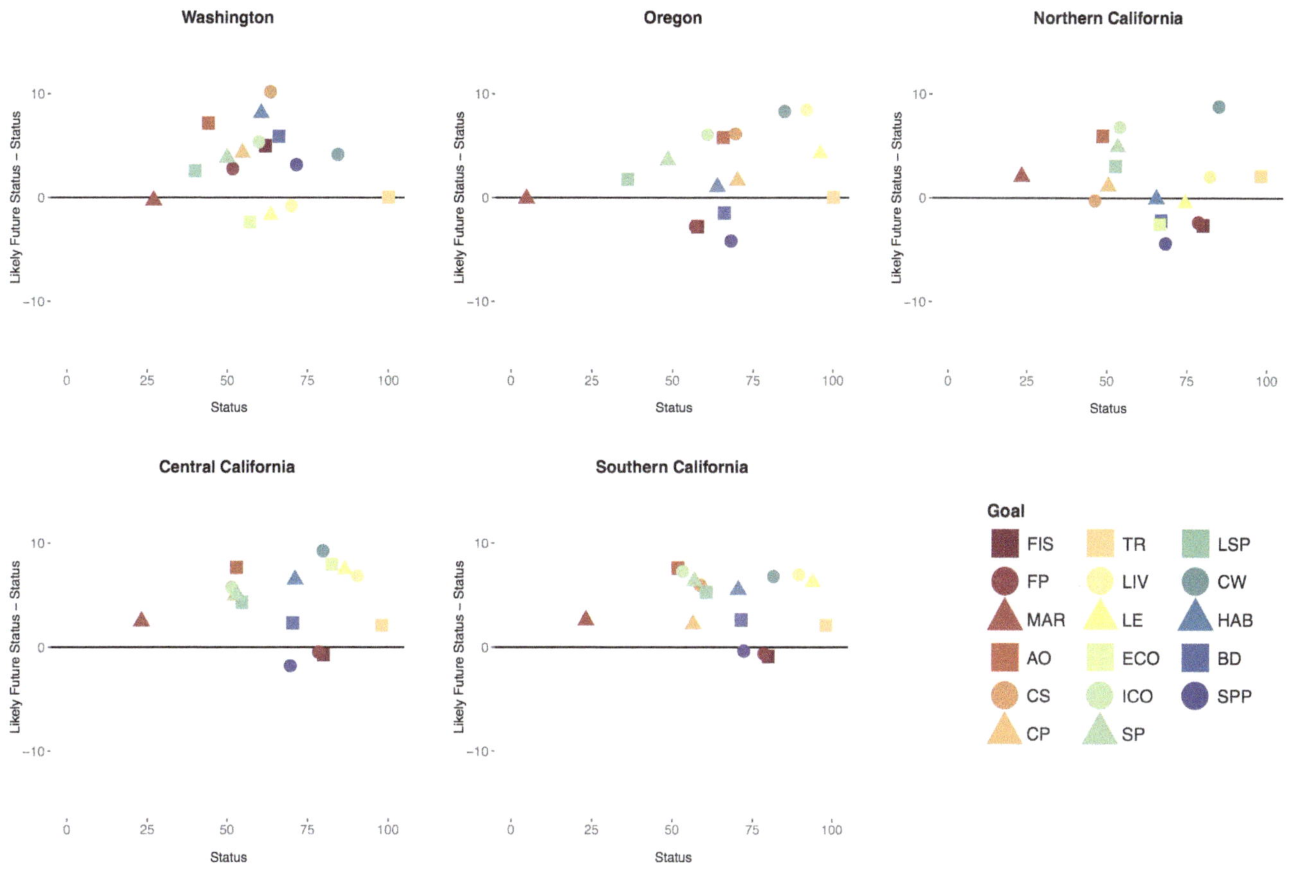

Figure 3. Current status versus the difference between likely future and current status for each goal and sub-goal within each sub-region. Values above the y-axis indicate the likely future status is greater than the current status. Note that y-axis is scaled −10 to 10.

special places sub-goal, in addition to evaluating protected areas inland and within 3 nm of the coast, we included a third zone (3 to 200 nm offshore) because we assume assessment of places offshore as well as nearshore is important to people in this region.

We conducted a number of analyses to assess how results were affected by various assumptions and data constraints (File S1), three of which we focus on here. First, to assess how results changed when goals were modeled differently, we calculated the regionally-modified goals using methods from the global study for comparison, when possible. Second, we assessed how alternate reference points modified goal scores, in particular for the mariculture sub-goal (see File S1). Finally, we assessed the consequences of using empirically-derived unequal weights, elicited from regional experts representing a diverse cross-section of stakeholders [26], for combining goals into a single Index score (Table S11 in File S1). Experts weighted sense of place and clean waters goals highest, and three to four times more heavily than the tourism & recreation and coastal livelihoods & economies goals, the two lowest-weighted goals (Table S11 in File S1).

Although we were able to estimate past status values for all goals and sub-goals except tourism and recreation, iconic species, and species diversity (which had insufficient time series; see File S1), most pressure and resilience metrics were not available for past time periods, precluding calculation of the 'likely future status', and thus the overall Index scores, for past years.

Scenario analyses

To further explore how the Index could be used within typical regional-scale decision contexts and to illustrate how the Index responds to typical management actions, we simulated several scenarios and recalculated the overall Index. The intent of this analysis was not to model precise changes but rather to illustrate expected types and relative magnitudes of change across goals. Rather than being prescriptive, these scenarios were chosen to illustrate how one can use the Index to explore consequences of management decisions. We recognize that realistic implementation would require engagement with decision-makers, normative decisions about management goals, fine-tuning of assumptions, and model-based simulations of future conditions. Our heuristic scenarios assessed what scores would be if 1) regulations had been adopted 5 years ago that successfully reduced land-based runoff of nutrients and pollutants each by 25%, 2) habitat restoration activities had been successfully implemented such that coastal wetlands and sand dunes were increased in extent by 10%, and 3) the Marine Life Protection Act (MLPA) process in California had not occurred and the currently-existing network of MPAs therefore had not been established within the state. For this third scenario we implemented three successive versions intended to measure changes over time in how the system, and thus the Index, would respond (for details see File S1).

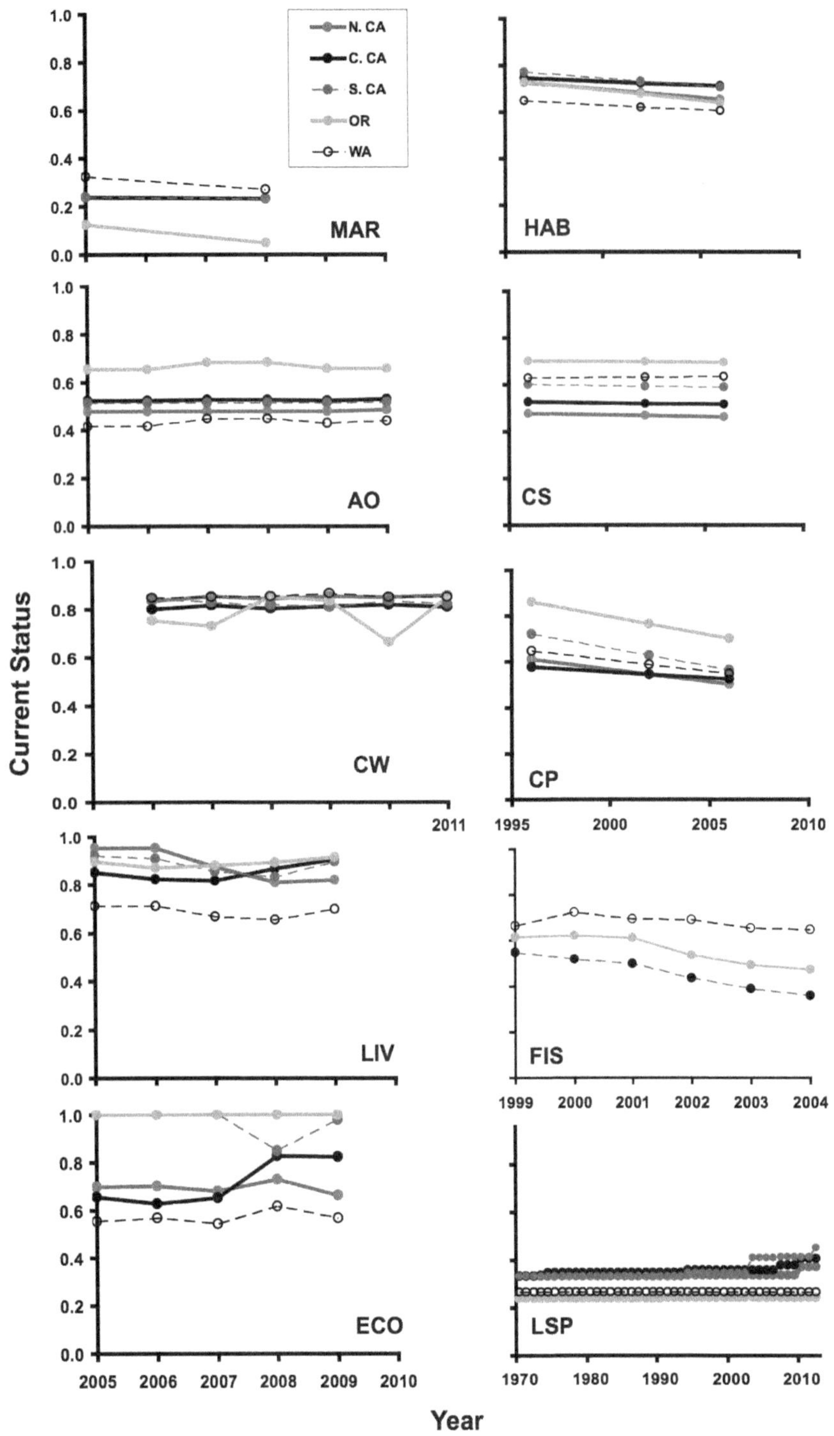

Figure 4. Time series of current status scores for goals and sub-goals with available historical data. Note different time scales on x-axes in right-hand plots. Plots are for the habitat sub-goal (HAB), carbon storage (CS), coastal protection (CP), artisanal fishing opportunities (AO), mariculture sub-goal (MAR), clean waters (CW), fisheries sub-goal (FIS), coastal livelihoods sub-goal (LIV), coastal economies sub-goal (ECO), and the lasting special places sub-goal (LSP). FIS could not be assessed for sub-regions within California and so a single state-level result is presented in that case.

Results

Overall the U.S. west coast scored 71 out of 100. Washington scored lowest of all sub-regions (65), with increasingly higher scores in Northern California (67), Central California (71), Southern California (73), and Oregon (74; Fig. 2). Goal scores varied from 22 (mariculture) to 99 (tourism and recreation), with tourism and recreation and clean waters scoring highest, and carbon storage, coastal protection, lasting special places and mariculture scoring lowest for all 5 sub-regions (see also Table S33 in File S1). Despite biophysical and socioeconomic differences

Table 3. Changes in goal scores for which sensitivity analyses were conducted using alternate methods for calculating goal status.

Region	Fisheries		Mariculture					Tourism and Recreation		Artisanal Fishing Opportunity	
	Orig without data poor stocks	**Alt with data poor stocks**	**Orig federal target**	**Alt1 spatial ref; global target**	**Alt2 spatial ref; national target**	**Alt3 temporal ref**	**Alt4 production function**	**Orig temporal ref**	**Alt spatial ref**	**Orig regional model**	**Alt global model**
Washington	64	55	27	1	81	80	10	100	55	47	61
Oregon	56	54	5	0.1	6	0	1	100	41	69	61
California	79	65		0.1	9	96	19	99	91		
N. California			24							51	59
C. California			24							57	58
S. California			25							55	59

Results are reported for each sub-region separately when possible. 'Orig' is the original approach used for reporting main results; 'Alt' is the alternative approach used for sensitivity analyses. Separate analyses for each sub-region within California were only possible for the artisanal fishing opportunity goal and for the original mariculture goal; results for other cases are reported for all of California as a single value. For mariculture we tested four different alternate reference points. See File S1 for details.

among sub-regions, overall Index scores for the sub-regions were within a 7 point range. Differences in scale, available data, and goal methodologies for several goals preclude direct quantitative comparison of these results to global scores [14].

For nearly every goal in each of the five sub-regions, likely near-term future scores were greater than those for the present (Fig. 3). Although the likely future state is strongly influenced by the recent trend (see Eq. 2), and the recent trend for many goals was (slightly) negative, the likely future status is also influenced by the balance between resilience measures and cumulative pressures, and in many cases resilience was greater than pressures (Table S34 in File S1). In only 24% of cases (11 of 45) were likely future scores worse, namely coastal livelihoods & economies in Washington; fisheries and species biodiversity in Oregon; fisheries, species biodiversity, carbon storage, and coastal livelihoods & economies in Northern California; and species biodiversity and fisheries in Central and Southern California. The potential future declines in the health of the fisheries (food provision) and species (biodiversity) sub-goals, despite significant resources being committed to their improvement in the region, is largely due to recent declines creating a negative trend (see below, Fig. 4, and Table S34 in File S1).

The likely future status does not tell the whole story, however. For goals and sub-goals with sufficient data to calculate past values of the current status, different patterns emerged. Habitat-based goals, notably habitat diversity and coastal protection, showed declines (8–17% respectively) across all sub-regions over the past ten years (Fig. 4). Coastal livelihoods and economies showed initial small declines in some sub-regions but recent recovery in many cases (Fig. 4). Because these values are standardized to remove broader economic patterns, this result suggests stronger effects of the global recession that began in 2008 and slower economic recovery in marine sectors compared to other sectors. Lasting special places showed recent improvements, in large part because of California's MPA initiatives, while remaining goals showed little recent change (Fig. 4). Because risk status of most species is rarely assessed more than once, we were not able to calculate past status scores for species diversity or iconic species sub-goals.

Comparisons of results obtained for several goals when assessed with the previous global approaches versus the refined regional approaches showed important differences. For the tourism and recreation goal, scores from the regional analysis were considerably higher than those obtained applying the global model, most likely reflecting both use of more informative data on participation rates instead of international tourist arrival data and the choice of a local temporal reference point instead of an across-region spatial one (Table 3). For large countries with coasts spanning sizable

Table 4. Changes in Index scores for each subregion and the U.S. West coast with goals weighted equally or unequally based on regionally-specific, empirically-derived preferences (Halpern et al. 2013b).

Region	Equal	Unequal
Washington	65	66
Oregon	74	74
N. California	67	66
C. California	71	69
S. California	73	71
U.S. west coast	71	70

biophysical gradients or bordering different oceans, such as the United States, sub-national assessments such as the one here are likely to produce Index scores that differ from those derived from the national-level global assessment. In contrast, the choice of a very different approach to modeling the artisanal fishing opportunity goal had relatively small effects on resulting scores (Table 3). Changes in the mariculture reference point significantly increased scores relative to the global approach, with highly variable results using other methods for setting reference points (Table 3). Unequal goal weighting [26], which represents one example of how people value goals differently, produced lower Index scores for some sub-regions and higher scores for others (Table 4).

The three management scenarios showed goal score changes from about −12% up to +11% depending on the goal and scenario (Fig. 5). For example, because land-based pollution is a pressure on nearly every goal, hypothetical decreases in this stressor led to modest increases in most goals (scenario 1). Simulated habitat restoration had a relatively large effect on habitat-based goals, a result that was influenced by choice of habitat reference points (see File S1, scenario analysis section), but not on other goals (scenario 2). The three versions of scenario 3 illustrate how an initial action (or in this case hypothetical lack thereof) could have cascading effects across multiple goals that may lead to increases in some goals and decreases in others. In the example here, the hypothetical removal of MPAs decreased the lasting special places score, increased the food provision score (through increased fishing), and decreased scores for other goals as a result of increased fishing pressure.

Discussion

By calculating the Index for the U.S. West coast, we were able to take advantage of regional data and knowledge of the system to

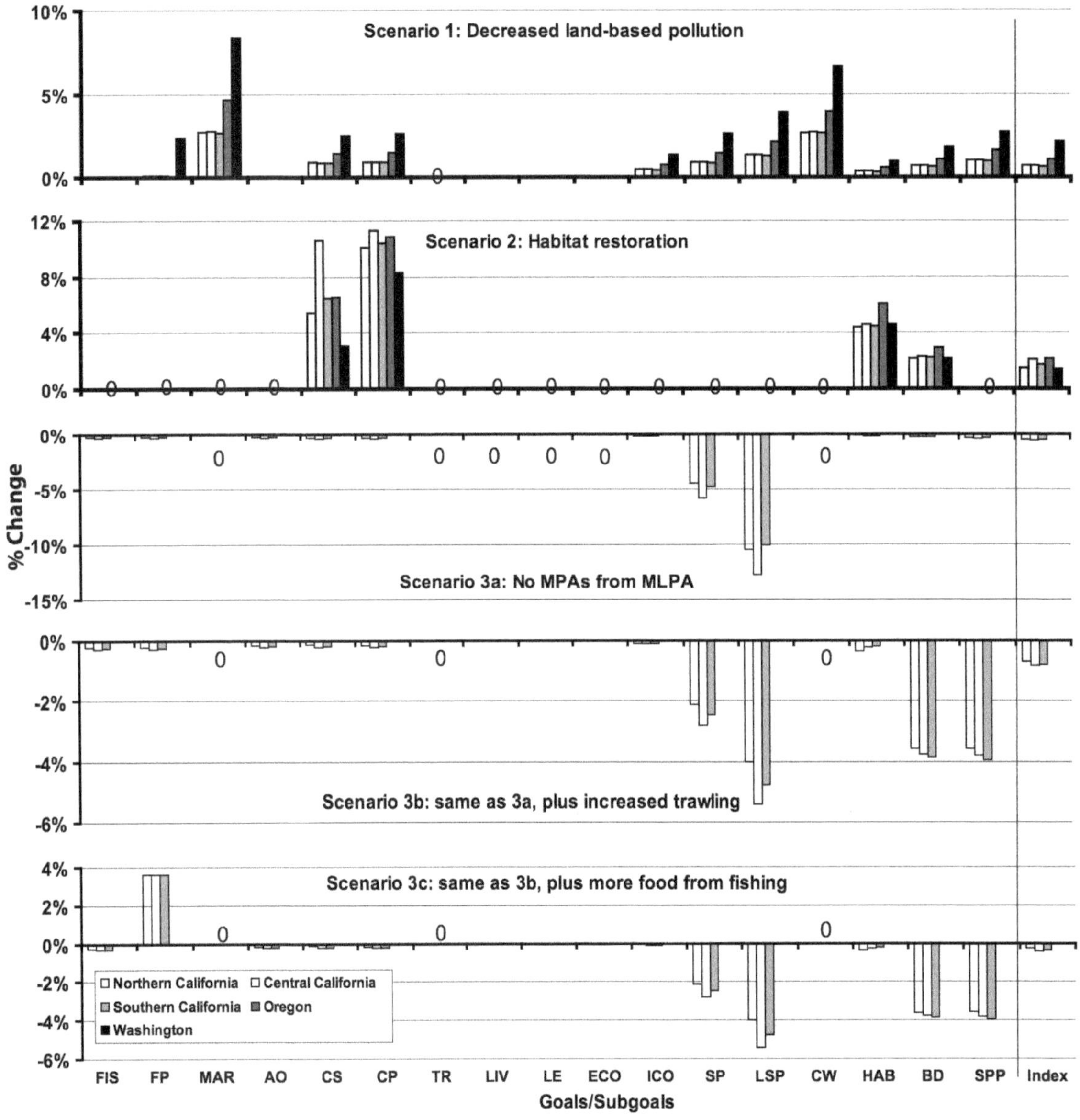

Figure 5. Scenario results as percent change in goal and Index scores for each sub-region. Goals with no change are indicated with a zero. Overall Index scores are on the far right, separated by the horizontal gray line. Note different scales on y-axes.

assess how particular goals and overall ocean health are faring at a regional scale, and whether conditions are getting better or worse. We found that current status scores for individual goals have gotten worse in the past decade or so (with the exception of recent improvements in lasting special places an livelihoods and economics for some sub-regions), but in most cases the near future looks better than the present (Figs. 3, 4). Assessment of the likely future status rewards the presence of regulatory and management measures; however, as data do not often exist on effectiveness of these measures, future estimates may be overly optimistic. In addition, the likely future status makes incorporates the potential impacts of climate change only as current climate-related pressures and not estimates of future conditions. Regardless, these differences highlight the importance of having time-series and maintaining on-going records of both ecological and governance information in order to understand likely future changes. The Index was designed explicitly to capture and quantify these different temporal components.

Spatial comparisons of sub-regional scores offered additional insights. Sub-regional scores had relatively small differences compared to the large range of scores globally in Halpern et al. [14]. This is understandable because biophysical and socioeconomic characteristics within the study region vary much less than among countries across the globe. Regional scores from this assessment were all higher than the score for the entire U.S. calculated from the global assessment. We cannot distinguish whether this difference stems from the use of different methods and data sources or if it suggests support for the widely-held view that the U.S. west coast is a relatively healthy and sustainably managed system. The overall U.S. score derived in the global study integrates scores from diverse coastal regions (i.e., Alaska, Hawaii, Gulf of Mexico, and east and west coasts of the U.S.) that vary historically, ecologically and in resource management actions.

Individual goal scores showed bigger sub-regional differences, but these differences were dampened when averaged with other goal scores to create overall Index scores (assuming equal weights for all goals; see File S1). For example, the fisheries sub-goal scored considerably higher in California (79; the Californian sub-regions could not be assessed separately for this goal due to the resolution of reported data) than in Oregon (56) and Washington (64; Fig. 2, Table S33 in File S1). This difference is in part due to differences in the dominant stock in the different regions; hake historically constituted roughly 30% of total catch in Oregon and Washington and are currently at low biomass and fishing effort levels (giving a score of 0.39 for the stock), whereas in California yellowfin tuna historically represented roughly 40% of total catch and are currently at ideal biomass and fishing effort levels (a score of 1.0).

The coastal livelihoods and economies sub-goals had very high scores for Central and Southern California and Oregon yet much lower scores for Northern California and Washington (Fig. 2, Table S33 in File S1). This goal uses a 'moving window' reference point, comparing each region to itself five years prior (while correcting for broader economic patterns, such as the global recession, that are independent of the condition of marine and coastal systems), based on the assumption that people mostly care about how they are doing economically relative to recent and local economic conditions. This technique avoids direct comparison, for example, of the absolute size of Southern California's coastal economy to that of Oregon. Consequently, Washington and Northern California scored lower because the largest sectors had significant declines in the last five years (in Washington, large declines occurred for jobs in tourism and transportation sectors and for revenue in tourism and living resources sectors; in Northern California, declines occurred for jobs and revenue in tourism), while such declines generally did not occur in the other regions.

Variation among sub-regions for the carbon storage goal (as well as the generally low scores for this goal for all sub-regions) is primarily due to the status of salt marsh habitats. Although salt marsh habitat loss has occurred throughout the U.S. west coast, this has been particularly severe in Central and Northern California. The exact values of the scores for this goal are highly dependent on the choice of reference point, which in this case is challenged by both practical and philosophical issues. Practically, few spatial data exist prior to the 1990s. Therefore, to set an ambitious yet realistic reference point, we relied on estimates of historical loss of these habitats from pre-industrial times and set the reference point to a fraction (50%) of this original extent. Philosophically, one must (subjectively) decide what serves as an ambitious yet realistic target (cf. [24]) for restoring this habitat, given that a great deal of potential ecosystems services were lost but also considering that the massive alteration driven by urbanization of estuarine systems is unlikely to be completely reversed.

Finally, differences among sub-region scores for the artisanal fishing opportunity goal were driven primarily by differences in coastal access, which together with economic factors and fish stock status determined this goal's score. Public access is provided along Oregon's entire coastline, leading to a higher score, whereas Washington and California both allow privately-owned access to the coast and have large stretches of restricted-access coastline. The choice of a common reference point for all sub-regions allows for direct comparison, but ultimately may not reflect sub-regional differences in management objectives, such as how local people want artisanal fishing to occur. If such differences are significant enough and the objective of local managers is primarily on managing at this finer scale, then a separate Index score, calculated with sub-regional best available data and locally-determined reference points, would be more appropriate.

Other goals are consistently low or high across all sub-regions but with results that may not immediately seem intuitive. For example, all sub-regions would like to increase coastal tourism, yet all scored nearly perfectly on this goal. One might hope that a different type of reference point could resolve this paradox. A functional relationship between people's values and the effects on the ecosystem caused by different levels of participation in coastal recreation would be ideal, as it could indicate what absolute levels of coastal tourism are both wanted by local communities and sustainable for local ecosystems. Unfortunately we do not have the information to construct one and it could differ in the different sub-regions. A spatial comparison reference point would probably not be appropriate as coastal communities in Oregon, for example, are very different biogeographically than those in Southern California, with local population density, weather and beach access being some of the place-specific factors affecting the number and frequency of people recreating in and around the ocean. Consequently, spatial comparison using a region-wide reference point would unfairly penalize one of the locations and would not be a useful indicator of ocean health. Currently no stated objectives for desired levels of tourism exist that could be used for an 'established target' reference point. This leaves temporal reference points as the best choice in this specific case, and, for reasons similar to economically-based temporal reference points for the livelihoods & economies goal, we used a moving-window reference point (i.e., that conditions are as good or better than they were 5 years ago). For all sub-regions, participation in coastal recreational activities has remained the same or increased in the past 5 years.

A similar issue of choice of reference point affected the scores for the mariculture sub-goal. We used an established target (increase mariculture by 350% from 2005 to 2020; [27]), which was based on socio-economic projections of seafood demand. This target produced relatively low scores, but our assessment of a range of other types of reference points showed that these scores are strongly dependent on choice of reference point (Table 3). Ideally we would have used a functional relationship reference point based on biophysical variables and societal preferences for how much available ocean space should be allocated to mariculture versus all the many other uses that currently exist and that will emerge in the near future. Unfortunately we currently have very little of these data, and so we relied on rough estimates of socially and ecologically desirable 'farmable area' in each sub-region. Although the estimate from Nash [27] is potentially arbitrary, a 350% increase in production is not unreasonable from an environmental perspective (i.e., very little area is currently dedicated to shellfish farms and the production of shellfish species has a minimal environmental impact), such that the socially/economically desirable reference point reported by federal managers seems reasonable using SMART principles for setting reference points [24]. A production function based on these parameters would likely lead to higher mariculture scores, as the reference points (i.e., targets) would likely be lower. Uncertainty in what these target values should be remains an important gap in our current understanding.

Lessons learned for regional assessments

Although from a global perspective the U.S. west coast is a relatively data-rich location, data from the recent past were largely lacking, and historical data even more so. Most notably these gaps include habitat condition (current quality, and historical and current extent), conservation status of most species, fisheries stock assessments, historical levels of human pressures on ecosystems, and the nature and effectiveness of regulatory measures – data gaps common nearly everywhere [28–29]. The process of pulling together the information necessary to calculate the Index serves as a means to systematically evaluate where key gaps remain. Such gaps are a perpetual challenge for managers and policy makers. Prioritizing efforts to fill those gaps remains critical. The assessment here also highlights the need for new or continued assessment of pressures and resilience measures, not only of status variables, for effective assessment of overall ocean health. Our study offers a valuable starting point, or baseline, for future assessments in this region, but only by filling key data gaps will we gain the ability to determine trends in overall ocean health, a critical need for ecosystem-based management [30]. The sooner that other regions can begin comprehensive and repeatable assessments of ocean health, the better equipped they will be to make fully informed and strategic resource management decisions.

A key challenge for any assessment of ecosystem health is to detect meaningful and significant change in condition. Ideally one can then attribute that change to natural versus anthropogenic drivers of change, although such attribution is notoriously difficult. Along the U.S. west coast efforts to detect and attribute change face the challenge of distinguishing between broad shifts in the ecosystem due to natural climate variability on timescales of interannual (such as El Nino-driven changes) to multi-decadal (such as Pacific Decadal Oscillations; PDO) from longer-term trends driven by human impacts such as climate change. In the Index most goals document the cumulative effect of natural and anthropogenic change but do not explicitly attribute the underlying cause. For example, the biodiversity, carbon storage, coastal protection and clean water goals all have fixed targets that are independent of the cause of change, and the human-focused goals of coastal livelihoods & economies and tourism & recreation are more indirectly affected by such natural variation and are also independent of the cause of change. The fisheries sub-goal of food provision is one case where attribution of change is important, as fisheries management benefits from knowing the cause(s) of mortality. In this case, stock assessment models are usually refined to adjust the reference points and assessments in consideration of known changes in key oceanographic conditions (i.e., adjusting B_{MSY} to account for what is sustainable for a given oceanographic regime). By using information from local stock assessments, the index is adjusted for these effects whenever local information and modeling tools allow.

Importantly, and perhaps unsettling to some, is the reality that assessing something as diverse and comprehensive as 'ocean health' requires accommodating regional values and perspectives of the people that are part of the coastal ecosystems being assessed. Within the Ocean Health Index, this means that some conceptual aspects of implementing the Index are inherently subjective. Although the Index framework provides guidelines that can help adapt models to available regional data, using simpler models or different proxies when necessary [14], it cannot prescribe which available regional data sets are preferred. Nor does the framework dictate the most appropriate models to use or choices of reference points, proxy data or goal weights. For example, the way we modified models or reference points used in the global assessment [14] for regional use in the food provision, tourism & recreation, and artisanal fishing opportunities goals (see File S1) has important consequences for the resulting goal scores (Table 3). These adaptations highlight the flexibility of the Index to incorporate different perspectives on how goals should be assessed, as reflected by debates on how fisheries were modeled [31–34]. The Index can also accommodate a different set of goals if they better reflect what local communities value, although we posit that the ten goals currently defined within the Index are sufficiently broad to capture a vast majority of values. To some this flexibility may seem to come at the cost of comparability or objectivity, but we argue that any local assessment of ecosystem health faces similar challenges of accounting for local variables and community values. No indicator is exempt from such subjective decisions. The Index's framework, however, requires one to identify, justify and track such assumptions explicitly and thus fosters careful and well-documented assessment of the sensitivity of results to such decisions.

Policy implications

The Index was explicitly designed to help inform decision-making by providing a comprehensive, comparable, and quantitative assessment of the range of components that drive overall ocean health. As with any decision support tool, the scale of assessment should match the scale of decision-making [35]. Our assessment here is thus most valuable to regional-scale (e.g., West Coast Governors Alliance, [2]; California Current LME, [36] and state-level (e.g., California's Marine Life Protection Act, [37]) decision-making. Decisions at smaller scales (such as Puget Sound or San Francisco Bay) would be best informed by recalculating the Index using best available local-scale information where possible. To help support application of the Index in these (and other) processes in the future, we have developed a software tool (www.ohi-science.org) that allows people to explore Index results as well as recalculate (or calculate anew) scores as new data become available. A key strength, and challenge, of the Index is that it requires an explicit statement of all assumptions and assignment of specific targets for each ocean health goal. The strength lies in providing stakeholders and decision-makers a forum to articulate

Table 5. Comparative summary of assessment tools and methods that have been applied to regions of the US West coast.

	OHI	IEA	PSP	FEP	CalCOFI
Ecological system assessed explicitly	no	yes	yes	yes	yes
Social system assessed explicitly	no	yes	no	no	no
Integrated assessment of socio-ecological systems	yes	no	no	no	no
Scalable to sub-regional level	yes	yes	yes	no	yes
Includes scenario analyses	yes	yes	no	no	no
Part of PFMC process	no	yes	no	yes	yes
Part of WCGA process	no	yes	no	no	no
Addresses most/all sectors	yes	yes	yes	yes	no
Combines all sectors into an overall quantitative assessment	yes	no	no	no	no
Reference points are explicitly delineated	yes	no	no	no	no

OHI, IEA and FEP methods have been applied to the entire west coast; PSP and CalCOFI are sub-regional assessments but are included for comparative purposes. Attributes which all or none of the methods achieve are not included in this table. See legend below for definition of acronyms.
Legend: OHI = Ocean Health Index; IEA = Integrated Ecosystem Assessment; PSP = Puget Sound Partnership; FEP = PFMC Fisheries Ecosystem Plan annual reports; PFMC = Pacific Fisheries Management Council; WCGA = West Coast Governors Agreement; CalCOFI = California Cooperative Oceanic Fisheries Investigations.

their reference points and assumptions, while leveraging their values and knowledge, and a means to disentangle and clearly define their multiple, interacting objectives. The challenge arises from the practical (e.g., data constraints) and political (e.g., managing expectations, achieving consensus) process of making these important decisions, and the inherent sensitivity of Index scores to these choices [24]. The Index offers a tool to engage stakeholders and decision-makers in these difficult but necessary discussions, while also helping agencies fulfill their mandates.

For example, the ability to use scenarios to evaluate the likely consequences of any particular management action for overall ocean health provides a powerful decision-support tool, but requires additional assumptions and decisions about how things will likely change in the future. We illustrated such a process with several heuristic scenarios (Fig. 5) intended to show how the Index could inform regional-scale decision-making on issues such as land-use regulations and MPA creation. Scenarios intended to inform decision-making at these or smaller scales in the future would benefit from vetting model assumptions through a planning process, and require that the Index be applied at the relevant spatial scale. Although hypothetical, the scenarios demonstrate several key aspects of the Index relevant to decision makers: 1) it responds quickly to management actions, giving initial 'credit' for those actions, and then further responds over time as the system (social, economic, and ecological) changes; 2) tradeoffs inherent in many decisions are captured by the Index (either explicitly as they are built into the Index or implicitly as they would emerge after management actions); and 3) the Index allows one to compare very different management actions in a transparent and quantitative way across different sub-regions, thus supporting strategic decision-making. The magnitude of expected change in the Index will necessarily be related to the scale of management action relative to the scale of assessment.

Such scenario analyses are also a key way that the Index can be used to explore potential implications of climate change on ocean health. As with the other scenario examples, because the Index does not model the future it cannot predict future ocean health. Instead, dynamic process models can be used to simulate ecological and social conditions, and then these results can be fed in as input parameters for calculating an alternate Index score. In this case, the Index can be used to indicate the likely overall ocean health in the future under status quo conditions and a changing climate. Additional management scenarios could then be layered on top of those outputs to better understand the likely effect of climate change on future ocean health.

Scenario analyses also illustrate how the Index can be used to identify and understand tradeoffs among goals. Some of these known tradeoffs are built into the architecture of the Index, for example in how increased (sustainable) fishing produces higher scores for food provision but lowers other goals due to its negative pressure on them. Other more complex, emergent tradeoffs become visible only when the Index is measured over time and one can track how goal scores change in similar or opposite directions. Because of the complexity of ecosystem responses, full attribution of a change in one goal causing a change in another goal is difficult, but such patterns can provide insight on where to direct further exploration of such possible tradeoffs. The ability to calculate past status scores, and then correlate changes in the Index with past management actions, illustrates a key way it can be used to assess management effectiveness. If the Index were adopted as a management tool, recalculating scores regularly could reveal whether management actions had the intended effect on both overall ocean health and particular goals. This objective demonstrates the power (and necessity) of having a quantitative, repeatable, transparent and comprehensive method for assessment.

The process of adapting the Index to finer geographic scales highlights its flexibility but also the limits to comparability of Index scores across scales. Most decision making focuses on optimizing outcomes for a region of interest (e.g., a particular country, or a state within a country), regardless of how other regions are performing, such that adapting the Index to the best available regional information is appropriate and ideal. However, it is human nature to ask how one is doing relative to others, and that desire for comparability can lead to misunderstanding of Index results if the comparisons are made across assessments at different scales (e.g., global version regional). Here we focus on results within and among U.S. west coast regions and minimize comparisons to global results for the U.S. for these exact reasons.

Many other assessment frameworks and tools have been applied to the U.S. west coast to evaluate different aspects of its health (see Table 5 for a summary of several prominent ones). Although it is instructive to compare the approaches to understand their strengths and weaknesses, it is important to note that each method was developed and applied for specific purposes, such that direct comparison among them is not always appropriate. Integrated

Ecosystem Assessments (IEAs), Fisheries Ecosystem Plans (FEPs) and CalCOFI reports are all part of the Pacific Fisheries Management Council's (PFMC) decision process, such that those assessments are directly affecting and assessing management actions, however the Ocean Health Index is too new to have had a chance to be vetted and potentially included in the PFMC process. All of the other methods directly and explicitly assess the ecological and biophysical aspects of the system, whereas in the Ocean Health Index these assessments are not separately available because they are combined within integrated socio-ecological indicators. On the other hand, for this reason the Index is currently the only method to offer a fully integrated assessment. Most of the methods assess the full range of sectors active in the region, but the Ocean Health Index generally combines them together into overall goal measures rather than tracking individual sectors separately. Finally, the Ocean Health Index makes explicit the process of defining and setting quantitative reference points that establish when goals are fully achieved, whereas the other methods tend to rely on expert judgment and informal evaluations.

Another important policy implication of applying the Index is to help prioritize data collection and primary research efforts. Most monitoring focuses on biological impacts without connecting them explicitly to benefits that people want and need. The Index framework, by explicitly showing the connection between societal goals and the ability of the system to provide those goals, highlights the importance of collecting ecological, social, institutional, and economic data to monitor and inform management, and motivates all stakeholders to strive for a more sustainable human-ocean system.

Application of the Ocean Health Index to the US west coast not only provided an assessment of ocean health for the region but also guidance on the opportunities and challenges in applying and adapting the general Index framework to a regional setting. In the relatively data-rich US west coast, we were able to take advantage of the best available knowledge and information and make sub-regional assessments, sub-regions that share some ecological and socio-economic aspects but also show many differences that are important for defining management strategies. Such sub-regional assessments are likely to be important in most regions of the world. In particular, this downscaled, regional application of the Index offers a means and a medium for conversations among disparate marine use sectors by providing measures of diverse aspects of ocean health in a common currency.

Acknowledgments

Thanks to Dan Ovando for help exploring options for calculating the status of data poor stocks in the region and Judith Kildow and Charles Colgan for help with livelihoods data. Dave Fluharty and an anonymous reviewer provided very helpful comments.

Author Contributions

Conceived and designed the experiments: BSH CL SKK KLM JFS. Performed the experiments: BSH CL DH BDB. Analyzed the data: BSH CL CS DH BDB. Wrote the paper: BSH CL CS DH BDB SCD SKK KLM AAR JFS.

References

1. MSFD (2008) Marine Strategy Framework Directive. Directive 2008/56/EC of the European Parliament and Council establishing a framework for community action in the field of marine environmental policy. http://eurlex.europa.eu/LexUriServ/LexUriServ.do?uri = OJ:L:2008:2164:0019:0040:EN:PDF.
2. WCGA (2006) West coast governors' agreement on ocean health. Agreement on Ocean Health Action Plan.
3. McLeod KL, Leslie HM (2009) Ecosystem-based management for the oceans. Island Press, Washington, DC.
4. COA (1998) Commonweath of Australia. Australia's Oceans Policy.
5. DFO (2002) Directorate, Fisheries and Oceans Canada. Canada's Oceans Strategy, Ottawa, Ontario.
6. Obama B (2010) Executive Order 13547: Stewardship of the ocean, our coasts, and the great lakes, Washington DC.
7. NMS (2004) National Marine Sanctuary Program. A Monitoring Framework for the National Marine Sanctuary System.
8. PFMC (2013) Pacific Fishery Management Council: *Pacific Coast Fishery Ecosystem Plan for the U.S. Portion of the California Current Large Marine Ecosystem – Public Review Draft, February 2013*. (Document prepared for the Council and its advisory entities.) Pacific Fishery Management Council, 7700 NE Ambassador Place, Suite 101, Portland, OR, 97220-1384.
9. CDFG (1999) California Department of Fish and Game. Marine Life Protection Act. Fish and Game Code Sections 2850–2863.
10. PSP (2008) Puget Sound Partnership. Puget Sound Action Agenda: Protecting and restoring the Puget Sound ecosystem by 2020, Olympia, Washington.
11. WCEBM (2010) West Coast EBM Network. Community-Based Management of Coastal Ecosystems: Highlights and Lessons of Success from the West Coast Ecosystem-Based Management Network. www.westcoastebm.org/WestCoastEBMNetwork_EBMGuide_June2010.pdf
12. Halpern BS, Diamond J, Gaines S, Gelcich S, Gleason M, et al. (2012a) Near-term priorities for the science, policy and practice of Coastal and Marine Spatial Planning (CMSP). Marine Policy 36:198–205.
13. Tallis H, Levin PS, Ruckelshaus M, Lester SE, McLeod K, et al. (2010) The many faces of ecosystem-based management: Making the process work today in real places. Marine Policy 34:340–348.
14. Halpern BS, Longo C, Hardy D, McLeod KL, Samhouri JF, et al. (2012b) An index to assess the health and benefits of the global ocean. Nature 488:615–620.
15. Karieva P, Marvier M, Lalasz R (2011) Conservation in the Anthropocene. Breakthrough Journal 2.
16. Karieva P, Marvier M (2012) What is conservation science? BioScience 62: 962–969.
17. Soulé M (2013) The "new conservation". Conservation Biology 27: 895–897.
18. Folke C, Carpenter SR, Walker BH, Scheffer M, Chapin III FS, et al. (2011) Reconnecting to the biosphere. Ambio doi:10.1007/s13280-011-0184-y;
19. Liu J, Dietz T, Carpenter SR, Alberti M, Folke C, et al. (2007) Complexity of coupled human and natural systems. Science 317: 1513–1516.
20. Walker B, Salt D (2006) Resilience thinking: sustaining ecosystems and people in a changing world. Washington, DC: Island Press.
21. Spalding M, Fox HE, Allen GR, Davidson N, Ferdana ZA, et al (2007) Marine ecoregions of the world: a bioregionalization of coastal and shelf areas. BioScience 57: 573–583.
22. Samhouri JF, Kim S, Zhang CI, Fogarty M (2014) Marine ecosystem-based management in temperate systems. In: Marine ecosystem-based management, Fogarty MJ, McCarthy JJ, editors. Vol. 16 of *The Sea*. Harvard University Press, Boston. pp. 325–367
23. MEA (2005) Millennium Ecosystem Assessment. Ecosystems and human well-being: synthesis report.
24. Samhouri JF, Lester SE, Selig ER, Halpern BS, Fogarty MJ, et al. (2012) Sea sick? Setting targets to assess ocean health and ecosystem services. Ecosphere 3: article 41.
25. Halpern BS, Longo C, Stewart Lowndes JS, Best BD, Frazier M, et al (in review) Patterns and emerging trends in global ocean health. PLoS ONE
26. Halpern BS, Longo C, McLeod KL, Cooke RG, Fischhoff B, et al. (2013b) Elicited preferences for components of ocean health in the California Current. Marine Policy 42:68–73.
27. Nash CE (2004) Achieving policy objectives to increase the value of the seafood industry in the United States: the technical feasibility and associated constraints. Food Policy 29:621–641.
28. Pereira HM, Cooper DH (2006) Towards the global monitoring of biodiversity change. Trends In Ecology & Evolution 21:123–129.
29. Rands MRW, Adams WM, Bennun L, Butchart SHM, Clements A, et al. (2010) Biodiversity conservation: challenges beyond 2010. Science 329:1298–1303.
30. Francis T B, Levin PS, Harvey CJ (2011) The perils and promise of future analysis in marine ecosystem-based management. Marine Policy 35:675–681.
31. Branch TA, Hively DJ, Hilborn R (2013) Fisheries assessments are biased in Ocean Health Index. Nature 495: E5–D6.
32. Halpern BS, Gaines S, Kleisner K, Longo C, Pauly D, et al. (2013a) Reply to Branch et al. Nature 495: E7.
33. Kleisner KM, Longo C, Coll M, Halpern BS, Hardy D, et al. (2013) Exploring patterns of seafood provision revealed in the global Ocean Health Index. Ambio DOI 10.1007/s13280-013-0447-x

34. Pauly D, Hilborn R, Branch TA (2013) Fisheries: does catch reflect abundance? Nature 494:303–306.
35. Cash DW, Moser SC (2000) Linking global and local scales: Designing dynamic assessment and management processes. Global Environmental Change 10: 109–120.
36. Aquarone MC, Adams S (2008) XIV-44 California Current: LME#3. In: K. Sherman K, Hempel G, editors. The UNEP large marine ecosystem report: A perspective on changing conditions in LMEs of the world's regional seas, UNEP Regional Seas Report and Studies No. 182. United Nations Environment Programme, Nairobi, Kenya. pp. 593–603.
37. CCC (1997) California Coastal Commission, California Coastal Commission Strategic Plan. Available: http://www.coastal.ca.gov/strategy.html. Accessed 2014 Feb. 10.

17

Quantifying Fish Assemblages in Large, Offshore Marine Protected Areas

Nicole A. Hill[1]*, Neville Barrett[1], Emma Lawrence[2], Justin Hulls[1], Jeffrey M. Dambacher[3], Scott Nichol[4], Alan Williams[5], Keith R. Hayes[3]

1 Institute for Marine and Antarctic Studies, University of Tasmania, Hobart, Tasmania, Australia, **2** Digital Productivity Flagship, Commonwealth Scientific and industrial Research Organisation (CSIRO), Brisbane, QLD, Australia, **3** Digital Productivity Flagship, Commonwealth Scientific and industrial Research Organisation (CSIRO), Hobart, Tasmania, Australia, **4** Geoscience Australia, Canberra, ACT, Australia, **5** Oceans and Atmosphere Flagship, Commonwealth Scientific and industrial Research Organisation (CSIRO), Hobart, Tasmania, Australia

Abstract

As the number of marine protected areas (MPAs) increases globally, so does the need to assess if MPAs are meeting their management goals. Integral to this assessment is usually a long-term biological monitoring program, which can be difficult to develop for large and remote areas that have little available fine-scale habitat and biological data. This is the situation for many MPAs within the newly declared Australian Commonwealth Marine Reserve (CMR) network which covers approximately 3.1 million km^2 of continental shelf, slope, and abyssal habitat, much of which is remote and difficult to access. A detailed inventory of the species, types of assemblages present and their spatial distribution within individual MPAs is required prior to developing monitoring programs to measure the impact of management strategies. Here we use a spatially-balanced survey design and non-extractive baited video observations to quantitatively document the fish assemblages within the continental shelf area (a multiple use zone, IUCN VI) of the Flinders Marine Reserve, within the Southeast marine region. We identified distinct demersal fish assemblages, quantified assemblage relationships with environmental gradients (primarily depth and habitat type), and described their spatial distribution across a variety of reef and sediment habitats. Baited videos recorded a range of species from multiple trophic levels, including species of commercial and recreational interest. The majority of species, whilst found commonly along the southern or south-eastern coasts of Australia, are endemic to Australia, highlighting the global significance of this region. Species richness was greater on habitats containing some reef and declined with increasing depth. The trophic breath of species in assemblages was also greater in shallow waters. We discuss the utility of our approach for establishing inventories when little prior knowledge is available and how such an approach may inform future monitoring efforts within the CMR network.

Editor: David Mark Bailey, University of Glasgow, United Kingdom

Funding: This work was undertaken for the Marine Biodiversity Hub, a collaborative partnership supported through funding from the Australian Government's National Environmental Research Program (NERP). NERP Marine Biodiversity Hub partners include the Institute for Marine and Antarctic Studies, University of Tasmania; CSIRO, Geoscience Australia, Australian Institute of Marine Science, Museum Victoria, Charles Darwin University and the University of Western Australia. The funders had no role in study design, data collection and analysis, decision to publish, or preparation of the manuscript.

Competing Interests: The authors have declared that no competing interests exist.

* Email: Nicole.Hill@utas.edu.au

Introduction

Marine ecosystems face pressures from a range of sources, including pollution, fishing, habitat destruction, invasive species and a changing climate [1]. These pressures threaten the essential processes and resources that marine ecosystems provide, such as climate regulation, nutrient cycling and the provision of food, as well as threatening their intrinsic natural and cultural value [2]. As a consequence, management strategies have been adopted to safeguard marine ecosystems and their associated biodiversity. These strategies include the establishment of Marine Protected Areas (MPAs) which offer varying levels of protection according to their zoning or IUCN designation [3]. Countries such as the United States, Australia, South Africa, New Zealand, Kenya and the Philippines have a long history of establishing MPAs [4,5]. However, the absence of a truly global system of MPAs prompted the Convention of Biological Diversity [6] and the Jakarta Mandate [7] to provide a framework and renewed international impetus for protecting marine biodiversity. This framework has resulted in a significant increase in recent years in the declaration of MPAs worldwide, with many forming part of large networks or systems of MPAs [8].

The increasing global commitment to the sustainability of healthy marine ecosystems is encouraging, but the declaration alone of MPAs does not ensure the conservation of biodiversity or ecosystems [4]. Management authorities must: clearly articulate the objectives of individual MPAs and MPA networks, both of which may contain multiple IUCN zones; formulate and enforce management plans [9] and periodically revisit and evaluate their management plans, which in turn requires that they establish a long-term biological monitoring program [4]. Developing monitoring programs for individual MPAs within networks of MPAs is often a difficult task. There is generally a mismatch between the spatial scales of information used to design MPA networks and those required to manage individual MPAs. Inventories of the types and distribution of habitats, communities and species within individual MPAs are often required before monitoring programs can be designed. In addition, for offshore or difficult to access MPAs, there are significant logistical and statistical challenges to working in remote environments.

Australia, a signatory to the CBD, provides a good example of the challenges involved in ultimately developing long-term monitoring programs in large MPA networks. In 2012, the Australian government announced an expansion of its MPA network to include 33 new MPAs. This makes a total of 60 MPAs covering approximately 3.1 million square kilometres, divided between six planning bioregions [10]. Although called the "Commonwealth Marine Reserve" or "CMR" network, areas with IUCN zones ranging from la (Sanctuary Zone) to VI (General Use Zone) are distributed throughout the network. From this point forward we adopt the above terminology and refer to MPAs within this network as reserves regardless of their IUCN zoning. All of the reserves in the CMR are however, located in offshore waters, most are large and encompass a broad depth range, and many are remote.

The Australian CMR network aims to meet the principles of being Comprehensive, Adequate and Representative (the CAR principles) [11] and utilised the national Integrated Marine and Coastal Regionalisation of Australia (IMCRA; [12]). The IMCRA bio-regionalisation is based on breaks in the distribution of demersal fish communities (together with some physical datasets) and is the best available continental-scale regionalisation of Australian marine fauna. Hence it was the most appropriate mechanism for delineating management regions and informing the placement of reserves. When designing biological monitoring programs for individual reserves however, much finer scale information is needed on the spatial distribution of important habitats (such as reefs) and the biological communities they contain. For many reserves in the new network this level of information is simply unavailable. It is therefore necessary to provide a robust inventory of the biological communities and key species within the reserves, delineate their spatial distribution and quantify the relationship between these distributions and key environmental drivers to enable the development of monitoring programs that will ultimately evaluate the effectiveness of each reserve.

Inventory, and ultimately monitoring, of large, deep reserves is challenging and relatively expensive. It requires vessels large enough to conduct surveys safely in offshore waters, together with a suite of non-extractive survey methods that can be deployed at depth, and sampling designs that are efficient and flexible enough to accommodate multiple (and potentially changing) objectives. A suite of non-extractive survey methods are currently available including multibeam sonar (MBS) for characterising the seafloor and associated habitat types [13], and image-based methods, such as towed camera systems, cameras attached to autonomous or remotely operated vehicles and baited cameras, for observing benthic fauna. Here we trialled MBS and Baited Remote Underwater Video (BRUVs), deployed within a spatially balanced design, for inventorying demersal fish in the multiple use zone (IUCN VI) of one reserve in the Australian network, the Flinders CMR. We use BRUVs because of their demonstrated ability to survey fish in deep waters [14] and because they have been extensively used in Australian inshore environments for monitoring and other ecological applications [15–17]. We trialled a spatially-balanced (i.e. evenly spread out in space) design known as the Generalised Random- Tessellation Stratified (GRTS) design [18] because it is flexible and provides a good way to obtain a representative sample that respects the spatial distribution of habitats and communities in the target population [19].

In this study we: identify patterns in demersal fish assemblages across the continental shelf (a multiple-use zone, IUCN VI) of the Flinders CMR; quantify these patterns in relation to environmental gradients; examine the spatial distribution of assemblages across the reserve; and describe how our approach and the information gained may be useful to the development of a long-term biological monitoring program. We focus on fish communities because: broad-scale distributional records of demersal fish were instrumental in the marine bioregionalisation that underpins the design of the reserve network [20]; many species are of recreational and commercial value and are therefore likely to respond to current and future management actions within the reserve; and demersal fish are key components of benthic ecosystems.

Methods

Survey Area

The Flinders Commonwealth Marine Reserve (CMR) was established in 2007 and lies about 25 km offshore of northern Tasmania (Fig. 1). The reserve is 26,975 km^2 in size and extends as a west-east corridor from 35 m to > 3,000 m water depths. It consists of two zones: a multiple use zone (IUCN Category VI) that covers the majority of the continental shelf and slope; and a marine national park zone (IUCN Category II) that extends from the continental slope to the edge of the reserve at the 200 nautical mile limit of Australia's Exclusive Economic Zone (Fig. 1 inset). Our study area is the continental shelf, part of the multiple use zone (IUCN Category VI), where activities that impact on benthic habitats are prohibited, including demersal trawling, Danish seining and scallop dredging, or subject to permit requirements, for example mining activities [21]. Benthic habitats on the Flinders CMR shelf consist of sediment plains with patches of low profile and sand-inundated reefs, and steep rocky outcrops where canyon heads incise the shelf break (Lawrence, unpublished data). Within this environment, reefs and rocky outcrops have been identified as features likely to contain enhanced benthic diversity [21].

The Flinders CMR shelf straddles two provincial biogeographical regions: the Southeast Shelf Transition in the north and the Tasmanian Shelf Province in the South [12]. This region is characterised by variable but generally high exposure and strong tidal currents, especially in shallow areas between Flinders Island and the Tasmanian mainland [22,23]. The region is also influenced by southwards incursions of the East Australian Current (EAC) which brings warmer waters on to the shelf in summer [24]. The flora of the region is moderately rich and contains species common in cold temperate waters as well as low abundances of species common to warmer temperate waters [20].

Sampling design and methods

In contrast to the Flinders CMR continental slope, which has been comprehensively mapped using multibeam sonar (from which seafloor habitats have been inferred), very little spatially explicit habitat or biological information is available for the shelf. This represents a challenge for designing targeted sampling programs. Interpolated bathymetry for the Flinders shelf exists, but is at a gridded resolution of 250 m [25] (Fig. 1) which does not allow the identification of fine scale features of interest on the shelf, such as reefs. As a consequence of our limited *a priori* knowledge of the area, the inventory and description of fish assemblages

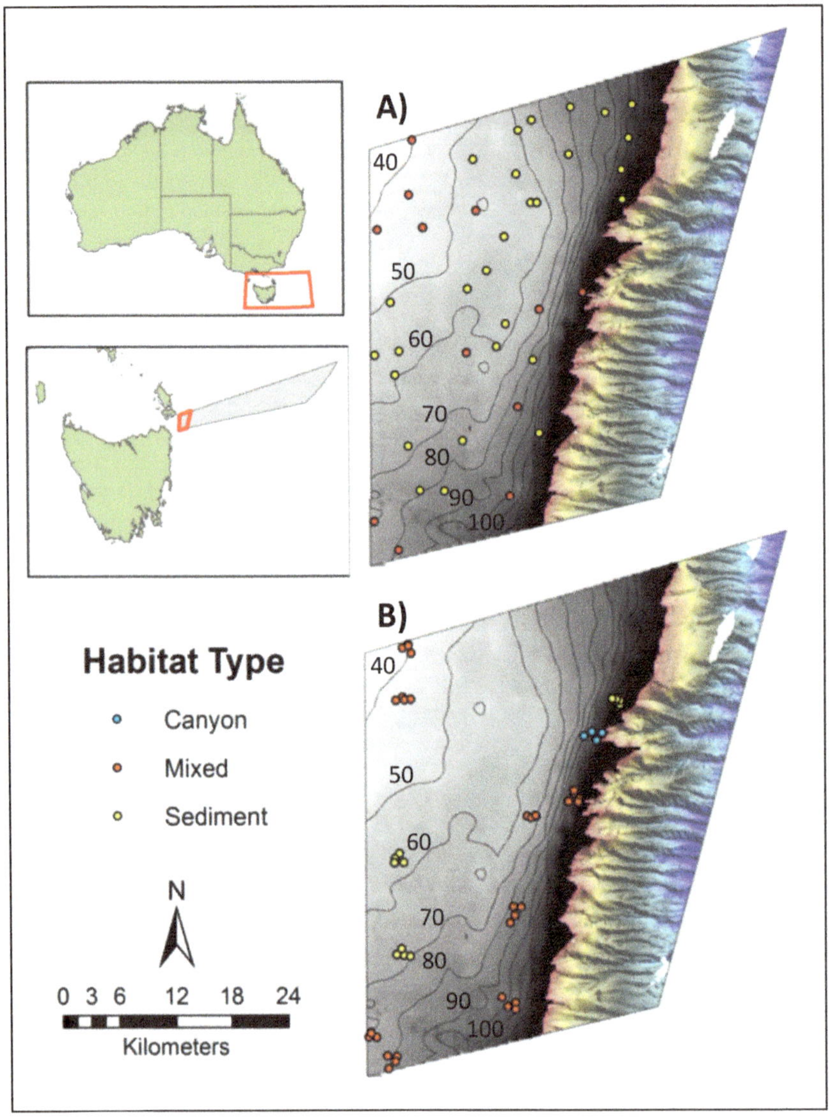

Figure 1. Location of the Flinders Commonwealth Marine Reserve (CMR) in Tasmania, Australia. Panels A and B show the CMR's Multiple Use Zone (IUCN VI) where survey work was conducted. The grey zone is the continental shelf (less than 200 m depth) where there was little pre-existing mapping data. Coarse bathymetry data (gridded at 250 m horizontal resolution) sourced from Geoscience Australia is shown with 10 m contour intervals overlain. The coloured area to the right shows the relatively steep and highly incised upper continental slope that had been mapped previously with multibeam sonar extending from 200 to 1500 m. A) Location of the 40 sites surveyed for habitat type in phase one of the sampling program. B) Location of the clustered sites where Baited Remote Underwater Videos (BRUVs) were deployed in phase two. Sites are coloured according to the broad habitat type: sediment (yellow); mixed, low-profile reef and sediments (red); canyon head (blue) recorded during phase one of sampling.

presented here forms one component of a multi-objective and multi-phase study within the Flinders CMR. Additional objectives of the broader study include examining the distribution and estimating the area of the different habitat types within the reserve, as well as quantifying benthic invertebrate communities using imagery captured by towed video and AUV, and these results will be reported elsewhere. To accommodate the objectives of the different components of the study, we used the probabilistic design, Generalised Random –Tessellation Stratified sampling technique (GRTS; [18]). GRTS ensures sampling sites are well spread out across the survey area (spatially-balanced), a desirable property for spatial sampling that enhances estimation efficiency and representativeness [18]. Other common sampling designs include simple or stratified random sampling or systematic sampling. However, simple random sampling often clumps sites [19] and stratified random sampling (based on habitat in our case) was not possible without more detailed knowledge of the study area. Systematic sampling prevents clumping but an accurate design-based variance estimator does not exist; a factor that was important for other aspects of the broader study [18,26]. provide a good discussion of the relative merits of various spatial sampling strategies. Another property of GRTS is that it produces an ordered list of sample sites and any set of consecutive sampling sites also maintain spatial balance, offering flexibility to adaptively change sample sizes in the field, if for example, sampling takes less time than expected. To date GRTS has primarily been used in natural resource assessment and monitoring in the United States e.g. [27].

Because of our lack of knowledge on the distribution of habitat types, we implemented a two-phase sampling program. In phase one, the distribution of habitat types across the reserve was quantified by characterising the first forty, 200 m square, sites from a master list of GRTS sites (essentially an ordered list of all possible GRTS sites in the study area. See [28] for a detailed discussion on the use of master lists) as either 'soft' sediments or 'mixed', low profile patchy reef using MBS and footage obtained from a drop camera (Fig. 1; Lawrence, unpublished data). In phase two of the sampling program, the subject of this paper, we sampled the fish assemblages using BRUVs. Ideally we would have revisited all forty of the phase one GRTS sites. The large size of the study area and the soak time required for each BRUV deployment however, meant that it was not possible to do so in the ship time available. Instead, we sampled clusters of sites surrounding a subset of the phase one GRTS- sites. In phase two, we sampled mixed reef habitats more intensively than sediment habitats because shelf reef systems within the reserve are recognised as an important biodiversity feature and a *priori* we expected that reefs would harbour a greater diversity of assemblage types than sediments. We utilised the fact that consecutive GRTS sites maintain spatial balance and selected (from the ordered list of phase one sites) the first eight mixed habitat sites and the first three sediment sites of the forty phase one sites as the basis of the phase two clusters (Fig. S1. shows the site numbers of the forty phase ones sites and lists those used as the basis of phase two clusters.) This amounts to selecting phase one sites for the BRUV sampling that are spatially balanced *within* each of the phase one habitat types, mixed reef and sediments, and presumably therefore also spread across the environmental gradients in the region. In addition, a site was added near the head of a canyon, to ensure representation of another identified feature of the reserve which occupies a relatively small proportion of space and is therefore less likely to be sampled in a probabilistic sampling design (Fig. 1B). Clusters of sites within 1 km of the subset of phase one sites were selected again using the GRTS master list, resulting in five sites per cluster (Fig. 1B).

At each site, one BRUV was deployed using systems and deployment conditions widely used in Australia [29]. Each stereo-BRUV consisted of a frame that houses a stereo camera pair, a bait bag attached to an arm within field of view of the cameras, a diode for synchronising imagery between camera pairs. In water deeper than approximately 70 m, a light with a blue filter was also attached [15]. Stereo- BRUVs were baited with 1 kg of crushed pilchards and deployed for 60 min (soak time). Appropriate ethics (University of Tasmania Animal Ethics Permit: A12514) and fieldwork (Australian Government Director of National Parks Approval of Research Activities in the Southeast Commonwealth Marine Reserve Network. Ref No 07/10622) approvals were obtained for this work.

Analysis

Imagery collected using stereo-BRUVs was scored using standard metrics including scoring the maximum number of fish occurring in any one frame for each species (MaxN) [15,30]. Scoring was completed with Event Measure software [31].

To determine the types of fish assemblages present at sites sampled on the Flinders CMR shelf, fuzzy clustering was performed on multivariate fish composition data using the 'cluster' package [32] in R [33]. Multivariate data were square root transformed to reduce the influence of the most abundant species and a Bray Curtis dissimilarity matrix generated. The number of fuzzy clusters was set to six, following preliminary analysis using several different clustering methods, and the membership exponent (which controls the 'fuzziness' of clusters) set to 1.3 after examining silhouette profiles of group membership.

A constrained ordination (Canonical Analysis of Principle Co-ordinates; CAP) [34] was used to test if the assemblage groups determined by fuzzy clustering were significantly different from each other (using a permutational test). In addition, jack-knife sampling was used to examine the overall and individual classification accuracy of the six groups. To enable the ecological interpretation of assemblages, the trophic level and habitat preference of species was defined. Species were assigned one of seven habitat preference categories: 1) pelagic; 2) wide-ranging demersal; 3) sediment associated; 4) reef and sediment associated; 5) reef and seagrass associated; 6) reef associated; and 7) demersal generalists (found in more than two habitats) based on species attributes described in [35,36]. The trophic level of species was extracted from FishBase [37]. Differences in the trophic level of species comprising the six groups was assessed with a one factor Analysis of Variance (ANOVA) after transforming data to satisfy the assumptions of ANOVA. To understand which species were the key contributors to each of the six assemblage groups, a SIMPER analysis was performed. SIMPER decomposes the Bray-Curtis dissimilarity matrix between all pairs within a group into the percentage contributions from each species [34].

Correlations between fish assemblages and their environment were examined against three easily quantifiable environmental variables that, from the literature, may influence benthic assemblages in our system; latitude, depth and the substratum type. At each cluster of five sites, only the central site had been surveyed with MBS and drop camera in phase one of the sampling program. Therefore, substratum type was derived from the BRUV field of view and characterised as either: 'sediment' if no hard substratum or no hard substratum-associated organisms (i.e. sponges etc.) were present; 'mixed' if some hard substratum was visible or if some hard substratum-associated organisms were present; and 'reef' if the majority of the field of view contained hard substratum or hard substratum-associated organisms. Latitude was taken from the recorded location of BRUV drops and depth was derived using the 250 m resolution bathymetry grid available from Geoscience Australia [25]. Relationships between assemblage groups and environmental variables were inferred by examining correlations with CAP axes and overlaying vectors of the environmental variables on CAP plots. Ordination and SIMPER analysis were performed in PRIMER + with PERMANOVA [34].

The influence of environmental variables in determining patterns in species richness (the number of species present) was examined using Generalised Linear Mixed Models (GLMMs). Observations were modelled using a Poisson distribution with a log link function and sampling cluster was included as a normally distributed random factor. Mixed models were fit in R [33] with penalised quasi-likelihood, using the glmmPQL function in the package MASS [38].

Results

Distribution of habitats across the CMR

The results of phase one sampling indicated that the majority (70%) of the 40 sites on the shelf were sedimentary habitats, and the remainder 'mixed' reef and sediment habitat. Mixed reef habitat occurred in the north-west, south-west, and near the edge of the shelf of the study area (Fig. 1A; Lawrence, unpublished data).

Table 1. List of species recorded in Baited Remote Underwater Video deployments across the continental shelf in the Flinders Commonwealth Reserve.

			Across CMR			Mean (SE) MaxN by group					
Family	Species name	Common name	Total MaxN	Max MaxN in any drop	Prevalence (% drops)	Group 1	Group 2	Group 3	Group 4	Group 5	Group 6
Blenniidae, Gobiidae, Tripterygiidae	*Blenniidae, Gobiidae, Tripterygiidae*	Blennies, Gobies, Triplefins	7	3	10	0.38 (0.18)	0.4 (0.31)	0	0	0	0
Callorhinchidae	*Callorhinchus milii*	Elephant shark	1	1	2	0	0.1 (0.1)	0	0	0	0
	Nemadactylus douglasii	Grey morwong	7	2	10	0	0.2 (0.13)	0.33 (0.24)	0.33 (0.33)	0	0
	Nemadactylus macropterus	Jackass morwong	379	43	59	0	10.7 (2.04)	23.22 (4.78)	0	5.25 (1.23)	0
Cyttidae	*Cyttus australis*	Silver dory	6	1	12	0	0.1 (0.1)	0	0	0.42 (0.15)	0
Diodontidae	*Diodon nicthemerus*	Globe fish	1	1	2	0	0	0	0	0.08 (0.08)	0
Gempylidae	*Thyrsites atun*	Barracouta	14	3	20	0.38 (0.37)	0.1 (0.1)	0.11 (0.11)	0	0.75 (0.22)	0
Gerreidae	*Parequula melbournensis*	Silverbelly	79	10	31	0	0.5 (0.27)	0	0	5.83 (0.6)	0.67 (0.67)
Heterodontidae	*Heterodontus portusjacksoni*	Port Jackson shark	4	1	8	0	0.3 (0.15)	0	0	0.08 (0.08)	0
Labridae	*Ophthalmolepis lineolatus*	Southern maori wrasse	1	1	2	0	0	0.11 (0.11)	0	0	0
	Pseudolabrus rubicundus	Rosy wrasse	72	13	25	0	1.4 (0.88)	0.67 (0.67)	0	4.33 (1.38)	0
Lamnidae	*Isurus oxyrinchus*	Shortfin Mako shark	1	1	2	0.12 (0.12)	0	0	0	0	0
Latridae	*Latris lineata*	Striped trumpeter	95	27	22	0	0.2 (0.13)	10.11 (3.47)	0.33 (0.33)	0	0
Loliginidae	*Sepioteuthis australis*	Southern calamari	21	3	29	0	0.2 (0.13)	0	0	1.5 (0.23)	0.17 (0.17)
Monacanthidae	*Acanthaluteres vittiger*	Toothbrush leatherjacket	15	3	18	0	0	0	0	1.25 (0.28)	0
	Meuschenia freycineti	Six-spine leatherjacket	3	1	6	0	0.2 (0.13)	0	0	0.08 (0.08)	0
	Meuschenia scaber	Velvet leatherjacket	97	11	53	0	3.3 (0.87)	0.78 (0.43)	0	4.25 (0.97)	1 (0.63)
	Meuschenia venusta	Stars and stripes leatherjacket	1	1	2	0	0.1 (0.1)	0	0	0	0
	Thamnaconus degeni	Degen's leatherjacket	471	89	41	0	1 (0.42)	0.11 (0.11)	0	36.92 (7.37)	2.83 (1.9)
Moridae	*Pseudophycis barbata*	Bearded cod	8	2	12	0	0.1 (0.1)	0.78 (0.28)	0	0	0
Mullidae	*Upeneichthys vlamingii*	Southern goatfish	16	8	16	0	0.3 (0.21)	0.11 (0.11)	0	1 (0.65)	0
Neosebastidae	*Neosebastes scorpaenoides*	Common gurnard perch	44	4	41	0	1.1 (0.31)	0	0	1.5 (0.42)	2.5 (0.5)
Ommastrephidae	*Ommastrephidae sp.*	Squid	9	2	12	0.5 (0.27)	0	0	0.83 (0.4)	0	0
Ostraciidae	*Aracana aurita*	Shaw's cowfish	7	2	12	0	0.1 (0.1)	0	0	0.5 (0.19)	0
Ostraciidae	*Aracana ornata*	Ornate cowfish	1	1	2	0	0.1 (0.1)	0	0	0	0
Palinuridae	*Jasus edwardsii*	Southern rocklobster	2	1	4	0	0	0.22 (0.15)	0	0	0
Paraulopidae	*Paraulopus nigripinnis*	Blacktip cucmberfish	47	12	22	2 (0.65)	0.1 (0.1)	0	5 (2.16)	0	0
Pinguipedidae	*Parapercis allporti*	Barred Grubfish	36	5	35	2.12 (0.52)	0.8 (0.25)	0.44 (0.44)	0.5 (0.34)	0	0.67 (0.42)
Platycephalidae	*Platycephalus bassensis*	Sand flathead	33	6	24	0.12 (0.12)	0	0	0.17 (0.17)	0.83 (0.42)	3.5 (0.96)
	Platycephalus richardsoni	Tiger flathead	21	4	25	0	0.2 (0.13)	0	1.67 (0.56)	0.17 (0.17)	1.17 (0.31)
Pristiophoridae	*Pristiophorus cirratus*	Longnose sawshark	2	1	4	0	0	0	0	0.08 (0.08)	0.17 (0.17)
Rajidae	*Dentiraja lemprieri*	Thornback skate	2	1	4	0	0	0	0	0.17 (0.11)	0

Table 1. Cont.

Family	Species name	Common name	Across CMR			Mean (SE) MaxN by group					
			Total MaxN	Max MaxN in any drop	Prevalence (% drops)	Group 1	Group 2	Group 3	Group 4	Group 5	Group 6
	Spiniraja whitleyi	Melbourne skate	9	1	18	0	0.3 (0.15)	0	0.33 (0.21)	0.33 (0.14)	0
Scyliorhinidae	*Cephaloscyllium laticeps*	Draughtboard shark	63	4	57	0.12 (0.12)	1.7 (0.5)	0.44 (0.34)	1.33 (0.49)	2.33 (0.31)	0.83 (0.48)
Sebastidae	*Helicolenus percoides*	Red gurnard perch	65	12	31	0.25 (0.25)	2.5 (0.81)	4.22 (1.28)	0	0	0
Serranidae	*Caesioperca lepidoptera/rasor*	Butterfly and Barber perch	140	60	14	0	6.1 (5.99)	2.11 (1.42)	0	5 (2.79)	0
Squalidae	*Squalus acanthias*	Spiny dogfish	1	1	2	0	0	0	0.17 (0.17)	0	0
	Squalus megalops	Shortnose spurdog	15	3	14	0.75 (0.49)	0	0	1.5 (0.43)	0	0
Triakidae	*Mustelus antarcticus*	Gummy shark	18	2	33	0.25 (0.16)	0	0	1.17 (0.17)	0.25 (0.13)	1
Triglidae	*Lepidoperca pulchella*	Tasmanian perch	11	5	10	0	0.1 (0.1)	1.11 (0.56)	0	0	0
	Lepidotrigla sp	Gurnards	3	1	6	0	0.1 (0.1)	0.11 (0.11)	0.17 (0.17)	0	0
Urolophidae	*Urolophus cruciatus*	Banded stingaree	2	1	4	0	0.1 (0.1)	0	0	0	0.17 (0.17)
	Urolophus paucimaculatus	Sparsely spotted stingaree	2	1	4	0	0.2 (0.13)	0	0	0	0
	Urolophidae sp.	Stingarees	4	1	8	0.12 (0.12)	0.2 (0.13)	0	0	0	0.17 (0.17)

The Total number of individuals recorded at Max N across all deployments, the greatest MaxN recorded in any single deployment and the prevalence (the percentage of deployments where a species was recorded) are presented as well as the mean (and standard error) MaxN for each assemblage group.

Table 2. Comparison between the number of sites assigned to each of the six groups using fuzzy clustering and Canonical Analysis of Principle coordinates (CAP).

	CAP Classified Group						Total	Correct
	1	**2**	**3**	**4**	**5**	**6**		**(%)**
Fuzzy Group 1	8	0	0	0	0	0	8	100
2	0	9	1	0	0	0	10	90
3	0	1	8	0	0	0	9	89
4	0	0	0	5	0	1	6	83
5	0	0	0	0	12	0	12	100
6	0	0	0	0	0	6	6	100

Jack knife validation in the CAP procedure was used to assess overall and group-wise accuracy of the six groups determined by fuzzy clustering. Classification accuracy of all groups is good, indicating that the groups determined by the fuzzy clustering are robust.

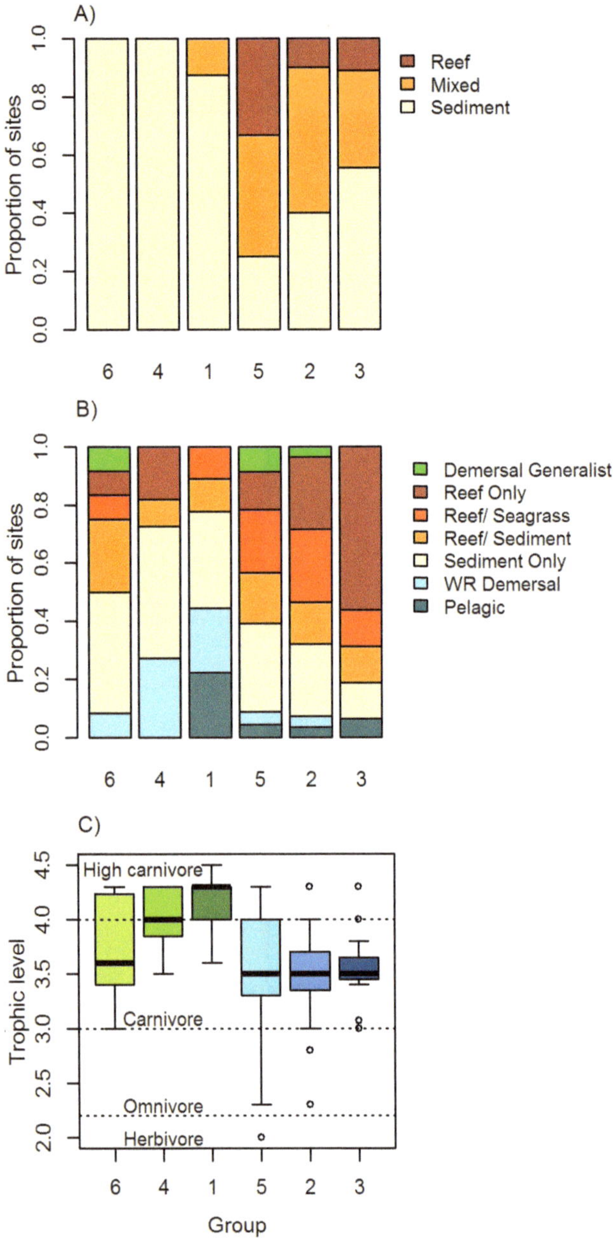

Figure 2. Characteristics of the six assemblage groups identified using fuzzy-clustering. A) Proportion of sites within each group classified as sediment, mixed or reef habitat based on the BRUVs footage; B) the habitat preference of species within each assemblage group; C) boxplot of the trophic level of species within each assemblage group; the box represents the 1st and 3rd quartiles, and circles denote potential outliers. In all panels, groups are ordered based on dominant habitat contained within groups (sediment versus mixed reef/reef) and from shallow to deep within each broad habitat type (based on spatial distribution of groups presented in Fig. 5) In plot C) sediment- associated groups are coloured green and reef-associated groups, blue.

Fish assemblages present in the CMR

Sixty BRUV deployments were completed in phase two, of which 51 yielded video footage of sufficient quality for scoring fish assemblage composition. Overall, 45 species were observed, with a total of 1,837 individuals recorded (at Max N) (Table 1). Of these species, 66% are endemic to Australia. Leatherjackets (Monacanthidae) were the most numerically abundant group, and

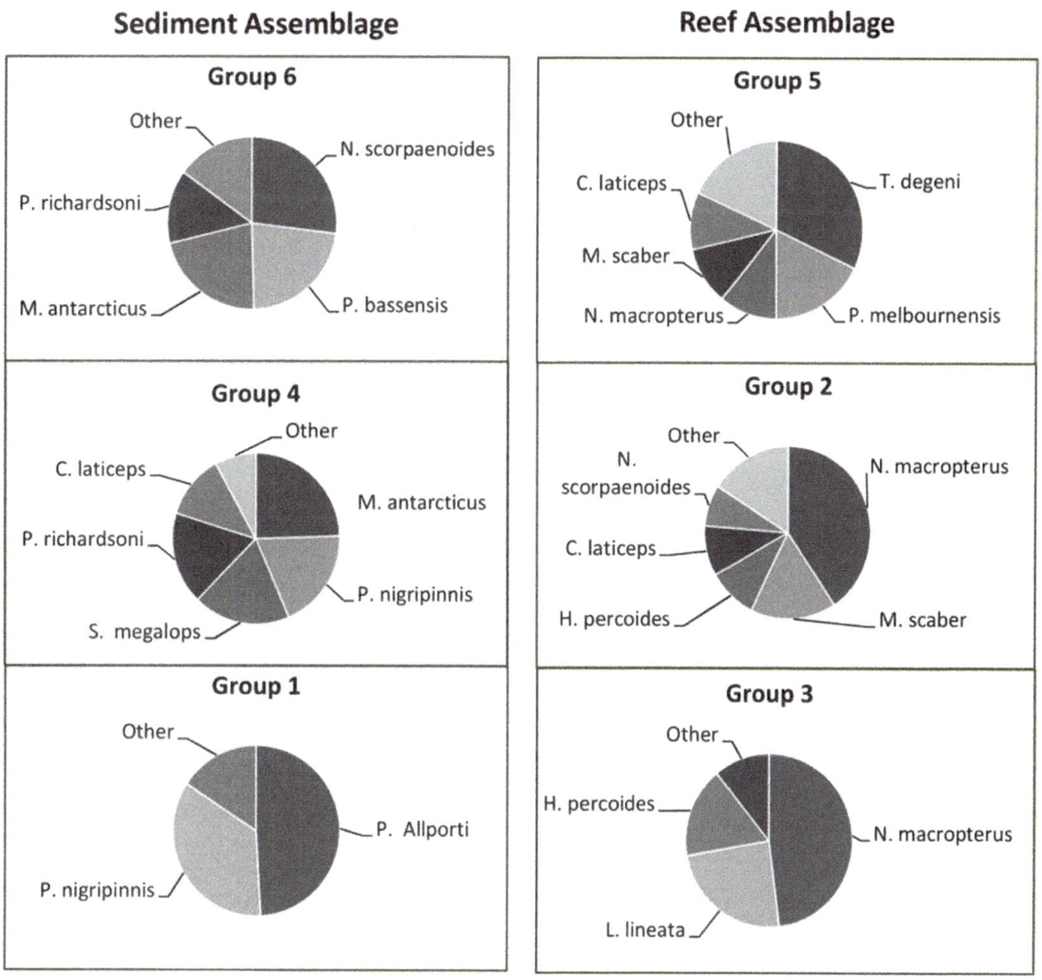

Figure 3. Species contributing up to 80% towards the similarity of the predominantly sediment- and predominantly reef-associated groups identified using fuzzy clustering.

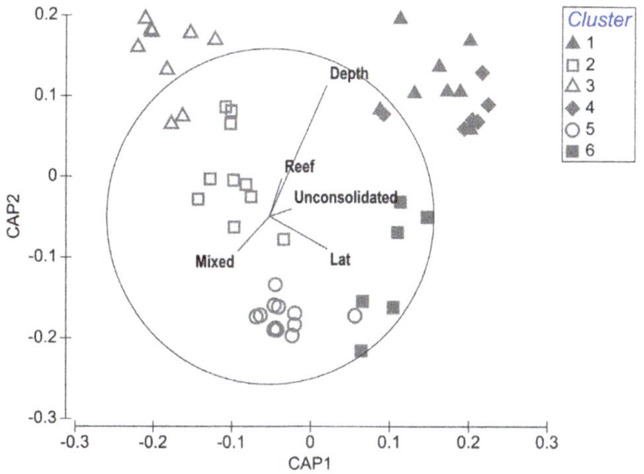

Figure 4. Canonical Analysis of Principle coordinates (CAP) plot discriminating sites based on fuzzy clustering. Each site was assigned to the cluster for which it had the highest probability of membership. Sites are shown as symbols: filled grey symbols represent sand-associated assemblages and open symbols represent reef-associated assemblages. Vectors of all environmental variables are overlaid and are proportional to their correlation with either CAP axis one or two.

comprised mostly *Thamnaconus degeni* (Degan's leatherjacket) and *Meuschenia scaber* (velvet leatherjacket) with, respectively, a total of 471 and 97 individuals recorded and MaxN of 89 and 11. *Nemadactylus macropterus* (jackass morwong), and *Caesioperca Lepidoptera and C. razor* (butterfly and barber perch) were the next most abundant species (379 and 140 individuals recorded and MaxN of 89 and 60). Other abundant species included *Latris lineata* (striped trumpeter), *Parequula melbournensis* (silverbelly), *Pseudolabrus rubicundus* (rosy wrasse), *Helicolenus percoides* (reef ocean perch), and *Cephaloscyllium laticeps* (draughtboard shark). The five most prevalent species were *N. macropertus* (59% of sites), *C. laticeps* (57% of sites), *M. scaber* (53% of sites), *Neosebastes scorpaenoides* (common gurnard perch; 41% of sites) and *T. degeni* (41% of sites).

The structure of fish assemblages surveyed was adequately represented by six groups. The overall classification accuracy of the six assemblage groups was high (94%) and a majority of sites (82%) were also assigned to groups with a high degree of confidence (greater than 0.85 probability of membership). Groups four and three were the least certain, but still had a high classification accuracy of 83% and 89%, respectively (Table 2). In addition, these six groups were statistically meaningful with significantly different multivariate community composition (P = 0.001).

Table 3. Results of Poisson GLMM for species richness recorded in BRUV deployments.

Factor	Co-efficient	Std Error	df	t-value	p-value
Depth	−0.277	0.067	36	−4.1489	0.0002
Mixed Habitat	0.330	0.092	36	3.575	0.0010
Reef Habitat	0.382	0.121	36	3.169	0.0031

Predictor variables have been scaled and centered and co-efficients are on the scale of the link function (log). Spatial clusters introduced as part of the sampling design were included as a random effect and the effects of habitat type (sediment, mixed or reef) are presented relative to sediment habitat.

The six assemblage groups were broadly aligned with habitats observed in the BRUV footage. Sites belonging to three groups (groups 1, 4, 6) contained exclusively or predominately sediment habitat (Fig. 2A) and contained greater numbers of species with a preference for sediment habitats. However, species with a strong affinity for reef were also found in the predominantly sediment-dominated habitat group, with groups 4 and 6 containing disproportionately more reef species than suggested by the available habitat (Fig. 2B). Sediment- associated assemblages (particularly groups 1 and 4) also contained a relatively high proportion of wide-ranging demersal species. Groups 2, 3, and 5 contained moderate to high proportions of mixed reef and reef habitat with the highest proportion of reef-associated species observed in group 3 (Fig. 2A, B).

Species representing a range of trophic levels were recorded in BRUV footage. Carnivores were the most numerous species, as expected given the use of baits. Even so, the average trophic level of species differed between the six assemblage groups (df = 5, F = 5.46, P<0.001; Fig. 2C). Groups 1 and 4 (sediment groups) contained species with significantly higher trophic level (higher carnivores) than groups 2 and 3 (mixed-reef) which contained primarily carnivores. The remaining two groups (5 and 6) contained species from a relatively wider range of trophic levels (Fig. 2C) with group 5 represented by the largest range in values and the only herbivore observed.

The characteristic species of each assemblage group (making the highest contributions to within-group similarity) re-enforced the broad habitat-related patterns described above (Fig. 3). Sediment-associated species such as flathead (*Platycephalus bassensis, Neoplatycephalus richardsoni*) and sharks (*Mustelus antarcticus, Squalus megalops*) as well as the common gurnard perch (*Neosebastes scorpaenoides*) and cucumber fish (*Paraulopus nigripinnis*) were characteristic of groups 4 and 6. Group 1 also consisted of sediment-associated fish (eastern- barred grubfish - *Parapercis allporti*- as well as *P. nigripinnis*), but was differentiated from groups 4 and 6 by relatively low abundances and very few species. Of the predominantly mixed-reef associated groups, groups 2 and 3 were both dominated by jackass morwong (*Nemadactylus macropterus*). Group 2 was also composed of a variety of others species including the velvet leatherjacket (*Meuschenia scaber*), reef ocean perch (*Helicolenus percoides*) and draughtboard sharks (*Cephaloscyllium laticeps*). Group 3 was further characterised by relatively high abundances of striped trumpeter (*Latris lineata*) as well as reef ocean perch. Group 5 was differentiated from other groups by large numbers of leatherjackets (primarily *Thamnaconus degeni*), but was also characterised by silverbelly (*Parequula melbournensis)* and jackass morwong.

Relationships with environmental variables

Environmental factors correlated well with the fish assemblage groupings (Fig. 4). Depth delineated groups most strongly (a correlation with CAP axis 2 of 0.78) and groups 1, 4 and 3 were generally found at greater depth. Substratum type also influenced assemblages, with sediment-associated groups (1, 4 and 6) falling on the right hand-side of the CAP plot, confirming the patterns seen in the composition of species above. Latitude was also moderately correlated with axis one (0.35) and appears to separate the mixed/reef associated groups 2, 3 and 5.

Species richness was also correlated with environmental factors. Richness decreased with depth and mixed and reef substratum also supported more species than sediments (Table 3). Latitude however, was not correlated with species richness (Table 3). This resulted in spatial patterns as depicted in Fig. 5, where assemblages are generally richer towards the shallow, western side of the reserve.

Spatial distribution of fish assemblages across CMR

Fish assemblages showed distinct spatial patterns across the Flinders CMR (Fig. 5). When combined with depth, the assemblages can be broadly categorised as: shallow, reef-associated (group 5); intermediate reef-associated (group 2); deeper, reef-associated (group 3); shallow sediment-associated (group 6); and deeper, reef-associated (groups 1 and 4). While sites within the same spatial cluster often belonged to the same fish assemblage, this was not always the case. Clusters of sites close to the shelf edge and in the south of the reserve were often more heterogeneous, composed of two to three different assemblages within a 1 km radius.

Discussion

This study trialled the use of BRUVs and a spatially- balanced survey design to provide the first quantitative description of demersal fish species and assemblages occurring the multiple use zone of the Flinders CMR shelf. Non-extractive sampling with BRUVs identified characteristic species (including jackass morwong, leatherjackets, draughtboard sharks, reef ocean perch and common gurnard perch), several species of commercial interest (including gummy sharks, reef ocean perch, striped trumpeter and jackass morwong) and species from trophic groups ranging from omnivores through to higher predators. While many of the species present in the study area are widely distributed across southern and/or south-eastern Australia, the majority are endemic to Australian waters. This highlights the unique biodiversity represented in temperate Australia, primarily as a result of its relatively isolated evolutionary history [39,40], and reinforces the global importance of conserving representative fauna.

Spatial and environmental patterns in fish assemblages

Six assemblages with a distinct spatial pattern were observed at sites across the reserve. Many species contributed to more than one assemblage and differences in the relative abundance of

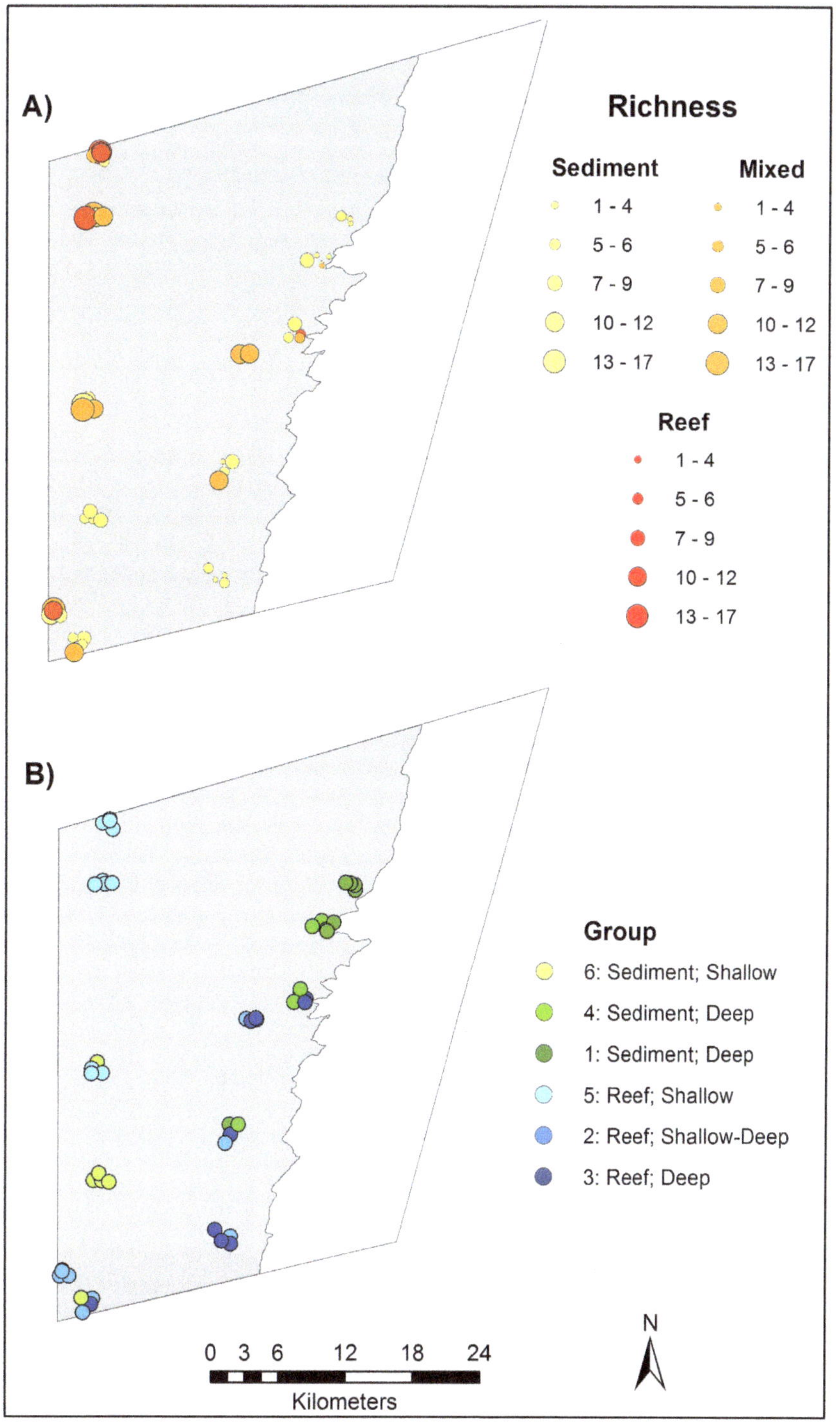

Figure 5. Spatial distribution of A) species richness B) fish assemblage groups on the Flinders CMR shelf (shaded grey). Increasing size of symbols in A) indicate increasing species richness. Symbols are colour coded according to the observed substratum type in BRUV footage: yellow = sediment; orange = mixed; red = reef. Assemblages in B) are coded by colour, with predominantly sediment-associated assemblages coloured green and predominantly reef-associated assemblages coloured blue.

species defined the differences between several assemblages. Assemblages were structured by depth and habitat type, and could be heterogeneous at relatively small scales (i.e. within a sampling cluster).

Depth-related assemblage transitions were most distinct and occurred in waters greater than approximately 80 m for sediment-associated assemblages and approximately 60 m for reef-associated assemblages, which is largely consistent with other studies on the southeast Australian shelf [41]. While some species were found across almost the full depth range of the study area (e.g. jackass morwong and the flathead, *Platycephalus richardsoni*) several species characterised particular substratum and depth zones. Leatherjackets (*T. degeni* in particular) for example, were characteristic of shallow reef sites; striped trumpeter were indicative of deeper, reef-associated assemblages; and cucumber fish were indicative of deeper, sediment-associated assemblages. These results accord with other studies where depth has a major influence on the distribution of fish assemblages [17,42].

As well as influencing assemblage composition, depth also influenced the diversity and ecology of assemblages. Species richness generally declined with depth consistent with observations elsewhere [39,43,44] and the trophic range of species occurring in shallow assemblages was generally greater than deeper assemblages. In reef-associated assemblages for example, larger trophic ranges of the shallow water assemblage were due to the co-occurrence of species such as leatherjackets (many of which are omnivorous) and draughtboard sharks (which are carnivores). Similarly, other studies have noted trophic-related changes with depth, such as increases in body size [42] and a transition towards higher –order predators [34].The combination of assemblage level patterns and changes in ecology and diversity with depth as seen here, have been used to suggest that the edge of the continental shelf forms an important faunal break. Consequently the shelf and slope environments should be considered separate management units [8,31] and not subsumed within the same management zone as currently occurs within the Flinders CMR.

Although a ubiquitous pattern, the processes involved in the depth structuring of benthic fish assemblages can be varied. In the shelf environments of the Flinders CMR, we suggest these are partly attributable to the strong tidal currents which affect the shallow, western section on the reserve [22]. In this region, low-profile reefs are dominated by tall biogenic structures with flexible forms such as sea whips and erect branching sponges (personal observation) that are indicative of high energy environments. This 3D biogenic structure is much reduced towards the deeper, eastern extent of the study area. These differences in reef structure [45], as well as physical limitations posed by increased current velocities [46], may have a large influence on the assemblages found in shallow regions of the reserve. Other physical and oceanographic conditions also co-vary with depth. For example, light availability decreases with depth, which affects primary productivity and therefore the range of food sources available to fish and appears to correspond with the observed changes in the trophic composition of assemblages. In addition, the oceanography in this region is complex, driven largely by inter-annual and inter-seasonal variability in the southward penetration of the warm-water, East Australian Current (EAC). The EAC can incur on the shelf or form fronts at the shelf break [24] and affects processes such as recruitment [47] and primary productivity [48] that are in turn likely influence the composition of fish assemblages, particularly on the outer shelf. Regardless of the proximal processes involved, the depth structuring of assemblages observed in the study area suggests that future sampling would benefit by breaking the region into shallow and deep strata around the 70 m depth mark.

Habitat or substratum type was also influential in discriminating between the assemblages observed in the Flinders CMR shelf. Despite preferentially sampling clusters of sites surrounding our initial mixed/reef sites, a large number of all sites surveyed contained only sediments, at least within the field of view of the BRUVs. This occurred because low-profile shelf reefs in the CMR are small scale features (tens of meters) and their distribution is highly patchy. Sediment sites were generally less species rich than those found on mixed or reef substrata. However, the composition of these sites varied enough to form three distinct assemblage groups, primarily on the basis of their depth distribution. The lower diversity of sediment assemblages has been attributed to lower habitat complexity with fewer niches to support co-existing species [49]. Still, sediment-associated assemblages contained some species with affinity for reef habitats. This may be because reef habitat was present outside the BRUVs field of view but within the swimming distance of fishes resulting in a 'halo' effect where species with an affinity for reef and sediments co-occur [49]. In subtropical shallow reefs this halo can extend 200 m from the reef edge [49]. The importance of local substratum in structuring fish assemblages highlights the need to develop comprehensive, high resolution habitat maps for the reserve. Assembling these habitat maps for marine reserves will require a commitment to strategic mapping of the seabed using multibeam sonar, so that over time complete coverage is achieved. For the continental shelf, this mapping could take years to decades [50]. In the meantime, management will need to adopt spatially representative sampling such as presented in this study to provide habitat and biodiversity inventories and to underpin monitoring programs.

Geomorphic features of the seafloor, such as submarine canyons, can substantially alter water flows, enhance local productivity [51] [52] and influence fish assemblage patterns [53]. Canyons are a conservation feature of the Flinders CMR and several canyons intersect the shelf edge. Canyons were represented in our sampling by a cluster of sites near a canyon head in the north of the survey area. Assemblages adjacent to canyon heads were not however, unique to canyon heads and were also found in other deeper sections of the shelf. Neither were these assemblages particularly species rich. Despite our preliminary findings, a better understanding of the ecological significance of canyon heads in this system will require a more targeted and intensive sampling strategy within and around the canyon head systems, e.g. [51].

Implications for Monitoring and Management

An objective of the Flinders CMR – and the larger Southeast Marine Reserve Network of which it is part - is to protect the biodiversity, natural and cultural values of the region. These include "representative examples of the ecosystems, communities and habitats" associated with the Tasmanian Shelf and Southeast Shelf Transition Provinces [21]. While these values are defined in broad terms from the national bioregionalisation, these have yet to be translated into inventories of conservation values, with specific metrics for monitoring their status. This study is a first step towards developing both an inventory of biological values, and elements of an effective biological monitoring program, including relevant environmental influences, for fish communities represented in the shelf component of the Flinders CMR. We suggest that representative demersal fish communities include those found on dynamic, low- profile reefs in both shallow (<70 m) and deep (70–160 m) areas of the shelf as well as those found on sediments in the same depth strata, and here we have characterised each of these assemblages. Several prevalent commercial and recreational species, including striped trumpeter and jackass morwong, which are likely to be responsive to management actions may be good candidates as indicator species and this work will provide the basis for further research into deriving specific monitoring targets.

We used a non-standard sampling design, GRTS, to survey fish assemblages and to satisfy the objectives of the broader survey of the Flinders CMR. Due to the cost and logistics associated with sampling in offshore and/or remote environments, the use of survey designs that are efficient, representative and flexible enough to accommodate multiple objectives is likely to become increasingly important. This is particularly true when simultaneously considering the inventory and monitoring needs of multiple reserves within networks. In the present study, the GRTS sampling ensured that clusters of reef and sediment sites sampled with BRUVs are well spread out across the reserve (within the distribution of each habitat type) and consequently encompass broad spatial and environmental gradients. This in turn provides confidence in the patterns observed and provides a better basis for additional or repeated sampling in a monitoring context. The site-ordered sampling of GRTS has the potential disadvantage of using ship time inefficiently if there is a large cumulative distance between deployments. This would have been the case in our study, but was circumvented by clustering samples around a subset of original sites. Conversely, the site ordering of GRTS allows dynamic altering of samples according to situations in the field, while maintaining spatial balance within strata.

The GRTS methodology will also be useful when choosing samples for future sampling, typically as part of an adaptive monitoring program. The master list of samples, that encompasses all possible sampling locations [28], can be subset in different ways by varying the inclusion probabilities for factors such as habitat types or depth, to accommodate new research questions or changing management objectives. While such changes lead to higher or lower sampling rates for different habitats or strata, the set of sites selected always maintains spatial balance within the strata. Inclusion probabilities are taken into account when deriving representative estimates of the occurrence, proportions or abundance of metrics or indicators across the entire sampling area [18]. Once suitable indicator species and metrics have been selected, the GRTS master sample can provide the basis for rotating panel designs [28] for repeat sampling that aims to detect changes in the abundance of these species, and possibly also major shifts in the distribution of assemblage types. Another advantage of GRTS in this respect, is its use of a local neighbourhood variance estimator, which is generally unbiased and more precise than variance estimators from other types of sampling designs [26]. Future sampling within the context of monitoring however, will need to incorporate comparable areas outside of the multiple use management zone of the Flinders CMR. In addition sufficient sample sizes will be needed to provide enough power to distinguish between natural variability and biological trends in order to be able to assess whether current management controls are effective in achieving their stated objectives.

The use of baited systems to survey fish communities has sometimes been criticised because of the bias towards sampling scavenging and predatory species [54]. However, all fish sampling methods have selectivity biases, and alternatives for sampling deep continental shelf fishes (e.g. trawls, meshnets and traps [41]) are extractive (and some destructive to the habitats the reserve is intended to protect). We recognise that by using BRUVs we have preferentially observed a subset of species that is attracted to, or undeterred by, baits. However, BRUVs data have been shown to clearly discriminate between fish assemblages in a variety of environmental settings [43] and have better statistical power to detect spatial and temporal changes in assemblage structure and abundances of individuals than unbaited methods [54]. In the future, we may complement BRUVs with other non-extractive video techniques (e.g. forward looking cameras on remotely or autonomously operated vehicles) to capture other components of the community. Never-the-less, in our present study BRUVs successfully observed a broad range of species, including key ecological species and potential indicator species. Standardised methodologies, in combination with robust sampling designs, will be essential for ongoing monitoring and assessment across multiple reserves in Australia's reserve network. Finally, the challenges we outline for the inventory and long-term monitoring of large and remote marine reserves are not unique to the Australian reserve network, nor exclusively to marine reserves. Therefore the approach that we have demonstrated may be usefully applied in other systems and under other management scenarios.

Supporting Information

Figure S1 The location of the forty GRTS sites sampled across the Flinders CMR shelf in phase one. Sites are coloured according to the broad habitat type: sediment (yellow); mixed, low-profile reef and sediments (red); canyon head (blue) recorded during phase one of sampling. Sites are labelled with their GRTS site number which, when sampled in order, represents a spatially balanced sample. The table shows the subset of sites sampled as the basis of clusters in the phase two sampling with BRUVs. These sites are the first three GRTS sites classified as sediment and the first eight classified as mixed reef.

Data S1 Matrix of the species observed at each BRUV deployment. Environmental variables used in analyses are also included.

Data S2 Attributes of species used in analysis, including trophic level, habitat preference and endemicity. Summary statistics on the prevalence and abundance of species across the survey area are also given.

Acknowledgments

The authors would like to thank the crew of *Challenger*, Matt Francis and Tim Green, for their expertise and dedication during fieldwork, Richard Zavalas for assistance with fieldwork and video scoring, Johnathan Kool for assistance with fieldwork, and Vanessa Lucieer for assistance in the planning phase. SN publishes with permission of the Chief Executive Officer of Geoscience Australia.

Author Contributions

Conceived and designed the experiments: NAH NB EL JMD SN KRH. Performed the experiments: NAH NB JH SN. Analyzed the data: NAH NB EL JH KRH. Contributed reagents/materials/analysis tools: NAH NB EL JH SN AW KRH. Wrote the paper: NAH NB EL JMD SN AW KRH.

References

1. Halpern BS, Walbridge S, Selkoe KA, Kappel CV, Micheli F, et al. (2008) A global map of human impact on marine ecosystems. Science 319: 948–952.
2. Halpern BS, Longo C, Hardy D, McLeod KL, Samhouri JF, et al. (2012) An index to assess the health and benefits of the global ocean. Nature: 615–620.
3. Day J, Dudley N, Hockings M, Holmes G, Laffoley D, et al. (2012) Guidelines for applying the IUCN Protected Area Management Categories to Marine Protected Areas. Gland, Switzerland: IUCN. 36 p.
4. Kemp J, Gibson R, Atkinson R, Gordon J, Hughes R, et al. (2012) Measuring the performance of spatial management in marine protected areas. Oceanography and Marine Biology: An Annual Review 50: 287–314.
5. Gubbay S, editor (1995) Marine Protected Areas: principles and techniques for management. London: Chapman and Hall.
6. CBD (1992) Convention on Biological Diversity. 5 June 1992 Rio de Janerio (Brazil).
7. CBD (1995) The Jakarta Mandate on Marine and Coastal Biological Diversity. Decisions of the Second Meeting of the Conference of the Parties to the Convention on Biological Diversity. Jakarta, Indonesia, 6–17 November 1995: UNEP.
8. Spalding MD, Meliane IN, Milam A, Fitzgerald C, Hale LZ (2013) Protecting marine spaces: Global targets and changing approaches. Ocean Yearbook Online 27: 213–248.
9. Edgar GJ, Stuart-Smith RD, Willis TJ, Kininmonth S, Baker SC, et al. (2014) Global conservation outcomes depend on marine protected areas with five key features. Nature advance online publication.

10. Barr LM, Possingham HP (2013) Are outcomes matching policy commitments in Australian marine conservation planning? Marine Policy 42: 39–48.
11. ANZECC (1998) Guidelines for establishing the National Representative System of Marine Protected Areas. Canberra: Australian and New Zealand Environment and Conservation Council (ANZECC) Task Force on Marine Protected Areas.
12. Commonwealth of Australia (2006) A Guide to the Integrated Marine and Coastal Regionalisation of Australia Version 4.0. In: Heritage DotEa, editor. Canberra, Australia.
13. Lucieer V, Lamarche G (2011) Unsupervised fuzzy classification and object-based image analysis of multibeam data to map deep water substrates, Cook Strait, New Zealand. Continental Shelf Research 31: 1236–1247.
14. Bailey DM, King NJ, Priede IM (2007) Cameras and carcasses: historical and current methods for using artificial food falls to study deep-water animals. Marine Ecology Progress Series 350: 179–191.
15. Willis TJ, Babcock RC (2000) A baited underwater video system for the determination of relative density of carnivorous reef fish. Marine and Freshwater Research 51: 755–763.
16. Langlois TJ, Harvey ES, Meeuwig JJ (2012) Strong direct and inconsistent indirect effects of fishing found using stereo-video: Testing indicators from fisheries closures. Ecological Indicators 23: 524–534.
17. Harvey ES, Cappo M, Kendrick GA, McLean DL (2013) Coastal fish assemblages reflect geological and oceanographic gradients within an Australian zootone. Plos One 8.
18. Stevens DL, Olsen AR (2004) Spatially balanced sampling of natural resources. Journal of the American Statistical Association 99: 262–278.
19. Olsen AR, Kincaid TM, Payton Q (2012) Spatially balanced survey designs for natural resources. In: Gitzen RA, Millspaugh JJ, Cooper AB, Licht DS, editors. Design and Analysis of Long-term Ecological Monitoring Studies. Cambridge, England: Cambridge University Press. pp. 126–150.
20. Commonwealth of Australia (2006) A Guide to the Integrated Marine and Coastal Regionalisation of Australia Version 4.0. Canberra, Australia: Department of the Environment and Heritage. 16 p.
21. Director of National Parks (2013) South-east Commonwealth Marine Reserves Network management plan 2013–23. Director of National Parks, Canberra.
22. Fandry C (1983) Model for the three-dimensional structure of wind-driven and tidal circulation in Bass Strait. Marine and Freshwater Research 34: 121–141.
23. National Tidal Centre (2014) Tide predictions for Australia, South Pacific and Antarctica.
24. Harris G, Nilsson C, Clementson L, Thomas D (1987) The water masses of the East Coast of Tasmania: Seasonal and interannual variability and the influence on phytoplankton biomass and productivity. Australian Journal of Marine and Freshwater Research 38: 569–590.
25. Whiteway TG (2009) Australian Bathymetry and Topography Grid. Canberra, Australia: Geoscience Australia. 46 p.
26. Stevens DL, Olsen A (2003) Variance estimation for spatially balanced samples of environmental resources. Environmetrics 14: 593–610.
27. Dambacher JM, Jones KK, Larsen DP (2009) Landscape-level sampling for status review of great basin redband trout. North American Journal of Fisheries Management 29: 1091–1105.
28. Larsen D, Olsen A, Stevens D (2008) Using a Master Sample to Integrate Stream Monitoring Programs. Journal of Agricultural, Biological, and Environmental Statistics 13: 243–254.
29. Harvey ES, Newman SJ, McLean DL, Cappo M, Meeuwig JJ, et al. (2012) Comparison of the relative efficiencies of stereo-BRUVs and traps for sampling tropical continental shelf demersal fishes. Fisheries Research 125–126: 108–120.
30. Cappo M, De'ath G, Speare P (2007) Inter-reef vertebrate communities of the Great Barrier Reef Marine Park determined by baited remote underwater video stations. Marine Ecology Progress Series 350: 209–221.
31. Seager J (2014) Transect Measure. 2.30 ed: SeaGIS.
32. Maechler M, Rousseeuw P, Struyf A, Hubert M, Hornik K (2013) cluster: Cluster Analysis Basics and Extensions. R package. 1.14.4 ed.
33. R Development Core Team (2008) R: A language and environment for statistical computing. Vienna, Austria: R Foundation for Statistical Computing.
34. Anderson MJ, Gorley RN, Clarke KR (2008) PERMANOVA+ for PRIMER: Guide to software and statistical methods. Plymouth, UK: PRIMER-E. 214 p.
35. Edgar GJ (2008) Australian Marine Life: the plants and animals of temperate waters. New Holland, Sydney. Sydney: Reed New Holland.
36. Gomon MF, Bray DJ, Kuiter RH (2008) Fishes of Australia's Southern Coast. Chatswood, Australia: Reed New Holland.
37. (2014) FishBase. In: Froese R, Pauly D, editors. Available: www.fishbase.org: World Wide Web electronic publication.
38. Venables WN, Ripley BD (2002) Modern Applied Statistics with S. New York: Springer.
39. Last PR, White WT, Gledhill DC, Pogonoski JJ, Lyne V, et al. (2011) Biogeographical structure and affinities of the marine demersal ichthyofauna of Australia. Journal of Biogeography 38: 1484–1496.
40. Phillips J (2001) Marine macroalgal biodiversity hotspots: why is there high species richness and endemism in southern Australian marine benthic flora? Biodiversity and Conservation 10: 1555–1577.
41. Williams A, Bax N (2001) Delineating fish-habitat associations for spatially based management: an example from the south-eastern Australian continental shelf. Marine and Freshwater Research 52: 513–536.
42. Fitzpatrick BM, Harvey ES, Heyward AJ, Twiggs EJ, Colquhoun J (2012) Habitat specialization in tropical continental shelf demersal fish assemblages. PLoS ONE 7: e39634.
43. Zintzen V, Anderson MJ, Roberts CD, Harvey ES, Stewart AL, et al. (2012) Diversity and composition of demersal fishes along a depth gradient assessed by Baited Remote Underwater Stereo-Video. PLoS ONE 7: e48522.
44. Chatfield BS, Van Niel KP, Kendrick GA, Harvey ES (2010) Combining environmental gradients to explain and predict the structure of demersal fish distributions. Journal of Biogeography 37: 593–605.
45. Gratwicke B, Speight MR (2005) The relationship between fish species richness, abundance and habitat complexity in a range of shallow tropical marine habitats. Journal of Fish Biology 66: 650–667.
46. Fulton CJ, Bellwood DR (2004) Wave exposure, swimming performance, and the structure of tropical and temperate reef fish assemblages. Marine Biology 144: 429–437.
47. Bruce BD, Evans K, Sutton CA, Young JW, Furlani DM (2001) Influence of mesoscale oceanographic processes on larval distribution and stock structure in jackass morwaong (*Nemodactlus macropertus:* Cheilodactylidae). ICES Journal of Marine Science 58: 1072–1080.
48. Bax NJ, Burford M, Clementson L, Davenport S (2001) Phytoplankton blooms and production sources on the south-east Australian continental shelf. Marine and Freshwater Research 52: 451–462.
49. Schultz AL, Malcolm HA, Bucher DJ, Smith SDA (2012) Effects of reef proximity on the structure of fish assemblages of unconsolidated substrata. Plos One 7.
50. Kloser RJ, Williams A, Butler A (2006) Exploring surveys of seabed habitats in Australia's deep ocean using remote sensing – needs and realities. In: Todd BJ, Greene HG, editors. Mapping the Seafloor for Habitat Characterization. pp. 93–109.
51. Currie DR, McClatchie S, Middleton JF, Nayar S (2012) Biophysical factors affecting the distribution of demersal fish around the head of a submarine canyon off the Bonney Coast, South Australia. Plos One 7.
52. Vetter EW, Smith CR, De Leo FC (2010) Hawaiian hotspots: enhanced megafaunal abundance and diversity in submarine canyons on the oceanic islands of Hawaii. Marine Ecology 31: 183–199.
53. Leathwick JR, Elith J, Francis MP, Hastie T, Taylor P (2006) Variation in demersal fish species richness in the oceans surrounding New Zealand: an analysis using boosted regression trees. Marine Ecology Progress Series 321: 267–281.
54. Harvey E, Cappo M, Butle RJ, Hall N, Kendrick G (2007) Bait attraction affects the performance of remote underwater video stations in assessment of demersal fish community structure. Marine Ecology Progress Series 350: 245–254.

Horizontal and Vertical Distribution of Marine Virioplankton: A Basin Scale Investigation Based on a Global Cruise

Yantao Liang[1,2,3], Li Li[1,2], Tingwei Luo[1,2], Yao Zhang[1,2], Rui Zhang[1,2]*, Nianzhi Jiao[1,2]*

1 State Key Laboratory of Marine Environmental Science, Xiamen University, Xiamen, China, **2** Institute of Marine Microbes and Ecospheres, Xiamen University, Xiamen, China, **3** Qingdao Institute of BioEnergy and BioProcess Technology, Chinese Academy of Sciences, Qingdao, China

Abstract

Despite the fact that marine viruses have been increasingly studied in the last decade, there is little information on viral abundance and distribution on a global scale. In this study, we report on a global-scale survey covering the Pacific, Atlantic, and Indian Oceans on viral distribution using flow cytometry. Viruses were stained with the SYBR Green I, which targets only dsDNA viruses. The average viral abundance was $1.10\pm0.73\times10^7$ ml^{-1} in global surface oceans and decreased from the areas with high chlorophyll concentration (on average, $1.47\pm0.78\times10^7$ ml^{-1}) to the oligotrophic subtropical gyres (on average, $6.34\pm2.18\times10^6$ ml^{-1}). On a large-spatial-scale, viruses displayed significant relationships with both heterotrophic and autotrophic picoplankton abundance, suggesting that viral distribution is dependent on their host cell abundance. Our study provided a basin scale pattern of marine viral distributions and their relationship with major host cells, indicating that viruses play a significant role in the global marine ecosystem.

Editor: Andrew C. Singer, NERC Centre for Ecology & Hydrology, United Kingdom

Funding: This work was supported by the National Key Basic Research Program of China (grant No. 2013CB955700), the SOA project (GASI-03-01-02-05), and the 863 Program (2012AA092003). R.Z. was supported by NSFC (41376132) and the Fundamental Research Funds for the Central Universities (2012121052). The funders had no role in study design, data collection and analysis, decision to publish, or preparation of the manuscript.

Competing Interests: The authors have declared that no competing interests exist.

* Email: ruizhang@xmu.edu.cn (RZ); jiao@xmu.edu.cn (NJ)

These authors contributed equally to this work.

¤ Current address: Department of Biology and Chemistry, City University of Hong Kong, Kowloon, Hong Kong SAR

Introduction

The interests in marine viruses have increased because they are the most abundant group of biological entities in the world's oceans. The estimated 10^{30} viruses in the ocean contain more carbon than 75 million blue whales [1]. Every second, approximately 10^{23} viral infections occur in the ocean [2]. These infections are a substantial source of mortality in a range of organisms including predominantly autotrophic and heterotrophic microbes. The prokaryotic mortality caused by viral lysis is equal to or sometimes exceeds grazing-induced mortality in marine environments [3,4]. This process of lytic activities causes the release of dissolved organic carbon (DOC), preventing its flow to higher trophic levels through a "viral shunt" [5,6]. A previous study indicated that 6–26% of the photosynthetically fixed carbon is "shunted" to the dissolved organic matter (DOM) pool by viral lysis of cells in marine pelagic systems [5]. In addition, the prokaryotic cell components released by viruses might be one of the most important sources of recalcitrant dissolved organic matte (RDOM) in the microbial carbon pump (MCP), which is a conceptual framework of long-term carbon storage in the global ocean [7]. In general, viruses modulate the transfer and transformation of carbon and nutrients in marine microbial food webs and play an important role in biogeochemical cycles and energy flow in the ocean and directly or indirectly affect the global ecosystem.

Over the past two decades, viral ecology has been studied in a wide range of marine habitats including estuary, coastal sea, open sea and extreme environments such as deep oceans [3]. These studies indicate that viruses are truly ubiquitous and their impact on the microbial community can be variable. However, our understanding of virioplankton in the global ocean is still limited, partially due to methodological inconsistency. Currently, there are three major techniques for the numeration of viral particles in environmental samples: transmission electron microscopy (TEM), epifluorescence microscopy (EFM) and flow cytometry (FCM) [3,8]. Although the data sets obtained by different methods may be positively correlated, they are incomparable with each other. It has been reported that EFM produces significantly higher viral numbers than TEM [9,10]. Even using the same technique, different protocols cause significant effects on the accuracy of virus enumeration [8,11]. Thus, the methodological discrepancy among different studies might to be one of the major difficulties for estimating marine viral abundance and determining the ecological implications on the global scale. In addition, there are still arguments regarding the factors controlling viral distribution and

Table 1. Mean values (± standard deviations) of microbial parameters in the surface waters of the study areas.

Provinces	Parameters[a]	Proc*	Syn*	Euk*	HetProk*	Prokaryotes*	Picoplankton*	HFV*	LFV*	Total viruses*	HTP %*	HVEukR*	LVProkR*	VPR*
Coastal/shelf/ upwelling	Abundance	$1.32\pm0.90\times10^5$	$3.44\pm3.52\times10^4$	$7.50\pm4.70\times10^3$	$8.73\pm3.96\times10^5$	$1.02\pm0.36\times10^6$	$1.02\pm0.36\times10^6$	$2.08\pm0.92\times10^6$	$1.26\pm0.72\times10^7$	$1.47\pm0.78\times10^7$	15.01 ± 4.70	372 ± 279	12.84 ± 5.80	14.93 ± 6.46
Gyres	Abundance	$9.26\pm7.74\times10^4$	$3.46\pm2.72\times10^3$	$1.00\pm0.70\times10^3$	$4.59\pm0.90\times10^5$	$5.55\pm1.23\times10^5$	$5.56\pm1.23\times10^5$	$6.25\pm3.29\times10^5$	$5.71\pm1.99\times10^6$	$6.34\pm2.18\times10^6$	9.78 ± 3.63	793 ± 597	10.59 ± 3.99	11.68 ± 4.21

[a]Unit of abundance: cells ml^{-1}. Abbreviations: Proc, *Prochlorococcus*; Syn, *Synechococcus*; Euk, picoeukaryotes; HetProk, heterotrophic prokaryotes; HFV, higher fluorescence virus; LFV, low fluorescence virus; HTP, the percent of HFV to total viruses; HVEukR, the ratio of HFV and picoeukaryotes abundance; LVProkR, the ratio of LFV and prokaryotic abundance; VPR, the ratio of total viruses and picoplankton abundance.
*indicated the parameters for Coastal/shelf/upwelling and gyres are statistically different (independent *t* test, P<0.01).

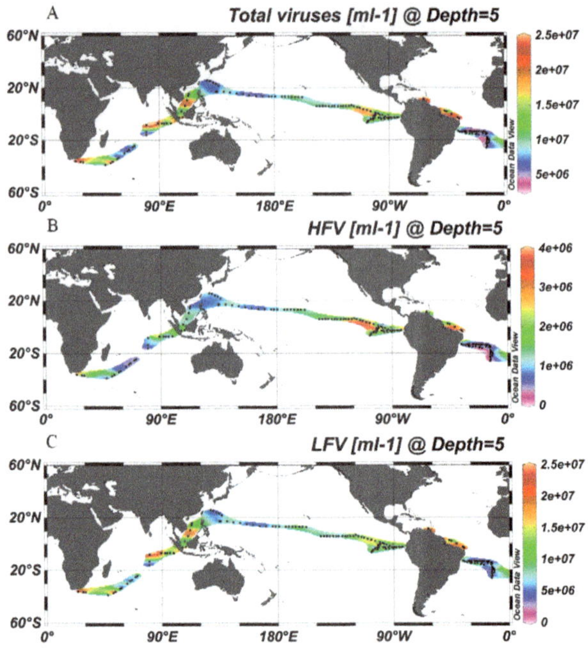

Figure 1. A global distribution pattern of the total viral abundance (A), HFV abundance (B) and LFV abundance (C). The map was generated with Ocean Data View software [40]. Abbreviations: HFV, high fluorescence viruses; LFV, low fluorescence viruses.

which factor plays a more important role on the large spatial scale [12–16]. Moreover, reports of vertical variation of viral abundance in the whole water column, especially in the deep sea, are few [13,15,17–21]. In this study, we performed a systematic survey of the abundances of viruses and their major host cells (picoplankton = heterotrophic prokaryotes + *Prochlorococcus* + *Synechococcus* + picoeukaryotes, Fig. S1) using flow cytometry during a global expedition cruise covering various coastal and oceanic environments. The correlations between viral abundance and host cell abundance or environmental factors were also investigated.

Materials and Methods

Study area and sampling

The samples were collected from a global expedition (December 2010–December 2011) aboard *R/V Ocean No. 1* (Fig. S2). The South China Sea samples were taken during December 2010, the Indian Ocean samples were taken from December 2010 to January 2011, the Atlantic Ocean samples were taken from March to June 2011 and the Pacific samples were taken from July to December 2011. The samples were collected from the vertical profile at 7 to 23 depths from epipelagic (0–200 m), mesopelagic (200–1000 m), and bathypelagic (1000–3150 m) zones at 22 stations including 4 stations in the Indian ocean, 13 stations at the south Atlantic gyre and 5 stations at the eastern equatorial Pacific upwelling area (No specific permissions were required for these locations/activities). The samples (1.5 ml) for flow cytometry analysis were fixed with glutaraldehyde (final concentration: 0.5%), flash-frozen in liquid nitrogen and stored at −80°C prior to analysis [8].

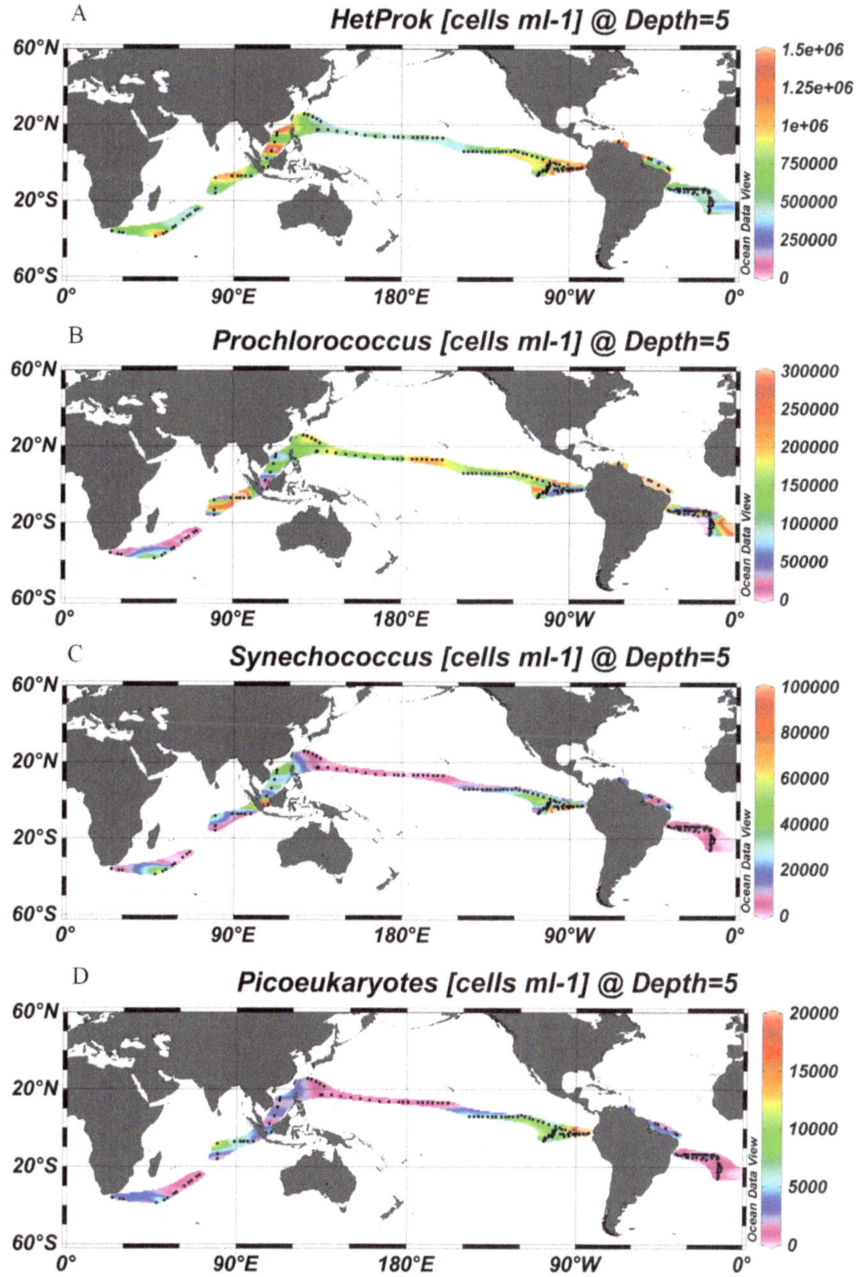

Figure 2. A global distribution pattern of the heterotrophic prokaryotic cell abundance (A), *Prochlorococcus* (B), *Synechococcus* (C) and picoeukaryotes abundance (D). The map was generated with Ocean Data View software [40].

Viral and host cell counting using flow cytometry

Viruses, heterotrophic prokaryotes and autotrophic picoplankton were analyzed on an Epics Altra II flow cytometer (Beckman Coulter, USA) with a 306C-5 argon laser (Coherent, USA). Virus enumeration was performed according to the method of Brussaard [8]. Once thawed at 37°C, the samples were diluted 5 to 50 times in 0.02-μm filtered Tris-EDTA buffer (pH = 8, Sigma-Aldrich). After staining with the SYBR Green I (1/20000 final concentration, Molecular Probes, which targets only dsDNA viruses), the diluted samples were heated for 10 min in the dark at 80°C and then cooled for 5 min prior to analysis. The samples were run at a flow rate of 0.1–1 ml h^{-1} (using HARVARD PHD2000 APPARATUS). For the enumeration of viruses, we set the discriminator to green fluorescence and set all parameters on logarithmic amplification. The typical settings on an Epics Altra II flow cytometer are forward scatter (PMT4, for the accurate detection of the weak signal of forward scatter of picoplankton, we changed the PMT4 as the forward scatter detector) = 400, side scatter (PMT1) = 590, green fluorescence (PMT2) = 720, orange fluorescence (PMT3) = 500, and red fluorescence (PMT5) = 910. The viruses were discriminated on the basis of the green fluorescence and side scatter signal. Usually, two subgroups of viruses were observed in the cytograms (Fig. S3A–C). The group with higher fluorescence was called the HFV group, and the group

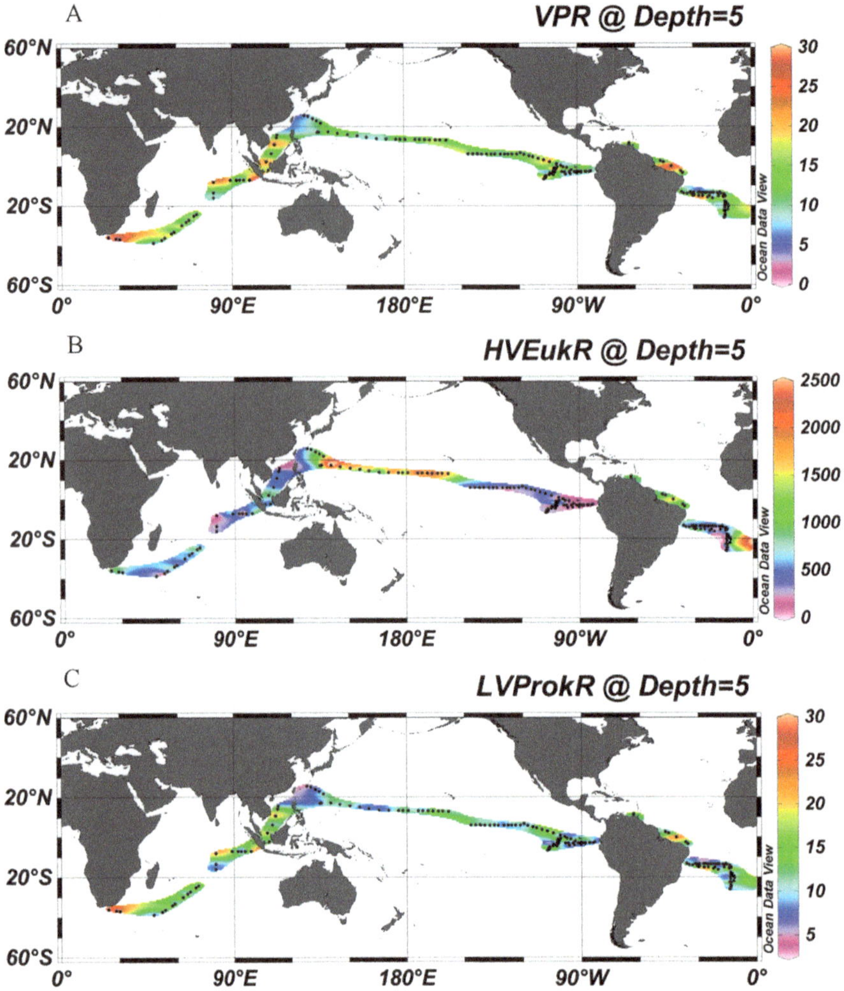

Figure 3. A global distribution pattern of the VPR (A), HVEukR (B) and LVProkR (C). The map was generated with Ocean Data View software [40]. Abbreviations: VPR, the ratio of total viral abundance and picoplankton abundance (heterotrophic prokaryotes + *Prochlorococcus* + *Synechococcus* + picoeukaryotes); HVEukR, the ratio of high fluorescence viral abundance and picoeukaryotes abundance; LVProkR, the ratio of low fluorescence viral abundance and prokaryotic abundance (heterotrophic prokaryotes + *Prochlorococcus* + *Synechococcus*).

with lower fluorescence was called the LFV group; total viruses = HFV + LFV.

Heterotrophic prokaryote enumeration was performed according to the method of Marie et al. [22]. The samples were stained with the SYBR Green I (1/10000 final concentration, Molecular Probes) for 15 min in the dark at room temperature prior to analysis. The samples were run at a flow rate of 0.1–1 ml h^{-1}. For the enumeration of heterotrophic prokaryotes, we set the discriminator to green fluorescence and set all parameters on logarithmic amplification. The typical settings on an Epics Altra II flow cytometer are PMT4 = 430, PMT1 = 400, PMT2 = 550, PMT3 = 640, and PMT5 = 1000. Heterotrophic prokaryotes were identified in plots of red fluorescence vs. green fluorescence (Fig. S3D–E). The autotrophic picoplankton abundance was determined according to the method of Jiao et al. [23]. We set the discriminator to red fluorescence and all parameters on logarithmic amplification. The typical settings on an Epics Altra II flow cytometer are PMT4 = 400, PMT1 = 350, PMT2 = 570, PMT3 = 750, and PMT5 = 1000. Autotrophic picoplankton are identified in plots of side scatter vs. red fluorescence and orange fluorescence vs. red fluorescence (Fig. S3F–G). Fluorescent microspheres (Molecular Probes Inc.) with a diameter of 1 μm were added to all samples as an internal standard. The data were analyzed with EXPOTM32 MultiCOMP software (Beckman Coulter, USA).

The ratios of a different viral group and their possible corresponding host cell abundance were calculated, including the ratio of HFV and picoeukaryotic abundance (HVEukR), the ratio of LFV and prokaryotes abundance (prokaryotes abundance = heterotrophic prokaryotes + *Prochlorococcus* + *Synechococcus* abundance, LVProkR) and the ratio of total viruses and total picoplankton abundance (total picoplankton abundance = picoeukaryotes + prokaryotes abundance, VPR).

Statistical analysis

An analysis of variance (ANOVA) and independent *t*-test were employed to compare the differences in parameters using the SPSS (18.0) software package (SPSS Inc., Chicago, IL, USA). A Pearson correlation analysis was applied to assess the degree of correlation among the investigated parameters. A linear regression

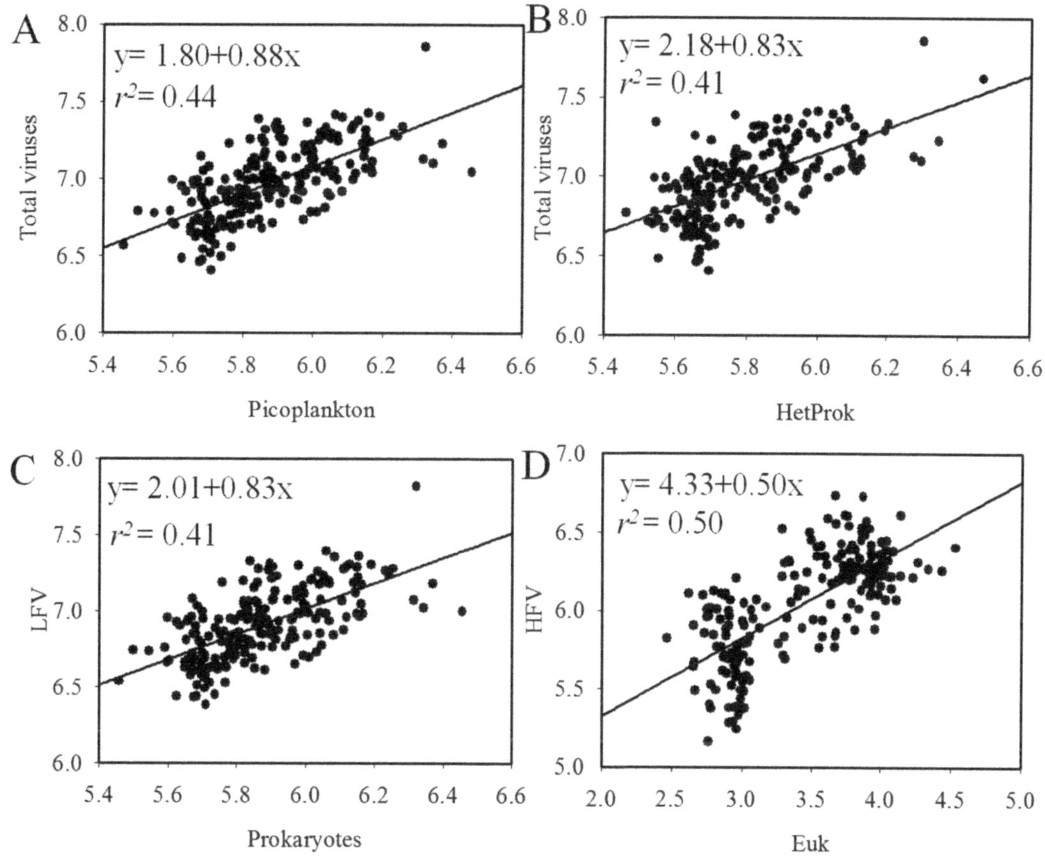

Figure 4. Relation between the abundance of total viruses and picoplankton (A), total viruses and HetProk (B), LFV and prokaryotes (C) and HFV and Euk (D) in surface waters. The fitting function used is: y = ax+b, all the data were log transformed. Abbreviations: HetProk, heterotrophic prokaryotes; LFV, low fluorescence viruses; HFV, high fluorescence viruses; Euk, picoeukaryotes.

analysis for log-transformed viral abundance and log-transformed host cell abundance was performed using Sigma-Plot 10.0. The distance-based multivariate analysis for a linear model using forward selection (DISTLM forward) was applied to test the relationships between viral abundance and biotic and abiotic environmental parameters [24].

Results and Discussion

Horizontal distribution pattern of viral abundance and its association with host cells

A total of 206 samples from surface water of the global ocean and 252 discrete samples from 22 depth profiles were collected and analyzed by FCM. Viruses were stained with the SYBR Green I.

Table 2. Results of the multivariate multiple regression analysis with forward selection (DISTLM forward) to explain the variability in viral abundance in the surface waters of coastal/shelf/upwelling and oceanic gyre water.

Provinces	*Selected variables*	*Pseudo-F*	P	r^2	*Cumulative*
Global ocean (n = 205)	HetProk	110.1725	0.0001	0.3518	0.3518
	Proc	4.7114	0.0279	0.0148	0.3666
Coastal/shelf/Upwelling (n = 115)	HetProk	16.6441	0.0010	0.1284	0.1284
	Euk	7.9945	0.0140	0.0581	0.1865
Gyres (n = 90)	Proc	12.5465	0.0030	0.1248	0.1248
	Syn	21.5631	0.0010	0.1738	0.2986

Abbreviations: Proc, *Prochlorococcus*; Syn, *Synechococcus*; Euk, picoeukaryotes; HetProk, heterotrophic prokaryotes.
The response variable was log-transformed and the resulting data were converted into Euclidian distance similarities matrices. The Pseudo-F and the *P*-values were obtained by permutation (n = 999).

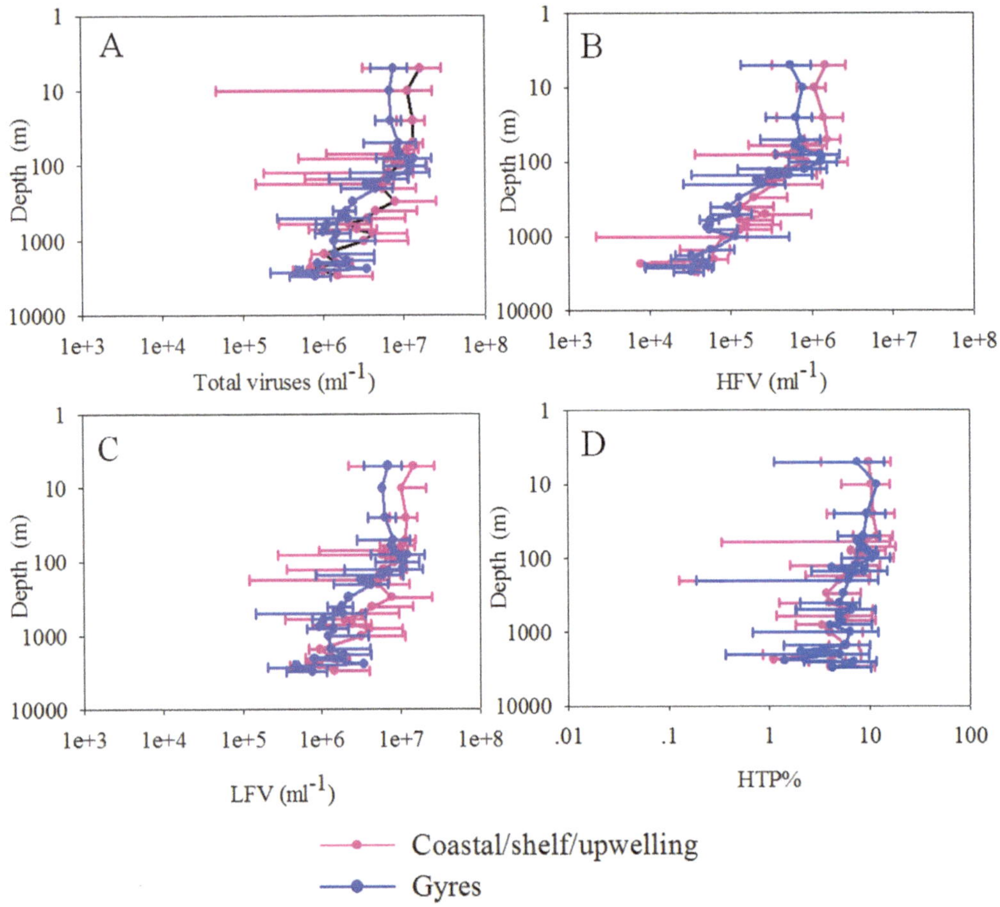

Figure 5. Depth profiles of total viral abundance (A), HFV abundance (B), LFV abundance (C) and HTP% (D) in the whole water column (0–3000 m) of the 22 vertical-sampling stations, respectively. The solid line is the average value, while the error bars represent 95% confidential intervals. The water depths of samples were considered as 2500 m when the bottom depths were between 2500 and 2700 m and as 3000 m when the depths were between 2900 and 3150 m. Abbreviations: HFV, high fluorescence viruses; LFV, low fluorescence viruses. HTP%, the percent of HFV to total viruses.

Although SYBR Green I targets only dsDNA viruses [8] and SYBR Gold would also be able to detect ssDNA and RNA viruses [25], Brussaard conducted a detailed evaluation of the two sensitive nucleic acid stains and recommended SYBR Green I for viral analysis [8]. Currently, SYBR Green I is widely used for viral enumeration [13–15,17,18,21,26]. To obtain comparable results to those of other studies, we stained the viral samples with SYBR Green I. Generally, the viral abundance of surface water varied from 2.54×10^{6} to 7.15×10^{7} ml^{-1} and was higher ($1.47\pm0.78\times10^{7}$ ml^{-1}) in the coastal/shelf waters and the presumed ocean upwelling waters with higher surface chlorophyll concentration (>0.07 mg m^{-3}) [27] than ($6.34\pm2.16\times10^{6}$ ml^{-1}) in the subtropical oceanic gyres (surface chlorophyll concentration ≤0.07 mg m^{-3}, $P<0.001$, independent *t*-test, Table 1, Fig. 1A) [28]. The distribution patterns of host cells (e.g., heterotrophic prokaryotes, *Synechococcus* and picoeukaryotes) were similar to that of the total viral abundance (Pearson correlation, $P<0.001$, Fig. 1A, 2A, and 2C–D).

Based on FCM analysis, the low DNA-content LFV group accounted for approximately 90% of the total viruses in the surface waters and averaged $1.26\pm0.72\times10^{7}$ and $5.71\pm1.99\times10^{6}$ ml^{-1} in the coastal/shelf/upwelling (CSU) areas and the subtropical gyres, respectively ($P<0.001$, independent *t* test, Table 1, Fig. 1B). The high DNA-content HFV group is usually considered to comprise viruses of autotrophic eukaryotes because of the higher genome sizes and consistent positions in the FCM biplots with algal viruses [29,30]. In our investigation, the CSU waters had a higher HFV abundance and percent of total viruses ($2.08\pm0.92\times10^{6}$ ml^{-1} and 15.01±4.70%) than the gyre waters ($6.25\pm3.29\times10^{5}$ ml^{-1} and 9.78±3.63%, $P<0.001$, independent *t*-test, Table 1, Fig. 1C).

The absolute values of latitude were negatively correlated with the abundance of HFV, total viruses, *Prochlorococcus*, *Synechococcus*, picoeukryotes, and heterotrophic prokaryotes (Pearson correlation, $r=-0.489$, -0.189, -0.216, -0.342, -0.555 and -0.291, $P<0.01$, respectively), which is similar with the results that have been reported in the north Atlantic Ocean [13]. They found that the latitude contributed to the variation of total viral abundance in the epipelagic layer, although the contribution of latitude was low ($r^2=0.05$; $P=0.04$).

The ratio of total viruses and total picoplankton abundance (VPR) varied from 4.95 to 34.94 in the study areas and showed the same pattern as viral abundance. The CSU waters had higher VPR (14.93±6.46) than the subtropical gyres (11.68±4.21) (P<

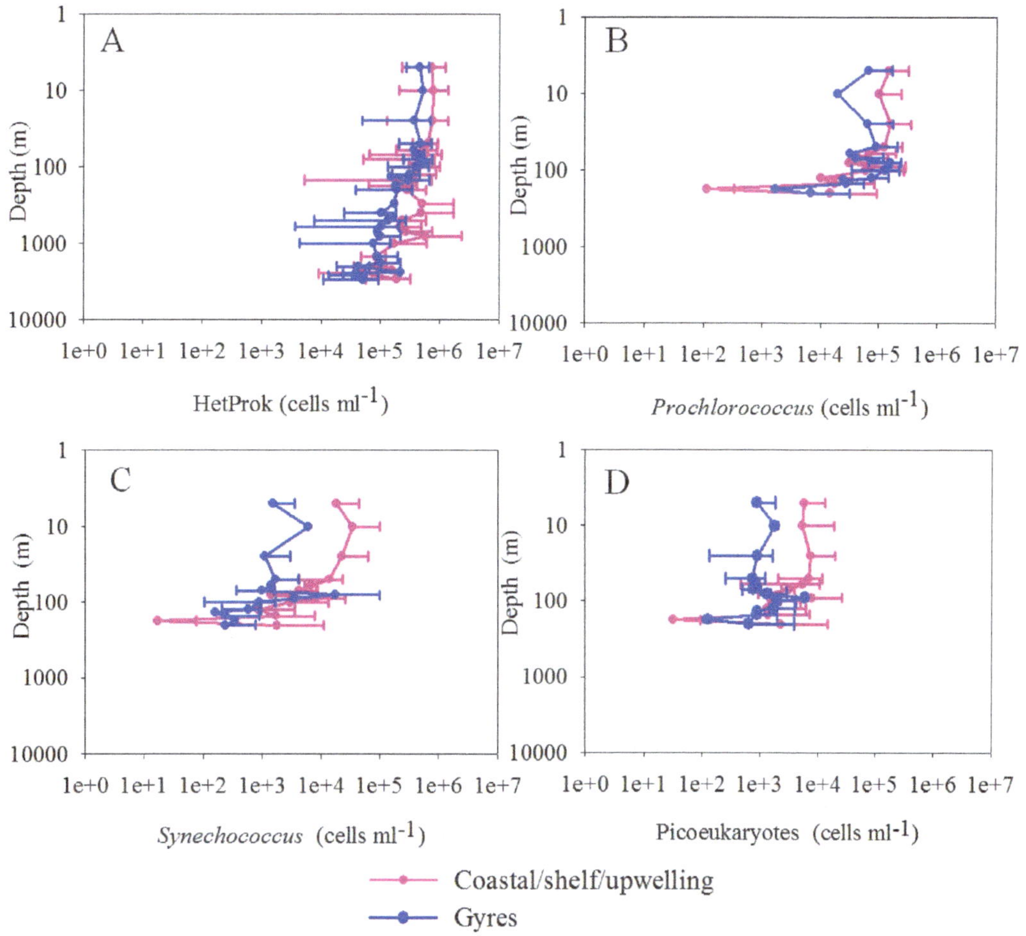

Figure 6. Depth profiles of heterotrophic prokaryotic abundance (A), *Prochlorococcus* abundance (B) *Synechococcus* abundance (C) and picoeukaryotes abundance (D) in the whole water column (0–3000 m) of the 22 vertical-sampling stations, respectively. The solid line is the average value, while the error bars represent 95% confidential intervals. The water depths of samples were considered as 2500 m when the bottom depths were between 2500 and 2700 m and as 3000 m when the depths were between 2900 and 3150 m.

0.001, independent *t*-test, Table 1, Fig. 3A), which is in agreement with the studies in the Adriatic Sea and the Southern Ocean [16,31]. LVProkR was also higher in the CSU waters (12.84±5.80) than the subtropical gyres (10.59±3.99) (P<0.001, independent *t*-test, Table 1, Fig. 3B). These results suggest that hosts sustain greater numbers of viruses under the environments favoring fast growth and high productivity. It has been reported that the increases in viral production might be due, in part, to a higher burst size in the high-nutrient waters [3]. Meanwhile, high autotrophic host abundance may induce a higher concentration of algal viruses and cyanophages, which could also result in a higher VPR [32]. In our study, the observed low VPR in the oligotrophic gyres agreed with the results reported in the low-nutrient marine environment [9,33]. The reasons might include the low contact rates between viruses and prokaryotes because of the low abundance of host cells, the presumably non-lytic life strategies of viruses (including chronic infections, lysogeny and pseudolysogeny), and the lower frequency of prokaryotes with mature phage in such environments [34]. HVEukR was higher than the VPR and LVProkR and varied from 127 to 3084 in the surface waters (P<0.001, independent t-test), which is similar to the reported burst sizes of eukaryotic viruses (ranged from 72–100,000 $cell^{-1}$, P<0.001, one-way ANOVA) [29,30]. The distribution pattern of HVEukR was different from the VPR and higher in the subtropical gyres (793±597) than in the CSU waters (372±279) (P<0.001, independent t-test, Table 1, Fig. 3A–C).

Over the entire sampling area, viral abundance displayed significant correlation with heterotrophic prokaryotic abundance (linear regression $r^2 = 0.41$, Pearson correlation, P<0.001, Fig. 4B), which is consistent with most previous studies [16,35,36] and suggests a tight coupling between heterotrophic prokaryotic and viral concentrations. The total picoplankton abundance (heterotrophic prokaryotes + *Prochlorococcus* + *Synechococcus* + picoeukayotes) has a slightly better explanation of total viral abundance than heterotrophic prokaryotic abundance (linear regression $r^2 = 0.44$, Pearson correlation, P<0.001, Fig. 4A–B). This indicated that viral distribution was dependent on both heterotrophic and autotrophic host cell abundance in the surface ocean. These observations were in agreement with the results of the central Pacific transects and the northwestern Sargasso Sea, which determined that autotrophic host cell abundance might contribute significantly to the variation in viral abundance [12,14]. The multivariate multiple regression analysis (DISTLM-*forward*) indicated that heterotrophic prokaryotic and

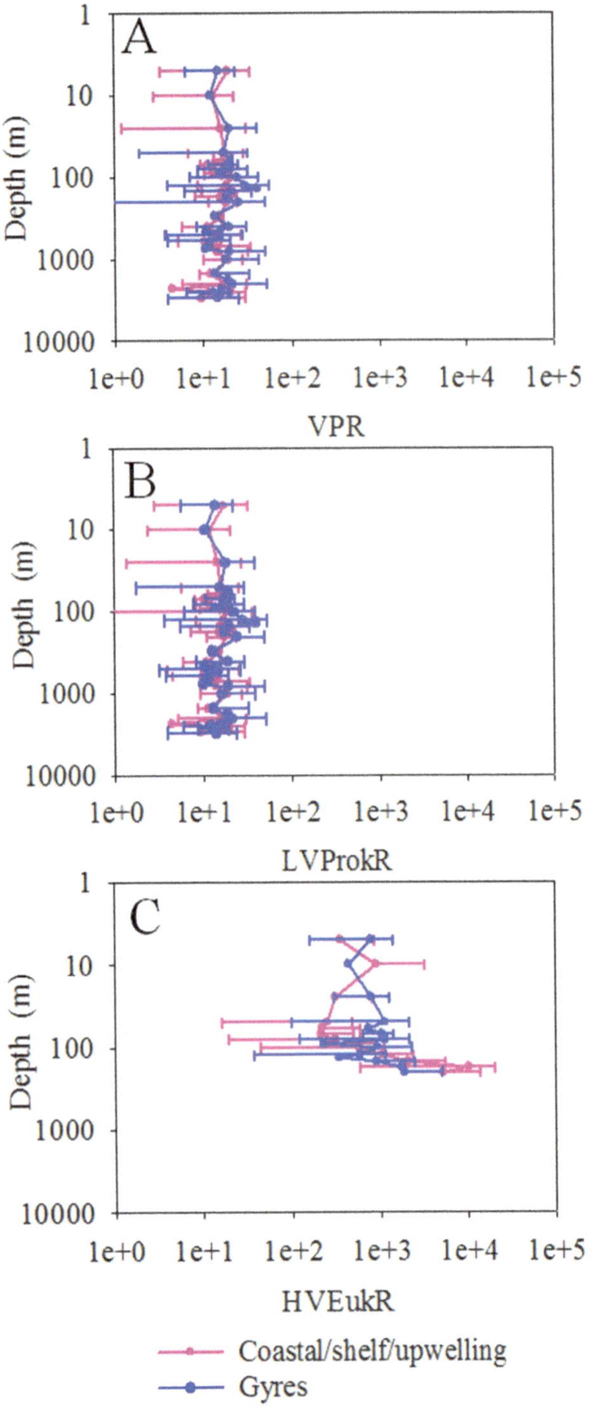

Figure 7. Depth profiles of VPR (A), LVProkR (B) and HVEukR (C) in the whole water column (0–3000 m) of the 22 vertical-sampling stations, respectively. The solid line is the average value, while the error bars represent 95% confidential intervals. The water depths of samples were considered as 2500 m when the bottom depths were between 2500 and 2700 m and as 3000 m when the depths were between 2900 and 3150 m. Abbreviations: VPR, the ratio of total viral abundance and picoplankton abundance (heterotrophic prokaryotes + *Prochlorococcus* + *Synechococcus* + picoeukaryotes); HVEukR, the ratio of high fluorescence viral abundance and picoeukaryotes abundance; LVProkR, the ratio of low fluorescence viral abundance and prokaryotic abundance (heterotrophic prokaryotes + *Prochlorococcus* + *Synechococcus*).

Prochlocococcus abundance together explained 36% of the variability in viral abundance in the surface waters of the global ocean (Table 2 & S1). This further suggested that viral distribution was dependent on both heterotrophic and autotrophic host cell abundance in the surface ocean.

Vertical distribution pattern of viral abundance and its association with host cell abundance

The abundance of viruses significantly decreased with depth from an average of $8.94\pm4.69\times10^6$ ml^{-1} in the epipelagic waters to $1.11\pm0.78\times10^6$ ml^{-1} in the bathypelagic waters (ANOVA on rank, P<0.001, Fig. 5A), which is comparable to the abundance reported in the Atlantic and Pacific Oceans and the northwestern Mediterranean Sea [13,15,17–19,21,37]. The abundance of high-fluorescence viruses significantly decreased with depth (linear regression $r^2=0.64$, Pearson correlation, P<0.001, Fig. 5B and 8B) and the percentage of high-fluorescence viruses decreased from an average of 8.34±3.21% in the epipelagic layer to 4.80±2.03% (ANOVA on rank, P<0.001, Fig. 5D) in the bathypelagic layer, which is in agreement with previous studies from the Atlantic Ocean [13,15]. The LFV abundance decreased with depth from an average of $8.25\pm4.12\times10^6$ ml^{-1} in the epipelagic waters to $1.07\pm0.77\times10^6$ ml^{-1} in the bathypelagic waters (ANOVA on rank, P<0.001, Fig. 5C). Comparatively, the prokaryotic abundance decreased by more than an order of magnitude in the vertical profiles (ANOVA on rank, P<0.001, Fig. 6A).

The VPR slightly decreased with depth from an average of 19.0±8.2 over all the stations in the epipelagic layer to 16.2±7.9 in the bathypelagic layer (ANOVA on rank, P<0.05, Fig. 7A). The highest VPR was usually displayed between 70 and 200 m in the Atlantic oceanic gyre water (average: 25.3±9.9, range: 12.4–47.6, Fig. 7A). The vertical distribution pattern of LVProkR is similar to that of VPR and slightly decreased with depth from an average of 17.5±7.7 over all the stations in the epipelagic layer to 15.5±7.8 in the bathypelagic layer (ANOVA on rank, P>0.05, Fig. 7A–B). These results are in agreement with the results reported in the northwestern Mediterranean Sea, the Arctic and the western Pacific [18,37–39]. Similar and slightly decreasing or increasing trends of VPR from the epipelagic layer to the deep sea waters were observed in the eastern and northern basin of the Atlantic [13,15,19], central Pacific [17] and the Mediterranean Sea [18,20,26]. Furthermore, VIEukR was relatively stable in the epipelagic zone of gyre waters (980±574) and increased with depth in the relatively eutrophic CSU waters (Fig. 7C) due to the different distribution of picoeukaryotes in these two environments (Fig. 6D).

Viral abundance was negatively related to depth (linear regression $r^2=0.61$, Pearson correlation, P<0.001, Fig. 8A) and positively related to heterotrophic prokaryotic abundance (linear regression $r^2=0.81$, Pearson correlation, P<0.001, Fig. 8D). In the epipelagic waters, the highest viral abundance usually displayed between 50 and 100 m in the oceanic gyre waters and above 50 m in the highly primary productive upwelling waters (Fig. 5A). This roughly matched the depth of the subsurface chlorophyll maxima (corresponding to the peak value of *Prochlorococcus*, *Synechococcus* and picoeukaryotes abundance) observed in our study (Fig. 6B–D). The subsurface peaks in virioplankton abundance around the depth of the chlorophyll maxima were observed previously [12,35,37]. This finding could be due to the relatively high virus loss due to sunlight in the surface area and a high viral production rate because of the peak of host cell abundance and productivity at chlorophyll maxima layers. The results of the correlation analysis and DISTLM-*forward* analysis

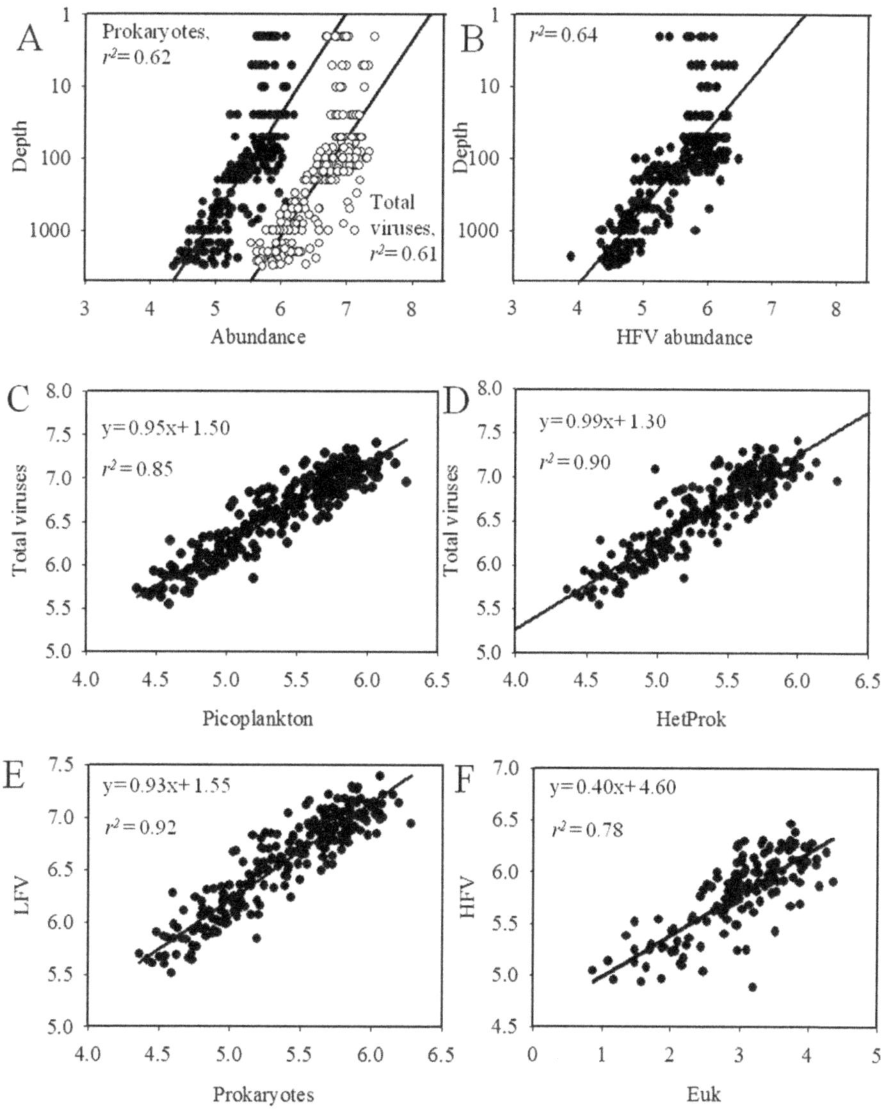

Figure 8. Relation between the abundance of prokaryotes and total viruses versus depth (A), LFV versus depth (B), total viruses versus picoplankton (C), total viruses versus HetProk (D), LFV and prokaryotes (E) and HFV abundance and Euk (F) in the whole water column of 22 vertical-sampling stations. The fitting function used is: y = ax+b, all the data were log transformed. Abbreviations: LFV, low fluorescence viruses; HFV, high fluorescence viruses; HetProk, heterotrophic prokaryotes; Euk, picoeukaryotes.

also showed that a vertical variation of viral abundance in the epipelagic waters was dependent on both heterotrophic and autotrophic host cell abundance. Significant positive correlations were found between the viral abundance and the cell concentration of *Prochlorococcus*, *Synechococcus*, picoeukaryotes and heterotrophic prokaryotes (linear regression $r^2 = 0.44$, $r^2 = 0.37$, $r^2 = 0.55$, and $r^2 = 0.46$, respectively; Pearson correlation, $P < 0.001$). DISTLM-*forward* analysis showed that the variables explaining most of the variability (57%) in viral abundance in the epipelagic zone were the abundance of heterotrophic prokaryotes, *Prochlorococcus*, *Synechococcus* and picoeukaryotes and depth (Table 3 & S1). In the meso- and bathypelagic waters, viral abundance was significantly and positively correlated with heterotrophic prokaryotic abundance (linear regression $r^2 = 0.73$, Pearson correlation, $P < 0.001$). DISTLM-*forward* analysis showed that heterotrophic prokaryotic abundance explained 54% and 30% of the variability in viral abundance in the meso- and bathypelagic waters, respectively, and that depth contributed to another 3% of the variation in viral abundance in the mesopelagic waters (Table 3 & S1).

Altogether, our study suggested that the horizontal and vertical distributions of virioplankton at large spatial scales are controlled by dynamic host-virus relationships. Virioplankton play an active and important role in marine ecosystems and biogeochemical cycling in all of the global ocean, including meso- and bathypelagic oceans. Correlative analyses of the abundances of total viruses and their potential host cells suggest possible controls on marine viral dynamics. Further phylogenetic quantification of phages, experimental determination of viral pressure on lineages of each host and additional modeling exercises using these data are required to resolve the drivers of the spatial variability in total

Table 3. Results of the multivariate multiple regression analysis with forward selection (DISTLM forward) to explain the variability in viral abundance throughout the water column (total) and in specific depth layers in different geographic regions.

Provinces	*Selected variables*	*Pseudo-F*	P	r^2	*Cumulative*
Global ocean (n = 252)	Depth	381.1793	0.0010	0.6039	0.6039
	HetProk	172.8139	0.0010	0.1623	0.7662
	Proc	33.3812	0.0010	0.0277	0.7939
	Euk	5.2376	0.0220	0.0043	0.7982
	Syn	7.5410	0.0060	0.0060	0.8042
Epi (n = 152)	HetProk	106.2202	0.0010	0.4146	0.4146
	Proc	28.6149	0.0010	0.0943	0.5089
	Euk	8.5936	0.0040	0.0270	0.5358
	Depth	7.4071	0.0070	0.0223	0.5581
	Syn	4.1951	0.0410	0.0123	0.5704
Meso (n = 56)	HetProk	62.2510	0.0010	0.5355	0.5355
	Depth	4.1708	0.0460	0.0339	0.5694
Bathy (n = 44)	HetProk	17.9534	0.0010	0.2995	0.2995
Coastal/shelf/upwelling (n = 120)	HetProk	168.3646	0.0010	0.5879	0.5879
	Depth	100.0706	0.0010	0.1900	0.7779
	Euk	24.4152	0.0010	0.0386	0.8165
	Syn	3.2953	0.0700	0.0051	0.8216
	Proc	5.1984	0.0290	0.0078	0.8294
Epi (n = 74)	HetProk	80.4848	0.0010	0.5278	0.5278
	Depth	20.4735	0.0010	0.1057	0.6335
	Euk	14.3137	0.0010	0.0622	0.6957
	Syn	11.9527	0.0010	0.0449	0.7406
Meso (n = 28)	HetProk	43.7601	0.0010	0.6273	0.6273
Gyres (n = 132)	HetProk	245.4824	0.0010	0.6538	0.6538
	Depth	76.5997	0.0010	0.1290	0.7828
	Proc	20.5376	0.0010	0.0300	0.8128
Epi (n = 78)	Proc	84.3087	0.0010	0.5259	0.5259
	Euk	5.7289	0.0120	0.0336	0.5596
	Depth	3.9734	0.0420	0.0224	0.5820
Meso (n = 28)	Depth	9.5536	0.0020	0.2687	0.2687
Bathy (n = 26)	HetProk	40.7493	0.0010	0.6293	0.6293

Abbreviations: Proc, *Prochlorococcus*; Syn, *Synechococcus*; Euk, picoeukaryotes; HetProk, heterotrophic prokaryotes.
The response variable was log-transformed and the resulting data were converted into Euclidian distance similarities matrices. The Pseudo-F and the *P*-values were obtained by permutation (n = 999).

viruses and their ecological effects on the biogeochemical cycles in the global ocean.

Supporting Information

Figure S1 The hierarchical tree of how the samples fractionated.

Figure S2 Sampling stations of the global cruise. The blue points represent the stations where surface samples were collected. The red points represent the stations where vertical profile samples were collected. The yellow oval circles indicated the gyre areas (surface chlorophyll concentration ≤0.07 mg m^{-3}).

Figure S3 Side scatter versus green fluorescence obtained for a natural viral sample (A), 0.02 µm filtered TE buffer (B), a pure-cultured marine roseophage RDJL Phi 1 lytically infecting *Roseobacter denitrificans* OCh114 (C), a natural heterotrophic prokaryotic sample (D), and red fluorescence versus green fluorescence obtained for the natural heterotrophic prokaryotic sample (E); all were stained with SYBR- Green I. Side scatter versus red fluorescence (F) and orange fluorescence versus red fluorescence (G) obtained for a natural autotrophic picoplankton sample. One micron fluorescent beads were added as an internal reference. Abbreviations: HFV, high fluorescence viruses; LFV, low fluorescence viruses; HetProk, heterotrophic prokaryotes; AutoProk, autotrophic prokaryotes; Proc, *Prochlorococcus*; Syn, *Synechococcus*, Euk, picoeukaryotes.

Table S1 The parameters and their numbers for the DISTLM forward analysis in the Table 2 & 3.

Acknowledgments

We would like to thank the captain and crews of the *R/V Dayang No. 1* for their support and the splendid atmosphere on board the ship. The authors would like to thank Kai Tang, Fan Zhang, Fei Zhang, Jia Sun, Keshao Liu and Tao Jiang for their valuable help during the sampling collection. We are grateful to Yawei Luo for his valuable comments and suggestions.

Author Contributions

Conceived and designed the experiments: RZ NJ. Performed the experiments: YL LL TL. Analyzed the data: YL TL RZ. Contributed reagents/materials/analysis tools: YL TL RZ. Contributed to the writing of the manuscript: YL LL YZ RZ NJ.

References

1. Suttle CA (2005) Viruses in the sea. Nature 437: 356–361.
2. Suttle CA (2007) Marine viruses–major players in the global ecosystem. Nat Rev Microbiol 5: 801–812.
3. Wommack KE, Colwell RR (2000) Virioplankton: viruses in aquatic ecosystems. Microbiol Mol Biol Rev 64: 69–114.
4. Fuhrman JA (1999) Marine viruses and their biogeochemical and ecological effects. Nature 399: 541–548.
5. Wilhelm SW, Suttle CA (1999) Viruses and nutrient cycles in the sea. Bioscience 49: 781–788.
6. Fuhrman J (1992) Bacterioplankton roles in cycling of organic matter: the microbial food web; Falkowski PG, and Woodhead, A. D., editor. New York: Plenum Press. 361–383 p.
7. Jiao N, Herndl GJ, Hansell DA, Benner R, Kattner G, et al. (2010) Microbial production of recalcitrant dissolved organic matter: long-term carbon storage in the global ocean. Nat Rev Microbiol 8: 593–599.
8. Brussaard CP (2004) Optimization of procedures for counting viruses by flow cytometry. Appl Environ Microbiol 70: 1506–1513.
9. Alonso M, Jimenez-Gomez F, Rodriguez J, Borrego J (2001) Distribution of virus-like particles in an oligotrophic marine environment (Alboran Sea, Western Mediterranean). Microb Ecol 42: 407–415.
10. Weinbauer M, Suttle C (1997) Comparison of epifluorescence and transmission electron microscopy for counting viruses in natural marine waters. Aquat Microb Ecol 13: 225–232.
11. Wen K, Ortmann AC, Suttle CA (2004) Accurate estimation of viral abundance by epifluorescence microscopy. Appl Environ Microbiol 70: 3862–3867.
12. Parsons RJ, Breitbart M, Lomas MW, Carlson CA (2012) Ocean time-series reveals recurring seasonal patterns of virioplankton dynamics in the northwestern Sargasso Sea. ISME J 6: 273–284.
13. De Corte D, Sintes E, Yokokawa T, Reinthaler T, Herndl GJ (2012) Links between viruses and prokaryotes throughout the water column along a North Atlantic latitudinal transect. ISME J 6: 1566–1577.
14. Yang Y, Motegi C, Yokokawa T, Nagata T (2010) Large-scale distribution patterns of virioplankton in the upper ocean. Aquat Microb Ecol 60: 233–246.
15. De Corte D, Sintes E, Winter C, Yokokawa T, Reinthaler T, et al. (2010) Links between viral and prokaryotic communities throughout the water column in the (sub) tropical Atlantic Ocean. ISME J 4: 1431–1442.
16. Corinaldesi C, Crevatin E, Del Negro P, Marini M, Russo A, et al. (2003) Large-scale spatial distribution of virioplankton in the Adriatic Sea: testing the trophic state control hypothesis. Appl Environ Microbiol 69: 2664–2673.
17. Yang Y, Yokokawa T, Motegi C, Nagata T (2014) Large-scale distribution of viruses in deep waters of the Pacific and Southern Oceans. Aquat Microb Ecol 71: 193–202.
18. Winter C, Kerros M-E, Weinbauer MG (2009) Seasonal and depth-related dynamics of prokaryotes and viruses in surface and deep waters of the northwestern Mediterranean Sea. Deep Sea Res Pt I 56: 1972–1982.
19. Parada V, Sintes E, van Aken HM, Weinbauer MG, Herndl GJ (2007) Viral abundance, decay, and diversity in the meso- and bathypelagic waters of the North Atlantic. Appl Environ Microbiol 73: 4429–4438.
20. Magagnini M, Corinaldesi C, Monticelli LS, De Domenico E, Danovaro R (2007) Viral abundance and distribution in mesopelagic and bathypelagic waters of the Mediterranean Sea. Deep Sea Res Pt I 54: 1209–1220.
21. Li Y, Luo T, Sun J, Cai L, Liang Y, et al. (2014) Lytic viral infection of bacterioplankton in deep waters of the western Pacific Ocean. Biogeosciences 11: 2531–2542.
22. Marie D, Partensky F, Vaulot D, Brussaard C (1999) Enumeration of phytoplankton, bacteria, and viruses in marine samples. Curr Protocols Cytom 11: 1–15.
23. Jiao N, Yang Y, Koshikawa H, Watanabe M (2002) Influence of hydrographie conditions on picoplankton distribution in the East China Sea. Aquat Microb Ecol 30: 37–48.
24. Anderson M, Ford R, Honeywill C, Feary D (2004) Quantitative measures of sedimentation in an estuarine system and its relationship with intertidal soft-sediment infauna. Mar Ecol-Prog Ser 272: 33–48.
25. Chen F, Lu J, Binder BJ, Liu Y, Hodson RE (2001) Application of digital image analysis and flow cytometry to enumerate marine viruses stained with SYBR Gold. Appl Environ Microbiol 67: 539–545.
26. Magiopoulos I, Pitta P (2012) Viruses in a deep oligotrophic sea: Seasonal distribution of marine viruses in the epi-, meso- and bathypelagic waters of the Eastern Mediterranean Sea. Deep Sea Res Pt I 66: 1–10.
27. Fiedler PC, Philbrick V, Chavez FP (1991) Oceanic upwelling and productivity in the eastern tropical Pacific. Limnol Oceanogr 36: 1834–1850.
28. Polovina JJ, Howell EA, Abecassis M (2008) Ocean's least productive waters are expanding. Geophys Res Lett 35: L03618.
29. Brussaard CP (2004) Viral control of phytoplankton populations–a review. J Eukaryot Microbiol 51: 125–138.
30. Van Etten JL, Lane LC, Dunigan DD (2010) DNA viruses: the really big ones (giruses). Annu Rev Microbiol 64: 83–99.
31. Brussaard C, Timmermans K, Uitz J, Veldhuis M (2008) Virioplankton dynamics and virally induced phytoplankton lysis versus microzooplankton grazing southeast of the Kerguelen (Southern Ocean). Deep Sea Res Pt II 55: 752–765.
32. Maranger R, Bird D, Juniper S (1994) Viral and bacterial dynamics in Arctic sea ice during the spring algal bloom near Resolute, N. W. T., Canada. Mar Ecol-Prog Ser 111: 121–127.
33. Hwang CY, Cho BC (2002) Virus-infected bacteria in oligotrophic open waters of the East Sea, Korea. Aquat Microb Ecol 30: 1–9.
34. Weinbauer MG, Fuks D, Peduzzi P (1993) Distribution of viruses and dissolved DNA along a coastal trophic gradient in the northern Adriatic Sea. Appl Environ Microbiol 59: 4074–4082.
35. Cochlan WP, Wikner J, Steward GF, Smith DC, Azam F (1993) Spatial distribution of viruses, bacteria and chlorophyll a in neritic, oceanic and estuarine environments. Mar Ecol-Prog Ser 92: 77–77.
36. Jiao N, Zhao Y, Luo T, Wang X (2006) Natural and anthropogenic forcing on the dynamics of virioplankton in the Yangtze river estuary. J Mar Biol Assoc UK 86: 543–550.
37. Hara S, Koike I, Terauchi K, Kamiya H, Tanoue E (1996) Abundance of viruses in deep oceanic waters. Mar Ecol-Prog Ser 145: 269–277.
38. Boehme J, Frischer M, Jiang S, Kellogg C, Pichard S, et al. (1993) Viruses, bacterioplankton, and phytoplankton in the southeastern Gulf of Mexico: distribution and contribution to oceanic DNA pools. Mar Ecol-Prog Ser 97: 1–10.
39. Arıstegui J, Gasol JM, Duarte CM, Herndl G (2009) Microbial oceanography of the dark ocean's pelagic realm. Limnol Oceanogr 54: 1501–1529.
40. Schlitzer R (2010) Ocean Data View. http://odv.awi.de.

19

Unusual Circulation Patterns of the Rias Baixas Induced by Minho Freshwater Intrusion (NW of the Iberian Peninsula)

Magda Catarina Sousa[1], Renato Mendes[1], Ines Alvarez[1,2], Nuno Vaz[1], Moncho Gomez-Gesteira[2], João Miguel Dias[1]*

1 CESAM, Departamento de Física, Universidade de Aveiro, Campus de Santiago, Aveiro, Portugal, **2** EPhysLab (Environmental Physics Laboratory), Universidade de Vigo, Facultade de Ciencias, Ourense, Spain

Abstract

The Minho River, situated 30 km south of the Rias Baixas, is the most important freshwater source flowing into the Western Galician coast (NW of the Iberian Peninsula). The buoyancy generated by the Minho estuarine plume can reverse the normal circulation pattern inside the Rias Baixas affecting the exchange between the Rias and the ocean, changing the input of nutrients. Nevertheless, this inversion of the circulation patterns is not a well-monitored phenomenon. The only published results based on *in situ* data related to the presence of the Minho River plume inside the Rias de Vigo and Pontevedra correspond to an event measured on spring 1998. In this case unexpectedly higher inflow surface current velocities were found at the Ria de Pontevedra, located further away from Minho River. Thus, the main aim of this study is to research the main factors inducing this unusual pattern on the circulation of the Rias de Vigo and Pontevedra. A numerical model implementation of MOHID previously developed, calibrated, and validated for this coastal area was used. Several scenarios were performed in order to explain the individual effect of the Minho River, rivers discharging into each Rias, and estuarine morphology changes. According to the model results, the Minho River discharge is a key factor in the establishment of the negative circulation, while small rivers inside the Rias slightly attenuate this circulation. The negative circulation was stronger in Ria de Pontevedra independently of the distance of this coastal system from the Minho River mouth, showing that morphologic estuarine features are the main factor justifying the different local circulation patterns.

Editor: Vanesa Magar, Centro de Investigacion Cientifica y Educacion Superior de Ensenada, Mexico

Funding: RM has been supported by the Portuguese Science Foundation (FCT) through a doctoral grant (SFRH/BD/79555/2011). IA and NV are supported by the Ramon y Cajal and Ciência 2008 programs, respectively. This work has been supported by FCT and by the European Union (COMPETE, QREN, FEDER) in the frame of the research project Pest (C/MAR/LA0017/2013) and partially supported by Xunta de Galicia under projects "Programa de Consolidación e Estructuración de Unidades de Investigación Competitivas: Grupos de Referencia Competitiva" (GR2013-001) co-funded by European Regional Development Fund (FEDER), EM2013/003, and "Galician Internal Basin Paleoclimatic Sedimentary Archives: Transport, Provenances, Hydrology and reliability (GALINCLIMARCH - CGL2010-16688/BTE). The funders had no role in study design, data collection and analysis, decision to publish, or preparation of the manuscript.

* Email: joao.dias@ua.pt

Introduction

Major rivers inject freshwater onto the adjacent shelf where mixing of these river plumes takes place, affecting the transport and transformation of dissolved and particulate materials in the coast. The coastal waters freshening due to river runoff is one of the mechanisms that control the density inside the estuaries located up north or south of the river mouth, modifying the circulation and the macronutrients concentrations available for production [1–3]. Numerous studies conducted over diverse coastal systems around the world have revealed a major interest in the ecological consequences of freshwater intrusion on nearby estuaries (e. g. Columbia River [1], Delaware River [3] or Mississippi River [2]). These studies showed that the river plumes from major rivers produce currents in the surface layers of neighbor estuaries, affecting local species and marsh plants.

The NW coast of the Iberian Peninsula is characterized by the presence of several embayments, usually called "Rias". The term "ria" defines a coastal shape correspondent to re-entrance, resultant from the submersion by the sea of the terminal zone of a fluvial network [4]. The Rias Baixas (Figure 1) are located in the northern limit of the Eastern North Atlantic Upwelling System [5]. They are heavy populated ecosystems of enormous potential, both economical (e.g. exploration of marine resources: fisheries, aquaculture and fishing industry) and social (e.g. tourism: beaches and natural beauty). Wind driven coastal upwelling is the main recognized local source of primary production, supporting the high fishery and aquaculture yields [6,7].

Located 30 km southward, the Minho River is the most important river flowing near this coastal region. It has a catchment area of 17080 km^2 and an annual average discharge of 300 m^3s^{-1}. The monthly average discharge oscillates between 100 m^3s^{-1} in

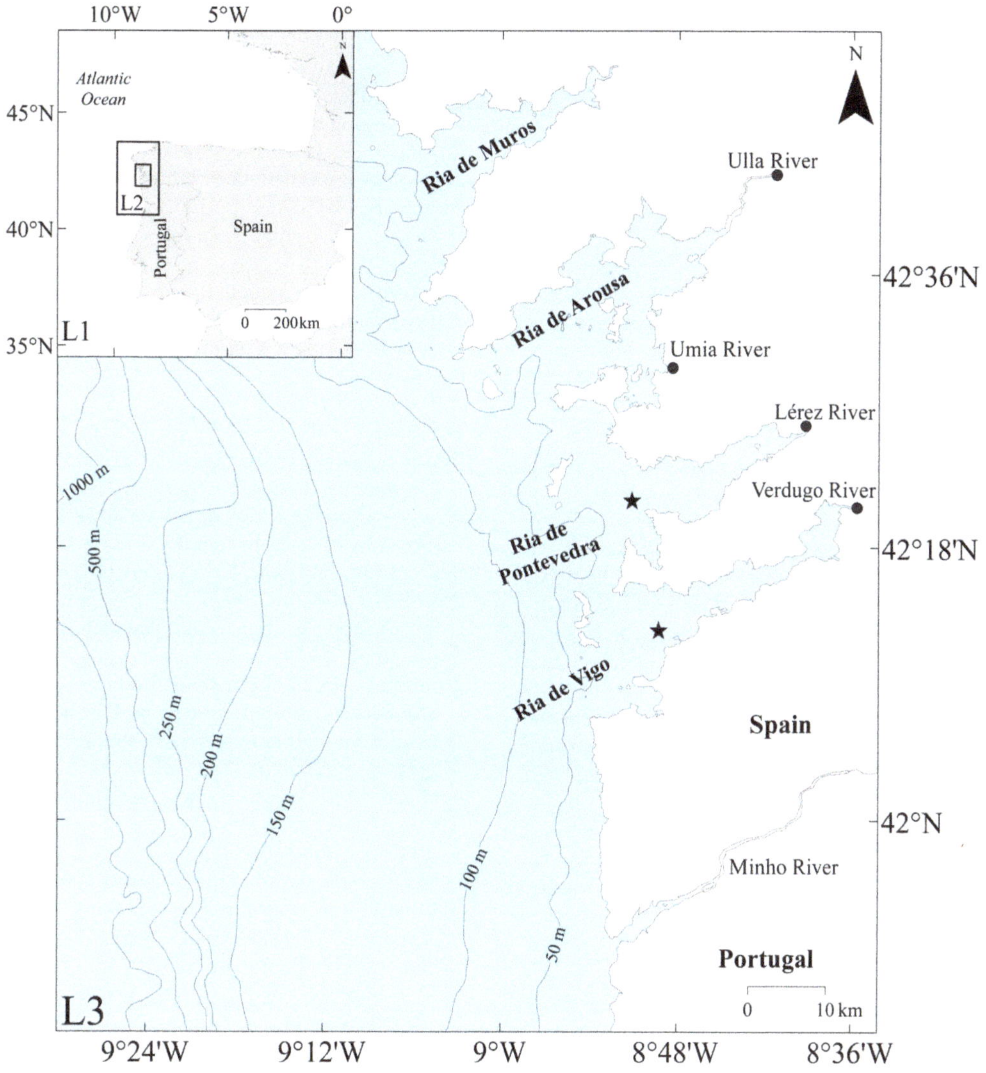

Figure 1. Nested grids and Rias Baixas map.

August and 800 m^3s^{-1} in February [8]. Essentially, during high river discharge and southerly wind conditions, its buoyant plume generates an important salinity decrease at the Rias Baixas mouth, reversing the normal circulation pattern [9–11]. This reverse in the circulation pattern (negative circulation) is characterized by the intrusion of less dense coastal waters at the surface, which accumulate within the Rias and finally flow towards the ocean through the bottom [9,10,12–14]. This inversion of the normal circulation patterns is not a well-documented phenomenon. The infrequent sampling in the inner and outer part of the Rias complicates the characterization of such events through analysis of field measurements. In fact, the only published results related to the presence of the Minho River plume inside the Rias Baixas correspond to an event measured on spring 1998 when a high Minho River discharge was reported as well as southerly wind conditions crucial to spread the river plume towards the Rias Baixas, generating a negative/reverse circulation inside them.

Alternatively, the use of numerical models constitutes a reliable and useful tool to analyze this type of events. The numerical models are used worldwide in order to quantify and understand different processes occurring in coastal waters [15–19]. They allow to investigate hypothetical scenarios (e.g. changes in wind and river discharge), which observations alone cannot provide. In addition, they afford the capability to study processes independently. Using a numerical model, Sousa et al. [14] reproduced the experimental campaign conducted on spring 1998 inside the Rias Baixas. They observed that under northward winds, a continuous moderate Minho River discharge (>700 $m^3\ s^{-1}$) is enough to produce the negative circulation pattern in Rias de Vigo and Pontevedra, reducing the importance of the existence of specific events of high river runoff. However, unexpectedly, higher inflow surface current velocities were found at the Ria located further away from Minho River (at Ria de Pontevedra rather than at Ria de Vigo, which is close to Minho River)). However, the explanation of these velocity differences was not evaluated in this previous study.

Considering these issues the present study aims to research the main reasons that induce this unusual pattern in the circulation in Rias de Vigo and Pontevedra on May 1998 event, through the application of a numerical model using a downscaling approach [11,14]. Several hypotheses were formulated to justify this pattern, namely the impact of different freshwater inflows at the head of each Ria, comparing to the impact of the Minho River and the Rias morphology, that have different width and depth and mouths with different features that may constrain the water exchange with the shelf. To test these hypotheses and explain the different current velocities observed at both Rias were defined several numerical scenarios. In the first, the real conditions corresponding to May 1998 are considered. The second and third scenarios are used in order to evaluate the influence of the Minho River and the rivers flowing into the Rias. The influence of estuarine morphology is evaluated in the last scenario, interchanging the location of the Rias de Vigo and Pontevedra.

The paper is organized as follows. In the second section, the description of study area is presented. The third part presents the general overview of the numerical model as well as the numerical experimental design. Results and discussion can be found in the fourth part. Finally, conclusions are drawn in the fifth part.

Study area

The study region comprises only two Galician Rias: the Rias de Vigo and Pontevedra, which are the closest estuarine regions located up north of the Minho Estuary.

The Ria de Vigo, which is the most meridional of the Rias Baixas is located around 30 km northwards of the Minho River mouth (Figure 1). This Ria is 32.5 km long, presenting an NE-SW direction, 1 km wide in its inner part (NE) and 10 km at the mouth of the Ria (SW). The mean width and depth are 4.8 km and 21 m, respectively. The mouth of this Ria is divided in two entrances defined by a set of islands, which allow the connection to the open sea. The northern entrance is 2.8 km wide and has a maximum depth of 25 m and the southern one is 5.1 km wide and 45 m depth. The Verdugo River is the main tributary of freshwater into Ria de Vigo with a catchment area of 333 km^2. Its annual mean annual discharge is 13 m^3s^{-1} with a greatest seasonal variability (values range from 120 m^3s^{-1} in winter to 1 m^3s^{-1} in summer) [20].

The Ria de Pontevedra is located northward of Ria de Vigo (Figure 1). This Ria is oriented in the NE-SW direction. The mean depth and width are 31 m and 3.8 km, respectively. It is also connected to the ocean by two entrances. The northern one is narrower (3.6 km) and shallower (15 m), while the southern mouth is wider (7.3 km) and deeper (60 m). The Lérez River is the main freshwater source that flows into Ria de Pontevedra. This river has a catchment area of 450 km^2 and an annual mean discharge of 27.5 m^3s^{-1}. The monthly mean discharge oscillates between 2 and 80 m^3s^{-1} and follows a similar pattern to the rainfall, as dams do not control its runoff.

Concerning the tidal forcing, both Rias are estuarine-like systems that have a semi-diurnal tidal regime. The tidal range varies between 2 and 4 m (mesotidal regime) and a significantly low Form Number of 0.25 is found [21]. The Rias behave normally as partially mixed estuaries, with positive residual circulation and two layer circulation pattern, with surface outflow and inflow on the bottom [22–24], although a reverse circulation may be found as previously referred.

Methodology

Numerical model

In this work was used the model MOHID (www.mohid.com), previously implemented and validated for the Rias Baixas by Sousa et al. [14].

The model implementation consists in a set of three-one-way nested models (Figure 1). The first domain is a barotropic tidal driven model using the FES2004 (Finite Element Solution) global solution as forcing [25]. It includes the whole Iberian Peninsula coast (L1, Figure 1), with a horizontal resolution of 0.06°. The ocean boundary conditions were given in cascade starting at the first level. Therefore, the 2D barotropic model is only used to predict the external tidal conditions necessary to feed the L2 (Figure 1) baroclinic model with surface elevation.

The second and third domains are baroclinic models (L2 and L3, Figure 1), being the second domain a coarse representation (horizontal resolution of 0.02°) of the Rias Baixas and adjacent coastal area, while the third includes the same area with a higher resolution (horizontal resolution of 0.005°) and it is directly coupled to L2 at the open boundaries. A z-level vertical discretization was adopted, with L2 and L3 having 46 and 42 vertical layers, where the bottom 39 (L2) and 35 (L3) were in Cartesian coordinates and the top 10 m were 7 sigma coordinate layers.

To obtain the initial ocean stratification, L2 and L3 were forced at the open boundaries with monthly mean climatologies profiles from Levitus [26,27].

The surface boundary condition was imposed using hourly high-resolution results from the Weather Research and Forecasting model with a spatial resolution of 4 km. At the surface, heat fluxes were imposed with parameterizations similar to those described by Chapra [28]. The sensible and latent heat fluxes were calculated using the Bowen and Dalton laws, respectively [28]. For the bottom boundary condition, shear friction was imposed, assuming a velocity logarithmic profile.

The freshwater inputs inside Rias Baixas and the Minho estuary outflow were considered as landward boundary conditions (L3). This last was imposed offline in L3 as momentum, water and mass discharges to the coastal model. This outflow was computed with a 0.0065°–0.001° horizontal resolution estuarine model for the inner part of the Minho River [14].

The model was run from November 1997 to May 1998 (the first six-months were considered as spin-up).

A more detailed description of this numerical implementation can be found in Sousa et al. [14].

Model scenarios

Four simulations were carried out in order to evaluate the main factors that induce the unusual pattern on May 1998 (Table 1). In the first scenario, the real conditions corresponding to May 1998 were considered (called reference scenario hereafter). The second simulation was run without Minho River discharge during the whole simulation period (called without Minho River scenarios hereafter), but including the freshwater discharges inside each Ria. In the third scenario, the influence of the rivers flowing into the Rias was analyzed. Thus, no discharges from rivers inside the Rias were considered (called without inside rivers scenario hereafter), but was considered the Minho River discharge. With respect to the influence of estuarine morphology, a new bathymetric grid was built interchanging the position of the entire estuarine areas in study, from the mouth to the easternmost grid point of both Rias (inclusive the islands) (Figure 2). This new grid was considered in the last scenario (called interchanged Rias scenario hereafter), where the Ria de Pontevedra is nearest to Minho River instead of the Ria de Vigo (Figure 2b). Despite this interchanging, each Ria preserves its freshwater input, i.e. Verdugo River discharges to the Ria de Vigo and Lérez River to the Ria de Pontevedra.

Three rivers were imposed as land boundaries for the simulations period: Minho, Verdugo, and Lérez (Figure 1). The Minho River discharge shows an atypical pattern with high values during early May (Figure 3). The maximum value is about 1600 m^3s^{-1}. After this date, a continuous decrease of the river discharge is observed. The Verdugo and Lérez Rivers present a similar pattern with a small runoff decrease in late May, except

Table 1. Model runs overview.

Name	*Minho River discharge*	*Verdugo River discharge*	*Lérez River discharge*
Reference	X	X	X
Without Minho river	-	X	X
Without Inside river	X	-	-
Interchanged Rias	X	X	X

between 29 April and 3 May when the Verdugo River shows high values (250 m^3s^{-1}, Figure 3).

To accomplish the main objective of this work, horizontal maps of the surface density were taken in the instant that Minho River plume reaches the Rias de Vigo and Pontevedra. They were analyzed for the reference and interchanged Rias scenarios. Moreover, the temporal evolution of vertical estuarine circulation at stations located at mouths of Rias de Vigo (~30 m deep) and Pontevedra (~35 m deep) (Figure 1, stars) were calculated for the same scenarios. The residual current was also computed for all scenarios by time averaging the hourly transient values from current velocities at mouth of both Rias (Figure 1, stars) over a period of 13 hours, corresponding to the plume intrusion.

Results and Discussion

In this section, the impact of the Minho River plume and the Rias de Vigo and Ponetevedra main freshwater sources in the establishment of negative circulation in the Rias will be evaluated. In addition, the impact of changes in the position of the Rias (e.g. estuarine morphology *versus* distance to Minho River) on estuarine circulation will be evaluated for the period of May 1998.

Real conditions in May 1998

Surface density maps show a northward spread of the Minho River plume reaching the Ria de Vigo on 11 May (Figure 4a). During the following days, the northward displacement of the plume continued and the plume intrusion is observed in Ria de Pontevedra on 13 May (Figure 4b). This situation generates an unusual surface density pattern influencing the physical properties within the Rias, reversing the normal density pattern characterized by the presence of saltier water at the outer part of the estuary and freshwater at the middle-inner part. These unusual inflows of freshwater are critical from an ecological point of view, affecting the exchange between the Rias and the ocean, and therefore changing the normal input of nutrients [9,10]. Thus, this intrusion into the Rias can impose a control on new production within the estuary environment and changes in the mortality of halophyte organisms inside an estuary [29]. Changes in estuarine density gradients can also affect higher trophic levels, particularly the distribution of fishes in estuaries [2]. These consequences demonstrate the high interest in characterizing these types of events in detail.

In terms of estuarine circulation, the existence of an important external freshwater supply at the mouth of an estuary can generate

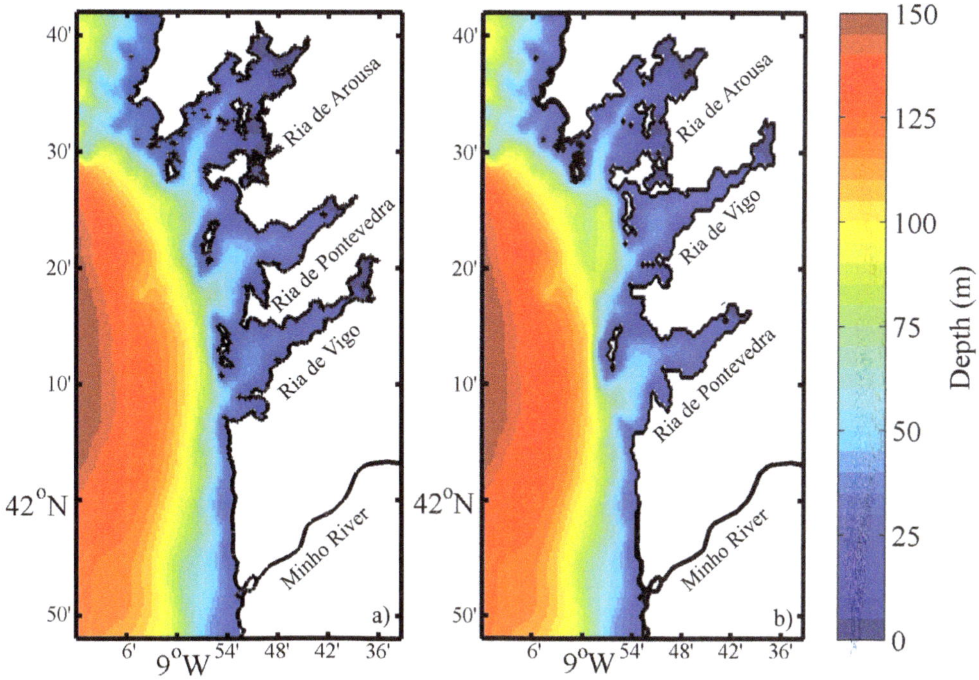

Figure 2. Reference (a) and interchanged (b) bathymetric grids. To simplify visual comparison the Verdugo and Lérez rivers were multiplied by 5.

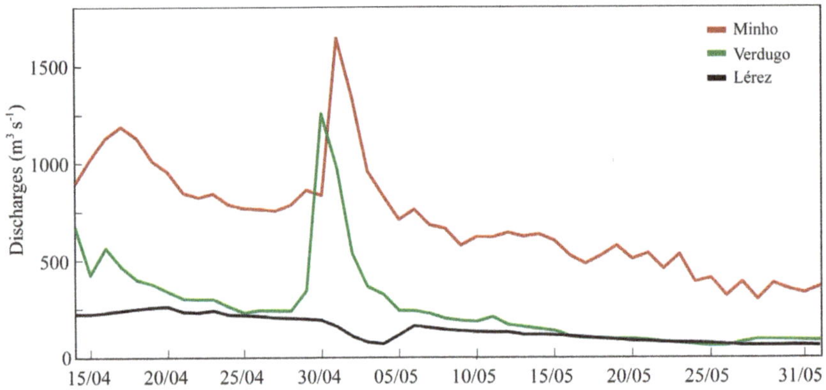

Figure 3. Time series of Minho, Verdugo and Lérez River discharges from 15 April to 31 May 1998.

the reversal of the normal estuarine circulation [14]. Thereby, the temporal evolution of vertical estuarine circulation at stations located at mouths of Rias de Vigo and Pontevedra (Figure 1, stars) between 11 and 16 May 1998 is also studied through the analysis of the along axis velocity component vertical profiles (Figure 5). In Ria de Vigo (Figure 5a), between 11 and 13 May, a negative circulation (upstream circulation) at upper layers (0–8 m) is observed. There is a landward water movement with highest velocities (0.25 m s^{-1}) between 0–2 m depth and a seaward movement at bottom layers. Positive surface velocities are more intense than the negative bottom ones, corresponding to the intrusion of the Minho estuarine outflow into the Ria de Vigo [14].

A similar pattern is recognized in Ria de Pontevedra (Figure 5b), although the surface water moving from ocean to land is observed later (13–15 May). In addition, this unusual circulation is more intense (0.30 m s^{-1}) than in Ria de Vigo, with positive values until 10 m depth. In fact, around 13–14 May surface water moving landward was observed during a tidal cycle. Freshwater supplied by the Minho River generates this unusual circulation, which tends to stop water exchange between the Rias and the shelf, increasing its residence time and consequently changing the water quality [10]. This inversion on estuarine circulation may introduce into the Rias dinoflagellate blooms, which are usually generated in coastal waters, decreasing the abundance of marine species in this area [30].

Due to the ecological importance of estuaries and other coastal ecosystems, external freshwater intrusion inside estuaries has also been studied worldwide, as usually as consequence of remote river plumes spreading. As an example, Monteiro and Largier [31] and Hickey et al. [32] showed that external sources of freshwater can

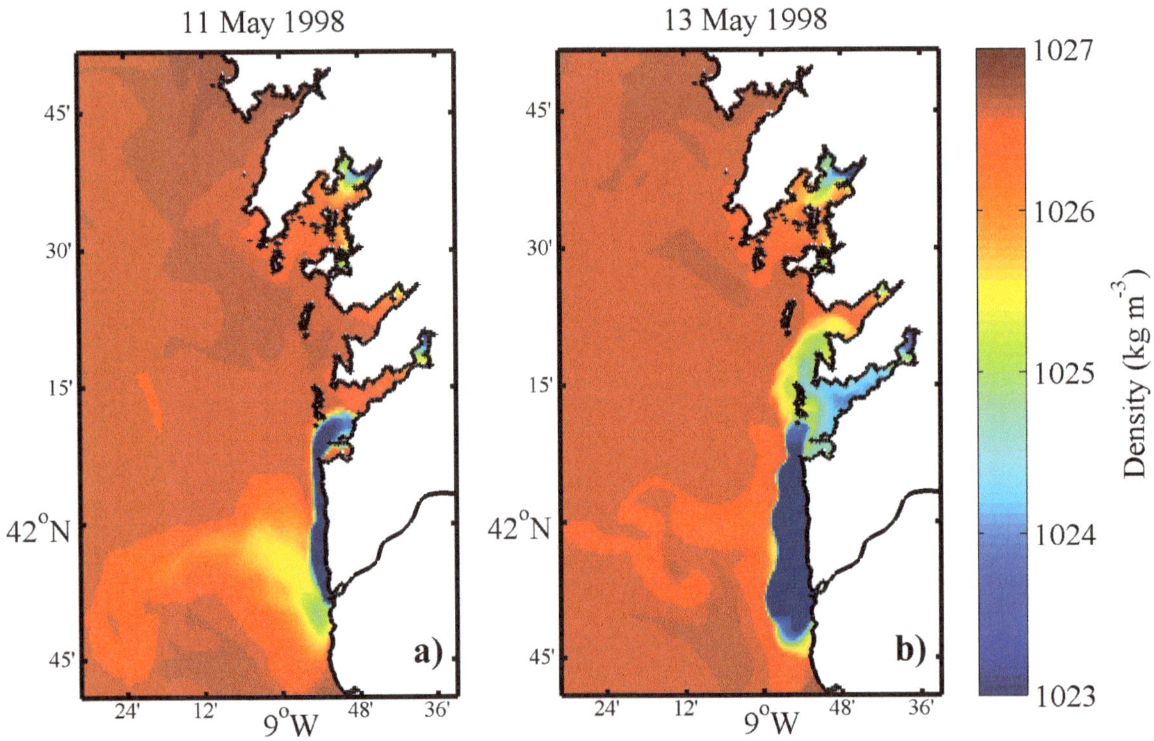

Figure 4. Surface density maps, considering the reference scenario.

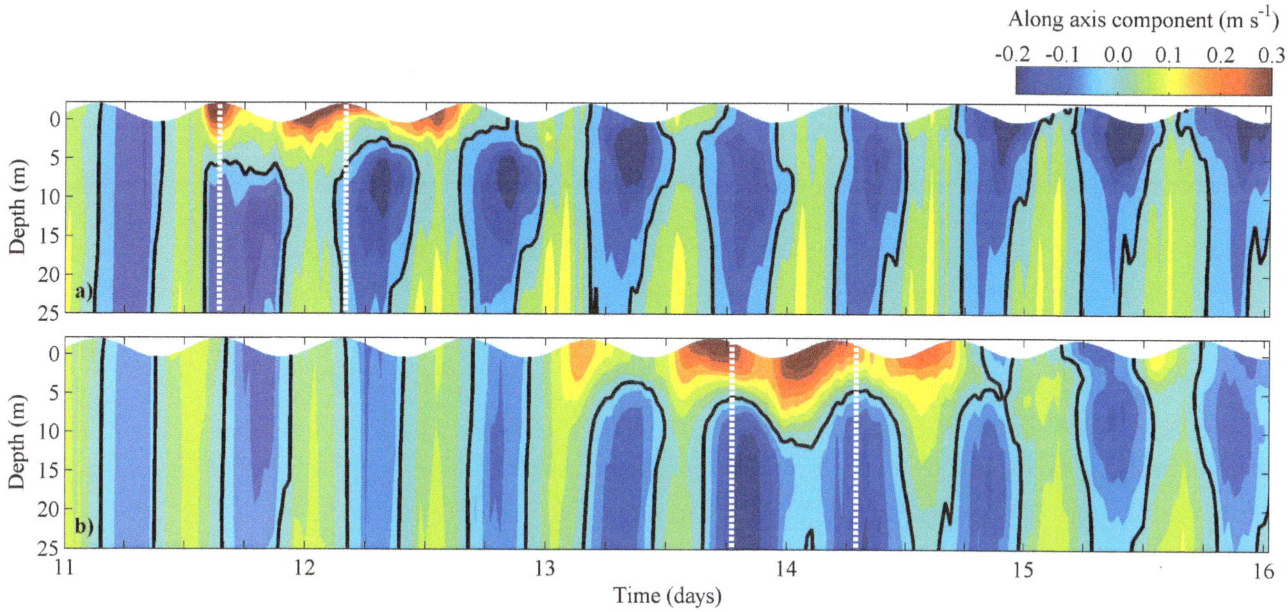

Figure 5. Temporal evolution of vertical profiles of along velocity component at stations located at mouths of Rias de Vigo (b) and Pontevedra (d), considering the reference scenario. Black line corresponds to 0 m s^{-1}.

have an higher impact on the mean baroclinic circulation inside the Saldanha and Willapa bays than usual estuarine processes (ex: changes in river flow or neap-spring variation in mixing). These results are consistent with the results presented here for the Rias de Vigo and Pontevedra.

Effect of Minho River and rivers flowing into the Rias

Residual circulation is crucial in order to determine the long-term transport in estuaries and therefore a key parameter in the dynamical behavior during the intrusion of the Minho estuarine plume on Rias Baixas. Thus, the residual currents induced by the plume intrusion are presented in Figure 6 for the reference, without Minho River, and inside rivers scenarios. The pattern of the vertical structure of velocity is very similar in both Rias concerning the reference scenario (Figure 6, red line). The inversion of circulation occurs at the same depth (~10 m). Both in the surface and in the bottom, the highest velocities are observed at the Ria located further away from Minho River (Ria de Pontevedra). Without Minho River discharge (Figure 6, grey line), lower residual velocities are observed for both Rias, showing that the Minho River discharge is the main responsible for the inversion of the circulation pattern and for the non-tidal estuarine transport of dissolved and particulate materials inside the Rias. For the third scenario (removing the rivers flowing into the Rias, Figure 6, blue line), the residual current profile is similar to the one corresponding to the reference simulation. A greater difference is observed on the surface layer for both Rias. The velocity is higher (about 0.02 m s^{-1}) than for the reference scenario, showing that if rivers flowing into the Rias are neglected, the Minho estuarine plume would affect more significantly the surface circulation of the Rias. These results also demonstrate that the unusual higher surface current velocities found at the Ria de Pontevedra are not induced by the internal freshwater discharges characteristics.

These results show that the Minho River discharge plays an important role in the establishment of the negative circulation inside the Rias. On the contrary, the small rivers that flow through the interior of the Rias do not influence significantly the inversion of the circulation. These findings are in accordance with the results reported for other regions worldwide, as Willapa Bay [1], where under favorable downwelling winds, the plume from Columbia River forces an axial gradient reversal inside the bay.

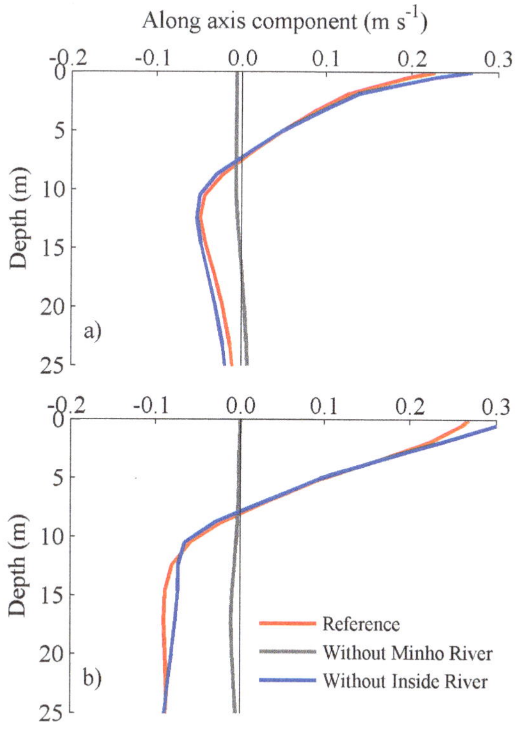

Figure 6. Residual current profiles during the plume intrusion (dashed lines, Figure 4) at stations located at mouths of Rias de Vigo (a) and Pontevedra (b), considering the different scenarios.

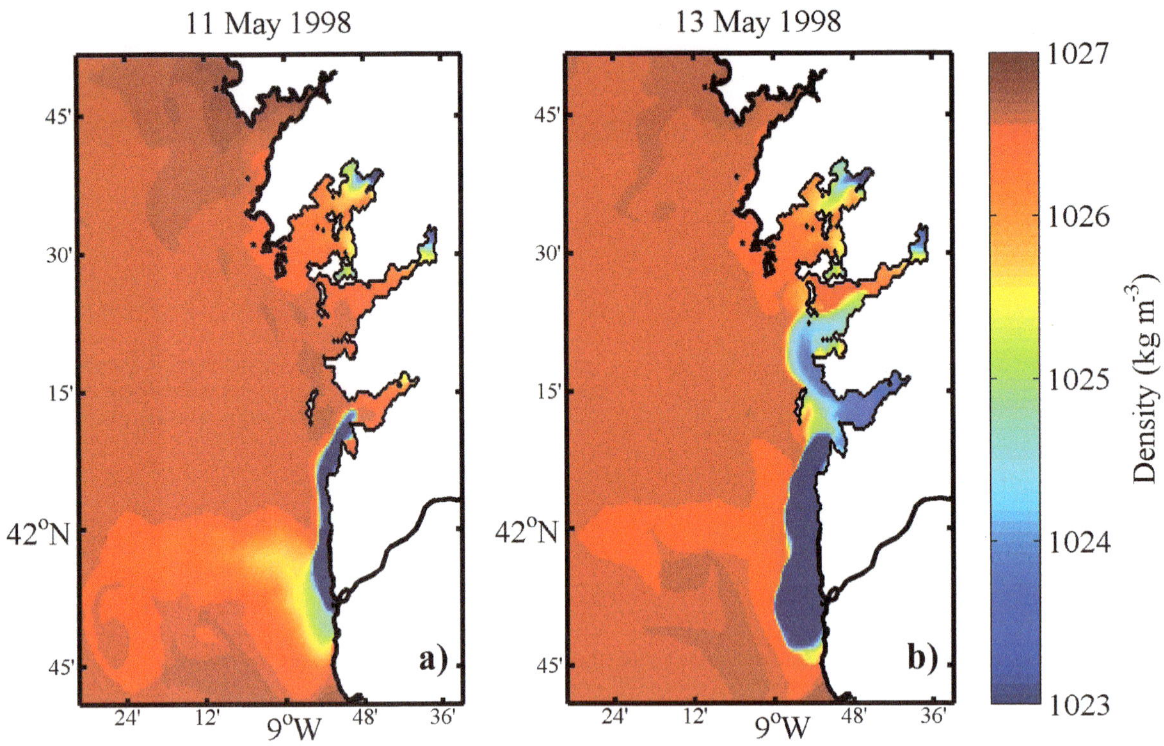

Figure 7. Surface density maps, considering the interchanged Rias scenario.

Effect of estuarine morphology

Comparing the results for each Ria independently, it is observed that the integrated residual currents, both at surface and bottom, are greater in Ria de Pontevedra (that is further way of Minho River) than in Ria de Vigo for all scenarios. For example, in the reference scenario (Figure 6, red line), the surface velocity in Ria de Pontevedra is 0.18 m s^{-1} (Figure 6b) while in Ria de Vigo is 0.12 m s^{-1} (Figure 6a). This could (or not) be related to differences in the morphology of each Ria (width and depth as well as different mouth geometries). In order to better analyze this situation the influence of the estuary morphology was evaluated by interchanging the location of Rias de Vigo and Pontevedra

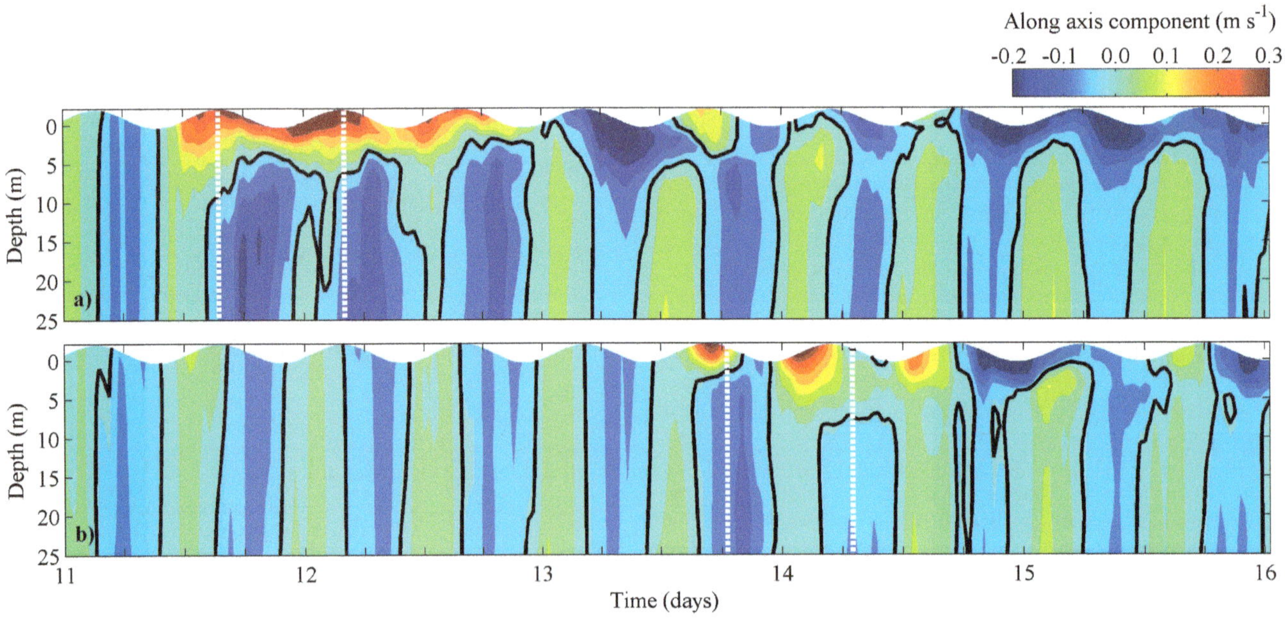

Figure 8. Temporal evolution of vertical profiles of along velocity component at stations located at mouths of Rias de Pontevedra (a) and Vigo (a), considering the interchanged Rias scenario. Black line corresponds to 0 m s^{-1}.

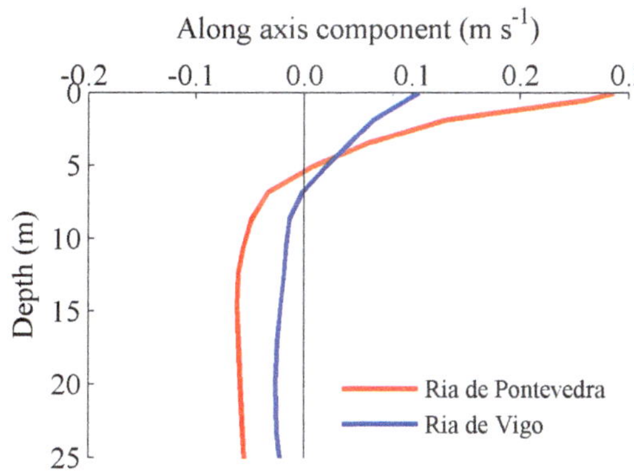

Figure 9. Residual current profiles during the plume intrusion (dashed lines, Figure 7) at stations located at mouths of Rias de Pontevedra and Vigo, considering the interchanged Rias scenario.

(Figure 2b). After that, the methodology adopted for the results presentation in this scenario is analogous to those shown previously.

Generally, the surface density maps are very similar to the reference scenario, with the plume intrusion reaching the Rias at the same period (Figures 7a and 7b).

For this scenario, the temporal evolution of the vertical profiles of along axis velocity component at the mouth of Rias de Pontevedra and Vigo is also calculated (Figures 8a and 8b). The pattern found for Ria de Pontevedra (Figure 8a) is very similar to that found for the real scenario (Figure 5b), with positive velocities at surface layers (0.25 m s^{-1}) and negative values near bottom (–0.10 m s^{-1}), corresponding to the intrusion of the Minho estuarine outflow. In Ria de Vigo (Figure 8d), the surface velocity is weaker (0.15 m s^{-1}) than for the reference scenario (Figure 5a), and in this case the surface water moving landwards only lasts one tidal cycle. In this scenario, the negative circulation is also stronger (in terms of velocity intensity) at Ria de Pontevedra (Figure 8a), indicating that the morphologic estuarine changes are particularly linked to the intensity of the circulation pattern. This is corroborated by the values of the residual currents generated during the plume intrusion in this scenario (Figure 9). Surface velocities are twice higher for Ria de Pontevedra (0.09 m s^{-1}) than for Ria de Vigo (0.04 m s^{-1}), showing that the estuarine morphology influences the intensity of the negative circulation.

As expected, these results reveal that the greater density gradients are visible in the Ria located close to the Minho River mouth for all studied scenarios (including the interchanged one). However, in terms of residual circulation, the morphology of each Ria become a critical factor, since residual flows are highly influenced by depth, width and mouth geometry of an estuary [33,34]. Here, the Ria de Pontevedra always presents higher velocities than the Ria de Vigo, showing the importance of the estuarine morphology. The Ria de Pontevedra is slightly narrower and deeper than Ria de Vigo. As the water content averages for both Rias is similar (3.12 km^3 for Ria de Vigo and 3.45 km^3 for Ria de Pontevedra, Alvarez et al. [35]), the morphology entrances is the crucial factor the plume propagation inside the Rias. Regarding only the Ria de Pontevedra, the residual surface velocity in the interchanged scenario is slightly higher (0.28 m s^{-1}) than the reference one (0.26 m s^{-1}). This difference can be attributed to the higher buoyancy forcing (>density gradients) which squeezes the plume to the surface, increasing the horizontal current velocity towards the interior of the Ria. Warrick and Stevens [36] also showed in their work in Elwha River in the Washington coast that the topographic features influence the buoyant plume behavior, affecting the coastal circulation and sediment transport.

Conclusions

The main aim of this work was to research the main reasons that induce unusual higher inflow surface current velocities at the Ria located further away from Minho River when the normal estuarine circulation is reversed, through a numerical model MOHID. The individual effect of the Minho River discharge, rivers tributaries flowing into the Rias and estuarine morphology in the Rias Baixas circulation was analyzed simulating several scenarios.

The results obtained from this study have shown that the freshwater intrusion of the Minho River into Rias de Vigo and Pontevedra was the main responsible for the negative circulation pattern and unusual density gradient in May 1998.

Without rivers flowing into the Rias, the Minho estuarine plume affects more significantly the Rias surface circulation, showing that small rivers inside Rias slightly attenuate the negative circulation.

Independently of the location of both Rias, the negative circulation was always stronger in Ria de Pontevedra, showing that the proximity with the Minho River is not the main factor justifying this pattern. Instead, the morphologic estuarine characteristics determine the unusual circulation pattern found.

In summary, the model predictions reproduce accurately the circulation pattern on May 1998 event. The obtained model results indicated that the Minho River plays an important role in the density patterns in Rias de Vigo and Pontevedra and that their morphology is the main factor for the establishment of the intensity of the circulation pattern inversion.

Author Contributions

Conceived and designed the experiments: MCS MGG JMD. Performed the experiments: MCS RM. Analyzed the data: MCS RM IA NV MGG JMD. Contributed reagents/materials/analysis tools: MCS RM IA NV. Wrote the paper: MCS RM IA NV MGG JMD.

References

1. Banas NS, Hickey BM, MacCready P, Newton JA (2004) Dynamics of Willapa Bay, Washington: A highly unsteady, partially mixed estuary. Journal of Physical Oceanography 34: 2413–2427.
2. Das A, Justic D, Inoue M, Hoda A, Huang HS, et al. (2012) Impacts of Mississippi River diversions on salinity gradients in a deltaic Louisiana estuary: Ecological and management implications. Estuarine Coastal and Shelf Science 111: 17–26.
3. Wong KC, Lu X (1994) Low-Frequency Variability in Delaware Inland Bays. Journal of Geophysical Research-Oceans 99: 12683–12695.
4. Nonn H (1966) Les régions cotiéres de la Galice (Espagne). Etude géomorphologique [PhD thesis]: Pub. Fac. Lettres. Univ. Strasbourg. 591 p.
5. Wooster WS, Bakun A, Mclain DR (1976) Seasonal Upwelling Cycle Along Eastern Boundary of North-Atlantic. Journal of Marine Research 34: 131–141.
6. Tenore KR, Alonsonoval M, Alvarezossorio M, Atkinson LP, Cabanas JM, et al. (1995) Fisheries and Oceanography Off Galicia, Nw Spain - Mesoscale Spatial and Temporal Changes in Physical Processes and Resultant Patterns of Biological Productivity. Journal of Geophysical Research-Oceans 100: 10943–10966.
7. Alonso-Perez F, Ysebaert T, Castro CG (2010) Effects of suspended mussel culture on benthic-pelagic coupling in a coastal upwelling system (Ria de Vigo, NW Iberian Peninsula). Journal of Experimental Marine Biology and Ecology 382: 96–107.

8. Rio-Barja FX, Rodríguez-Lestegás F (1996) Os ríos, in as augas de Galicia. Consello de Cultura.
9. Alvarez I, deCastro M, Gomez-Gesteira M, Prego R (2006) Hydrographic behavior of the Galician Rias Baixas (NW Spain) under the spring intrusion of the Mino River. Journal of Marine Systems 60: 144–152.
10. deCastro M, Alvarez I, Varela M, Prego R, Gomez-Gesteira M (2006) Mino River dams discharge on neighbor Galician Rias Baixas (NW Iberian Peninsula): Hydrological, chemical and biological changes in water column. Estuarine Coastal and Shelf Science 70: 52–62.
11. Sousa MC, Vaz N, Alvarez I, Dias JM (2013) Effect of Minho estuarine plume on Rias Baixas: numerical modeling approach. Journal of Coastal Research SI65: 2059–2064.
12. deCastro M, Gomez-Gesteira M, Alvarez I, Prego R (2004) Negative estuarine circulation in the Ria of Pontevedra (NW Spain). Estuarine Coastal and Shelf Science 60: 301–312.
13. Mourino C, Fraga F (1982) Hydrography of the Ria of Vigo - 1976-1977 - Abnormal Influence of the Mino River. Investigacion Pesquera 46: 459–468.
14. Sousa MC, Vaz N, Alvarez I, Gomez-Gesteira M, Dias JM (2014) Modeling the Minho River plume intrusion into the Rias Baixas (NW Iberian Peninsula). Continental Shelf Research 85: 30–41.
15. Choi BJ, Wilkin JL (2007) The effect of wind on the dispersal of the Hudson River plume. Journal of Physical Oceanography 37: 1878–1897.
16. García-Berdeal I, Hickey BM, Kawase M (2002) Influence of wind stress and ambient flow on a high discharge river plume. Journal of Geophysical Research-Oceans 107.
17. Jurisa JT, Chant R (2012) The coupled Hudson River estuarine-plume response to variable wind and river forcings. Ocean Dynamics 62: 771–784.
18. Marques WC, Fernandes EH, Monteiro IO, Moller OO (2009) Numerical modeling of the Patos Lagoon coastal plume, Brazil. Continental Shelf Research 29: 556–571.
19. Vaz N, Mateus M, Plecha S, Sousa MC, Leitao PC, et al. (2014) Modeling SST and chlorophyll patterns in a coupled estuary-coastal system of Portugal: the Tagus case study. Journal of Marine Systems. doi: 10.1016/j.jmarsys.2014.1005.1022.
20. Rios AF, Perez FF, Fraga F (1992) Water Masses in the Upper and Middle North-Atlantic Ocean East of the Azores. Deep-Sea Research Part a-Oceanographic Research Papers 39: 645–658.
21. Varela RA, Roson G, Herrera JL, Torres-Lopez S, Fernandez-Romero A (2005) A general view of the hydrographic and dynamical patterns of the Rias Baixas adjacent sea area. Journal of Marine Systems 54: 97–113.
22. Prego R, Dale AW, deCastro M, Gomez-Gesteira M, Taboada JJ, et al. (2001) Hydrography of the Pontevedra Ria: Intra-annual spatial and temporal variability in a Galician coastal system (NW Spain). Journal of Geophysical Research-Oceans 106: 19845–19857.
23. Fraga F, Margalef R (1979) Las rias gallegas. Estudio y Explotacion del mar en Galicia. Santiago de Compostela: 101–121.
24. Prego R, Fraga F (1992) A Simple-Model to Calculate the Residual Flows in a Spanish Ria - Hydrographic Consequences in the Ria-De-Vigo. Estuarine Coastal and Shelf Science 34: 603–615.
25. Lyard F, Lefevre F, Letellier T, Francis O (2006) Modelling the global ocean tides: modern insights from FES2004. Ocean Dynamics 56: 394–415.
26. Locarnini RA, Mishonov AV, Antonov JI, Boyer TP, Garcia HE, et al. (2010) World Ocean Atlas 2009, Volume 1: Temperature. Washington, DC: U.S. Government Printing Office. 184 p.
27. Antonov JI, Seidov D, Boyer TP, Locarnini RA, Mishonov AV, et al. (2010) World Ocean Atlas 2009, Volume 2: Salinity. Washington, DC: U.S. Government Printing Office. 184 p.
28. Chapra SC (1997) Surface Water Quality Modeling. New York: McGraw-Hill.
29. Genio L, Sousa A, Vaz N, Dias JM, Barroso C (2008) Effect of low salinity on the survival of recently hatched veliger of Nassarius reticulatus (L.) in estuarine habitats: A case study of Ria de Aveiro. Journal of Sea Research 59: 133–143.
30. Sordo I, Barton ED, Cotos JM, Pazos Y (2001) An inshore poleward current in the NW of the Iberian peninsula detected from satellite images, and its relation with G-catenatum and D-acuminata blooms in the Galician Rias. Estuarine Coastal and Shelf Science 53: 787–799.
31. Monteiro PMS, Largier JL (1999) Thermal stratification in Saldanha Bay (South Africa) and subtidal, density-driven exchange with the coastal waters of the Benguela upwelling system. Estuarine Coastal and Shelf Science 49: 877–890.
32. Hickey BM, Zhang X, Banas N (2002) Coupling between the California Current System and a coastal plain estuary in low riverflow conditions. Journal of Geophysical Research-Oceans 107.
33. Lee J, Valle-Levinson A (2012) Influence of bathymetry on hydrography and circulation at the region between an estuary mouth and the adjacent continental shelf. Continental Shelf Research 41: 77–91.
34. Valentim JM, Vaz N, Silva H, Duarte B, Cacador I, et al. (2013) Tagus estuary and Ria de Aveiro salt marsh dynamics and the impact of sea level rise. Estuarine Coastal and Shelf Science 130: 138–151.
35. Alvarez I, deCastro M, Gomez-Gesteira M, Prego R (2005) Inter- and intra-annual analysis of the salinity and temperature evolution in the Galician Rias Baixas-ocean boundary (northwest Spain). Journal of Geophysical Research-Oceans 110.
36. Warrick JA, Stevens AW (2011) A buoyant plume adjacent to a headland-Observations of the Elwha River plume. Continental Shelf Research 31: 85–97.

Permissions

All chapters in this book were first published in PLOS ONE, by The Public Library of Science; hereby published with permission under the Creative Commons Attribution License or equivalent. Every chapter published in this book has been scrutinized by our experts. Their significance has been extensively debated. The topics covered herein carry significant findings which will fuel the growth of the discipline. They may even be implemented as practical applications or may be referred to as a beginning point for another development.

The contributors of this book come from diverse backgrounds, making this book a truly international effort. This book will bring forth new frontiers with its revolutionizing research information and detailed analysis of the nascent developments around the world.

We would like to thank all the contributing authors for lending their expertise to make the book truly unique. They have played a crucial role in the development of this book. Without their invaluable contributions this book wouldn't have been possible. They have made vital efforts to compile up to date information on the varied aspects of this subject to make this book a valuable addition to the collection of many professionals and students.

This book was conceptualized with the vision of imparting up-to-date information and advanced data in this field. To ensure the same, a matchless editorial board was set up. Every individual on the board went through rigorous rounds of assessment to prove their worth. After which they invested a large part of their time researching and compiling the most relevant data for our readers.

The editorial board has been involved in producing this book since its inception. They have spent rigorous hours researching and exploring the diverse topics which have resulted in the successful publishing of this book. They have passed on their knowledge of decades through this book. To expedite this challenging task, the publisher supported the team at every step. A small team of assistant editors was also appointed to further simplify the editing procedure and attain best results for the readers.

Apart from the editorial board, the designing team has also invested a significant amount of their time in understanding the subject and creating the most relevant covers. They scrutinized every image to scout for the most suitable representation of the subject and create an appropriate cover for the book.

The publishing team has been an ardent support to the editorial, designing and production team. Their endless efforts to recruit the best for this project, has resulted in the accomplishment of this book. They are a veteran in the field of academics and their pool of knowledge is as vast as their experience in printing. Their expertise and guidance has proved useful at every step. Their uncompromising quality standards have made this book an exceptional effort. Their encouragement from time to time has been an inspiration for everyone.

The publisher and the editorial board hope that this book will prove to be a valuable piece of knowledge for researchers, students, practitioners and scholars across the globe.

List of Contributors

Ari J. S. Ferreira, Rania Siam, João Ahmed Moustafa, Ahmed Sayed, Mohamed A. Ghazy, Hazem Sharaf, Amged Ouf, Alyaa M. Abdel-Haleem, Eman Ramadan and Hamza El- Dorry
Department of Biology and the Science and Technology Research Center, School of Sciences and Engineering, The American University in Cairo, Cairo, New Cairo, Egypt

C. Setubal
Departamento de Bioquímica, Instituto de Química, Universidade de São Paulo, São Paulo, Brazil

Felipe S. Chambergo
Escola de Artes, Ciências e Humanidades, Universidade de São Paulo, São Paulo, Brazil

Adam S. Dawe, Intikhab Alam, Heikki Lehvaslaiho, John A. C. Archer, Boris R. Jankovic and Vladimir B. Bajic
Computational Bioscience Research Center, King Abdullah University of Science and Technology, Thuwal, Kingdom of Saudi Arabia

AndréAntunes
Institute for Biotechnology and Bioengineering, Centre of Biological Engineering, University of Minho, Portugal

Ulrich Stingl
Red Sea Research Center, King Abdullah University of Science and Technology, Thuwal, Kingdom of Saudi Arabia

Mitchell Sogin
Josephine Bay Paul Center, Marine Biological Laboratory, Woods Hole, Massachusetts, United States of America

Fabiana Saporiti, Lisette Zenteno, Alex Aguilar and Luis Cardona
Department of Animal Biology and Institut de Recerca de la Biodiversitat (IRBio), Faculty of Biology, University of Barcelona, Barcelona, Spain

Stuart Bearhop
Centre for Ecology and Conservation, School of Biosciences, University of Exeter, Cornwall Campus, Penryn, United Kingdom

Laura Silva, Damián G. Vales and Enrique A. Crespo
Laboratory of Marine Mammals, Centro Nacional Patagónico - Consejo Nacional de Investigaciones Científicas y Tícnicas (CENPAT-CONICET), Puerto Madryn, Argentina

Carolyn K. Robb
School of Environment and Management, Royal Roads University, Victoria, British Columbia, Canada

Carlotta Molfese and Jason M. Hall-Spencer
Marine Biology and Ecology Research Centre, Plymouth University, Plymouth, United Kingdom

Doug Beare
Worldfish, Batu Maung, Penang, Malaysia

Ursula Witte
Oceanlab, University of Aberdeen, Aberdeen, United Kingdom

Niels Jobstvogt
Oceanlab, University of Aberdeen, Aberdeen, United Kingdom
Aberdeen Centre for Environmental Sustainability (ACES), University of Aberdeen, Aberdeen, United Kingdom

Michael Townsend
National Institute of Water and Atmospheric Research (NIWA), Hamilton, New Zealand

Nick Hanley
Department of Geography and Sustainable Development, University of St Andrews, St Andrews, United Kingdom

Timothy Hoellein, Miguel Rojas, Adam Pink, Joseph Gasior and John Kelly
Department of Biology, Loyola University Chicago, Chicago, Illinois, United States of America

Ellen Kenchington, Camille Lirette and Lindsay Beazley
Bedford Institute of Oceanography, Department of Fisheries and Oceans, Dartmouth, Nova Scotia, Canada

Francisco Javier Murillo
Departamento de Zooloxía e Antropoloxía Física, Facultade de Bioloxía, Universidade de Santiago de Compostela, Santiago de Compostela, Spain

Mar Sacau
Instituto Español de Oceanografía, Centro Oceanográfico de Vigo, Programa de Pesquerías Lejanas, Vigo, Spain

Mariano Koen-Alonso, Neil Ollerhead and Vonda Wareham
Northwest Atlantic Fisheries Centre, Department of Fisheries and Oceans, St. John's, Newfoundland and Labrador, Canada

Andrew Kenny
Centre for Environment, Fisheries and Aquaculture Science, Lowestoft, Suffolk, United Kingdom

Sandip Mandal, Christopher Locke, Mamoru Tanaka and Hidekatsu Yamazaki
Department of Ocean Sciences, Tokyo University of Marine Science and Technology, Minato-ku, Tokyo, Japan

Erik Olsen
Institute of Marine Research, Bergen, Norway

David Fluharty
School of Marine and Environmental Affairs, University of Washington, Seattle, Washington, United States of America

Alf Håkon Hoel
Institute of Marine Research, Tromsø, Norway

Kristian Hostens and Ellen Pecceu
Animal Sciences, Aquatic Environment and Quality, Bio-environmental Research, Institute for Agricultural and Fisheries Research (ILVO), Oostende, Belgium

Frank Maes
Maritime Institute, Ghent University, Ghent, Belgium

David D. Huff
nstitute of Marine Sciences, University of California Santa Cruz, Santa Cruz, California, United States of America

Rebecca R. Miller
Institute of Marine Sciences, University of California Santa Cruz, Santa Cruz, California, United States of America
Fisheries Ecology Division, Southwest Fisheries Science Center, National Marine Fisheries Service, National Oceanic and Atmospheric Administration, Santa Cruz, California, United States of America

John C. Field, Don E. Pearson and Alec D. MacCall
Fisheries Ecology Division, Southwest Fisheries Science Center, National Marine Fisheries Service, National Oceanic and Atmospheric Administration, Santa Cruz, California, United States of America

Jarrod A. Santora
Center for Stock Assessment Research, University of California Santa Cruz, Santa Cruz, California, United States of America

Isaac D. Schroeder
Environmental Research Division, Southwest Fisheries Science Center, National Marine Fisheries Service, National Oceanic and Atmospheric Administration, Pacific Grove, California, United States of America

Meisha Key
California Department of Fish and Wildlife, c/o, National Marine Fisheries Service, Southwest Fisheries Science Center, Santa Cruz, California, United States of America

Juan Moreno Navas, Lea-Anne Henry and Sebastian J. Hennige
Centre for Marine Biodiversity and Biotechnology, Heriot-Watt University, Edinburgh, United Kingdom

J. Murray Roberts
Centre for Marine Biodiversity and Biotechnology, Heriot-Watt University, Edinburgh, United Kingdom
Center for Marine Science, University of North Carolina, Wilmington, North Carolina, United States of America
Scottish Association for Marine Science, Scottish Marine Institute, Oban, United Kingdom

Peter L. Miller
Remote Sensing Group, Plymouth Marine Laboratory, Plymouth, United Kingdom

Mariana G. Bender and Sergio R. Floeter
Departamento de Ecologia e Zoologia, Universidade Federal de Santa Catarina, Floriano´ polis, SC, Brazil

Gustavo R. Machado, Cassiano Monteiro-Netto and Carlos E. L. Ferreira
Departamento de Biologia Marinha, Universidade Federal Fluminense, Niteroi, RJ, Brazil

Paulo Joséde Azevedo Silva
Fundac¸ão Instituto de Pesca, Arraial do Cabo, RJ, Brazil

Osmar J. Luiz
Department of Biological Sciences, Macquarie University, Sydney, NSW, Australia

Kathleen Schwerdtner Máñez
Department of Social Sciences, Leibniz Center for Tropical Marine Ecology, Bremen, Germany
Asia Research Center, Murdoch University, Murdoch, Western Australia, Australia

Poul Holm
Trinity Long Room Hub Arts and Humanities Research Institute, Trinity College Dublin, Dublin, Ireland

Louise Blight
World Wildlife Fund (WWF) -Canada, Vancouver, British Columbia, Canada

Marta Coll
Renewable Marine Resources Department, Institute of Marine Science, Barcelona, Spain
Laboratoire Écosystémes Marins Exploités, SéteCedex, France

Alison MacDiarmid
Marine Ecology, National Institute of Water and Atmospheric Research, Kilbirnie, Wellington, New Zealand

Henn Ojaveer
Estonian Marine Institute, University Tartu, Tartu, Estonia

Bo Poulsen
Faculty of Social Sciences, Aalborg University, Aalborgø, Denmark

Malcolm Tull
Murdoch Business School, Murdoch University, Murdoch, Western Australia, Australia

Sébastien Personnic, Charles F. Boudouresque, Delphine Thibault-Botha, Mireille Harmelin-Vivien, Jean-Christophe Poggiale, Thierry Thibaut and Sandrine Ruitton
Aix-Marseille University, Mediterranean Institute of Oceanography (MIO), Universitéde Toulon, CNRS/ INSU, IRD, UM 110, Marseille, France

Patrick Astruch and Patrick Bonhomme
GIS Posidonie, Pytheas Institute, Aix-Marseille University, Marseille, France

Enric Ballesteros
Centre d'Estudis Avanc¸ats de Blanes - CSIC, Blanes, Spain

Sylvain Blouet
Aire marine protégée de la côte Agathoise, site natura 2000, Agde, France

Denise Bellan-Santini and Florent Renaud
Aix-Marseille University, Institut Méditerranéen de Biodiversitéet d'Ecologie (IMBE), UMR 7263, Station Marine d'Endoume, Marseille, France

Eric Feunteun
Museum National d'Histoire Naturelle, UMR 7208, Station Marine de Dinard, France

Gérard Pergent and Christine Pergent-Martini
Equipe Ecosystémes Littoraux, FRES 3041, University of Corsica, Corte, France

Jérémy Pastor
Universitéde Perpignan, Via Domitia, Centre de Formation et de Recherche sur les Environnements Méditerranéens, UMR 5110, Perpignan, France

Noel A. Pelland and Charles C. Eriksen
School of Oceanography, University of Washington, Seattle, Washington, United States of America

Jeremy T. Sterling and Rolf R. Ream
National Marine Mammal Laboratory, National Oceanographic and Atmospheric Administration, Seattle, Washington, United States of America

Mary-Anne Lea
Institute for Marine and Antarctic Studies, University of Tasmania, Hobart, Tasmania, Australia

Nicholas A. Bond
Joint Institute for the Study of the Atmosphere and Ocean, Seattle, Washington, United States of America

Craig M. Lee
Applied Physics Laboratory, University of Washington, Seattle, Washington, United States of America
School of Oceanography, University of Washington, Seattle, Washington, United States of America

Catherine Longo, Courtney Scarborough and Benjamin D. Best
National Center for Ecological Analysis and Synthesis, Santa Barbara, California, United States of America

Benjamin S. Halpern
National Center for Ecological Analysis and Synthesis, Santa Barbara, California, United States of America
Bren School of Environmental Science and Management, University of California, Santa Barbara, California, United States of America
Imperial College London, Silwood Park Campus, Ascot, United Kingdom

Darren Hardy
National Center for Ecological Analysis and Synthesis, Santa Barbara, California, United States of America
Digital Library Systems & Services, Stanford University, Stanford, California, United States of America

Scott C. Doney
Marine Chemistry & Geochemistry Department, Woods Hole Oceanographic Institution, Woods Hole, Massachusetts, United States of America

Steven K. Katona
Conservation International, Arlington, Virginia, United States of America

Karen L. McLeod
COMPASS, Oregon State University, Department of Zoology, Corvallis, Oregon, United States of America

Andrew A. Rosenberg
Union of Concerned Scientists, Cambridge, Massachusetts, United States of America

Jameal F. Samhouri
Conservation Biology Division, Northwest Fisheries Science Center, National Marine Fisheries Service National Oceanic and Atmospheric Administration, Seattle, Washington United States of America

Nicole A. Hill, Neville Barrett and Justin Hulls
Institute for Marine and Antarctic Studies, University of Tasmania, Hobart, Tasmania, Australia

Emma Lawrence
Digital Productivity Flagship, Commonwealth Scientific and industrial Research Organisation (CSIRO), Brisbane, QLD, Australia

Jeffrey M. Dambacher and Keith R. Hayes
Digital Productivity Flagship, Commonwealth Scientific and industrial Research Organisation (CSIRO), Hobart, Tasmania, Australia

Scott Nichol
Geoscience Australia, Canberra, ACT, Australia

Alan Williams
Oceans and Atmosphere Flagship, Commonwealth Scientific and industrial Research Organisation (CSIRO), Hobart, Tasmania, Australia

Yantao Liang
State Key Laboratory of Marine Environmental Science, Xiamen University, Xiamen, China
Institute of Marine Microbes and Ecospheres, Xiamen University, Xiamen, China
Qingdao Institute of BioEnergy and BioProcess Technology, Chinese Academy of Sciences, Qingdao, China

Li Li, Tingwei Luo, Yao Zhang, Rui Zhang and Nianzhi Jiao
State Key Laboratory of Marine Environmental Science, Xiamen University, Xiamen, China
Institute of Marine Microbes and Ecospheres, Xiamen University, Xiamen, China

Magda Catarina Sousa, Renato Mendes, Nuno Vaz and João Miguel Dias
CESAM, Departamento de Física, Universidade de Aveiro, Campus de Santiago, Aveiro, Portugal

Ines Alvarez and Moncho Gomez-Gesteira
CESAM, Departamento de Física, Universidade de Aveiro, Campus de Santiago, Aveiro, Portugal
EPhysLab (Environmental Physics Laboratory), Universidade de Vigo Facultade de Ciencias, Ourense, Spain

Index